Mathematics in Healthcare: Concepts and Applications

Mathematics in Healthcare: Concepts and Applications

Editor: Eden Paul

STATES
ACADEMIC PRESS
www.statesacademicpress.com

Published by States Academic Press
109 South 5th Street,
Brooklyn, NY 11249, USA
www.statesacademicpress.com

Mathematics in Healthcare: Concepts and Applications
Edited by Eden Paul

International Standard Book Number: 978-1-63989-346-1 (Hardback)

Cataloging-in-Publication Data

Mathematics in healthcare : concepts and applications / edited by Eden Paul.
p. cm.
Includes bibliographical references and index.
ISBN 978-1-63989-346-1
1. Medical care--Mathematical models. 2. Medicine--Mathematics. 3. Mathematics. I. Paul, Eden.
R853.M3 M38 2022
610.151 95--dc23

Contents

Preface

Mathematics in healthcare focuses on supporting important activities in a hospital such as measurement of a patient's status, evaluation of quality of services, calculating dosages, analyzing test results, taking temperatures, etc. The concepts of mathematics such as probability and statistics are used in validating the effectiveness of new drugs, procedures or estimating the survival rate of cancer patients who are undergoing treatments. The mathematical topics such as the mechanics of a moving body, the statistics governing the atoms in a gas or developing more efficient algorithms in computational understanding have varied applications in the healthcare sector. This book outlines the concepts and applications of mathematics in healthcare in detail. It presents researches and studies performed by experts across the globe. This book will serve as a valuable source of reference for graduate and post graduate students.

This book is a result of research of several months to collate the most relevant data in the field.

When I was approached with the idea of this book and the proposal to edit it, I was overwhelmed. It gave me an opportunity to reach out to all those who share a common interest with me in this field. I had 3 main parameters for editing this text:

1. Accuracy - The data and information provided in this book should be up-to-date and valuable to the readers.

2. Structure - The data must be presented in a structured format for easy understanding and better grasping of the readers.

3. Universal Approach - This book not only targets students but also experts and innovators in the field, thus my aim was to present topics which are of use to all.

Thus, it took me a couple of months to finish the editing of this book.

I would like to make a special mention of my publisher who considered me worthy of this opportunity and also supported me throughout the editing process. I would also like to thank the editing team at the back-end who extended their help whenever required.

Editor

1

Cancer Markers Selection using Network-Based Cox Regression: A Methodological and Computational Practice

Antonella Iuliano [1†], Annalisa Occhipinti [2*†], Claudia Angelini [1], Italia De Feis [1] and Pietro Lió [2]*

[1] Istituto per le Applicazioni del Calcolo "Mauro Picone," Consiglio Nazionale delle Ricerche, Naples, Italy, [2] Computer Laboratory, University of Cambridge, Cambridge, UK

Edited by:
Krasimira Tsaneva-Atanasova,
University of Exeter, UK

Reviewed by:
Guanglong Jiang,
Indiana University School of Medicine, USA
Anelia Horvath,
George Washington University, USA

***Correspondence:**
Antonella Iuliano
a.iuliano@na.iac.cnr.it;
Annalisa Occhipinti
ao356@cam.ac.uk

† Joint first authors.

International initiatives such as the Cancer Genome Atlas (TCGA) and the International Cancer Genome Consortium (ICGC) are collecting multiple datasets at different genome-scales with the aim of identifying novel cancer biomarkers and predicting survival of patients. To analyze such data, several statistical methods have been applied, among them Cox regression models. Although these models provide a good statistical framework to analyze omic data, there is still a lack of studies that illustrate advantages and drawbacks in integrating biological information and selecting groups of biomarkers. In fact, classical Cox regression algorithms focus on the selection of a single biomarker, without taking into account the strong correlation between genes. Even though network-based Cox regression algorithms overcome such drawbacks, such network-based approaches are less widely used within the life science community. In this article, we aim to provide a clear methodological framework on the use of such approaches in order to turn cancer research results into clinical applications. Therefore, we first discuss the rationale and the practical usage of three recently proposed network-based Cox regression algorithms (i.e., Net-Cox, AdaLnet, and fastcox). Then, we show how to combine existing biological knowledge and available data with such algorithms to identify networks of cancer biomarkers and to estimate survival of patients. Finally, we describe in detail a new permutation-based approach to better validate the significance of the selection in terms of cancer gene signatures and pathway/networks identification. We illustrate the proposed methodology by means of both simulations and real case studies. Overall, the aim of our work is two-fold. Firstly, to show how network-based Cox regression models can be used to integrate biological knowledge (e.g., multi-omics data) for the analysis of survival data. Secondly, to provide a clear methodological and computational approach for investigating cancers regulatory networks.

Keywords: cancer, Cox model, high-dimensionality, gene expression, network, regularization, survival

INTRODUCTION

Recent developments in high-throughput technology have produced a huge amount of multiple and diverse genome-scale data to deal with biological and clinical questions in cancer. For example, genomics, transcriptomics, and epigenomics information is nowadays publicly available for tens of different cancer cell lines from thousands of patients in The Cancer Genome Atlas (TCGA, http://cancergenome.nih.gov/). Mutations data over one million tumor samples are also reported in Cosmic (http://cancer.sanger.ac.uk/cosmic), the world's largest and most comprehensive resource for exploring the impact of somatic mutations. Other valuable databases include The Gene Expression Omnibus (GEO, http://www.ncbi.nlm.nih.gov/gds) among others. Such amount of data is likely to revolutionize genetics and biomedical cancer research, but a thorough integration of all these different types of information is necessary. Indeed, cancer is a "multi-factorial" disease caused by a combination of genetic, environmental, and lifestyle factors. Such factors play an important role in discovering prognostic and diagnostic cancer gene signatures opening a new way toward the so called "personalized medicine." The term refers to a new type of therapy that is essentially based on the features of each patient. For instance, the anticancer drug Cetuximab (Karapetis et al., 2008) inhibits cells proliferation by binding to the EGF receptor and, consequently, preventing activation of the downstream signaling pathway. However, it has been found that Cetuximab can work only if the K-RAS gene is not mutated. Another example is the anti-cancer drug Trastuzumab (Hudis, 2007), which is effective only in patients that highly express the human epidermal growth factor (HER2) at the cell surface, to which the antibody binds. These examples highlight the need of identifying stable and interpretable biomarkers able to predict patient survival and characterize a patient-personalized therapy. In addition, the knowledge of complex cancer processes and networks is important to optimize the use of technology within health care (Raghupathi and Raghupathi, 2014). By discovering associations within the data, big data analytics has the potential to improve care, save lives, and lower costs.

As a consequence, in the last years, there has been a growing interest in developing methods that integrate different genome-scale data into regression models for survival data to create a comprehensive view of human biology and disease (Wang et al., 2014). A popular used approach for the integration of genomic and clinical information is the Cox proportional hazard model (Cox, 1972). The main goal of such method is investigating the connection between gene expression data and survival information to predict cancer survival, assess cancer outcomes, and identify new gene markers. However, since gene expression data are usually characterized by a number of covariates p much larger than the sample size n, the traditional Cox model cannot be applied. Hence, several penalized Cox regression methods have been developed to identify core pathways and biomarkers involved in cancer progression, e.g., the Cox model based on Lasso penalty (Tibshirani, 1996, 1997; Gui and Li, 2005). Alternative penalized Cox regression models based on variable selection include the SCAD (Fan and Li, 2001), the adaptive Lasso (Zou, 2006), the elastic net model (Zou and Hastie, 2005; Simon et al., 2011a; Wu, 2012), and the Dantzig selector (Candes and Tao, 2007) among others. These methods are able to cope with the high-dimensionality of gene expression data, thus solving the "$p \gg n$" issue (Engler and Li, 2009). All these penalized models are statistically efficient in high-dimensional regression, but they perform poorly on data with high collinearity. Moreover, no biological knowledge is taken into account. Indeed, they are simply based on statistical frameworks completely ignoring biological regulatory network, protein–protein interaction (PPI), signaling pathways, and well-known relationships among genes. In such models, the lack of biological information produces instability in predictors reducing the predictive ability of the models. Hence, in order to provide more reliable and biologically meaningful results, the inclusion of *a-priori* biological knowledge into the models is mandatory. To address this issue, new penalized Cox methods based on the integration of genomic information have been recently proposed (Zhang et al., 2013; Gong et al., 2014; Sun et al., 2014). In such models, the genomic information is encoded by a network whose graph structure identifies a given relation (edges) between genes (nodes). The resulting Laplacian matrix is then integrated as penalty in the Cox regression models. In particular, the network can represent the correlation between genes (Zhang et al., 2013), KEGG pathways identification (Sun et al., 2014), functional interaction network (Huttenhower et al., 2009), or PPI. These Cox models based on *a-priori* biological network are called "network-based Cox regression."

The network-based Cox regression methods provide an efficient tool to perform Cox regression on high-dimensional data incorporating genes network information. In literature, there are some recent approaches that analyze different Cox methods. For instance, an accurate review of eight different methods that integrate network information into multi-variable Cox models is presented to study the risk prediction in breast cancer and the integrated Brier score is used as a performance measure (Fröhlich, 2014). However, the study performed enrichment analysis on the signatures genes selected by the compared models without showing any survival prediction analysis in terms of Kaplan–Meier curves. A network-based Cox regression model that explores gene-to-gene connections in multiple cancer datasets is also performed for maximizing the overall association of the sub-network with clinical outcomes (Martinez-Ledesma et al., 2015). A potential limitation of these conventional networks is that the edges only reflect the information of within-features or within-relations, and do not consider the association between features and outcomes, which may be useful in improving the predictive power. Therefore, an alternative network construction method for the outcome-guided gene-interaction network has to be introduced in order to improve the performance of survival analysis in network-based Cox regression (Jeong et al., 2015).

In this work, we present a methodological framework for the analysis of molecular and survival data through a cross-validated approach of network-based Cox regression algorithms (*Net-Cox*, *Adalnet*, and *fastcox*, see Section Methods). The method starts from the analysis of raw data and, through a cross-validated penalty approach, it guides the reader to the interpretation of the final results. As shown in **Figure 1**, the general steps of our approach are the following: (i) defining the biological question and the experimental design using microarray data, then integrating *a-priori* biological information using functional map of the human genome such as HEFalMp (Huttenhower et al., 2009) and KEGG; (ii) performing biological screening of the data for selecting relevant features through cross-validated penalization (Simon et al., 2011b); (iii) implementing network-based Cox regression models for the analysis of cancer-related genes; (iv) evaluating survival models to predict cancer patient prognosis and exploring cancer associated pathways. The presented approach provides a new methodological framework for the study and the interpretation of regression methods through gene-network and pathways analyses and it can be easily adapted to incorporate other network-based Cox regression algorithms.

A preliminary study for the comparison of penalized Cox models was presented in Iuliano et al. (2014), where the analysis was limited to cancer survival prediction using top ranked genes. No simulation studies, extensive pathways analysis or validation of the data were performed in that study. On the contrary, this article presents a more accurate and complete analysis based on a cross-validated approach (Simon et al., 2011b), the overall workflow (see **Figure 2**) that includes both simulation studies and novel real cancer datasets (see Section Data Analysis). Simulated data have been used to perform a statistical comparison of the methods in terms of sensitivity, specificity, number of selected genes, false positive rates, and Matthews correlation coefficient in two simulation settings with different genetic effects. On the other hand, real datasets analysis was performed to assess the relevance of the selected genes in the training dataset and to test the survival prediction accuracy of each model. Cross-validated Kaplan–Meier curves for survival analysis and pathway analysis were also computed (see Section Results). The novelty of the current study consists in the integration of a cross-validated approach (Simon et al., 2011b) to obtain an accurate survival prediction even when the number of cases is relatively small for an effective sample splitting (see **Figure 2**). Cross-validation methods have been largely applied in Cox regression models to estimate prediction errors and for model parameters tuning (Vasselli et al., 2003; Molinaro et al., 2005; Simon et al., 2011b). Some of the most relevant cross-validation approaches include leave-one-out cross-validation (LOOCV; Kearns and Ron, 1999), *k*-fold (Refaeilzadeh et al., 2009), and bootstrap algorithms (Kohavi, 1995). However, all these methods do not provide a good estimation if the data available are limited for an effective division in training and test sets. On the contrary, the cross-validation method used in our analysis (Simon et al., 2011b) is based on a re-sampling algorithm that allows an accurate prediction of the survival risk model regardless the data size. Therefore, in this work, we first present a novel statistical approach to infer pathway interaction networks from gene expression data that relies on a new mathematical

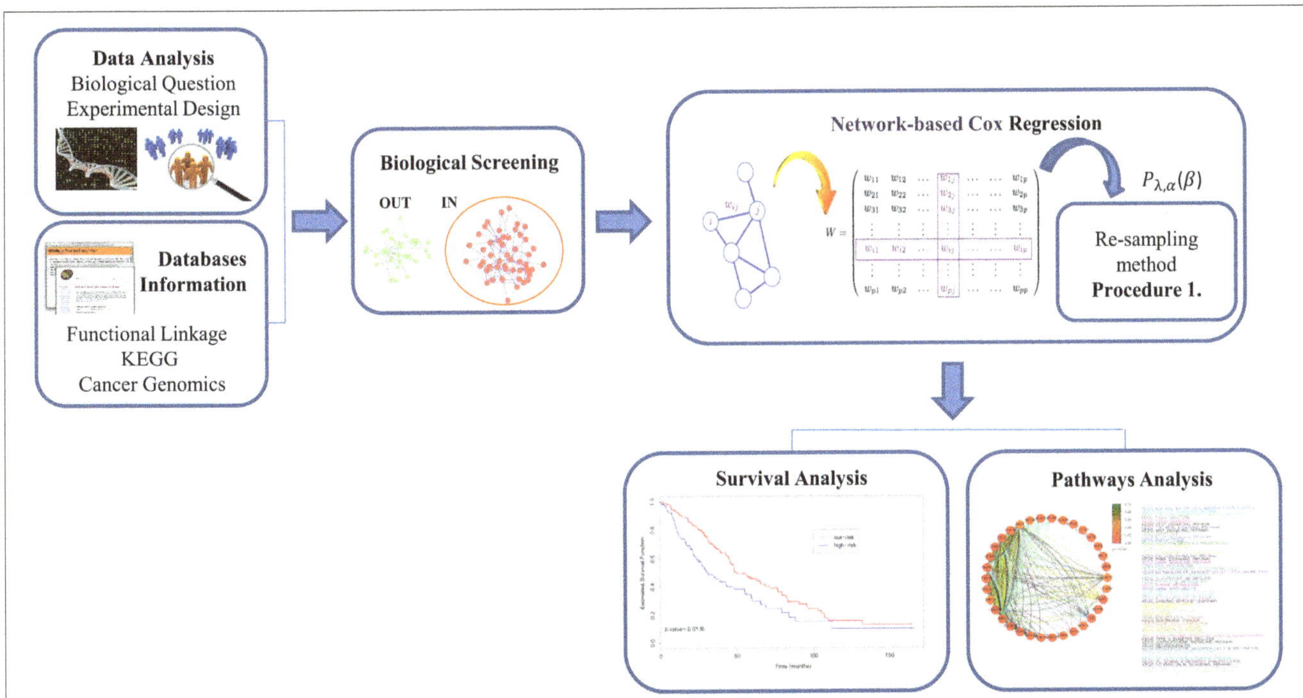

FIGURE 1 | The pipeline of network-based Cox models approach for cancer survival analysis in four general steps. (1) Define the biological question and the experimental design and then, integrate *a-priori* biological information using functional map of the human genome; (2) perform biological screening of the data in order to select IN variables to use in the analysis; (3) implement network-based Cox regression models with the integration of a re-sampling method based on a cross-validated approach; (4) apply survival analysis to predict cancer patients and pathway analysis to explore groups of genes associated to the disease.

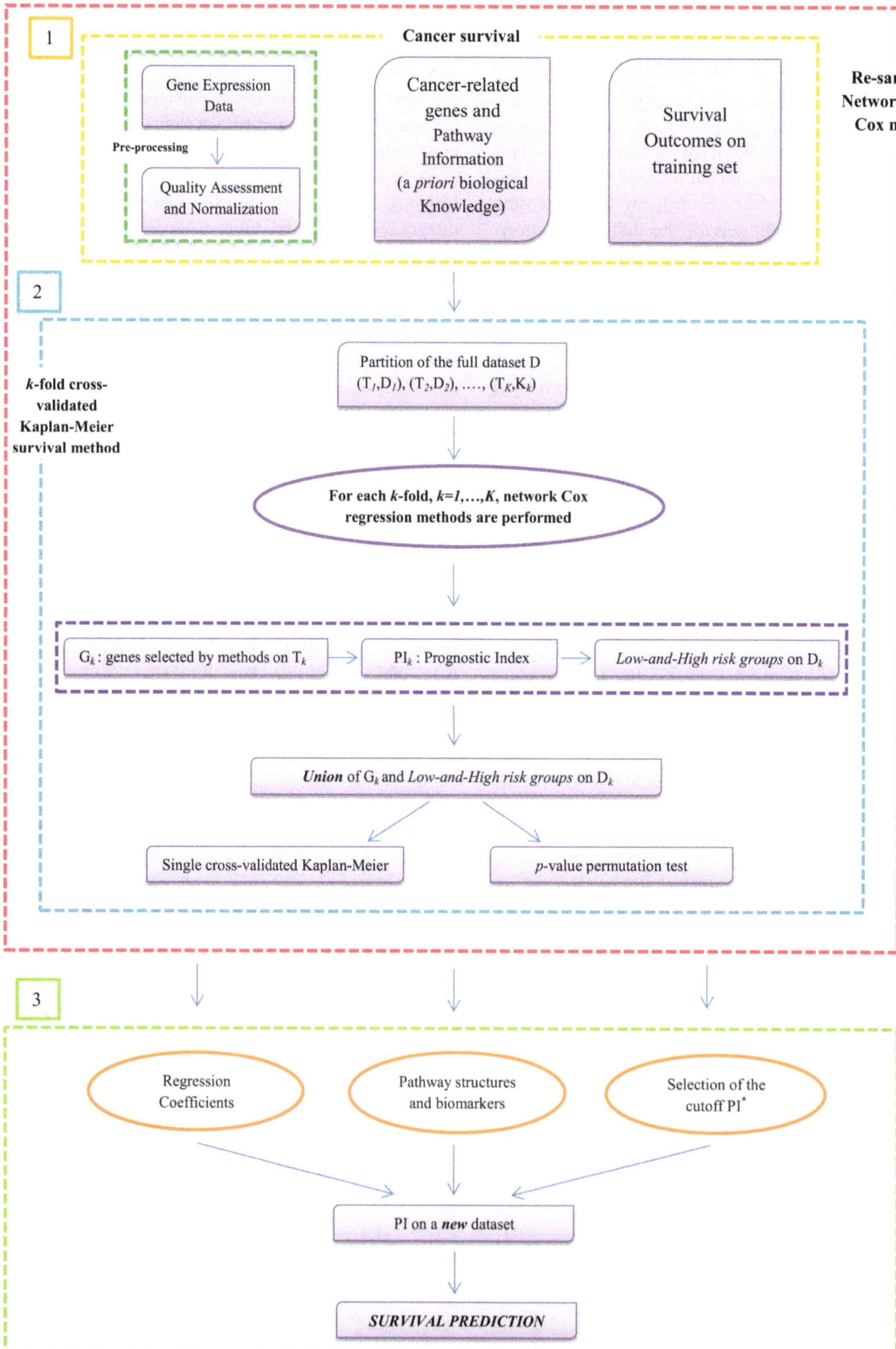

FIGURE 2 | Workflow of prognostic model building by using gene expression profile in cancer. The method starts from the analysis of raw data and, through a cross-validated penalty approach, it leads to the interpretation of the final results. Step (1) includes the input data for the survival analysis: gene expression data,

(Continued)

FIGURE 2 | Continued
cancer-related genes, pathway information, and overall survival (OS) times. Step (2) illustrates the novelty of the work based on a k-fold cross-validation Kaplan–Meier procedure by integrating network-regularized Cox models for selecting significant genes and pathways structures. The Prognostic Index (*PI*) has been used to divide the patients in high-risk and low-risk groups. Then, the union of these two groups is done to plot single cross-validated Kaplan–Meier curves and to calculate the *p*-value permutation test. Step (3) shows the survival prediction to test how well the models generalize across independent cancer datasets.

concept (based on the biological screening and network-based Cox regression methods) for understanding pathways' activity and relationships. Second, we provide a methodological strategy to researchers for the use of network-based Cox regression models in order to turn cancer research results into clinical applications.

METHODS

Network-Regularized Cox Regression Models

The Cox Proportional hazards model (Cox, 1972) is the most widely used model to describe the relationship between survival times and predictor covariates.

Given a sample of n subjects, let T_i and C_i be the survival time and the censoring time, respectively, for subject $i = 1, \ldots, n$. Let $t_i = \min\{T_i, C_i\}$ be the observed survival time and $\delta_i = I(T_i \leq C_i)$ the censoring indicator, where $I(\cdot)$ is the indicator function (i.e., $\delta_i = 1$ if the survival time is observed and $\delta_i = 0$ if the survival time is censored). We denote by $\mathbf{X}_i = (X_{i1}, \ldots, X_{ip})'$ the regression vector of p-variables for the ith subject (i.e., the gene expression profile of the ith patient over p genes). The survival time T_i and the censoring time C_i are assumed to be conditionally independent given $\boldsymbol{X}_i$. Furthermore, the censoring mechanism is assumed to be non-informative. The observed data can be represented by the triplets $\{(t_i, \delta_i, \boldsymbol{X}_i), i = 1, ..., n\}$. The Cox regression method assumes that the hazard function $h(t|\mathbf{X}_i)$, which is the risk of death at time t for the ith patient with gene expression profile $\boldsymbol{X}_i$, can be written as

$$h(t|\mathbf{X}_i) = h_0(t)\exp\left(X_i'\boldsymbol{\beta}\right)$$

where $h_0(t)$ is the baseline hazard and $\boldsymbol{\beta} = (\beta_1, \ldots, \beta_p)'$ is the column vector of the regression parameters.

In the classical setting, the regression coefficients are estimated by maximizing the Cox's log-partial likelihood

$$pl(\boldsymbol{\beta}) = \sum_{i=1}^{n} \delta_i \left\{ \mathbf{X}_i'\boldsymbol{\beta} - \log\left[\sum_{j \in R(t_i)} \exp(\mathbf{X}_j'\boldsymbol{\beta})\right]\right\}, \quad (1)$$

where t_i is the survival time (observed or censored) for the ith patient, $R(t_i)$ is the risk set at time t_i (i.e., the set of all patients who still survived prior to time t_i).

However, in the analysis of gene expression data, the number of genes p is usually larger than the sample size n and the standard Cox-model cannot be directly applied. To cope with the curse of dimensionality ($p \gg n$), a variety of penalization approaches have been proposed for achieving good prediction performance and easy interpretation of the data. Although these regularization methods induce sparsity into the solution by shrinking some estimates to zero, the biological relationship of gene expression profiles is not taken into account. Hence, in order to integrate information from molecular interactions between genes, network-based constrained methods for high-dimensional Cox regression have been introduced.

In this context, the regression coefficients are estimated by maximizing the penalized Cox's log-partial likelihood function

$$pl_{pen}(\boldsymbol{\beta}) = \sum_{i=1}^{n} \delta_i \left\{ X_i'\boldsymbol{\beta} - \log\left[\sum_{j \in R(t_i)} \exp(X_j'\boldsymbol{\beta})\right]\right\} - P_\lambda(\boldsymbol{\beta}), \quad (2)$$

where $P_\lambda(\boldsymbol{\beta})$ is a network-constrained penalty function on the coefficients $\boldsymbol{\beta}$.

Such penalty function describes the existing relationships among the covariates (genes) specified by a network $G = (V, E, W)$ (weighted and undirected graph), where $V = \{1, \ldots, p\}$ is the set of vertices (genes/covariates), an element (i, j) in the edge set $E \subset V \times V$ indicates a link between vertices i and j and $W = (w_{ij}), (i, j) \in E$ is the set of weights associated with the edges. These weights are usually used to represent the relations between genes in terms of gene–gene interaction, KEGG pathway analysis or PPI. Hence, the network structure plays an important role since it incorporates prior gene regulatory information often ignored.

The three regularized network-based Cox regression models used in our study are presented below and differ in the form of the penalty function $P_\lambda(\boldsymbol{\beta})$.

Net-Cox method

Net-Cox regression (Zhang et al., 2013) is an extension of the L_2-Cox model and uses the following penalty function

$$P_{\lambda,\alpha}(\boldsymbol{\beta}) = \lambda\left[\alpha \,\|\boldsymbol{\beta}\|_2^2 + (1-\alpha)\,\Phi(\boldsymbol{\beta})\right], \quad (3)$$

where $\lambda > 0$ and $\alpha \in (0, 1]$ are two regularization parameters in the network constraint. and

$$\Phi(\boldsymbol{\beta}) = \sum_{(i,j) \in E} w_{ij}(\beta_i - \beta_j)^2. \quad (4)$$

The penalty (3) consists of two terms: the first one is an L_2-norm of $\boldsymbol{\beta}$ that regularizes the uncertainty in the network constraint; the second term is a network Laplacian penalty $\Phi(\boldsymbol{\beta})$ that encourages smoothness among correlated gene in the network and encode prior knowledge from a network.

Given a normalized graph weight matrix W, we assume that co-expressed (related) genes are assigned similar coefficients by defining the cost term $\Phi(\boldsymbol{\beta})$ as reported in Equation (4). $\Phi(\boldsymbol{\beta})$

can be also written as $\Phi(\boldsymbol{\beta}) = \boldsymbol{\beta}'(\mathbf{I} - \mathbf{W})\boldsymbol{\beta} = \boldsymbol{\beta}'\bar{L}\boldsymbol{\beta}$ where $\bar{L}$ is a positive semi-definite matrix derived from network information (weight matrix $\mathbf{W}$) and $\mathbf{I}$ is an identity matrix. Hence, the objective function will result in a significant cost in the network if any pair of genes is connected by an high weight edge and the difference between their coefficients is large.

Note that to identify the signature genes classified by *Net-Cox*, which is a ridge regression based method, we create a consensus ranking of the relevant cancer genes.

AdaLnet Method

*Ada*ptive *L*aplacian *net* (Sun et al., 2014) is a modified version of a network-constrained regularization procedure for fitting linear models and for variable selection (Li and Li, 2008, 2010) where the predictors are genomic data with graphical structures. *AdaLnet* is based on prior gene regulatory network information, represented by an undirected graph for the analysis of gene expression data and survival outcomes.

Denoting with $d_i = \sum_{i:(i,j)\in E} w_{ij}$ the degree of vertex i, *AdaLnet* defines the normalized Laplacian matrix $\mathbf{L} = (l_{ij})$ of the graph G by

$$l_{i,j} = \begin{cases} 1, & \text{if } i = j \text{ and } d_i \neq 0, \\ -w_{ij}/\sqrt{d_i d_j}, & \text{if}(i,j) \in E, \\ 0, & \text{otherwise.} \end{cases} \tag{5}$$

Note that $\mathbf{L}$ is positive semi definite. The network-constrained penalty in Equation (2) is given by

$$P_{\lambda,\alpha}(\boldsymbol{\beta}) = \lambda\left[\alpha\, \|\boldsymbol{\beta}\|_1 + (1-\alpha)\,\Psi(\boldsymbol{\beta})\right], \tag{6}$$

with

$$\Psi(\boldsymbol{\beta}) = \sum_{(i,j)\in E} w_{ij}\left(\text{sign}(\tilde{\beta}_i)\beta_i/\sqrt{d_i} - \text{sign}(\tilde{\beta}_j)\beta_j/\sqrt{d_j}\right)^2. \tag{7}$$

Equation (6) is composed by two penalty terms. The first one is an L_1-penalty that induces a sparse solution, the second one is a quadratic Laplacian penalty $\Psi(\boldsymbol{\beta}) = \boldsymbol{\beta}'\tilde{L}\boldsymbol{\beta}$ that imposes smoothness of the parameters $\boldsymbol{\beta}$ between neighboring vertices in the network. Note that $\tilde{L} = S'LS$ with $S = \text{diag}(\text{sign}(\tilde{\beta}_1), \ldots, \text{sign}(\tilde{\beta}_p))$ and $\tilde{\boldsymbol{\beta}} = (\tilde{\beta}_1, \ldots, \tilde{\beta}_p)$ is obtained from a preliminary regression analysis. The scaling of the coefficients $\boldsymbol{\beta}$ respect to the degree allows the genes with more connections (i.e., the hub genes) to have larger coefficients. Hence, small changes of expression levels of these genes can lead to large changes in the response.

An advantage of using penalty (6) consists in representing the case when two neighboring variables have opposite regression coefficient signs, which is reasonable in network-based analysis of gene expression data. Indeed, when a transcription factor (TF) positively regulate gene i and negatively regulate gene j in a certain pathway, the corresponding coefficients will result with opposite sign.

Note that in *Net-Cox* and *AdaLnet*, λ is the parameter controlling the weight between the likelihood and the network constraint and $\alpha \in (0, 1]$ is the parameter weighting the network constraint.

Fastcox Method

The penalty function of *fastcox* (Yang and Zou, 2012) computes the solution paths of the elastic net penalized Cox's proportional hazards model (Wu, 2012). In this method the penalty function in Equation (2) is given by

$$P_{\lambda,\alpha}(\boldsymbol{\beta}) = \lambda\left[\alpha w\|\boldsymbol{\beta}\|_1 + \frac{1}{2}(1-\alpha)\|\boldsymbol{\beta}\|_2^2\right],$$

where the non-negative weights w allow a more flexible estimation. In particular, setting $w_j = 0$ implies no shrinkage and the variable j will be always included in the final model. Default is 1 for all variables. $\alpha \in (0, 1]$ is the elastic net trade off. This regularization technique is a combination of the lasso and ridge penalty that produce a sparse model (given by the L_1-penalty) with good prediction accuracy, while encouraging a grouping effect. It is worthy to note that this method does not include any gene network information. It has been used in our study to obtain pathways investigation and survival prediction from a relevant method that is simply based on statistical framework.

Tuning Parameters by Five-Fold Cross-Validation

For all the methods, we estimated the regularization parameters using cross-validation. Four-folds of data are used to build a model for validation on the fifth fold, cycling through each of the five-folds in turn. Then, the (λ,α) pair that minimizes the cross-validation log-partial likelihood (CVPL) are chosen as the optimal parameters. CVPL is defined as

$$CVPL(\lambda,\alpha) = -\frac{1}{n}\sum_{k=1}^{K}\{\ell(\hat{\boldsymbol{\beta}}^{(-k)}(\lambda,\alpha)) - \ell^{(-k)}(\hat{\boldsymbol{\beta}}^{(-k)}(\lambda,\alpha))\}, \tag{8}$$

where $\hat{\boldsymbol{\beta}}^{(-k)}(\cdot)$ is the estimate obtained from excluding the kth part of the data with a given pair of (λ,α), $\ell(\cdot)$ is the Cox log-partial likelihood on all the sample and $\ell^{(-k)}(\cdot)$ is the log-partial likelihood when the kth fold is left out (van Houwelingen et al., 2006).

General Algorithm: A Re-Sampling Method for Survival Prediction

The prediction capabilities of a given method are usually evaluated using a training set to select the markers and a testing set to measure the goodness of the prediction. In several cases training and test sets are obtained splitting a given dataset in two parts. However, findings could be over optimistic depending on the specific split. To further understand the role of the network information in cross-validation and to overcome the drawbacks of investigating only one split, each network-based model was validated with the re-sampling procedure suggested by Simon et al. (2011b). This method is based on a cross-validated estimate of the survival distribution of the risk groups and provide a more efficient use of data than fixed sample splitting (see **Figure 2**). The steps of the re-sampling algorithm for survival prediction are presented below.

Procedure 1: *k*-fold Cross-validated Kaplan–Meier survival method

1. The full dataset D is partitioned into K approximately equal parts $D_1, \ldots, D_K$.
   For each $k = 1, \ldots, K$
2. Set $T_k = D - D_k$ as the training set and D_k as the testing set.
3. Perform network-based Cox regression on T_k and select high-risk cancer genes G_k. Denote the parameter estimate by $\hat{\beta}_{T_k}$.
4. Calculate the prognostic index (PI) for each patient i_k in D_k as

$$PI_{i_k}^{D_k} = x'_{i_k}\hat{\beta}_{T_k},$$

   where x_{i_k} is the vector of gene expression value associated to the i_k-th patient into the k-fold. Each patient i_k in D_k is assigned into the *high/low-risk* group if its prognostic index $PI_{i_k}^{D_k}$ is above (or below) a fixed threshold PI^{*,T_k} defined adaptively on T_k.
5. All the patients classified as *low-and-high risk* in any of the folds are grouped together and a single Kaplan–Meier curve is computed as the union of the risk groups defined in each fold. The set of predictive genes is selected as the union of G_k, for $k = 1, \ldots, K$.
6. Compute the log-rank χ_0^2 statistic under the null hypothesis that survival is independent of expression profile.
7. Calculate a permutation p-value as follows:
   (*i*) from the m-th permutation data ($m = 1, \ldots, M$), compute the log-rank χ_b^2 statistic using the cross-validation procedure (1–6),
   (*ii*) compute the permutation p-value, $\hat{p}$, as

$$\hat{p} = M^{-1}\sum_{i=1}^{M} I(P_m \geq P_0).$$

For our analysis, the estimate $\hat{\beta}_{T_k}$ in step 4 was computed by using five-fold cross-validation (i.e., $K = 5$) to select the optimal tuning parameter values $(\hat{\lambda}_{T_k}, \hat{\alpha}_{T_k})$, that we used to fit the corresponding penalized function $P_{\hat{\lambda}_{T_k},\hat{\alpha}_{T_k}}(\hat{\beta}_{T_k})$ on T_k. In particular, we first set α to a sufficiently fine grid of values on [0, 1]. For each fixed α, λ was chosen from $\{10^{-5}, 10^{-4}, 10^{-3}, 10^{-2}, 10^{-1}, 1\}$ for *Net-Cox*, while it was set λ to a decreasing sequence of values λ_{max} to λ_{min} automatically chosen for *AdaLnet* and *fastcox*.

In step 5, we selected PI^{*,T_k} as the optimal cut-off in terms of PI^{D_k}. By using the $PI_{i_k}^{T_k}$, it was possible to split the patients in two subgroups, i.e., *high-risk* and *low-risk* prognosis groups. Thus, the patient i_k in T_k was assigned to the *high-risk* (or *low-risk*) group if his prognostic index $PI_{i_k}^{T_k}$ was above (or below) the quantile selected on a grid of given values that spans from 30 to 70%. The cut-off PI^{*,T_k} was chosen in correspondence to the lowest p-value in a log rank test on this grid.

In step 7, we set M equal to 500.

Survival Analysis

Network-based Cox regression model was used to discover significant variables, i.e., genes, correlated with death risk. Overall survival (OS) curves were estimated using the Cross Kaplan–Meier estimator and compared using the two-sided log-rank test as implemented in the R package *survival*. The statistical significance of the log-rank statistic related to the cross-validated Kaplan–Meier curves was obtained through a permutation distribution (Simon et al., 2011b) as described in the previous section. Permutation test was used to test the association between *high-risk* or *low-risk groups* and $p < 0.05$ were considered statistically significant. A simple scheme of the applied procedure for OS estimation is reported in **Figure 2**.

Furthermore, we also validated the predictive performance of the three methods using independent dataset for training and testing. In this context, we used the largest dataset as training set to identify the gene expression signatures (see **Figure 2**, step 2). Then, the second independent dataset was considered as test set in order to analyze the survival prediction of the models. We used Kaplan–Meier survival curves and log-rank test to perform the analysis (see **Figure 2**, step 3).

Pathway Analysis

We performed pathway analysis based on KEGG database and on the Human Experimental/Functional Mapper (Huttenhower et al., 2009). In particular, we focused on a gene–gene interaction analysis developing gene-networks that describe the relations between genes in terms of KEGG pathways. Each node in the network represents a gene and an edge between two nodes means that the two genes belongs to the same pathway. Different colors are used for different pathways. The color of each node indicates how strong is the relationship between the gene and the disease under analysis (ovarian and breast cancer; Huttenhower et al., 2009). The p-value chosen within the interval [0, 0.1] represents the node color intensity. Red color, that is $p = 0$, means that there is a high significant gene-disease relation, while green color, that is $p = 0.1$, means that not exist a relevant gene-disease relation.

Gene networks have been computed by considering only the *not isolated* genes in the intersection between KEGG pathways and the set of genes selected by each method. Given a set of genes G and the set of all the KEGG pathways K, we defined a gene g as *not isolated* if $G \cap K \not\supseteq \{g\}$. Namely, g is *not isolated* if there is at least another gene $g' \in G$ belonging to the same pathways of g.

Software

The methodological approach presented in **Figure 2** has been implemented as an integrative R script that allows to run the different algorithms under the same R environment. *Net-Cox*, which is a Matlab toolbox (http://compbio.cs.umn.edu/Net-Cox/), *AdaLnet*, available as an R code and sent us upon request and *fastcox*, which is an R package (http://code.google.com/p/fastcox/) were merged together by using *R.matlab*, https://cran.r-project.org/web/packages/R.matlab/index.html. The script also includes the implementation of the re-sampling permutation approach (Simon et al., 2011b) and the cross-validation method for parameters estimation. Both simulated and real data can be used to run the script which can be easily adapted for the integration of new Cox models.

For real data analysis, the microarray data were preprocessed using R packages available in Bioconductor. First, we selected from the initial dataset the genes that were more likely to be involved in cancer by using a functional map summarizing the most relevant interactions in the cancer area of interest (Huttenhower et al., 2009). Then, we used HEFalMp tool (Huttenhower et al., 2009) to build the genes network and identify the weight of the edges between the selected genes. Finally, *Net-Cox*, *AdaLnet*, and *fastcox* were implemented integrating a cross-validation method for selecting the optimal tuning parameters λ and α and a re-sampling based procedure (Simon et al., 2011b), see Procedure 1.

The scripts are available upon request from the first two authors.

DATA ANALYSIS

Simulation Scheme

We used the three methods in two different simulation settings (Wu and Wang, 2013; Sun et al., 2014) in order to investigate the performances and the properties of the three models and to facilitate the interpretation of results. We considered two scenarios that are likely to be encountered in genomic studies and we simulated gene expression data as network constrained. Both the two settings consist of 100 regulatory networks. Each regulatory network is composed by one transcription factor (TF) that regulates 10 genes resulting in a total of 1100 genes. Detailed settings are given below.

Scenario 1: Not-Overlapped Networks

The first setting simulates a scenario with not-overlapped networks, which means that the 100 regulatory networks are disjoint each other and each gene is linked to only one TF. Under this assumptions, the degree d_i of each TF = 10 and $d_i = 1$ for the regulated genes. The edges' weight $w_{ij} = 1$ between the TFs and their regulated genes, $w_{ij} = 0$ otherwise. The expression value of each TF was generated from a normal standard distribution. The expression values of the ten regulated genes were generated from a conditional normal distribution with positive correlation ($\rho = 0.7$) between the expression of five genes and the corresponding TF, and negative correlation ($\rho = -0.7$) for the remaining five genes. This simulates the activation or repression of each gene under the effect of the corresponding TF. The failure times were generated from the Cox model

$$\lambda(t|X) = \lambda_0(t)\exp\left(\sum_{j=1}^{88} \beta_j X_j\right)$$

which includes only s = 88 relevant genes (i.e., eight regulatory networks). The baseline hazard function $\lambda_0(t)$ was specified by a Weibull distribution with shape parameter 5 and scale parameter 2. Censoring times were generated from $U(2, 15)$ with a censoring rate of about 30%. The sample size was fixed at $n = 200$ and the simulation were replicated 100 times. In this setting of not-overlapped genes, the coefficients β_j, j=1, ..., 44 were generated from the uniform distribution $U(0.1, 1)$, while β_j, j=45, ..., 88 were generated from $U(-1.5, -0.1)$.

For each of the settings above, we quantified the noise as the mean between the variance of each transcription factor (TF) and the variance of the 10 corresponding regulated genes.

Scenario 2: Overlapped Networks

The second setting simulates a scenario with overlapped networks, where four regulatory networks (i.e., 44 genes) are connected to the other four networks. This mimics the fact that some genes can belong to different pathways regulating different biological processes, as often observed in cancer. For the sake of simplicity, we assume that all the genes (including the TF) in the networks P_3, P_4, P_5, and P_6 are connected to the genes in the remaining four network P_1, P_2, P_7, and P_8 which are maintained disjointed and independent each other. The expression values of the TFs and the regulated genes were generated from a multivariate normal distribution with $cov(X_i, X_j) = 0.5^{|i-j|}$. The coefficients β_j, $j = 1, \ldots, 22$, corresponding to $P1$ and $P2$, were generated from the uniform distribution $U(0.1, 0.5)$, the coefficients corresponding to the 44 common genes β_j, $j = 23, \ldots, 66$ were generated from $U(-0.1, 0.1)$ and the coefficients β_j, $j = 67, \ldots, 88$, corresponding to P_7 and P_8, were generated from the uniform distribution $U(-1, -0.5)$. Survival times were generated as reported in the first setting with the same censoring rate.

Statistical Measures

The performance of each method is summarized by four measures: sensitivity, specificity, number of genes selected, and the Matthews correlation coefficient (MCC). The *sensitivity or true positive rate (TPR)* and *specificity or true negative rate (TNR)* are given by

$$TPR = \frac{TP}{TP + FN}, \quad TNR = \frac{TN}{TN + FP},$$

where TP, TN, FP, and FN denote the numbers of true positives, true negatives, false positives, and false negatives, respectively. A test with high sensitivity (few false negative) has a low type II error rate, while a test with a high specificity (few false positive) has a low type I error rate. The number of genes selected refers to the genes identified as relevant by each method in the training set. The analysis of these genes gives information on prediction accuracy.

The Matthews correlation coefficient (MCC) is defined as

$$MCC = \frac{TP \times TN - FP \times FN}{\sqrt{(TP + FP)(TP + FN)(TN + FP)(TN + FN)}}.$$

The MCC measure is an global measure of accuracy, and a larger MCC indicates a better performance.

Real Data Applications

We applied the three network methods on different real datasets containing large-scale microarray gene expression measurements from ovarian and breast cancer including survival information (see **Table 1**) in order to facilitate the detection of molecular biomarker and pathway analysis with clinical utility.

TABLE 1 | Microarray Dataset Summary (OS = overall survival).

Datasets	Ref.	Sample number	Platform	Genes number	Survival data	Cancer type
GSE26712	Bonome et al., 2008	185	Affymetrix U133A	13104	OS	Ovarian
OV-TCGA	The Cancer Genome Atlas Research Network, 2011	578	Affymetrix U133A	13104	OS	Ovarian
GSE20685	Kao et al., 2011	327	Affymetrix U133Plus2	21686	OS	Breast
GSE7390	Desmedt et al., 2007	198	Affymetrix U133A	13718	OS	Breast

Ovarian Datasets

We downloaded the first ovarian dataset from NCBI Gene Expression Omnibus as raw .CEL files (Bonome: GSE26712). The data contain gene expression profiling for extensive set of 185 primary ovarian tumors untreated late-stage (III–IV) high-grade (2,3) patients hospitalized at the Memorial Sloan-Kettering Cancer Center between 1990 and 2003. The Affymetrix human U133A microarray platform was used. The second ovarian dataset, the ovarian TCGA, was downloaded from The Cancer Genome Atlas data portal (The Cancer Genome Atlas Research Network, 2011). It was obtained at the gene level (level 3) using the Affymetrix human U133A microarray from 578 samples. All patients were diagnosed with high-grade serous carcinoma and were in an advanced stage. We noted that such datasets are very similar in terms of type of patients, platforms, and cancer disease. Therefore, they can be also used for validation.

Breast Datasets

The breast cancer microarray datasets were downloaded from NCBI GEO database as raw .CEL files (Kao: GSE20685 and Desmedt: GSE7390). Gene expression profiling of the first dataset was conducted on fresh frozen breast cancer tissue collected from 327 patients diagnosed and treated between 1991 and 2004 at the Koo Foundation Sun-Yat-Sen Cancer Center. Hybridization targets were prepared from total RNA according to the Affymetrix U133 plus 2.0 platform. The second breast cancer dataset was chosen on gene expression profiling of frozen samples from 198 N—systemically untreated patients at the Bordet Institute. It was based on the Affymetrix U133 platform.

Preprocessing

All the raw files were processed and normalized by RMA package available in Bioconductor (Gentleman et al., 2004). Between arrays normalization was carried out by using the *preprocessCore* package available in Bioconductor (Gentleman et al., 2004). Survival data (OS, i.e., overall survival), censoring indicator and time to death, for each patients in every dataset were also given (**Figure 2**, step 1).

Cancer Genes and Related Functional Networks

Following our previous study (Iuliano et al., 2014), in order to better analyze real datasets, we first applied a biologically inspired size reduction of the dataset, then we built an *a-priori* network information for the type of cancer under investigation (see **Figure 2**, step 1). For a better focus on genes that are more likely to be relevant in cancer, we selected the high-risk cancer genes using the Human Experimental/Functional Mapper

TABLE 2 | Significant genes number selected using HEFalMp tool.

Datasets	Genes number
GSE26712	1068
OV-TCGA	1068
GSE20685	536
GSE7390	536

(Huttenhower et al., 2009), which is based on a regularized Bayesian integration system. This mapper provides a p-value for each gene describing the significance of the relation between the gene and the disease of interest (breast and ovarian cancer, respectively). In our analysis, we selected only the genes with $p < 0.05$. A summary of the final number of the genes selected from each dataset is reported in **Table 2**. The network matrices used to test the network-based Cox models in our analysis were also derived from the Human Experimental/Functional Mapper which provides maps describing the genes functional activity and interaction networks in over 200 areas of human cellular biology with information from 30,000 genome-scale experiments. This functional network summarizes information from a variety of biologically informative perspectives: prediction of protein function and functional modules, cross-talk among biological processes, and association of novel genes and pathways with known genetic disorders (Huttenhower et al., 2009). The edges of the network are weighted between [0, 1] and express the functional relation between two genes. Note that the functional linkage network includes more information than Human PPI, frequently used as the network prior knowledge. It is clear that taking into account such biological knowledge helps in identifying significant genes that are functionally related in order to obtain important results biologically interpretable.

In order to adapt the gene network to the different methods, the final weight matrix was slightly different from method to method. In particular, since *AdaLnet* requires a weight matrix consisting of 0 and 1, each matrix element was set equal to 0 (or 1) if the weight value was below (or above) a fixed threshold equals to 0.5. On the other hand, *Net-Cox* uses the original weight matrix as obtained in the original paper (Huttenhower et al., 2009).

RESULTS

In our study, we analyzed three network-based Cox regression methods described in Section Methods both on simulated

TABLE 3 | Simulation results for Not-Overlapped settings. Sensitivity, specificity, number of selected genes, false positive rates, and MCC were averaged over the 100 replications.

	Sensitivity	Specificity	No. genes	No. FP	MCC
Net-Cox					
No. genes = 44	0.240 (0.042)	0.977 (0.004)	44.000 (0.000)	22.910 (3.677)	0.300 (0.063)
No. genes = 88	0.489 (0.071)	0.956 (0.006)	88.000 (0.000)	44.940 (6.233)	0.445 (0.077)
No. genes = 176	0.737 (0.087)	0.890 (0.008)	176.000 (0.000)	111.180 (7.692)	0.464 (0.070)
AdaLnet					
General setting	0.444 (0.250)	0.792 (0.170)	249.360 (193.786)	210.330 (172.384)	0.190 (0.059)
No. genes ≤ 100	0.200 (0.085)	0.967 (0.021)	51 (27.256)	33.395 (21.227)	0.220 (0.064)
No. genes > 100	0.627 (0.160)	0.660 (0.099)	399 (113.254)	343.807 (100.118)	0.166 (0.041)
fastcox					
General setting	0.141 (0.117)	0.970 (0.037)	42.62 (46.613)	30.19 (37.833)	0.160 (0.082)
No. genes ≤ 10	0.017 (0.017)	0.999 (0.0002)	1.524 (1.486)	0.048 (0.216)	0.099 (0.07)
No. genes > 10	0.231 (0.063)	0.949 (0.036)	72.379 (40.331)	52.017 (36.492)	0.204 (0.054)

The table reports three consensus rankings for Net-Cox *obtained selecting 44, 88, and 176 genes. For* AdaLnet *and* fastcox, *we show the results related to the general setting, and the statistical measures obtained when the number of selected genes is higher (or lower) of a fixed threshold (threshold was set equal to 100 for* AdaLnet *and equal to 10 for* fastcox*). Standard deviation is reported in brackets.*

TABLE 4 | Simulation results for overlapped settings.

	Sensitivity	Specificity	No. genes	No. FP	MCC
Net-Cox					
No. genes = 44	0.156 (0.043)	0.970 (0.004)	44.000 (0.000)	30.240 (3.766)	0.175 (0.064)
No. genes = 88	0.288 (0.044)	0.938 (0.004)	88.000 (0.000)	62.620 (3.842)	0.227 (0.048)
No. genes = 176	0.386 (0.044)	0.860 (0.003)	176.000 (0.000)	142.010 (3.860)	0.182 (0.035)
AdaLnet					
General Setting	0.262 (0.178)	0.879 (0.144)	145.280 (160.666)	122.240 (145.679)	0.166 (0.067)
No. genes ≤ 100	0.141 (0.064)	0.977 (0.020)	35.635 (24.760)	23.206 (20.296)	0.196 (0.060)
No. genes > 100	0.467 (0.106)	0.713 (0.105)	331.973 (114.325)	290.865 (106.135)	0.114 (0.043)
fastcox					
General setting	0.098 (0.099)	0.974 (0.039)	34.55 (47.732)	25.89 (39.807)	0.134 (0.061)
No. genes ≤ 10	0.019 (0.015)	0.999 (0.0001)	1.679 (1.281)	0.0178 (0.134)	0.115 (0.065)
No. genes > 10	0.199 (0.061)	0.942 (0.040)	76.386 (45.224)	58.818 (40.830)	0.158 (0.044)

Sensitivity, specificity, number of selected genes, false positive rates and MCC were averaged over the 100 replications. The table reports three consensus rankings for Net-Cox *obtained selecting 44, 88, and 176 genes. For* AdaLnet *and* fastcox, *we show the results related to the general setting, and the statistical measures obtained when the number of selected genes is higher (or lower) that a fixed threshold (threshold was set equal to 100 for* AdaLnet *and equal to 10 for* fastcox*). Standard deviation is reported in brackets.Standard deviation is reported in brackets.*

and real data. Here, the major interest is the association of genomic features with clinical outcomes under specific scenarios. Simulation studies were based on two different biological scenarios and were introduced to show the performance of the selected network methods. While, real data analysis was performed in order to provide a better understanding of the outcomes in terms of predictive/prognostic biomarkers and to demonstrate their validity and clinical utility. In particular, we first investigated the three methods in terms of survival prediction performances and then, a pathway analysis was carried out focusing on the relevance in cancer of the selected genes.

It is important to note that the goal of this study is not to provide a rank list of the analyzed methods, but to present a accurate study for the identification of new cancer related genes and core pathways in order to make available such information to biomedical community in the form of a comprehensive methodological procedure (see **Figure 1**).

Simulation Studies

We analyze the performance of the three analyzed methods in two simulation settings where the number of relevant genes is fixed *a-priori* to 88 genes. The first setting simulates a scenario with not overlapped pathways, which means that each gene in the network belongs to only one pathway (not-overlapped pathways). The second setting represents a more realistic scenario with a set of genes shared among different pathways (overlapped pathways). In both cases, a five-fold cross validation was conducted on the full dataset in order to select the tuning parameters (λ, α) and to obtain the coefficient estimates

TABLE 5 | Optimal α cross-validated value calculated on the k training sets.

Datasets	*k* Partitions	Net-Cox		AdaLnet		fastcox	
		α	Genes selected	α	Genes selected	α	Genes selected
GSE26712	5	0.2	101	0.5	23	0.01	453
OV-TCGA	5	0.5	99	0.5	38	0.1	623
GSE20685	5	0.5	76	0.5	28	0.01	298
GSE7390	5	0.5	89	0.5	14	0.01	423

by using the three methods. The details of the simulation data are reported in Section Methods.

The performance of each method is summarized by several statistical measures: sensitivity, specificity, number of selected genes, false positive rates, and Matthews correlation coefficient (MCC). Simulation results for both the models are reported in **Tables 3**, **4**, respectively (standard deviation is reported in brackets). To analyze the signature genes identified by *Net-Cox*, which is a method based on ridge regression, we considered three different consensus rankings where the number of significant genes selected by the method was fixed to 44, 88, and 176 genes, respectively. The selected genes were classified in descending order according to the absolute value of the regression coefficients. On the other hand, to better highlight the variable selection performance of *AdaLnet* and *fastcox*, we split the 100 iterations in two groups based on the number of genes selected at each iteration. We fixed 100 genes as threshold for *AdaLnet* and 10 genes for *fastcox*, then we computed again the statistical measures based on the two groups.

In the not-overlapped setting, *Net-Cox* performed better than the other two methods as showed by the MCC, which provides an overall measure of accuracy. In particular, when considering 44 and 88 genes, the false positive rate in *Net-Cox* was 22.910 and 44.940, respectively, with MCC equals to 0.300 and 0.445. Sensitivity and specificity were, respectively, 0.240 and 0.977 in the first case, 0.489 and 0.956 in the second case study. When the number of selected genes was increased to 176, even if the false positive rate increased resulting in a lower specificity (0.890), the sensitivity reached its highest values producing the highest MCC (0.464).

Since the majority of the selected genes were irrelevant and both *AdaLnet* and *fastcox* resulted in sparse models, specificity was much higher than sensitivity and was comparable between the two variable selection methods. In particular, in the not-overlapped setting, *AdaLnet* selected in average 249.360 genes with a false positive rate equals to 210.330. Sensitivity and specificity were equal to 0.444 and 0.792 resulting in a MCC of 0.190. On the other hand, *fastcox* selected in average 42.62 genes with a false positive rate of 30.19. MCC was equal to 0.160 with sensitivity 0.141 and specificity 0.970.

AdaLnet had the best performance when the number of selected genes was below 100, while *fastcox* exhibit the best performance when the number of genes was above 10. This means that in the other cases the methods fail in the execution of the cross-validation (see Supplementary Image 1).

In the overlapped-pathways setting, *Net-Cox* obtained the highest MCC overall when considering 88 genes (MCC equals to 0.227) with a false positive rate equals to 62.620, sensitivity 0.288 and specificity 0.938. However, even if the specificity levels of the three consensus rankings were almost equal to the previous setting (specificity for 44, 88, and 176 genes equals to 0.970, 0938, and 0.860, respectively), in this setting *Net-Cox* sensitivity decreased resulting in lower MCC compared to the not-overlapped case (MCC for 44, 88, and 176 genes equals to 0.175, 0.227, and 0.182, respectively). *AdaLnet* and *fastcox* also reported lower MCCs compared to the not-overlapped setting (MCC equals to 0.166 in *AdaLnet* and 0.134 in *fastcox*). In particular, both *AdaLnet* and *fastcox* showed an higher specificity than before (0.879 and 0.974, respectively) but a lower sensitivity (0.262 and 0.098). Further analysis showed that *AdaLnet* had the highest MCC when the number of selected genes was below 100 (MCC 0.196), while *fastcox* had the highest MCC (0.158) when the number of selected genes was above 10, in accordance with the previous results (see Supplementary Image 2).

Real Data Analysis

In order to evaluate the performance of the three Cox models in terms of survival analysis, we used cross-validated Kaplan–Meier curves (Simon et al., 2011b) for overall survival (OS) both on ovarian and breast microarray studies (see **Figure 2**, step 2). Note that p-value was estimated within the same dataset but the cross-validation approach is used to correct over optimistic conclusions due to the lack of independence between samples.

Moreover, since the ovarian datasets are comparable in terms of types of patients, platforms and cancer disease, Kaplan–Meier curves and two-side log-rank test were used to estimate the survival time and stratify the low-risk and high-risk groups on the independent test set (see **Figure 2**, step 3).

Table 5 reports the number of genes selected by the three Cox regression methods for each OS and the optimal tuning parameter α. Interestingly, the optimal α was often equal to 0.5, indicating that there was a good balance between statistical constraints and network information. These results confirm that the network carries important information useful for improving survival analysis. Moreover, since *Net-Cox* is a method based on ridge regression, the genes are only shrunk and it is necessary to fix a threshold for selecting the most relevant cancer genes. Hence, within each fold, we ordered the genes according to the absolute value of the corresponding regression coefficients, then we considered the union of the top 50 genes selected in each fold.

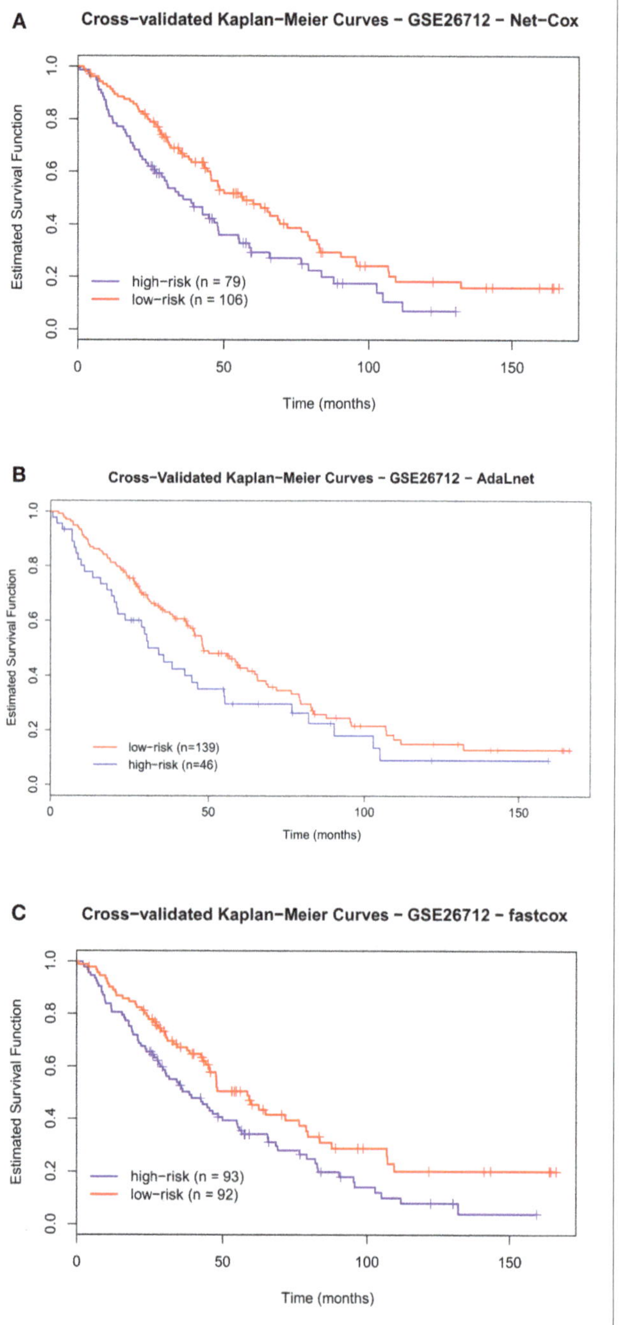

FIGURE 3 | Cross-validated Kaplan–Meier curves of the prognostic models on GSE26712 dataset. The patients are divided in *high-risk* and *low-risk groups* based on the pathways and genes selected by each methods for overall survival (OS). The survival probabilities of these two groups are compared using the log-rank test by using *Net-Cox* **(A)**, *AdaLnet* **(B)**, and *fastcox* **(C)**.

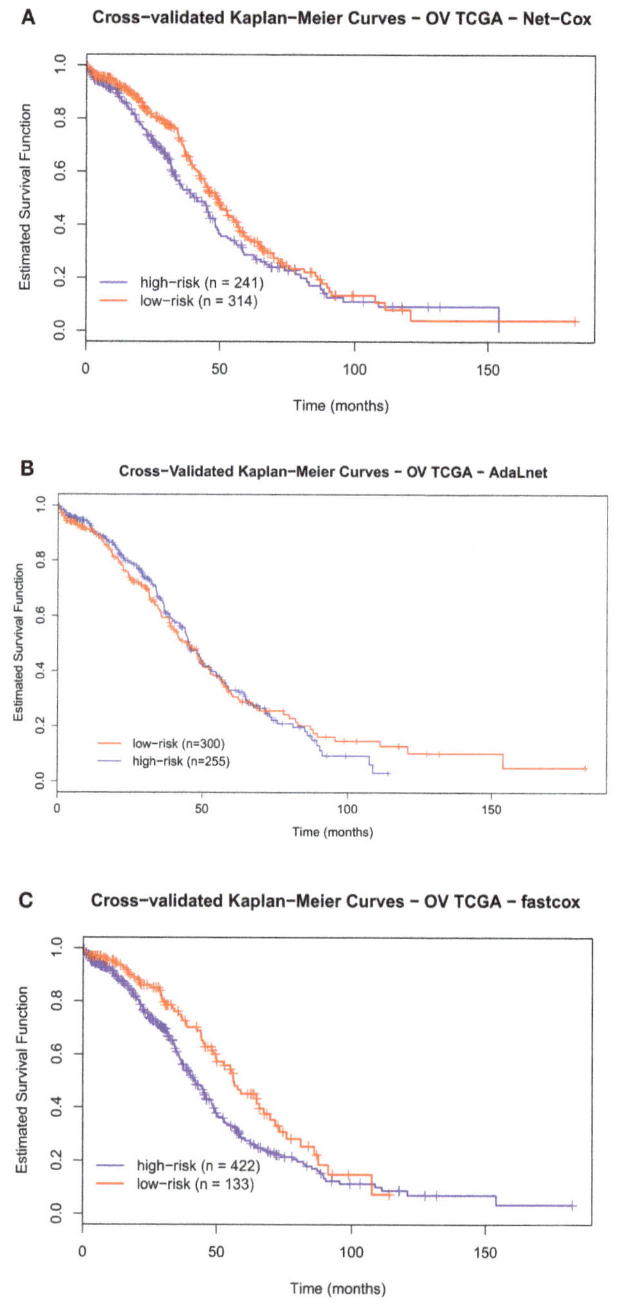

FIGURE 4 | Cross-validated Kaplan–Meier curves of the prognostic models on OV TCGA dataset. The patients are divided in *high-risk* and *low-risk groups* based on the pathways and genes selected by each methods for overall survival (OS). The survival probabilities of these two groups are compared using the log-rank test by using *Net-Cox* **(A)**, *AdaLnet* **(B)**, and *fastcox* **(C)**.

In the following, we present the main results obtained.

Results on the Ovarian Datasets

Figures 3, **4** show the cross-validated Kaplan–Meier curves for *high-and-low risk groups* patients selected in the ovarian datasets (Benome: GSE26712 and OV TCGA datasets, respectively). **Figure 3** shows that in the Bonome dataset the gap between the survival curves of the two risk groups in *Net-Cox* (**Figure 3A**) and *fastcox* (**Figure 3C**) is wider compared to *AdaLnet* (**Figure 3B**). In particular, in predicting survival probabilities, *fastcox* (permuted $p < 0.05$) seem to discriminate the risk groups better than *Net-Cox* and *AdaLnet* where the permuted $p > 0.05$. These

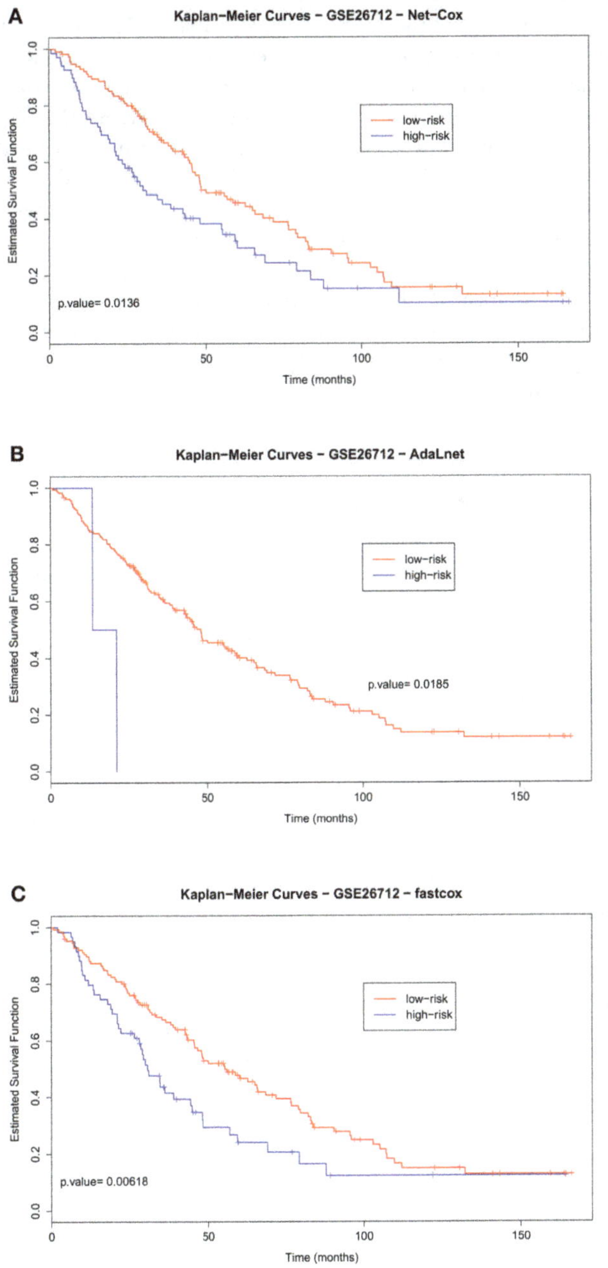

FIGURE 5 | Kaplan–Meier curves for validation test on GSE26712 ovarian dataset. The curves show the patients stratified by using the genes selected in the OV TCGA dataset by *Net-Cox*, *AdaLnet*, and *fastcox* [**(A)**, **(B)**, and **(C)**, respectively] with threshold $p < 0.05$.

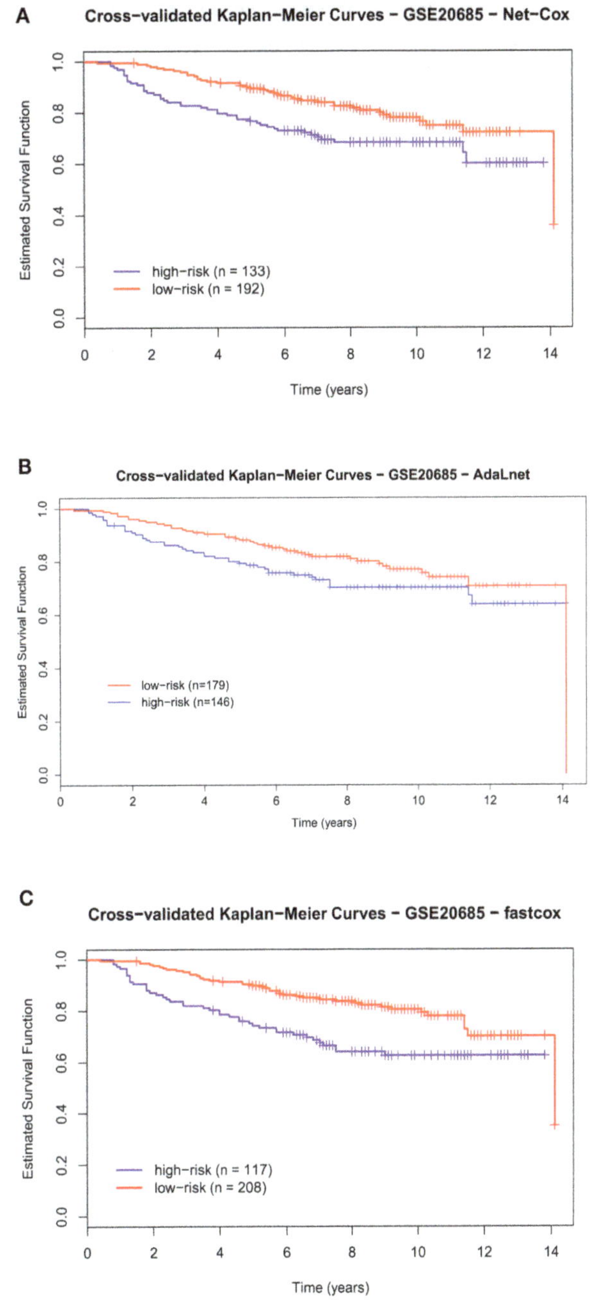

FIGURE 6 | Cross-validated Kaplan–Meier curves of the prognostic models on GSE20685 dataset. The patients are divided in *high-risk* and *low-risk groups* based on the pathways and genes selected by each methods for overall survival (OS). The survival probabilities of these two groups are compared using the log-rank test by using *Net-Cox* **(A)**, *AdaLnet* **(B)**, and *fastcox* **(C)**.

findings confirm the results previously obtained in Iuliano et al. (2014), in relation to the survival curves for each method. This was mainly due to the cross-validation approach used in this analysis to overcome the sample splitting problem with too small dataset.

On the other hand, in the OV TCGA dataset (**Figure 4**), the survival curves for *high-and-low risk* patients are not significantly separated. In particular, *fastcox* is the only method with a significant difference (permuted $p < 0.05$) in the OS between the *high-and-low-risk groups*.

Finally, to test the survival prediction across independent datasets, we used the ovarian OV TCGA dataset as training set, and the Benome dataset as the test set to predict the risk scores

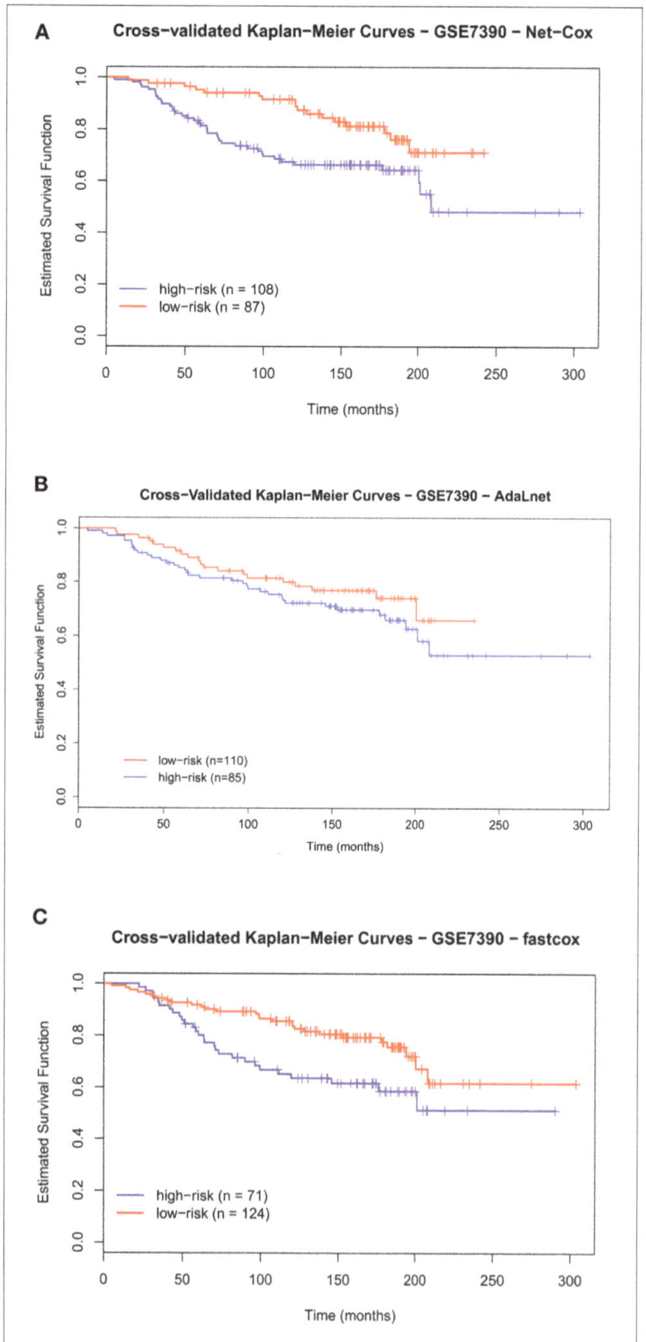

FIGURE 7 | Cross-validated Kaplan-Meier curves of the prognostic models on GSE7390 dataset. The patients are divided in *high-risk* and *low-risk groups* based on the pathways and genes selected by each methods for overall survival (OS). The survival probabilities of these two groups are compared using the log-rank test by using *Net-Cox* **(A)**, *AdaLnet* **(B)**, and *fastcox* **(C)**.

of the patients (see **Figure 2**, step 3). **Figure 5** shows the Kaplan–Meier curves for the two risk groups (high-and-low risk groups) in the Benome dataset obtained by *Net-Cox* (**Figure 5A**), *AdaLnet* (**Figure 5B**), and *fastcox* (**Figure 5C**). All the three methods gave a significant p-value at the 5% significance level (log-rank test, $p < 0.05$).

Results on Breast Datasets

Figures 6, **7** show the cross-validated Kaplan–Meier curves for *high-and-low risk groups* patients selected in the breast datasets (Kao: GSE20685 and Desmedt: GSE7390, respectively). In the Kao dataset, the permuted p-value related to **Figure 6A** (*Net-Cox*) and **Figure 6C** (*fastcox*) was smaller than 0.05, which means the *high-risk* and *low-risk* groups were significantly separated and the selected pathways and genes were related to survival times. In **Figure 6B** (*AdaLnet*), a patient of the high-risk group fell in the low-risk group and the permuted p-value is not significant.

We performed the same analysis for *high-and-low risk* patients in the Desmedt dataset. Also in this case, there was a significant difference in OS between the two risk groups as shown in **Figure 7A** (*Net-Cox*) and **Figure 7C** (*fastcox*) where the permuted p-value is smaller than 0.05. In **Figure 7B** (*AdaLnet*) the permuted p-value is not significant.

Identified Pathways

In this section, we present the results of the analysis in terms of KEGG pathways analysis based only on *not-isolated* genes (see section Methods for details). We report here only the networks related to *AdaLnet* and *Net-Cox* since all the networks related to *fastcox* have more than 100 node and 2000 edges and a clear visualization would not be possible. However, the lists of the genes selected by *fastcox* and the related pathways are reported in Supplementary Table 1 (ovarian datasets) and Supplementary Table 2 (breast datasets).

Figures 8, **9** show the gene-networks obtained for the Bonome dataset (GSE26712) built on the genes identified by *Net-Cox* and *Adalnet*, respectively. From the color of the nodes, we can infer that all the selected genes have a significant relation with ovarian cancer. Indeed, almost all the genes are close to red except for *AKT3* which has a p-value correlation equal to 0.039. Indeed, *AKT3* is usually involved in prostate and breast cancer (Nakatani et al., 1999). However, since it was selected both by *Net-Cox* and *fastcox*, a possible significant relation between *AKT3* and ovarian cancer could be inferred as indeed confirmed by literature (Liby et al., 2012). In particular, *AKT3* has a specific role in the genesis of ovarian cancer through modulation of G2-M phase transition (Cristiano et al., 2006). As showed in **Figure 8**, *AKT3* is also involved in many cancer pathways, such as *KEGG basal cell carcinoma, KEGG prostate cancer*, and *KEGG melangiogenesis*. It is worthy to note that this gene was also selected in our previous study (Iuliano et al., 2014) by all the analyzed methods and it was also involved in the same cancer related pathways. These findings confirm the importance of *AKT3* in ovarian cancer as confirmed indeed by literature (Cristiano et al., 2006).

In the Bonome dataset (GSE26712), *Adalnet* selected only two *not-isolated* genes (*RB1* and *BRCA2*) involved in two different cancer pathways (**Figure 9**). Both the genes have been frequently observed in epithelial ovarian cancer (Flesken-Nikitin et al., 2003; Dinulescu et al., 2005; Naora and Montell, 2005) and several studies report their stable correlation (Flesken-Nikitin et al., 2003; The Cancer Genome Atlas Research Network, 2011). Moreover, the strong interaction between *RB1* and the tumor protein *TP53* (Dong et al., 1997; Schuijer and Berns, 2003) has been identified by *Net-Cox* and *fastcox* (**Figure 8**).

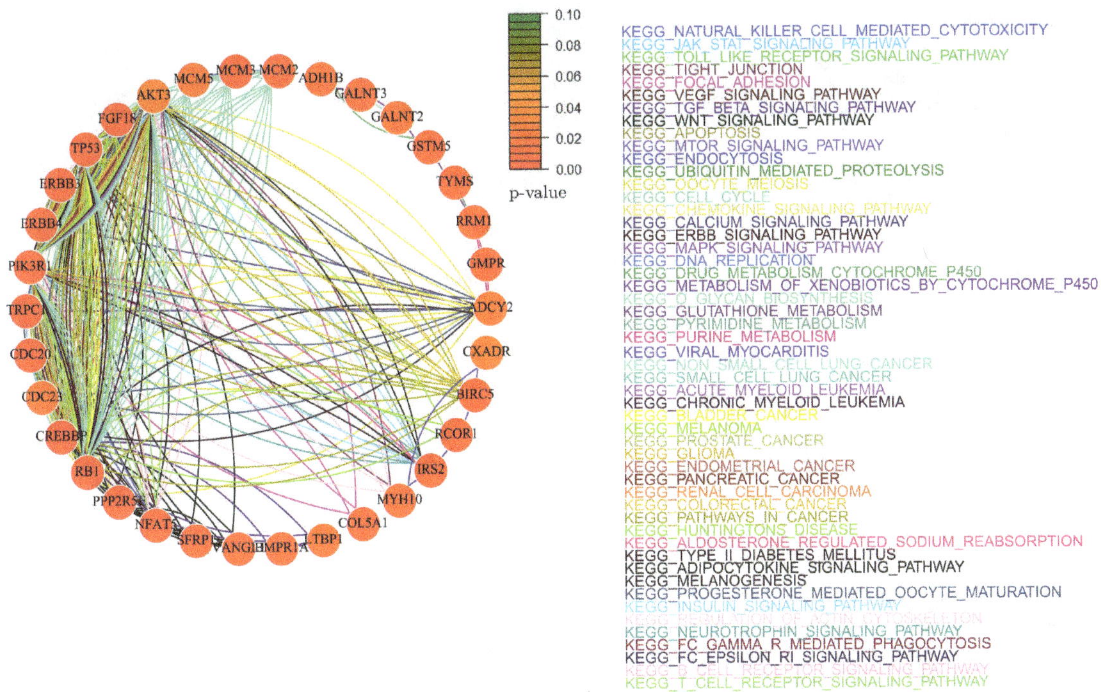

FIGURE 8 | Gene-network of not isolated genes selected by *Net-Cox* in the Bonome ovarian dataset (GSE26712). Each node represents a gene and an edge between two nodes means that the two genes belongs to the same pathway. Different colors are used for different pathways. The color of each node represents the *p*-value of the interaction between the gene and ovarian cancer (Huttenhower et al., 2009). Genes with $p > 0.10$ are represented in green.

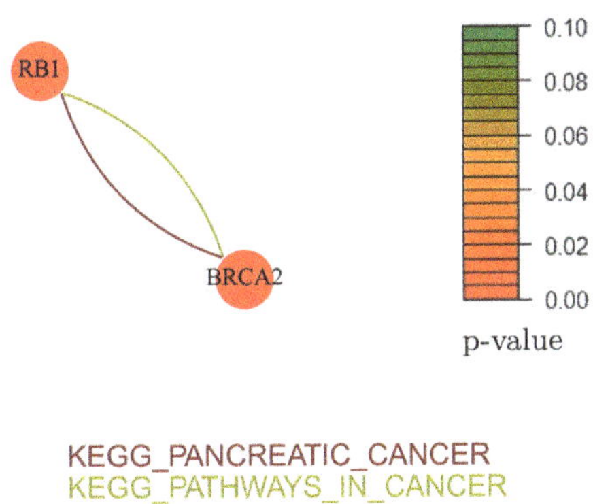

FIGURE 9 | Gene-network of not isolated genes selected by *Adalnet* in the Bonome ovarian dataset (GSE26712). Each node represents a gene and an edge between two nodes means that the two genes belongs to the same pathway. Different colors are used for different pathways. The color of each node represents the *p*-value of the interaction between the gene and ovarian cancer (Huttenhower et al., 2009). Genes with $p > 0.10$ are represented in green.

Figures 10, 11 show the gene-networks obtained for the OV TCGA ovarian dataset built on the genes identified by *Net-Cox* and *Adalnet*, respectively. As already observed in the Bonome dataset analysis, all the selected genes in the OV TCGA dataset resulted strongly correlated with ovarian cancer. Indeed, almost all the genes are close to red. The only gene with a slightly different color is *FZD3* which has a *p*-value of 0.049 and was selected by all the three methods. Hence, even if this gene has been mainly classified as gastric-cancer-related (Katoh, 2005), our results prove that it also has a relevant effect in ovarian cancer as confirmed by literature (Tapper et al., 2001). It is also important to note that other genes have been selected by all the three methods (i.e., *GMPR*, *ENPP1*, and *APC*). Such genes have been already classified as ovarian-related in cancer literature (Gayther et al., 1997; Kikuchi et al., 2007; Rikova et al., 2007), but, in our analysis, the pathways involved in such relation are also investigated. For example, while *GMPR* and *ENPP1* interact simply through the *KEGG purine metabolism* pathway, the *APC-FZD3* interaction involves three different pathways: *KEGG basal carcinoma*, *KEGG pathways in cancer*, and *KEGG wnt signaling pathway*.

It is worthy to note that some of the genes selected by the three methods (e.g., *NPY*, *COL5A1*, *EGFR*, and *FBL1*) have been already reported in literature (Zhang et al., 2013) where an analysis of subnetwork signatures in ovarian cancer based on Cox model is presented. Moreover, our approach selected new genes, such as *AKT3* and *RB1*, which are also related to ovarian cancer (Flesken-Nikitin et al., 2003; Cristiano et al., 2006). These results show that our findings are consistent with the previous ones including, at the same time, other gene signatures.

Figures 12, 13 report the gene-networks selected in the Kao dataset (GSE20685) by *Net-Cox* and *Adalnet*, respectively. *FGFR2* and *BCL2* were again selected in this dataset confirming the strong relevance of the two genes in breast cancer. Moreover,

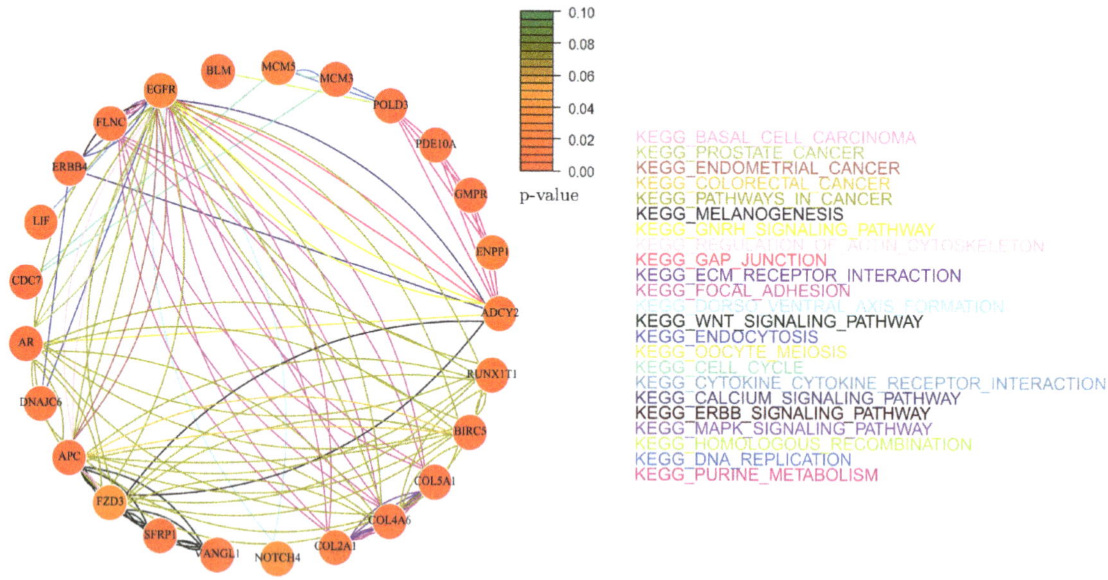

FIGURE 10 | Gene-network of not isolated genes selected by *Net-Cox* in the TCGA ovarian dataset. Each node represents a gene and an edge between two nodes means that the two genes belongs to the same pathway. Different colors are used for different pathways. The color of each node represents the *p*-value of the interaction between the gene and ovarian cancer (Huttenhower et al., 2009). Genes with $p > 0.10$ are represented in green.

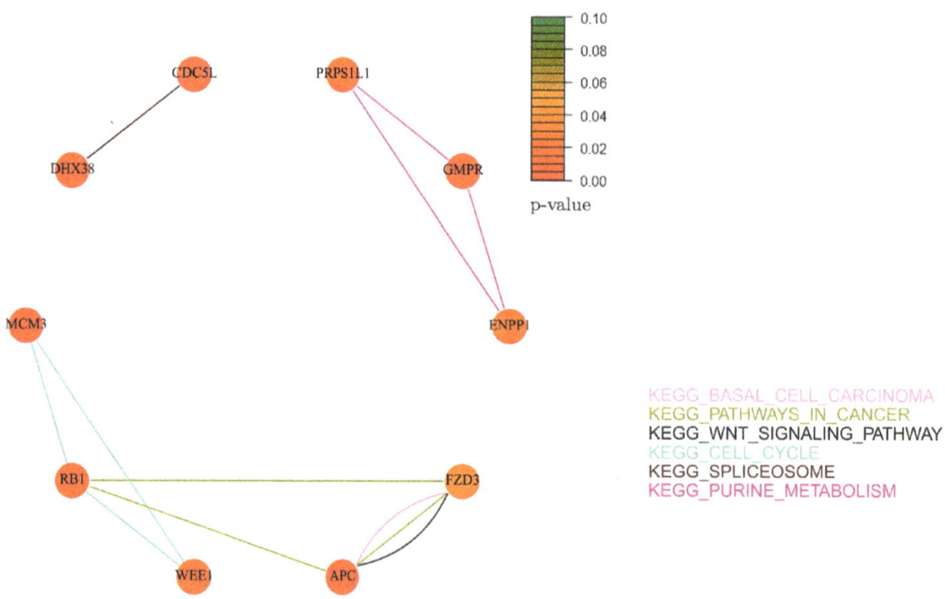

FIGURE 11 | Gene-network of not isolated genes selected by *Adalnet* in the TCGA ovarian dataset. Each node represents a gene and an edge between two nodes means that the two genes belongs to the same pathway. Different colors are used for different pathways. The color of each node represents the *p*-value of the interaction between the gene and ovarian cancer (Huttenhower et al., 2009). Genes with $p > 0.10$ are represented in green.

BRCA2 (Wooster et al., 1995) was selected by *Net-Cox* and *fastcox* confirming the accuracy of our analysis. It is also worthy to note that in all the breast cancer gene-networks the *KEGG prostate cancer* is always recurrent. This is mainly due to the common biomarkers between the two diseases (Yang et al., 1998; Mattie et al., 2006) and through our analysis new common biomarkers can be identified.

In the Desmedt dataset (GSE7390), all the genes selected by *Adalnet* were *isolated* and no network was built in this case. A list of the genes selected is reported in **Table 6. Figure 14**

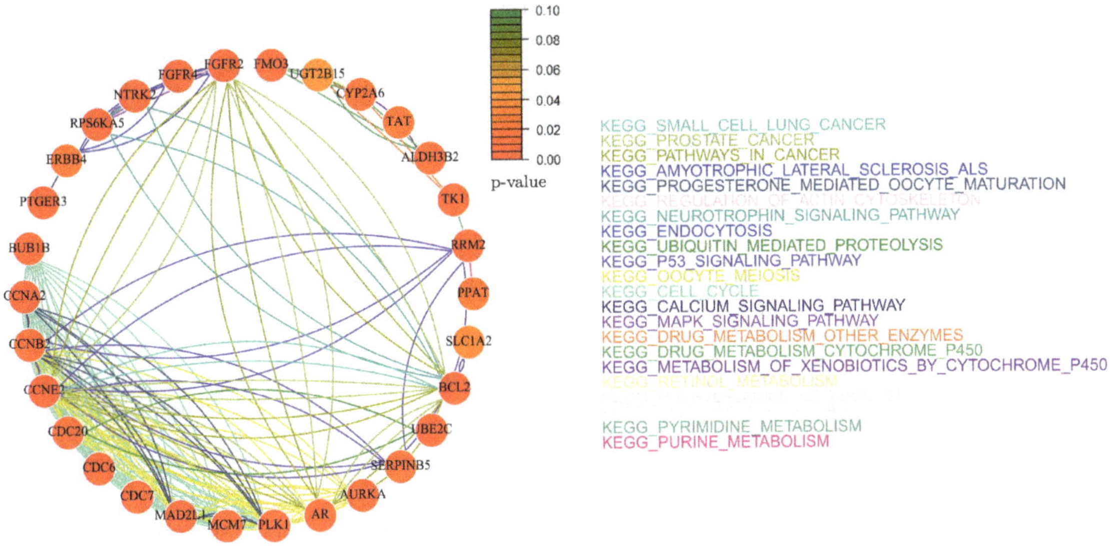

FIGURE 12 | Gene-network of not isolated genes selected by *Net-Cox* in the GSE20685 breast dataset. Each node represents a gene and an edge between two nodes means that the two genes belongs to the same pathway. Different colors are used for different pathways. The color of each node represents the *p*-value of the interaction between the gene and breast cancer (Huttenhower et al., 2009). Genes with $p > 0.10$ are represented in green.

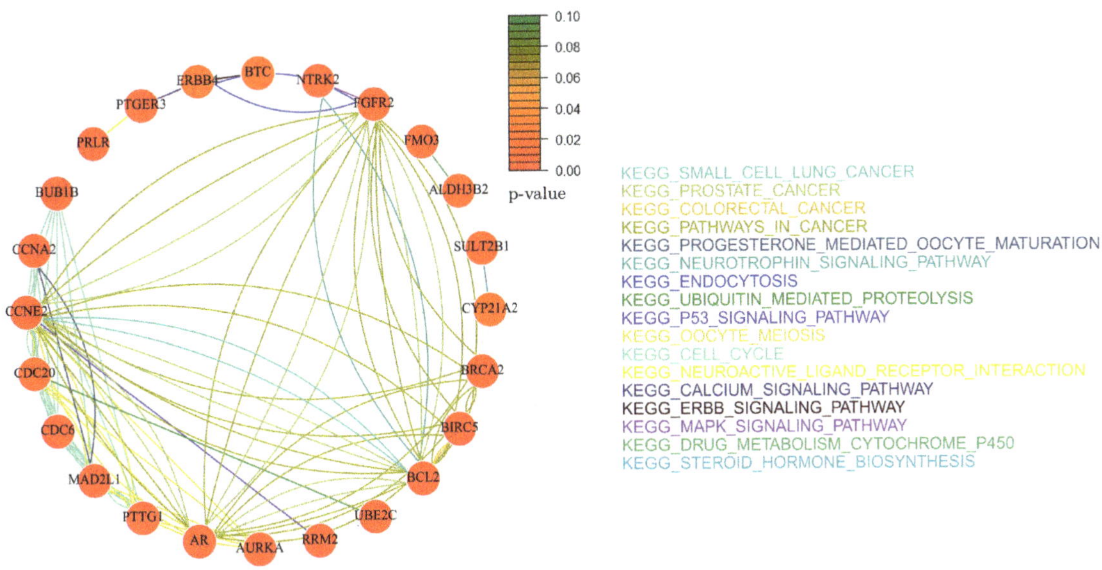

FIGURE 13 | Gene-network of not isolated genes selected by *Adalnet* in the GSE20685 breast dataset. Each node represents a gene and an edge between two nodes means that the two genes belongs to the same pathway. Different colors are used for different pathways. The color of each node represents the *p*-value of the interaction between the gene and breast cancer (Huttenhower et al., 2009). Genes with $p > 0.10$ are represented in green.

reports the gene-network related to the genes selected by *Net-Cox*. All the selected genes show a strong relation with the disease, such as *FGFR2* and *BCL2*, which were selected by both *Net-Cox* and *fastcox* and are involved in *KEGG prostate cancer* and in *KEGG pathways in cancer*. Both the genes are largely known as independent prognostic marker in breast cancer (Hunter et al., 2007; Thomadaki et al., 2007; Callagy et al., 2008). Both *Net-Cox* and *fastcox* selected *UGT2B15*, which has a breast-cancer-correlation $p = 0.049$. This gene has been usually involved in prostate cancer (Gsur et al., 2002), but recent works highlight its role also in breast cancer (Wegman et al., 2007).

In the analysis of the breast datasets, there was no overlap with our previous study (Iuliano et al., 2014). This was mainly due to the different datasets analyzed here potentially (different cancer subtype and different types of conditions) and to the more sophisticated procedures followed in this analysis. Indeed, in our previous work, we split the dataset in training and test set only

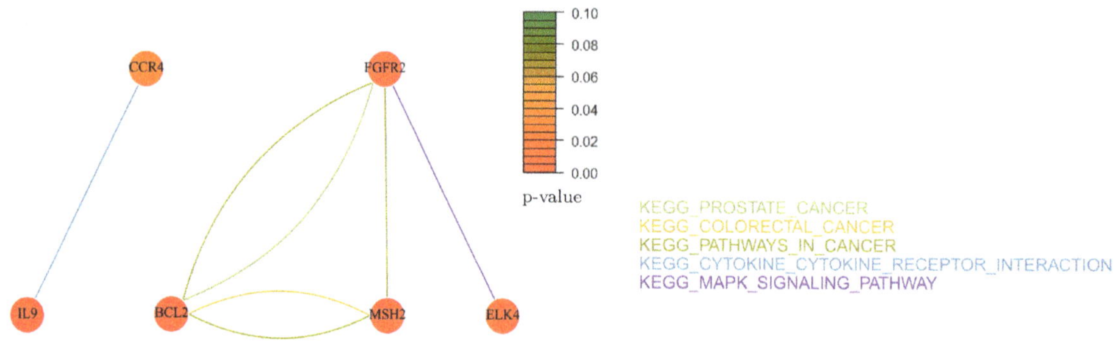

FIGURE 14 | Gene-network of not isolated genes selected by *Net-Cox* in the GSE7390 breast dataset. Each node represents a gene and an edge between two nodes means that the two genes belongs to the same pathway. Different colors are used for different pathways. The color of each node represents the p-value of the interaction between the gene and breast cancer (Huttenhower et al., 2009). Genes with $p > 0.10$ are represented in green.

TABLE 6 | List of genes selected by *Adalnet* in the breast dataset GSE7390.

Genes	p-values
BRCA1	0
GYPB	0.0489
MYBL2	0.0026
ADH6	0.0259
GHRHR	0.0007
GUCY2C	0.0323
PPP2R1B	0.0321
SLC1A2	0.0450
SLC12A3	0.0483
LIPF	0.0449
TRIP13	0.0001
PPM1E	0.0026
CEP152	0.0064
PSPC1	0.0475

The second column reports the breast-cancer correlation p-value of each gene accordingly (Huttenhower et al., 2009). All the selected genes resulted isolated *and no network was built in this case.*

once, while here we used a cross-validation procedure that is expected more robust results.

DISCUSSION AND CONCLUSIONS

A key issue in cancer survival analysis is uncovering the relation between gene expression profiles and cancer patients survival in order to identify biomarkers for disease diagnosis and treatment. In the last years, there has been a growing interest in methods that incorporate network information into classification algorithms for genes signature discovery. The main aims are to identify molecular biomarkers that reliably predict patient's response to therapy and to avoid ineffective treatment for reducing drug side-effects and associated costs. For this purpose, prognostic and diagnostic biomarker signatures need to be derived from omics data for various disease entities in order to offer useful methodological and practical strategy in research and clinical settings.

Here, we presented an extended methodological strategy for the analysis of gene signatures and survival prediction (see **Figure 1**). We integrated a new cross-validation method (Simon et al., 2011b) with the most recent network penalized Cox models (Yang and Zou, 2012; Zhang et al., 2013; Sun et al., 2014) to obtain an effective multi-splitting of the data and achieve an accurate survival prediction (see **Figure 2**). The analysis of the models was based both on simulated and real datasets in order to provide an accurate analysis in terms of statistical and biological investigation. Indeed, we showed that, given a number of variables not extremely high, all the analyzed methods were able to select the altered genes under different simulation settings. On the other hand, the analysis on real cancer datasets showed that through the integration of network information into Cox regression methods it is possible to identify cancer gene signatures with an accurate prognostic performance. Therefore, the contribution of this study is two-fold. Firstly, to obtain an integrative analysis of cancer genes networks and survival prediction. Secondly, to provide a computational and methodological framework for better investigating cancers regulatory networks and facilitating the management of patients in terms of prognosis, diagnosis and treatment.

The findings of this study have a number of important implications for future practice. Firstly, a practically appealing study based on a fast screening procedure (Fan and Lv, 2008; Fan et al., 2010) could be introduced in order to reduce the size of the feature space to a moderate scale. In fact, several types of screening procedures could be combined to integrate biological information into statistical screening analysis and provide more definitive understanding of the gene-regulatory networks. Secondly, the integration of clinical information and data from different omics (e.g., epigenomics or metabolomics) into the screening procedure could also provide a more accurate investigation and prevent the drawbacks of the current methods. Moreover, a more accurate biomarkers investigation could be performed using a number of high-quality binary PPIs available in literature (Rolland et al., 2014) where a proteome-scale map of the human binary interactome is compared to alternative network maps in order to give a deeper insight into genotype-phenotype relationships. Finally, it will be necessary to develop an user-friendly interface to turn this methodological framework into a practical tool.

AUTHOR CONTRIBUTIONS

AI and AO are joint first authors and both authors contributed equally. AI and AO prepared the computational codes and carried out all of the statistical analysis. CA, ID, and PL initiated and coordinated the work, guided the study design, supervised all data curation and analysis, and finalized all study conclusion. CA, ID, and PL are equal contributors. All the authors wrote, reviewed and approved the final manuscript.

ACKNOWLEDGMENTS

This research was partially supported by BioforIU Project and by EPIGEN Project. AO acknowledges the Qualcomm Research Scholarship and Cambridge Home and European Scholarship Scheme (CHESS). AI and AO would like to thank Prof. Pietro Lió for the hospitality at the Computer Laboratory in Cambridge and CA and ID for the hospitality at the IAC in Naples. We would like to thank also Prof. Hokeun Sun for sharing *AdaLnet* code and the reviewers for their insightful comments. PL received funding from the EU FP7-Health-F5-2012 under grant no. 305280 (MIMOmics).

REFERENCES

Bonome, T., Levine, D. A., Shih, J., Randonovich, M., Pise-Masison, C. A., Bogomolniy, F., et al. (2008). A gene signature predicting for survival in suboptimally debulked patients with ovarian cancer. *Cancer Res.* 68, 5478–5486. doi: 10.1158/0008-5472.CAN-07-6595

Callagy, G. M., Webber, M. J., Pharoah, P. D., and Caldas, C. (2008). Meta-analysis confirms BCL2 is an independent prognostic marker in breast cancer. *BMC Cancer* 8:153. doi: 10.1186/1471-2407-8-153

Candes, E., and Tao, T. (2007). The Dantzig selector: statistical estimation when *p* is much larger than *n*. *Ann. Stat.* 35, 2313–2351. doi: 10.1214/00905360600 0001523

Cox, D. R. (1972). Regression models and life-tables. *J. R. Stat. Soc. B Methodol.* 187–220.

Cristiano, B. E., Chan, J. C., Hannan, K. M., Lundie, N. A., Marmy-Conus, N. J., Campbell, I. G., et al. (2006). A specific role for AKT3 in the genesis of ovarian cancer through modulation of G2-M phase transition. *Cancer Res.* 66, 11718–11725. doi: 10.1158/0008-5472.CAN-06-1968

Desmedt, C., Piette, F., Loi, S., Wang, Y., Lallemand, F., Haibe-Kains, B., et al. (2007). Strong time dependence of the 76-gene prognostic signature for node-negative breast cancer patients in the transbig multicenter independent validation series. *Clin. Cancer Res.* 13, 3207–3214. doi: 10.1158/1078-0432.CCR-06-2765

Dinulescu, D. M., Ince, T. A., Quade, B. J., Shafer, S. A., Crowley, D., and Jacks, T. (2005). Role of K-ras and pten in the development of mouse models of endometriosis and endometrioid ovarian cancer. *Nat. Med.* 11, 63–70. doi: 10.1038/nm1173

Dong, Y., Walsh, M. D., McGuckin, M. A., Cummings, M. C., Gabrielli, B. G., Wright, G. R., et al. (1997). Reduced expression of retinoblastoma gene product (pRB) and high expression of p53 are associated with poor prognosis in ovarian cancer. *Int. J. Cancer*, 74, 407–415. doi: 10.1002/(SICI)1097-0215(19970822)74:4<407::AID-IJC8>3.0.CO;2-Z

Engler, D., and Li, Y. (2009). Survival analysis with high-dimensional covariates: an application in microarray studies. *Stat. Appl. Genet. Mol. Biol.* 8, 1–22. doi: 10.2202/1544-6115.1423

Fan, J., Feng, Y., and Wu, Y. (2010). "High-dimensional variable selection for cox's proportional hazards model," in *Borrowing Strength: Theory Powering Applications–A Festschrift for Lawrence D. Brown*, eds J. O. Berger, T. T. Cai, and I. M. Johnstone (Beachwood, OH: Institute of Mathematical Statistics), 70–86.

Fan, J., and Li, R. (2001). Variable selection via nonconcave penalized likelihood and its oracle properties. *J. Am. Stat. Assoc.* 96, 1348–1360. doi: 10.1198/016214501753382273

Fan, J., and Lv, J. (2008). Sure independence screening for ultrahigh dimensional feature space. *J. R. Stat. Soc. B Stat. Methodol.* 70, 849–911. doi: 10.1111/j.1467-9868.2008.00674.x

Flesken-Nikitin, A., Choi, K.-C., Eng, J. P., Shmidt, E. N., and Nikitin, A. Y. (2003). Induction of carcinogenesis by concurrent inactivation of p53 and Rb1 in the mouse ovarian surface epithelium. *Cancer Res.* 63, 3459–3463.

Fröhlich, H. (2014). Including network knowledge into Cox regression models for biomarker signature discovery. *Biom. J.* 56, 287–306. doi: 10.1002/bimj.201300035

Gayther, S. A., Mangion, J., Russell, P., Seal, S., Barfoot, R., Ponder, B. A., et al. (1997). Variation of risks of breast and ovarian cancer associated with different germline mutations of the BRCA2 gene. *Nat. Genet.* 15, 103–105. doi: 10.1038/ng0197-103

Gentleman, R. C., Carey, V. J., Bates, D. M., Bolstad, B., Dettling, M., Dudoit, S., et al. (2004). Bioconductor: open software development for computational biology and bioinformatics. *Genome Biol.* 5:R80. doi: 10.1186/gb-2004-5-10-r80

Gong, H., Wu, T. T., and Clarke, E. M. (2014). Pathway-gene identification for pancreatic cancer survival via doubly regularized Cox regression. *BMC Syst. Biol.* 8(Suppl. 1):S3. doi: 10.1186/1752-0509-8-s1-s3

Gsur, A., Preyer, M., Haidinger, G., Schatzl, G., Madersbacher, S., Marberger, M., et al. (2002). A polymorphism in the UDP-glucuronosyltransferase 2B15 gene ($D^{85}Y$) is not associated with prostate cancer risk. *Cancer Epidemiol. Biomarkers Prev.* 11, 497–498.

Gui, J., and Li, H. (2005). Penalized Cox regression analysis in the high-dimensional and low-sample size settings, with applications to microarray gene expression data. *Bioinformatics* 21, 3001–3008. doi: 10.1093/bioinformatics/bti422

Hudis, C. A. (2007). Trastuzumabmechanism of action and use in clinical practice. *N. Engl. J. Med.* 357, 39–51. doi: 10.1056/NEJMra043186

Hunter, D. J., Kraft, P., Jacobs, K. B., Cox, D. G., Yeager, M., Hankinson, S. E., et al. (2007). A genome-wide association study identifies alleles in FGFR2 associated with risk of sporadic postmenopausal breast cancer. *Nat. Genet.* 39, 870–874. doi: 10.1038/ng2075

Huttenhower, C., Haley, E. M., Hibbs, M. A., Dumeaux, V., Barrett, D. R., Coller, H. A., et al. (2009). Exploring the human genome with functional maps. *Genome Res.* 19, 1093–1106. doi: 10.1101/gr.082214.108

Iuliano, A., Occhipinti, A., Angelini, C., De Feis, I., and Lió, P. (2014). "Applications of network-based survival analysis methods for pathways detection in cancer," in *Computational Intelligence Methods for Bioinformatics and Biostatistics*, eds C. Di Serio, P. Liò, A. Nonis, and R. Tagliaferri (Springer), 76–88.

Jeong, H.-H., Kim, S. Y., Wee, K., and Sohn, K.-A. (2015). Investigating the utility of clinical outcome-guided mutual information network in network-based Cox regression. *BMC Syst. Biol.* 9:1. doi: 10.1186/1752-0509-9-S1-S8

Kao, K.-J., Chang, K.-M., Hsu, H.-C., and Huang, A. T. (2011). Correlation of microarray-based breast cancer molecular subtypes and clinical outcomes: implications for treatment optimization. *BMC Cancer* 11:143. doi: 10.1186/1471-2407-11-143

Karapetis, C. S., Khambata-Ford, S., Jonker, D. J., O'Callaghan, C. J., Tu, D., Tebbutt, N. C., et al. (2008). K-ras mutations and benefit from cetuximab in advanced colorectal cancer. *N. Engl. J. Med.* 359, 1757–1765. doi: 10.1056/NEJMoa0804385

Katoh, M. (2005). WNT/PCP signaling pathway and human cancer (review). *Oncol. Rep.* 14, 1583–1588. doi: 10.3892/or.14.6.1583

Kearns, M., and Ron, D. (1999). Algorithmic stability and sanity-check bounds for leave-one-out cross-validation. *Neural Comput.* 11, 1427–1453. doi: 10.1162/089976699300016304

Kikuchi, R., Tsuda, H., Kanai, Y., Kasamatsu, T., Sengoku, K., Hirohashi, S., et al. (2007). Promoter hypermethylation contributes to frequent inactivation of a putative conditional tumor suppressor gene connective tissue growth factor in ovarian cancer. *Cancer Res.* 67, 7095–7105. doi: 10.1158/0008-5472.CAN-06-4567

Kohavi, R. (1995). "A study of cross-validation and bootstrap for accuracy estimation and model selection," in *IJCAI*, Vol. 14 (Stanford, CA), 1137–1145.

Li, C., and Li, H. (2008). Network-constrained regularization and variable selection for analysis of genomic data. *Bioinformatics* 24, 1175–1182. doi: 10.1093/bioinformatics/btn081

Li, C., and Li, H. (2010). Variable selection and regression analysis for graph-structured covariates with an application to genomics. *Ann. Appl. Stat.* 4, 1498. doi: 10.1214/10-AOAS332

Liby, T. A., Spyropoulos, P., Buff Lindner, H., Eldridge, J., Beeson, C., Hsu, T., et al. (2012). Akt3 controls vascular endothelial growth factor secretion and angiogenesis in ovarian cancer cells. *Int. J. Cancer* 130, 532–543. doi: 10.1002/ijc.26010

Martinez-Ledesma, E., Verhaak, R. G., and Treviño, V. (2015). Identification of a multi-cancer gene expression biomarker for cancer clinical outcomes using a network-based algorithm. *Sci. Rep.* 5:11966. doi: 10.1038/srep11966

Mattie, M. D., Benz, C. C., Bowers, J., Sensinger, K., Wong, L., Scott, G. K., et al. (2006). Optimized high-throughput microrna expression profiling provides novel biomarker assessment of clinical prostate and breast cancer biopsies. *Mol. Cancer* 5:24. doi: 10.1186/1476-4598-5-24

Molinaro, A. M., Simon, R., and Pfeiffer, R. M. (2005). Prediction error estimation: a comparison of resampling methods. *Bioinformatics* 21, 3301–3307. doi: 10.1093/bioinformatics/bti499

Nakatani, K., Thompson, D. A., Barthel, A., Sakaue, H., Liu, W., Weigel, R. J., et al. (1999). Up-regulation of Akt3 in estrogen receptor-deficient breast cancers and androgen-independent prostate cancer lines. *J. Biol. Chem.* 274, 21528–21532. doi: 10.1074/jbc.274.31.21528

Naora, H., and Montell, D. J. (2005). Ovarian cancer metastasis: integrating insights from disparate model organisms. *Nat. Rev. Cancer* 5, 355–366. doi: 10.1038/nrc1611

Raghupathi, W., and Raghupathi, V. (2014). Big data analytics in healthcare: promise and potential. *Health Inf. Sci. Syst.* 2:3. doi: 10.1186/2047-2501-2-3

Refaeilzadeh, P., Tang, L., and Liu, H. (2009). "Cross-validation," in *Encyclopedia of Database Systems*, eds L. Liu and M. T. Özsu (New York, NY: Springer), 532–538. doi: 10.1007/978-0-387-39940-9_565

Rikova, K., Guo, A., Zeng, Q., Possemato, A., Yu, J., Haack, H., et al. (2007). Global survey of phosphotyrosine signaling identifies oncogenic kinases in lung cancer. *Cell* 131, 1190–1203. doi: 10.1016/j.cell.2007.11.025

Rolland, T., Taşan, M., Charloteaux, B., Pevzner, S. J., Zhong, Q., Sahni, N., et al. (2014). A proteome-scale map of the human interactome network. *Cell* 159, 1212–1226. doi: 10.1016/j.cell.2014.10.050

Schuijer, M., and Berns, E. M. (2003). TP53 and ovarian cancer. *Hum. Mutat.* 21, 285–291. doi: 10.1002/humu.10181

Simon, N., Friedman, J., Hastie, T., and Tibshirani, R., (2011a). Regularization paths for Cox's proportional hazards model via coordinate descent. *J. stat. Softw.* 39, 1–13. doi: 10.18637/jss.v039.i05

Simon, R. M., Subramanian, J., Li, M.-C., and Menezes, S. (2011b). Using cross-validation to evaluate predictive accuracy of survival risk classifiers based on high-dimensional data. *Brief. Bioinform.* 12, 203–214. doi: 10.1093/bib/bbr001

Sun, H., Lin, W., Feng, R., and Li, H. (2014). Network-regularized high-dimensional Cox regression for analysis of genomic data. *Stat. Sin.* 24:1433. doi: 10.5705/ss.2012.317

Tapper, J., Kettunen, E., El-Rifai, W., Seppälä, M., Andersson, L. C., and Knuutila, S. (2001). Changes in gene expression during progression of ovarian carcinoma. *Cancer Genet. Cytogenet.* 128, 1–6. doi: 10.1016/S0165-4608(01)00386-7

The Cancer Genome Atlas Research Network (2011). Integrated genomic analyses of ovarian carcinoma. *Nature* 474, 609–615. doi: 10.1038/nature10166

Thomadaki, H., Talieri, M., and Scorilas, A. (2007). Prognostic value of the apoptosis related genes BCL2 and BCL2L12 in breast cancer. *Cancer Lett.* 247, 48–55. doi: 10.1016/j.canlet.2006.03.016

Tibshirani, R. (1996). Regression shrinkage and selection via the lasso. *J. R. Stat. Soc. B Methodol.* 267–288. doi: 10.1002/(SICI)1097-0258(19970228)16:4<385::AID-SIM380>3.0.CO;2-3

Tibshirani, R. (1997). The Lasso method for variable selection in the cox model. *Stat. Med.* 16, 385–395.

van Houwelingen, H. C., Bruinsma, T., Hart, A. A., van't Veer, L. J., and Wessels, L. F. (2006). Cross-validated Cox regression on microarray gene expression data. *Stat. Med.* 25, 3201–3216. doi: 10.1002/sim.2353

Vasselli, J. R., Shih, J. H., Iyengar, S. R., Maranchie, J., Riss, J., Worrell, R., et al. (2003). Predicting survival in patients with metastatic kidney cancer by gene-expression profiling in the primary tumor. *Proc. Natl. Acad. Sci. U.S.A.* 100, 6958–6963. doi: 10.1073/pnas.1131754100

Wang, B., Mezlini, A. M., Demir, F., Fiume, M., Tu, Z., Brudno, M., et al. (2014). Similarity network fusion for aggregating data types on a genomic scale. *Nat. Methods* 11, 333–337. doi: 10.1038/nmeth.2810

Wegman, P., Elingarami, S., Carstensen, J., Stal, O., Nordenskjold, B., and Wingren, S. (2007). Genetic variants of CYP3A5, CYP2D6, SULT1A1, UGT2B15 and tamoxifen response in postmenopausal patients with breast cancer. *Breast Cancer Res.* 9:R7. doi: 10.1186/bcr1640

Wooster, R., Bignell, G., Lancaster, J., Swift, S., Seal, S., Mangion, J., et al. (1995). Identification of the breast cancer susceptibility gene BRCA2. *Nature* 378, 789–792. doi: 10.1038/378789a0

Wu, T. T., and Wang, S. (2013). Doubly regularized Cox regression for high-dimensional survival data with group structures. *Stat. Interface* 6, 175–186. doi: 10.4310/SII.2013.v6.n2.a2

Wu, Y. (2012). Elastic net for Coxs proportional hazards model with a solution path algorithm. *Stat. Sin.* 22:27. doi: 10.5705/ss.2010.107

Yang, G., Truong, L. D., Timme, T. L., Ren, C., Wheeler, T. M., Park, S. H., et al. (1998). Elevated expression of caveolin is associated with prostate and breast cancer. *Clin. Cancer Res.* 4, 1873–1880.

Yang, Y., and Zou, H. (2012). A cocktail algorithm for solving the elastic net penalized Coxs regression in high dimensions. *Stat. Sin.* 6, 167–173.

Zhang, W., Ota, T., Shridhar, V., Chien, J., Wu, B., and Kuang, R. (2013). Network-based survival analysis reveals subnetwork signatures for predicting outcomes of ovarian cancer treatment. *PLoS Comput. Biol.* 9:e1002975. doi: 10.1371/journal.pcbi.1002975

Zou, H. (2006). The adaptive lasso and its oracle properties. *J. Am. Stat. Assoc.* 101, 1418–1429. doi: 10.1198/016214506000000735

Zou, H., and Hastie, T. (2005). Regularization and variable selection via the elastic net. *J. R. Stat. Soc. B Methodol.* 67, 301–320. doi: 10.1111/j.1467-9868.2005.00503.x

2

Development of a Patient-Specific Multi-Scale Model to Understand Atherosclerosis and Calcification Locations: Comparison with *In vivo* Data in an Aortic Dissection

Mona Alimohammadi[1*†], *Cesar Pichardo-Almarza*[1†], *Obiekezie Agu*[2] *and Vanessa Díaz-Zuccarini*[1]

[1] *Mechanical Engineering, University College London, London UK,* [2] *Vascular Unit, University College Hospital, London, UK*

Edited by:
Zbigniew R. Struzik, The University of Tokyo, Japan

Reviewed by:
Umberto Morbiducci, Politecnico di Torino, Italy
Miguel Angel Martinez, University of Zaragoza, Spain

***Correspondence:**
Mona Alimohammadi
mona.alimohammadi.10@ucl.ac.uk

[†] *These authors have contributed equally to this work.*

Vascular calcification results in stiffening of the aorta and is associated with hypertension and atherosclerosis. Atherogenesis is a complex, multifactorial, and systemic process; the result of a number of factors, each operating simultaneously at several spatial and temporal scales. The ability to predict sites of atherogenesis would be of great use to clinicians in order to improve diagnostic and treatment planning. In this paper, we present a mathematical model as a tool to understand why atherosclerotic plaque and calcifications occur in specific locations. This model is then used to analyze vascular calcification and atherosclerotic areas in an aortic dissection patient using a mechanistic, multi-scale modeling approach, coupling patient-specific, fluid-structure interaction simulations with a model of endothelial mechanotransduction. A number of hemodynamic factors based on state-of-the-art literature are used as inputs to the endothelial permeability model, in order to investigate plaque and calcification distributions, which are compared with clinical imaging data. A significantly improved correlation between elevated hydraulic conductivity or volume flux and the presence of calcification and plaques was achieved by using a shear index comprising both mean and oscillatory shear components (HOLMES) and a non-Newtonian viscosity model as inputs, as compared to widely used hemodynamic indicators. The proposed approach shows promise as a predictive tool. The improvements obtained using the combined biomechanical/biochemical modeling approach highlight the benefits of mechanistic modeling as a powerful tool to understand complex phenomena and provides insight into the relative importance of key hemodynamic parameters.

Keywords: mathematical modeling, multiscale, atherosclerosis, patient-specific, aortic dissection, *in vivo* data

INTRODUCTION

Atherogenesis is a complex, multifactorial and systemic process; the result of a number of factors, each operating simultaneously at several spatial and temporal scales. The bewildering molecular and cellular complexity is well-described in Lusis' classical review more than a decade ago (Lusis, 2000), which highlights a plethora of biological mechanisms and gene associations, revealing an incredible etiological complexity. In addition to the biological components of the disease, atherosclerosis is also known to be related to mechanical stimuli on the vessel wall and hemodynamic parameters (Suo et al., 2006). Experimental evidence indicates that hemodynamic stimuli influence mechanotransduction and affect permeability (Davies, 1995). Increased permeability can lead to penetration and accumulation of lipoproteins (e.g., Low Density Lipoproteins–LDL) in the arterial wall and thus initiation of atherosclerosis. In recent years, much research has been conducted in order to draw correlations between hemodynamics and the atherogenic process (Peiffer et al., 2013a; Alimohammadi et al., 2015a). Certain hemodynamic parameters have been identified as key; these include flow distribution, pressure and wall shear stress (WSS) indices. Nevertheless, given the incredible complexity of the atherogenesis process, these hemodynamic analyses, on their own, have been inconclusive (Peiffer et al., 2013a) and a clear metric for plaque location remains elusive. Strong correlations between atherosclerotic disease and vascular calcification have been well-documented in the literature, including large cohort studies (Sangiorgi et al., 1998). Although the underlying molecular cause of calcification is unknown (Lanzer et al., 2014), the severity and extent of mineralization in calcification reflect atherosclerotic plaque burden (Demer and Tintut, 2008). Given the central role of inflammation in atherogenesis, an interesting possibility is that vascular mineral itself may initiate, promote, or perpetuate atherosclerosis by inducing inflammatory cytokines in monocytes that encounter and ingest hydroxyapatite crystals (Nadra, 2005).

Modeling and simulation have been used in a large number of studies in order to improve understanding of the role of hemodynamic variables in plaque formation. Although this work is mathematically elegant and can provide detailed insight into hemodynamics, it is disconnected from molecular research. Similarly, biomedical researchers often reduce the complexity of investigations of cardiovascular disease into manageable parts, for example, working on cell-lines or employing large-scale genome wide association studies (GWAS) to identify SNPs related to CVD (Tegner et al., 2006). However, such statistical genetic models have no mechanistic basis, and it is significant that Lusis highlighted the revival of functional studies in a relatively recent review (Lusis, 2012).

In this paper, a clear application of mathematics for healthcare will be made by unifying multi-mechanistic factors in the prediction of atherosclerosis location. The location of atherosclerotic plaque and vascular calcification will be investigated using a patient-specific biomechanical model of an aortic dissection (AD). From a physiological point of view, AD is a life threatening condition in which a tear forms in the wall of the aortic wall and blood splits the media layer, forming two lumina: the true lumen (TL) and false lumen (FL; Braverman, 2010). Blood flows from the TL into the FL via a primary tear and, in communicating dissections, returns to the TL via one or multiple tears downstream. The section of intima and media that separates the two lumina, called the intimal flap (IF), often stiffens over time due to fibrosis (Criado, 2011). Common comorbidities in patients with AD are atherosclerosis (Coady et al., 1999; Tsai et al., 2006) and inner wall calcification (de Jong et al., 2014). Additionally, patients suffering from AD usually have an elevated pulse pressure, which would likely be further increased in the presence of calcification, due to the reduction of vessel elastance (Demer and Tintut, 2008). The patient data used for this research showed an AD with significant atherosclerosis and calcification regions, consistent with the condition. Detailed hemodynamic characterization was achieved by using patient-specific dynamic boundary conditions representing the downstream vasculature, based on *in vivo* measurements collected for the same patient, treated in University College Hospital (UCH; Alimohammadi et al., 2014). Although complex flow simulations for this patient have been published previously, the work presented here, including the quantification and analysis of plaque and calcification areas as well as the multi-scale framework used in this context, is completely new.

The postulate of this research is that multiscale modeling and simulation can pave the way to study multi-mechanistic factors to explain disease in a cohesive modeling framework, which can integrate key markers at different biological scales and can provide insight into endothelial mechanotransduction, as well as potential predictive power in patient-specific analyses, compared to purely hemodynamic, biomechanical or biochemical approaches. We will use a virtual "follow-up" approach, combining a fluid-structure interaction (FSI) simulation model of a dissected aorta with a model of plaque formation, using a number of somewhat disparate indicators available in the literature, which will be described below. The patient-specific simulation results are then coupled to an endothelial permeability model following the three-pore approach (Olgac et al., 2008a), also recently used by Kim and Giddens (2015).

This paper is organized as follows: Section Methods presents the methods and simulation details. Results are presented in Section Results, including a clear, interpretable metric for atherogenic potential, which shows a better performance when compared to others based on "established" descriptors in the current literature. Simulation results of the model will be presented in and compared to atherosclerotic plaques and calcifications indicated in the original CT scans. The discussion, limitations and conclusions of this work will be presented in Section Discussion.

METHODS

Details of the Simulation

The specifics of the FSI simulation used for the present study are presented briefly in this section and are described in detail in Alimohammadi et al. (2015b). The patient gave informed oral consent and ethical approval was given by the National Research Ethics Service, UK, REC reference: 13/EM/0143. The fluid domain was reconstructed from CT scans of a female patient, starting at the ascending arch and extending down to the thoracic aorta, upstream of the iliac bifurcation. The supra-aortic branches were included in the model, but the visceral branches were not clearly resolvable from the CT data and so were omitted. The vessel wall was modeled by extruding the outer wall of the fluid geometry uniformly by 2.5 mm, based on reports of thickened aortic walls in hypertension (Malayeri et al., 2008), which occurs in ~70% of AD patients (Hagan et al., 2000; Khan and Nair, 2002). The intimal flap (IF), separating the TL and FL, was created by filling the gap between the two lumina and was 2.45 ± 0.34 mm (median ± median absolute deviation). Fluid and solid meshes were generated in ANSYS ICEM-CFD, and contained ~230,000 (with 7 prismatic layers at the wall) and 50,000 elements respectively.

Simulations were carried out using ANSYS mechanical and CFX. The vessel wall was modeled using the isotropic hyperelastic model of Raghavan and Vorp (2000). This model is comparable to a linear elastic model with a Young's modulus of 1 MPa, but displays a small amount of strain-stiffening. An external pressure of 52.5 mmHg (diastolic pressure in the descending aorta) was applied and at each of the outlets of the 3D domain, the solid geometry was restricted to planar motion about the center-point of the lumen, enabling expansion of the vessels.

At the inlet, a flow wave from another study of a patient with type-B aortic dissection was used (Karmonik et al., 2008), as such data was not available for the present patient. At the fluid outlets, three element Windkessel models were used, with parameters tuned to invasive pressure measurements on the patient, using an iterative technique described in a previous study (Alimohammadi et al., 2014). Blood was modeled as an incompressible fluid of density 1056 kg/m^3. Turbulence was modeled using the hybrid k-ε, k-ω shear stress transport (SST) turbulence model, with a 1% turbulence level at the inlet. In order to account for the shear-thinning properties of blood, the Carreau-Yasuda (CY) viscosity model was utilized with the parameters reported by Gijsen et al. (1999). The CY viscosity model includes infinite and zero shear viscosities, a characteristic time constant and two exponents without direct physical meaning to describe the viscosity-shear response of blood. The parameters of Gijsen et al. (1999) were fitted to a blood analog, which showed similar characteristics to blood samples. The viscosity predicted by this model is ~3.5 mPa s (a commonly used Newtonian viscosity) at 100/s, and decreases for higher shear rates to around 2.5 mPa s at 1000/s. At shear rates below 100/s, the viscosity increases, exceeding 10 mPa s for shear rates below 2/s.

Simulations were run with a time step of 5 ms, a periodic solution was achieved after two cycles and the third cycle was extracted for further analysis. In a previous study (Alimohammadi et al., 2015b), it was shown that the low shear rates in the slow flow regions of the distal and proximal FL led to significantly increased viscosity.

Shear Stress Indices (SSI)

In the permeability calculations described in Section Endothelial Permeability Model, one of the inputs is an index indicative of the shear stress condition that leads to increased permeability. The time-average wall shear stress (TAWSS) and oscillatory shear index (OSI; Ku et al., 1985) are two commonly used indices that are considered important for plaque formation. TAWSS describes the average magnitude of the shear stress and the OSI gives an indication of the directionality of the shear stress, yielding 0 for uniaxial flows and 0.5 when there is no preferential direction. Typically, in permeability models for atherogenesis, TAWSS is used as the shear index related to permeability.

A number of studies (Malek et al., 1999; Xiang et al., 2010; Chiu and Chien, 2011; Meng et al., 2014) have shown that regions with low average shear stress *combined* with highly oscillatory shear stress have increased endothelial permeability along with other pathological responses. More recently, Sáez et al. (2015) presented the 3D remodeling of endothelial cells as the combined effect of OSI and TAWSS in a computational framework, which fitted experimental works presented before in *in vitro* studies.

In light of the indications for increased permeability in low, oscillatory regions, we propose an index, HOLMES (Highly Oscillatory, Low MagnitudE Shear), given by:

$$HOLMES = TAWSS\,(0.5 - OSI) \tag{1}$$

This parameter is equivalent to half the reciprocal of relative residence time (RRT) which was previously identified as a potential index for combining these two characteristics (Himburg, 2004). The *HOLMES* indicator can be understood as a modified *TAWSS*, with the (0.5 − *OSI*) term further reducing the index in regions where the wall shear stress is both low in magnitude and oscillatory in nature. Additionally, HOLMES provides a conceptually alternative explanation, offering a linear (rather than reciprocal) index, proportional to WSS that intuitively corresponds to the observed effects of shear characteristics on endothelial permeability.

In the present study, we compare the efficacy of using the most-widely used hemodynamic/shear stress index (SSI) for atheroprone regions, i.e., TAWSS, with HOLMES, separate and in a mechanistic model, as shown in the sections below.

Endothelial Permeability Model

The endothelial permeability model proposed here is based on previous work describing the early stages of atherosclerosis, using a transport model of low density lipoprotein (LDL) from the artery lumen into the arterial wall, taking into account the effects of mechanical stimuli exerted by the blood flow on the endothelial cell layer and its pathways of volume and solute flux; see Díaz-Zuccarini et al. (2014) for more details. An excellent and recent analysis and use of this model along the same lines using time-average wall shear stress (TAWSS) has been recently published by Kim and Giddens (2015).

The endothelial layer is described with a three-pore modeling approach considering the contributions of the vesicular pathway, normal junctions, and leaky junctions. The fraction of leaky junctions is calculated as a function of the mechanical stimuli and is used in conjunction with the pore theory to determine the transport properties of this pathway.

The LDL transport equations are decomposed using three main penetration pathways: leaky junctions, normal junctions and vesicular pathways; so the bulk of volume flux (J_v) through the endothelial membrane is given by:

$$J_v = J_{v,lj} + J_{v,nj} \tag{2}$$

where $J_{v,lj}$ is the flux through leaky junctions and $J_{v,nj}$ is the flux through normal junctions.

The volumetric flux through leaky junctions ($J_{v,lj}$) is calculated using a modified version of the Kedem-Ketchalsky equations for membrane transport:

$$J_{v,lj} = L_{p,lj} \left(\Delta p_{end} - \sigma_d \Delta \Pi \right) \tag{3}$$

where $L_{p,lj}$ is the hydraulic conductivity, Δp_{end} is the pressure difference through the endothelium, σ_d is the osmotic reflection coefficient and $\Delta \Pi$ is the osmotic pressure. The value of Δp_{end} is estimated by subtracting the externally applied pressure from either the average pressure throughout the domain (uniform pressure gradient) or the spatially varying time average pressure at each wall location (time-averaged pressure gradient).

According to the three *pores* theory, solute flux (LDL flux in this case) only occurs through endothelial leaky cell junctions and vesicles:

$$J_s = J_{s,lj} + J_{s,v} \tag{4}$$

Assuming that the solute flux through the vesicular pathway ($J_{s,v}$) is 10% of the solute flux through the leaky junction pathway ($J_{s,lj}$) (Olgac et al., 2008b).

Finally, in the interest of brevity, Equation (5) shows a general form of the function used to calculate $J_{s,lj}$ which is proportional to the magnitude of $J_{v,lj}$ (more details about the calculation of $J_{s,lj}$ are shown elsewhere (Díaz-Zuccarini et al., 2014; Alimohammadi et al., 2015a):

$$J_{s,lj} = \phi \left(J_{v,lj}, Pi, Pe, \sigma, c_{lum}, c_{w,end}, c_{ave} \right) \tag{5}$$

where Pi is the diffusive permeability, Pe the modified Peclet number, c_{lum} and $c_{w,end}$ the LDL concentration in the lumen and the sub-endothelial layer respectively, c_{ave} is the mean endothelial concentration and σ the solvent drag coefficient.

Endothelial cell shape will affect the amount of leaky junctions. Experimental findings have shown that in areas of altered hemodynamics, endothelial cells do not have a typical cobblestone shape, but rather exhibit a more circular shape as well as increased permeability. Previous models used a relationship between endothelial permeability and local WSS based on the Endothelial Cell Shape Index (ECSI; Levesque et al., 1986). ECSI is related to the cellular shape and takes values from zero to one, i.e., a circle has an ECSI of one whilst a straight line has an ECSI of zero.

Some approaches based on ECSI calculated this variable as a function of WSS (for instance Olgac et al., 2008b used WSS values in steady state simulations). However, as previously mentioned, recent work shows that ECSI can be affected by other indices *as well*, such as OSI. Based on the seminal paper by Levesque et al. (1986), excellent work from Sáez et al. (2015), presents how different OSI and TAWSS modify the endothelial cell shape. This combined relationship will be key for the modeling work presented here.

In the present study, we consider the role of the two different shear stress indices (SSI), i.e., TAWSS and HOLMES, described in Section Shear Stress Indices (SSI).

For each SSI, *ECSI* is defined according to

$$ECSI = 0.380e^{-0.79SSI} + 0.225e^{-0.043SSI} \tag{6}$$

Figure 1 shows a comparison of normalized ECSI using Saez et al. results (Sáez et al., 2015), which are dependent on TAWSS and OSI, with the normalized ECSI using Equation (6) and HOLMES. Very similar behavior can be seen in both surfaces, showing how areas of low TAWSS and high OSI will increase the values of ECSI. Following this, we will use HOLMES as a combined index able to capture key haemodynamic features in the regions of interest for atherosclerotic plaque/calcification locations.

Leaky cells have high permeability to LDL, which can be linked to the magnitude of the SSI. Areas with high ECSI will be related

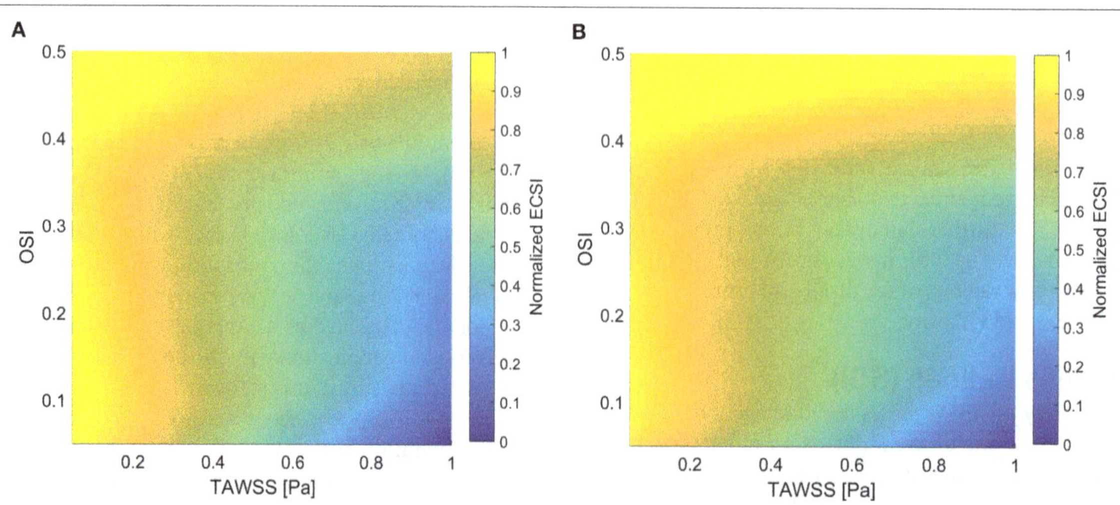

FIGURE 1 | Normalized ECSI as a function of TAWSS and OSI. (A) Using Saez et al. results (Sáez et al., 2015); **(B)** Using Equation (6) and HOLMES.

to a higher number of mitotic cells (MC) which are calculated as follows:

$$MC = 0.003739e^{14.75ECSI} \tag{7}$$

Assuming that within the endothelium the quantity of leaky mitotic cells is ~80.5%, which represents ~45.3% of the total number of leaky cells (LC) in that area (Tedgui and Lever, 1984a), the number of LC is calculated as:

$$LC = 0.307 + 0.805MC \tag{8}$$

The ratio of endothelium (ϕ) covered by LCs is calculated using

$$\phi = \frac{LC\,\pi R_{cell}^2}{unit\ area} \tag{9}$$

where R_{cell} is the radius of a single cell. Finally, the total hydraulic conductivity of the endothelial leaky junctions ($L_{p,lj}$) is defined as:

$$L_{p,lj} = \phi \cdot L_{p,slj} \tag{10}$$

where $L_{p,slj}$ is the hydraulic conductivity of a single leaky junction calculated as follows:

$$L_{p,slj} = \frac{w^2}{3\mu l_{lj}} \tag{11}$$

with w and l_{lj} being the half-width (20 nm) and the length (2 μm) of the leaky junctions and μ the viscosity term used for the estimation of the LDL penetration.

Estimating Plaque Location and Metrics Evaluation

Previous work (Alimohammadi et al., 2015a) has shown that calculation of the hydraulic conductivity in leaky junctions ($L_{p,lj}$) can be used to estimate the magnitude of LDL fluxes across the wall along the artery and identify atheroprone regions. However, as mentioned previously, other variables such as the volume flux $J_{v,lj}$ [which is closely related to the solute (LDL) flux, $J_{s,lj}$], may be a better metric to identify atheroprone areas. As shown in Equation (3), $J_{v,lj}$ is proportional to both $L_{p,lj}$ and the pressure drop across the endothelium Δp_{end} and thus, $J_{v,lj}$ may be more sensitive in the estimation of plaque locations. For a given metric ($J_{v,lj}$ or $L_{p,lj}$), the aorta was divided into subsections and the mapping of this variable was compared with the plaque/calcification observed in the same subsections from CT scans, as shown in **Figure 2**. Plaque areas are calculated as a percentage (%) of the total area covered for the selected section of the artery, with plaque defined as HU>220 in the clinical image (Isgum et al., 2004) and $L_{p,lj}$ >1.2 × 10^{-11} m^2/s/kg (Tedgui and Lever, 1984b). The ratio of the area coved in the model compared to the clinical image is reported as a percentage match.

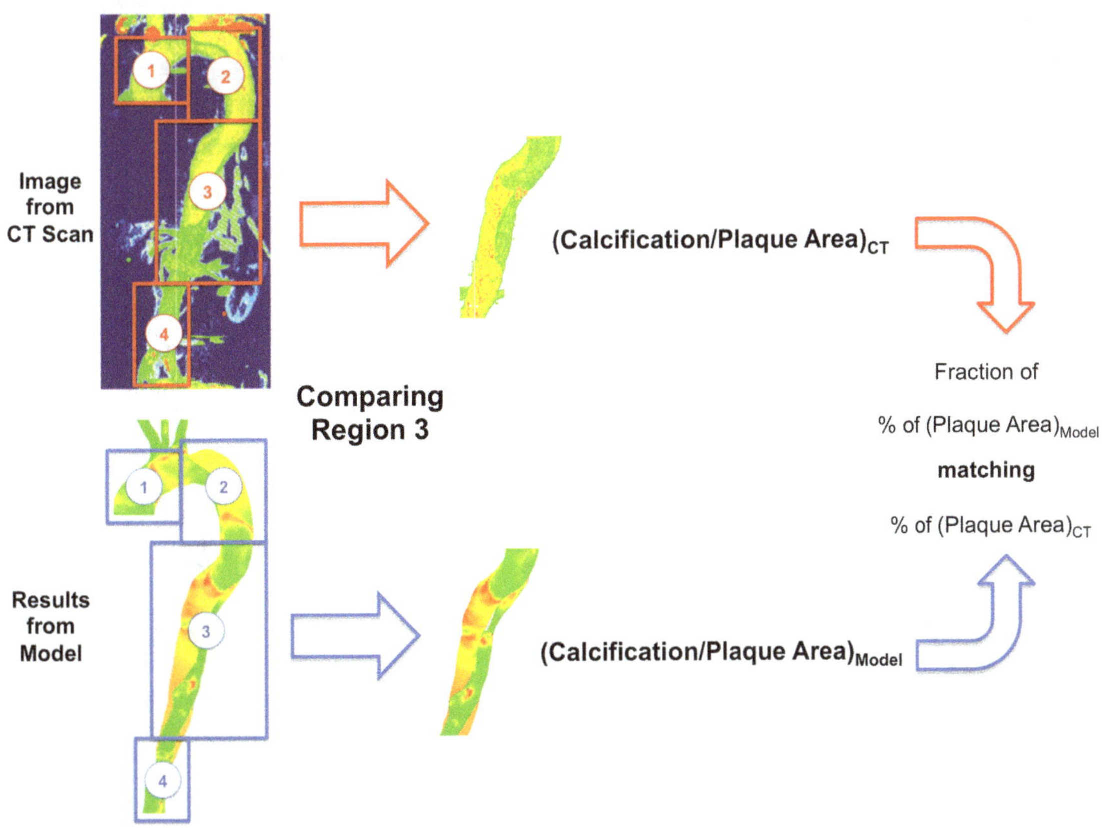

FIGURE 2 | Definition of the different regions in the aorta for comparison and analysis.

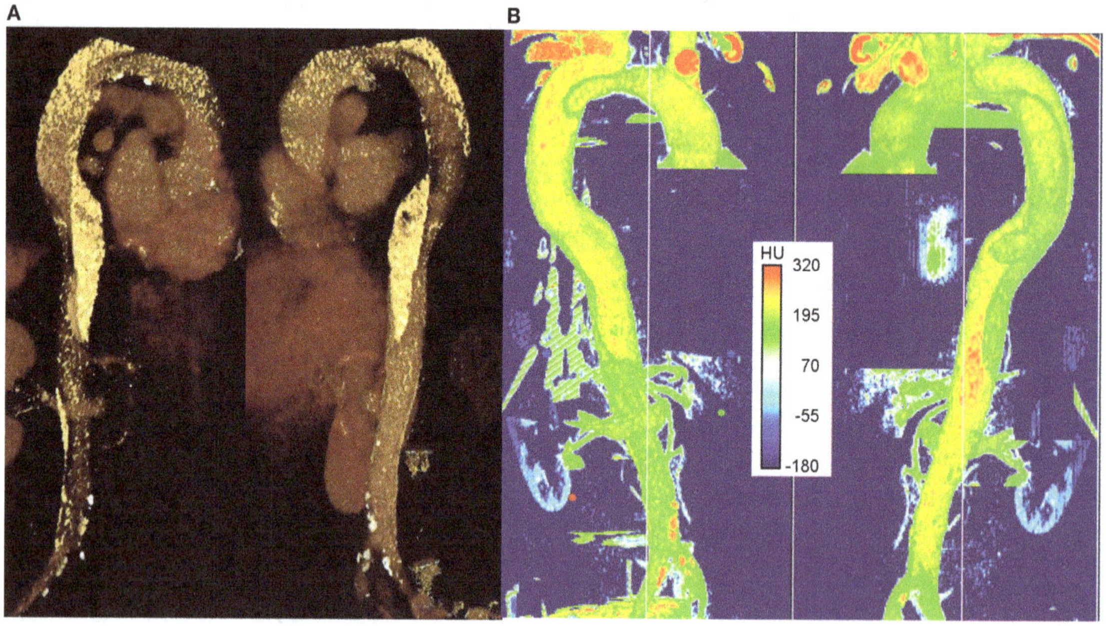

FIGURE 3 | The right anterior and left posterior views of the aorta. (A) 3D CT data visualized in Aquarius and **(B)** 3D LUT volume viewer in FIJI (Image J).

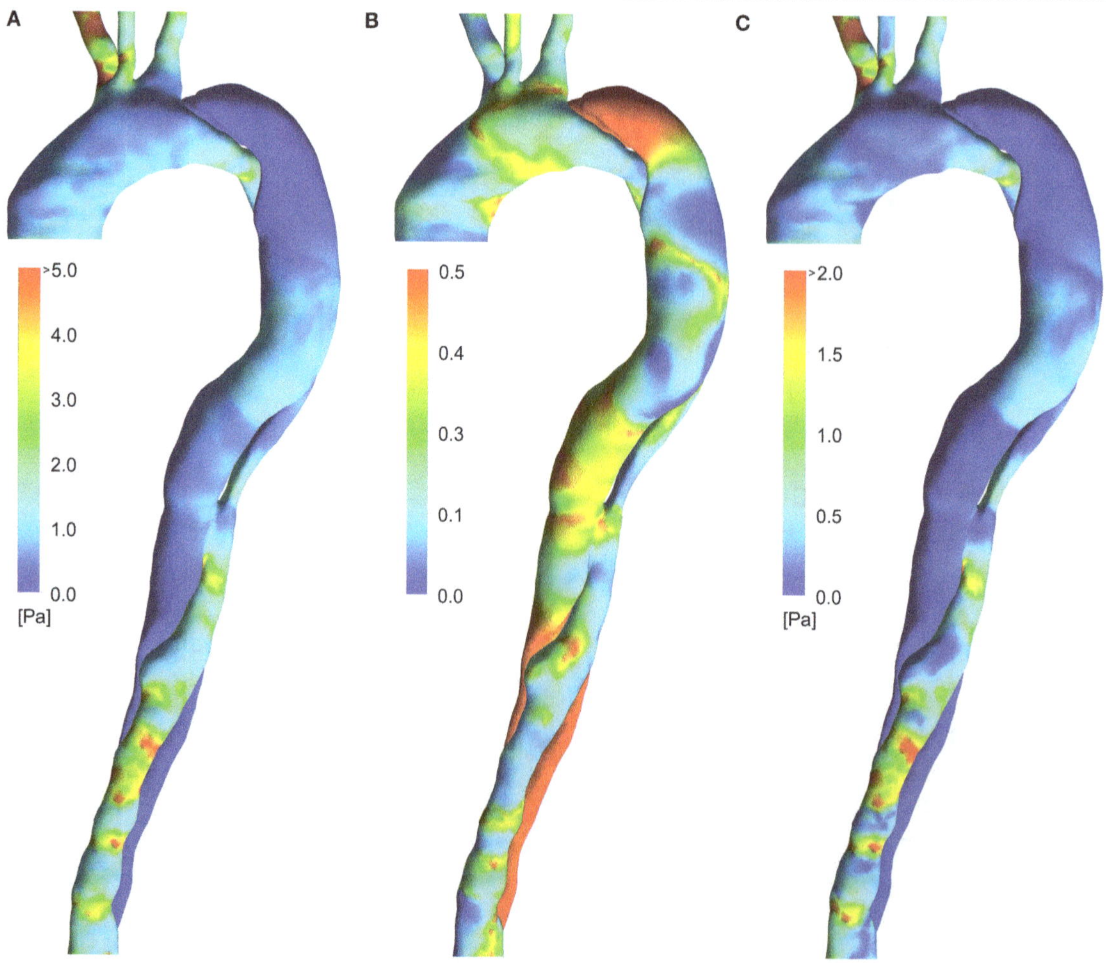

FIGURE 4 | Wall shear indices from the FSI simulations. (A) TAWSS, **(B)** OSI, **(C)** HOLMES.

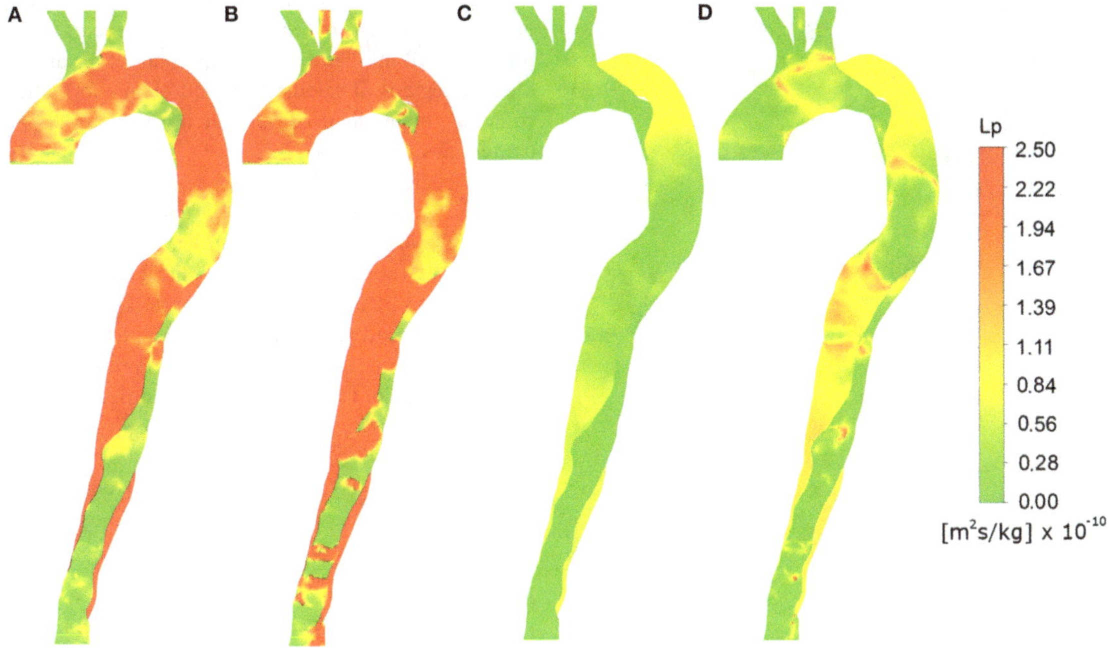

FIGURE 5 | Total hydraulic conductivity of leaky junctions ($L_{p,lj}$) values calculated using various parameter combinations from the FSI results shown in the right anterior view. (A) Constant plasma viscosity and TAWSS, **(B)** Constant plasma viscosity and HOLMES, **(C)** Non-Newtonian blood viscosity and TAWSS, **(D)** Non-Newtonian blood viscosity and HOLMES.

Comparison of the calcification/plaque area observed in the CT scans and those calculated by the chosen metric following the methodology shown here, was used to evaluate the efficacy of the different metrics and the role of some of the selected hemodynamic (viscosity and SSI) variables.

RESULTS

Figure 3A shows an image of the 3D CT image reconstruction in Aquarius (TeraRecon, USA), in which the highest intensity regions indicate plaques (white) or vessel wall calcification. **Figure 3B** shows an image of the CT scans reconstructed in the LUT volume viewer in Fiji (Schindelin et al., 2012), with intensities in the range 0–255. The colorbar shows Hounslow units (HU) from the CT scan. Although there is not a single threshold that can be considered to be a calcification, a value of 220 HU has been used previously as an indicator of calcified regions (Isgum et al., 2004). The plaques appear red and the calcifications yellow (it should be noted that the exact rendering of both panels in **Figure 2** is dependent on the distance from the first slide as there is no easy way to ensure the same depth in both pieces of software, and so the views may differ slightly).

As seen in both panels, a number of completely formed plaques can be observed in the lower edge of the aortic arch, between the two tears in FL and around the iliac bifurcation. Calcifications can be seen along the aortic arch proximal to the supraaortic branches, and in the proximal and distal FL. The existence and locations of the plaques and calcifications were confirmed and approved by the vascular surgeon who managed this patient.

Wall Shear Stress Indices

Figure 4 shows the distributions of the three SSI used in the present study. The TAWSS (**Figure 4A**) is moderate in the ascending aorta and in the FL between the two tears. Elevated regions of TAWSS can be observed in the visceral branches, at the coarctation and in the distal TL. The OSI (**Figure 4B**) shows scattered elevated regions throughout the domain, with particularly high values in the distal and proximal FL (which would not be captured using a rigid wall model; Alimohammadi et al., 2015b). The HOLMES index (**Figure 4C**) is similar to TAWSS, but lower in magnitude overall, and particularly in regions where the OSI is high, such as between the two tears and in the aortic arch.

In the following sub-sections, we systematically compare the indices and parameters used in the literature to indicate or predict locations of calcification and/or atherogenesis with our integrated model, the use of HOLMES and spatially varying viscosity.

Analysis of the Total Hydraulic Conductivity

As the volume flux is dependent on both hydraulic conductivity and pressure gradient across the endothelium, we first consider L_p alone.

Figures 4A,B show the total hydraulic conductivity of leaky junctions ($L_{p,lj}$) values calculated assuming a constant viscosity at the wall, equal to the plasma viscosity and TAWSS and HOLMES,

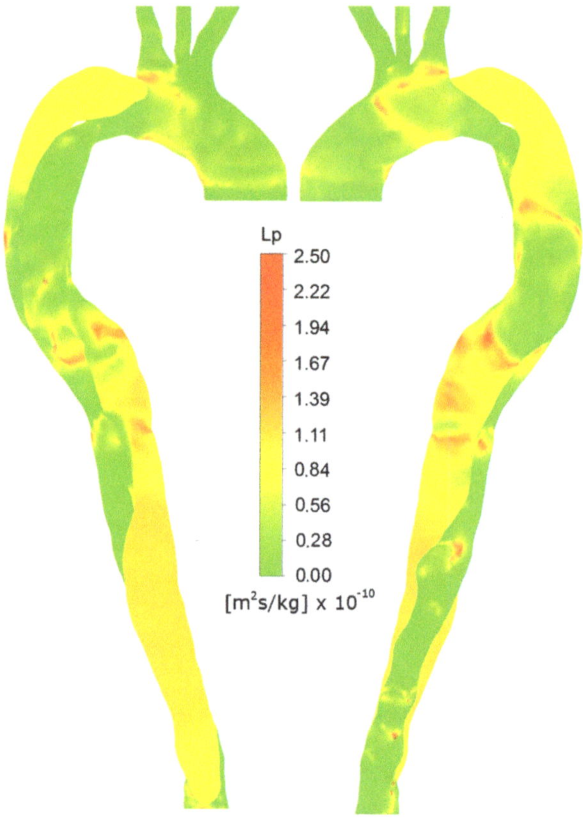

FIGURE 6 | Right anterior and left posterior views of $L_{p,lj}$, calculated using non-Newtonian viscosity and HOLMES.

respectively, as wall shear stress indices. **Figures 4C, 5D** show the same wall shear stress indices, but with the viscosity at the wall predicted based on the continuum assumption using the Carreau-Yasuda viscosity model. Note that all panels use the same color scale. Tedgui and Lever (1984b) reported $L_{p,lj}$ values of 1.2×10^{-11} m^2s/kg, corresponding to green values in the figure. Higher values indicate pathologically elevated $L_{p,lj}$. For both of the plasma viscosity models (**Figures 4A,B**), the estimated $L_{p,lj}$ values are extremely high, and no qualitative correlation can be observed when comparing these images to **Figure 3B**. Using the non-Newtonian viscosity model at the vessel wall and TAWSS to calculate $L_{p,lj}$ (**Figure 5C**), reduced the magnitude of the calculated $L_{p,lj}$ and resulted in a reasonable indication of the calcified/atheroprone locations in both the distal and proximal FL, although the extent of the calcified regions is shorter than those observed in **Figure 2** and the calcified region between the two tears is not identified. Additionally, this model did not identify any of the plaque locations. Comparing the $L_{p,lj}$ distribution predicted using HOLMES and non-Newtonian viscosity (**Figure 5D**) to the clinical images (**Figure 3B**), it can be seen that there is a good qualitative correlation between the two figures. **Figure 5** shows this $L_{p,lj}$ distribution in greater detail in both right anterior and left posterior views.

Under the assumption that plaque formation and calcification occur as a result of elevated $L_{p,lj}$, the correlation between **Figures 6, 3B** implies that this combination of variables is able to identify regions of calcification and plaque formation, and thus could be potentially used to predict further development of such pathologies.

The region of calcification with scattered small plaques observed in the FL between the two tears is captured in the $L_{p,lj}$ prediction in **Figure 6**. Similarly, the calcification of the proximal FL (both views) and the distal FL (left posterior view), are predicted by elevated $L_{p,lj}$ in these regions. The low $L_{p,lj}$ in the distal TL and ascending arch in the left posterior view also correlate with minimal calcification in **Figure 3B**. The plaques at the lower edge of the aortic arch, best observed in **Figure 3A** in the left posterior view, correspond to regions of elevated $L_{p,lj}$ in **Figure 6**.

Some regions of the image do not correlate, such as the low $L_{p,lj}$ in the right anterior view of the ascending arch, where there is calcification in the CT image.

Analysis of the Volume Flux

Figure 6 provides images of the predicted volume flux across the vessel wall, $J_{v,lj}$, using various combinations of wall shear indices and viscosity assumptions, along with either a uniform pressure gradient or a time-averaged pressure gradient. These figures can be compared with **Figure 3B** to analyze the hypothesis that calcification and plaque formation are better correlated with volume flux, $J_{v,lj}$, rather than $L_{p,lj}$ alone. Meyer et al. (1996) reported J_v values in the range $\sim$2–3 $\times 10^{-8}$ m/s, corresponding to green regions in the figure. Higher values indicate elevated solute flux.

None of the constant plasma models (**Figures 7A,B,E,F** produced $J_{v,lj}$ distributions that compared well with **Figure 3B**. **Figures 7C,D** using the uniform pressure gradient are qualitatively identical to **Figures 4C,D** as the uniform pressure distribution does not affect the relative values of $J_{v,lj}$ as compared to $L_{p,lj}$. When using the time averaged pressure gradient (**Figures 7G,H**), the distributions do not significantly change in qualitative terms, indicating that the prediction is more sensitive to $L_{p,lj}$, and thus appropriate models for wall shear stress and blood viscosity at the wall, than to the pressure gradient. **Figure 8**, showing the distribution of $J_{v,lj}$ using HOLMES, the non-Newtonian model and time averaged pressure distribution, is therefore qualitatively similar to **Figure 6**, although there are some differences in the magnitudes.

DISCUSSION

The results shown in Section Results are promising and thus warrant further investigation. They also open up a number of questions. Is the volume flux (which determines the magnitude of the solute flux) a metric for plaque location? If so, then what is required in order to estimate it correctly? As mentioned in the introduction, the question of plaque location remains elusive. There are presently no available tools to predict where this might occur, or indeed to sufficiently explain why. This may be because, if we take volume flux ($J_{v,lj}$) (or even hydraulic conductivity) as

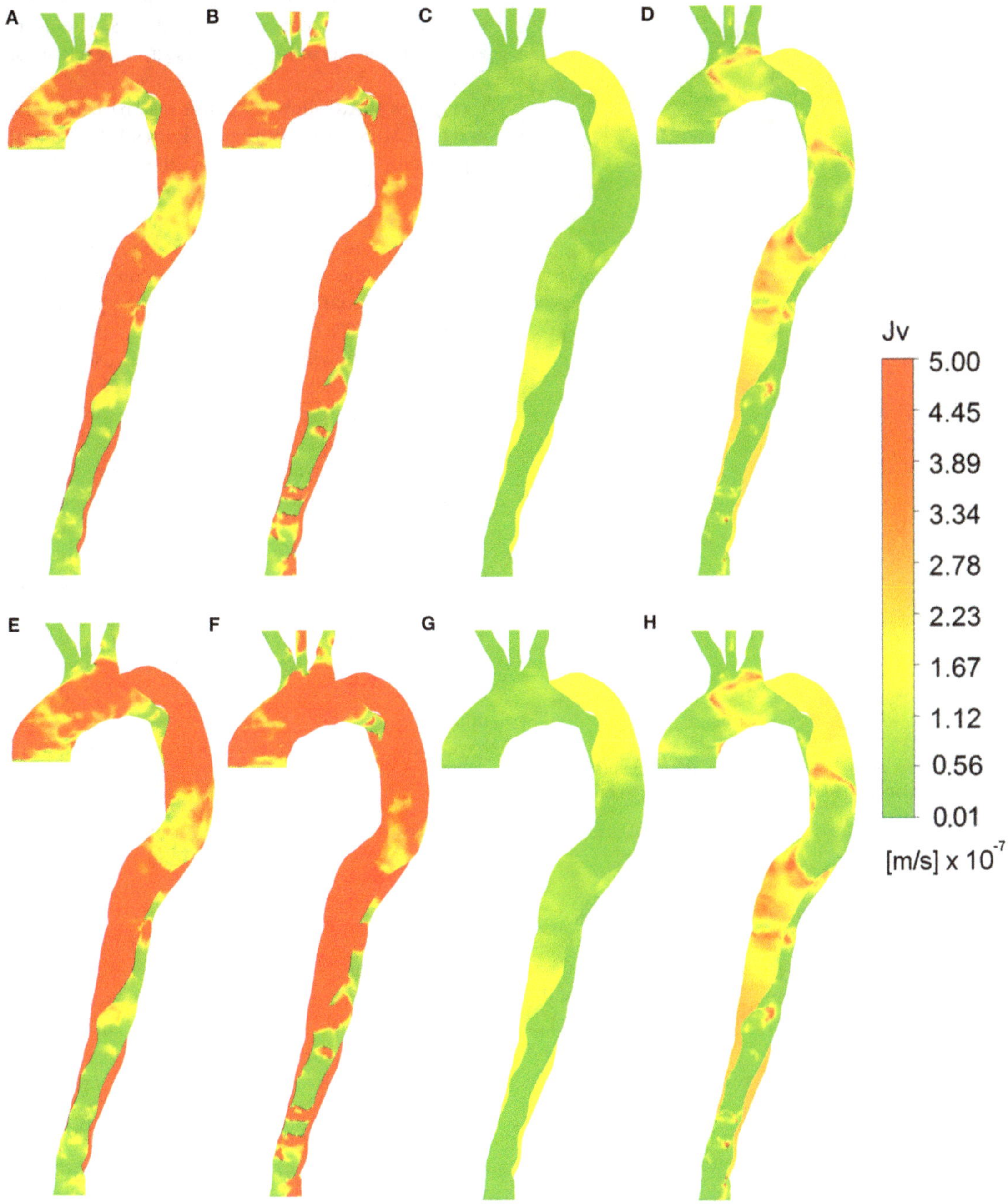

FIGURE 7 | Volume flux ($J_{v,lj}$) calculated using various parameters. Wall shear indices **(A)** TAWSS, **(B)** HOLMES with constant plasma viscosity and uniform pressure gradient. Wall shear indices **(C)** TAWSS, **(D)** HOLMES with non-Newtonian blood viscosity and uniform pressure gradient. **(E)** TAWSS, **(F)** HOLMES with constant plasma viscosity and time averaged pressure gradient. **(G)** TAWSS, **(H)** HOLMES with non-Newtonian blood viscosity, and time-averaged pressure gradient.

the metric for plaque location, its generation is multifactorial and complex and thus simplified models lack some of the key features that seem to impact plaque location (see Section Results). In the present study, strong correlations were only produced through the use of a multiscale model combining multiple hypotheses at the cellular and physiological level, as evidenced by the comparison of **Figures 5**, **7** with **Figure 3**. This complex approach involves the detailed formulation, extraction and combination of more than 11 different variables including the calculation of variable pressure differences along the arterial wall across the endothelium, the use of a non-linear model for blood viscosity and use of the HOLMES index to isolate regions of low, oscillatory shear. This formulation, making use of mechanistic models, also provides these metrics ($J_{v,lj}$ and $L_{p,lj}$) with strong

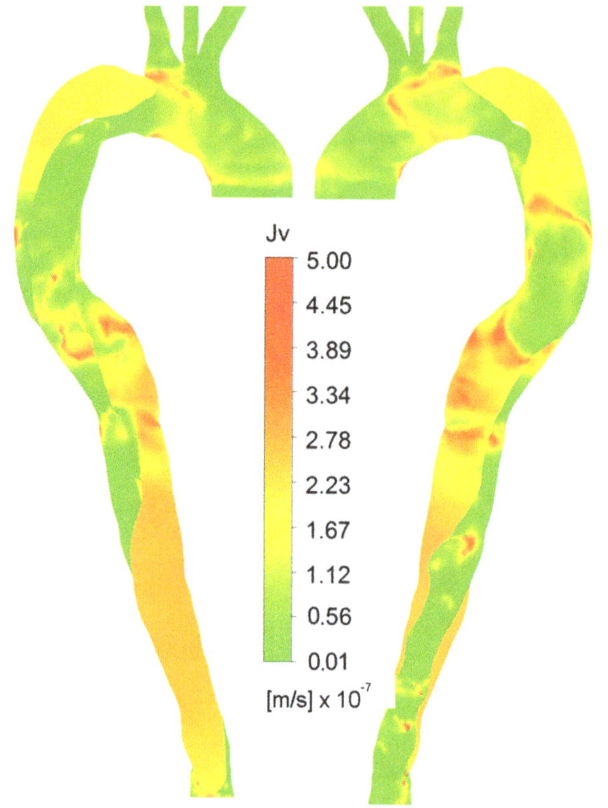

FIGURE 8 | Right anterior and left posterior views of $J_{v,lj}$, calculated using non-Newtonian viscosity, HOLMES and time-averaged pressure gradient.

interpretability and physical meaning. As previously explained, the origin and interpretation of HOLMES as a combined index is strongly rooted in the most recent, relevant literature (Sáez et al., 2015). The work presented in this paper exemplifies the power of mechanistic models as "hypothesis tester" and a key example of the use of mathematics and mechanistic models and formulations *to understand* biological processes. In fact, it can be can argued that *a posteriori hypotheses*, as a result of abandoning part of a priori thinking in the light of new observations, can pave the way for future studies (Erren, 2007). Here, an unambiguous formulation could enable others to follow up on the findings and modified conjectures to advance knowledge in this field. Despite some limitations (please see below), it remains a fact that the work presented here is strongly anchored in the state of the art and current knowledge in this area.

In addition to the clear qualitative improvement in the correlation between the CT scans and $L_{p,lj}$ and $J_{v,lj}$ distributions, we utilized a simple measure based on plaque areas, as shown in **Figure 9** for the descending aorta, a key region in the case of AD (as explained in Section Estimating Plaque Location and Metrics Evaluation).

When comparing $L_{p,lj}$ values in specific regions of the aorta, using the results shown in **Figure 5**, the calcified regions predicted by the model overlapped 80% of the calcified regions observed in the CT scans for the HOLMES model. No calcification was predicted using TAWSS. The next step was to analyze the results of $J_{v,lj}$ (given its stronger relationship with the LDL flux).

Figures 10A,C show $J_{v,lj}$ in this region calculated with TAWSS and HOLMES, respectively and **Figures 10B,D** show the calcified region only. **Figures 10E,F** show the equivalent figures for the CT scans. Comparing the proportional area of the calcified region between the two cases, 26 and 93% of the regions overlap for TAWSS and HOLMES, respectively. When calculated using HOLMES in the selected region (Region 3 in **Figure 2**), 45% of the visible arterial wall is predicted to be prone to calcification, compared to 48% of the visible arterial wall observed in the CT scan image. Applying the same analysis using HOLMES to the whole aorta predicts between 80 and 95% of overlapping. This analysis supports the *potential* predictive power of the proposed model.

From a hemodynamics point of view, the heterogeneous distribution of calcification/plaque formation will be influenced by disturbed flow. Wall shear indices such as TAWSS and OSI capture only partial aspects of the complex flow fields. The wall shear stress vector has both direction and magnitude at each moment throughout the cardiac cycle. TAWSS captures the average magnitude, but contains no information about directionality. OSI characterizes the variability in the directionality. However, a number of indices ("emerging" multidirectional predictors, as defined by Gallo et al., 2016) have been proposed and some of these might have potential as predictors of atherogenesis. Peiffer et al. (2013b) proposed the transverse wall shear stress, which is the temporal average of the component of the wall shear stress vector that is perpendicular to the time average wall shear stress vector. Morbiducci et al. (2015) defined the preferential direction as being relative to the direction of the vessel centerline and similar considered the temporal average, producing an additional index based on the temporal average of the ratio of these two components. Arzani and Shadden (2016) also introduced backward wall shear stress, considering the average of the negative instances of wall shear stress in the direction opposite to the centerline. In future studies, it will be important to explore the efficacy of these new indices.

In this paper, we have exclusively focused on the "established" predictors, for which the links between endothelial cell behavior and haemodynamics have been shown in the relevant literature as described above.

In the work presented here, the use of patient-specific data (including invasive haemodynamic measurements) and BCs, inclusion of wall motion and use of a non-Newtonian blood viscosity model are likely to improve the estimates of these hemodynamic variables and thereby increase the likelihood of an improved correlation. Importantly, as evidenced by **Figures 4, 7**, the interpretation of the role of these variables individually was unable to predict distributions of atherogenesis/calcification. It was only through a combination of HOLMES, a compound shear index, a non-Newtonian fluid viscosity and a sophisticated simulation model that this was achieved. In general, when extracting information related to values calculated at the wall, motion is extremely important. In this particular case and as

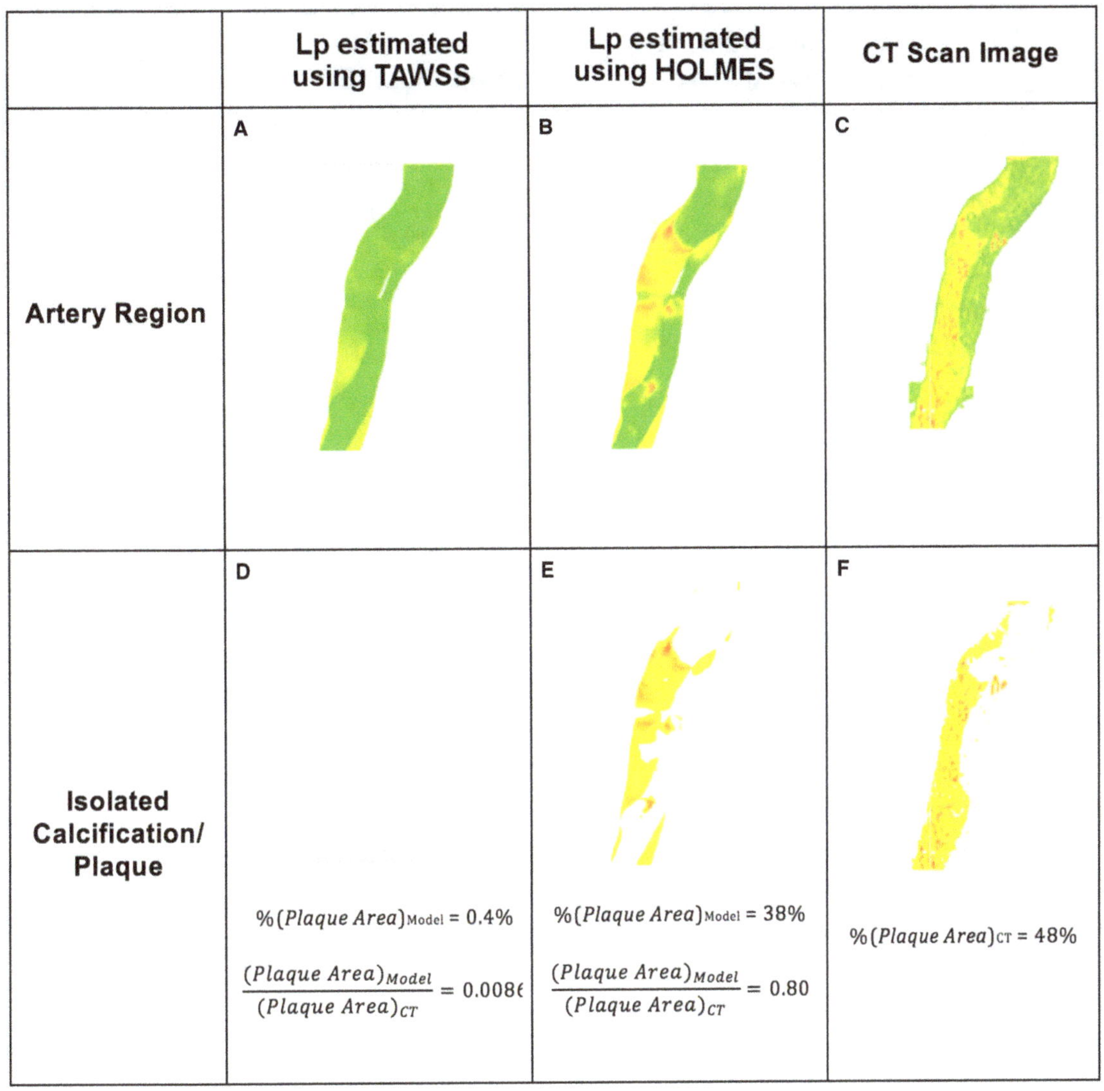

FIGURE 9 | Example of atherosclerotic plaque identification from simulations using hydraulic conductivity and comparison with 3D CT scans. (A,B) Section of the artery showing plaques based on $L_{p,lj}$; **(D,E)** Plaque extracted from **(A,B)**; **(C)** Section of the artery from CT scan; **(F)** Plaque extracted from **(C)**.

shown previously (Alimohammadi et al., 2015b), simulating wall motion is key, since the dynamic interactions between the intraluminal pressure-gradient, the vessel wall elasticity and the intimal flap motion play a critical role in accurately predicting haemodynamics in the false lumen. Moreover, in healthy aortae, rigid wall simulations have been reported to produce somewhat comparable distributions of WSS, albeit with overestimated magnitude (Brown et al., 2012; Reymond et al., 2013). In dissected aortae, the differences between the results of rigid wall (CFD only) and FSI simulations are significant (Alimohammadi et al., 2015b), which has a clear impact in the interpretation of the effect of mechanical stimuli on endothelial behavior in this case. These differences are enhanced by the complex, intertwining lumina, in which even small motion variations have a decisive effect. Hence, the use of FSI, while more time-consuming and technically challenging, is important.

Limitations

One limitation of the fluid dynamics simulation is the absence of the visceral arteries, which would alter the flow in the descending aorta. As previously stated, it was not possible to resolve these vessels from the CT scans, so they were omitted. Nonetheless, even in their absence, we believe the usefulness of the model was demonstrated.

A patient-specific inflow waveform was not available for the present study, and as such, a waveform from the literature was selected from a patient suffering from a similar type-B AD. The absence of a patient-specific inflow reduces the specificity to the patient of the present results however, given that the boundary conditions were tuned to patient-specific values using the same inflow, the hemodynamic environment predicted by the simulations is expected to be an appropriate representation of a patient with type-B AD. The flow waveform at the inlet was

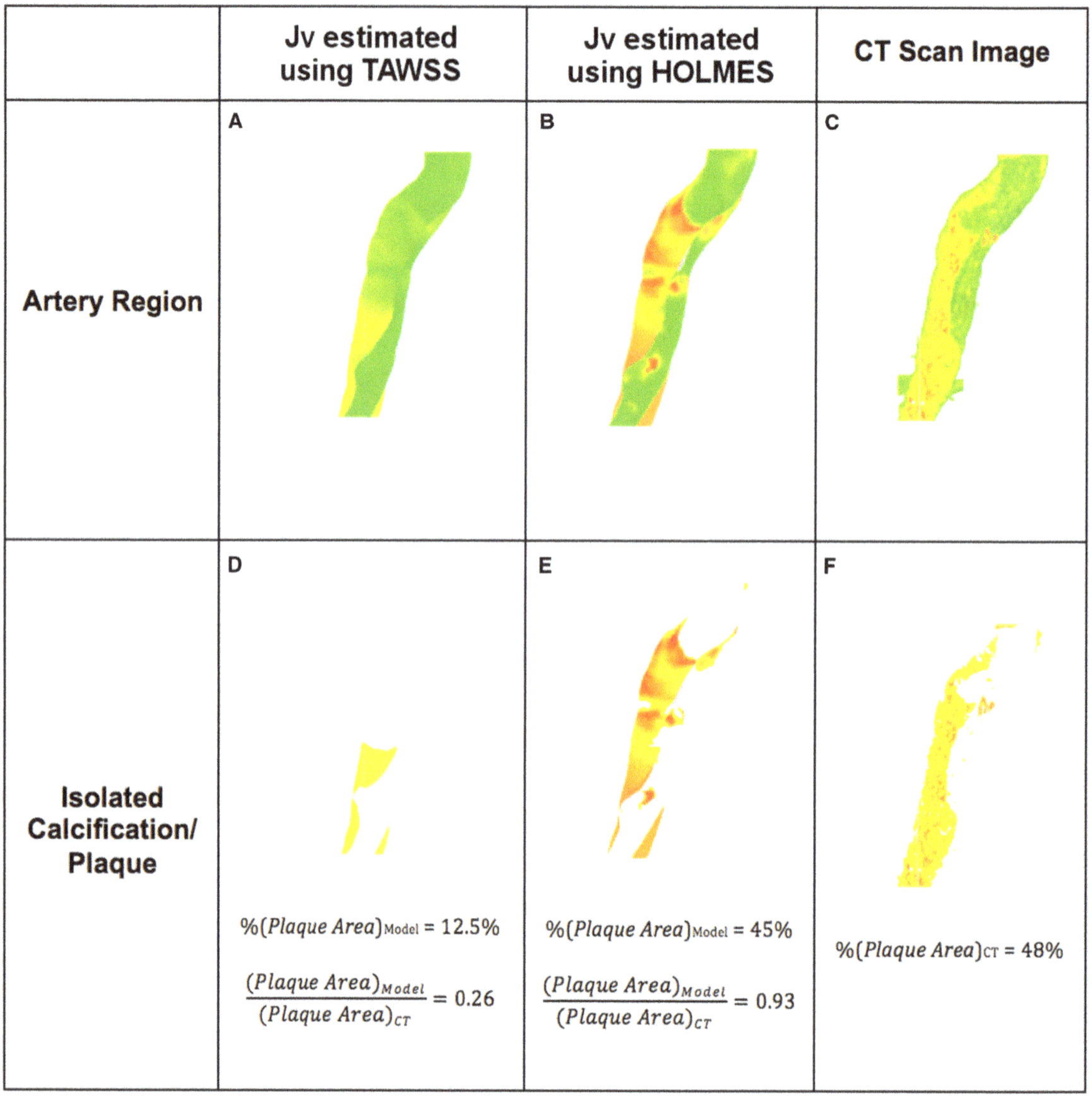

FIGURE 10 | Example of atherosclerotic plaque identification from simulations using solute flux and comparison with 3D CT scans. (A,B) Section of the artery showing plaques based on $J_{v,lj}$; **(D,E)** Plaque extracted from **(A,B)**; **(C)** Section of the artery from CT scan; **(F)** Plaque extracted from **(C)**.

applied as a uniform velocity, rather than mapping to parabolic or Womersley profiles. Whilst accurate axial velocity profile, as extracted using pcMRI can provide improved predictions of characteristics such as helical flow (Morbiducci et al., 2013), it is not clear that parabolic or Womersley profiles offer any improvement over a uniform velocity (Marzo et al., 2009; Campbell et al., 2012; Morbiducci et al., 2013).

The isotropic hyperelastic model of Raghavan and Vorp (2000) is a simplification of the true aortic wall properties, which are known to be anisotropic (Gasser et al., 2006). This model has been shown to provide improved predictions of stress distributions compared with a linear elastic model, but underestimates peak stresses relative to anisotropic models (Roy et al., 2014). However, it is important to note that wall stresses are not explicitly evaluated in the present study; additionally, fibrosis in the vessel wall alters wall properties over time in AD, and is thus highly patient-specific. Furthermore, wall motion was not captured as part of the clinical data collection. More detailed imaging as well as experimental data on vessel wall properties in AD is necessary for better, patient-specific simulations. Given these constraints, although we acknowledge that more complex models of vessel wall properties have been applied in AD and that this might result in local variations, in this case these are unlikely to yield improved accuracy in what is our ultimate goal, the quantification of their effect on hemodynamic parameters for individual patients.

Another limitation of this approach, which is fairly consistent throughout the related literature is the use of "visual maps," in order to establish qualitative correlations between the *in vivo* data and the simulation results. Recent studies address the issue of quantification of these type of results by using statistical methods to establish quantitative correlations and statistical significance of the variables analyzed (simulated) with respect to plaque location, obtained from *in vivo* data (Morbiducci et al., 2015; Gallo et al., 2016). This is clearly a welcome development that we will be keen to use for future, multi-patient studies. As it has been presented here, the focus of this paper is on the use of multi-scale modeling and simulation tools and the development of interpretable, physiologically-based metrics to understand plaque location.

The results presented in this paper when using the multi-criteria, patient-specific, multi-scale complex framework described here, compare well, qualitatively, to *in vivo* data and although there is no perfect match, this is to be expected, when considering the number of assumptions, simplifications and limitations as described (please see above). Moreover, as for patient-specific data, only anatomical information was fed into the model and hemodynamic variables were calculated based on invasive pressure measurements in order to properly characterize the flow, as described in Alimohammadi et al. (2014). Apart from the hemodynamic calculations previously described, it is important to mention that the authors have been extremely careful to inform the model with data and values taken straight from the literature, so, there is no unique estimation of parameters or "fitting" apart from appropriate characterization of flow variables, for the simulation presented here. It remains a striking feature of this approach that even taking this into account; the *correspondence* between observed and simulated calcified/atherosclerotic regions is high, can be explained and appears to be a marked improvement upon other standard methods reported in previous studies. It also offers a coherent, mechanistic explanation that is able to shine some light on combined mechanisms responsible for the location of atherosclerotic/calcification areas and to interpret them together and simultaneously. Although the focus of this paper is on the use of multi-criteria and multi-scale mathematical modeling to understand atherosclerosis and results are shown for one patient only, it is important to say that preliminary and very encouraging results from a different anatomical site with a simpler model, but also following a mechanistic approach have been published in Alimohammadi et al. (2015a). Equally, results from other simulations currently performed by our group (not shown here) using the approach developed in this work for different arteries, show similar effectiveness than the one discussed here, in terms of identification of calcification/atheroprone areas. We acknowledge nevertheless the need to test this approach in a small cohort of patients, taking into account the limitations and new developments described above. This is work that is already under way.

CONCLUSIONS

In the present study, we have presented a mechanistic, mathematical model of endothelial mechanotransduction to understand plaque location/calcification. The model is tested on a patient-specific case for which *in vivo* measurements were obtained at University College Hospital and a patient-specific biomechanical model produced (Alimohammadi et al., 2015a) and attempts to provide a clear, multi-factorial metric for plaque location, with strong physiological meaning and interpretation. The results from this model compared favorably with *in vivo* data and outperformed other well-established indices currently used in the literature. The model used an advanced FSI simulation, comprising patient specific dynamic outlet boundary conditions and non-Newtonian blood viscosity. This was coupled with an established model for atherogenesis in order to investigate the roles of various hemodynamic parameters on the development of calcified regions in the aortic wall.

As an input to the endothelial permeability model relating to shear stress, we hypothesized that regions of oscillatory, low magnitude shear stress would be susceptible to calcification, due to the known connection to increased permeability. We therefore, proposed the compound HOLMES shear index, which includes both magnitude and oscillatory characteristics and will thus emphasize oscillatory, low magnitude shear stress and found that it considerably improved the predictive power of the model over TAWSS-based analysis.

The role of the complex characteristics of the hemodynamics near the wall was investigated by hypothesizing that elevated blood viscosity near the wall would limit the convection of plasma into the vessel wall, and vice versa. It was found that when the spatially varying blood viscosity at the wall as estimated using an empirical non-Newtonian viscosity model, was used in the endothelial permeability model, the predicted $L_{p,lj}$ and $J_{v,lj}$ distributions bore increased resemblance to the observed regions of calcification.

This work provides a good example of the use of multiscale mathematical modeling to understand physiology. The promising results obtained from this approach warrant further investigation. Next steps will include studies of a larger number of patients to enable comparisons amongst patients and potentially statistical analyses, in order to investigate in detail the predictive power of the model.

AUTHOR CONTRIBUTIONS

MA and CP conceived the study and carried out simulations and wrote the manuscript. OA acquired the clinical data and VD conceived the study and wrote the manuscript.

FUNDING

EPSRC grant "Personalised Medicine Through Learning in the Model Space" (grant number EP/L000296/1). Leverhulme Trust Senior Research Fellowship "Exploring the Unknowable Using Simulation: Structural Uncertainty in Multiscale Models" (Fellowship number RF-446 2015-482).

REFERENCES

Alimohammadi, M., Agu, O., Balabani, S., and Díaz-Zuccarini, V. (2014). Development of a patient-specific simulation tool to analyse aortic dissections: assessment of mixed patient-specific flow and pressure boundary conditions. *Med. Eng. Phys.* 36, 275–284. doi: 10.1016/j.medengphy.2013.11.003

Alimohammadi, M., Pichardo-Almarza, C., Giulia Di, T., Stavroula, B., Agu, O., and Diaz-Zuccarini, V. (2015a). "Predicting atherosclerotic plaque locationin an iliac bifurcation using a hybrid CFD/biomechanical approach," in *Part II, Bioinformatics and Biomedical Engineering: Volume 9044 of the Series Lecture Notes in Computer Science*, eds F. Ortuño and I. Rojas (Springer Link), 594–606.

Alimohammadi, M., Sherwood, J. M., Karimpour, M., Agu, O., Balabani, S., and Díaz-Zuccarini, V. (2015b). Aortic dissection simulation models for clinical support: fluid-structure interaction vs. rigid wall models. *Biomed. Eng. Online* 14, 1661–1616. doi: 10.1186/s12938-015-0032-6

Arzani, A., and Shadden, S. C. (2016). Characterizations and correlations of wall shear stress in aneurysmal flow. *J. Biomech. Eng.* 138, 014503–014510. doi: 10.1115/1.4032056

Braverman, A. C. (2010). Acute aortic dissection: clinician update. *Circulation* 122, 184–188. doi: 10.1161/CIRCULATIONAHA.110.958975

Brown, A. G., Shi, Y., Marzo, A., Staicu, C., Valverde, I., Beerbaum, P., et al. (2012). Accuracy vs. computational time: translating aortic simulations to the clinic. *J. Biomech.* 45, 516–523. doi: 10.1016/j.jbiomech.2011.11.041

Campbell, I. C., Ries, J., Dhawan, S. S., Quyyumi, A. A., Taylor, W. R., and Oshinski, J. N. (2012). Effect of inlet velocity profiles on patient-specific computational fluid dynamics simulations of the carotid bifurcation. *J. Biomech. Eng.* 134, 051001–0510018. doi: 10.1115/1.4006681

Chiu, J. J., and Chien, S. (2011). Effects of disturbed flow on vascular endothelium: pathophysiological basis and clinical perspectives. *Physiol. Rev.* 91, 327–387. doi: 10.1152/physrev.00047.2009

Coady, M. A., Rizzo, J. A., and Elefteriades, J. A. (1999). Pathologic variants of thoracic aortic dissections. Penetrating atherosclerotic ulcers and intramural hematomas. *Cardiol. Clin.* 17, 637–657. doi: 10.1016/S0733-8651(05)70106-5

Criado, F. J. (2011). Aortic dissection: a 250-year perspective. *Tex. Heart Inst. J.* 38, 694–700.

Davies, P. F. (1995). Flow-mediated endothelial mechanotransduction. *Physiol. Rev.* 75, 519–560.

de Jong, P. A., Hellings, W. E., Takx, R. A. P., Išgum, I., van Herwaarden, J. A., and Mali, W. P. (2014). Computed tomography of aortic wall calcifications in aortic dissection patients. edited by Jeroen Hendrikse. *PLoS ONE* 9:e102036. doi: 10.1371/journal.pone.0102036

Demer, L. L., and Tintut,Y. (2008). Vascular calcification: pathobiology of a multifaceted disease. *Circulation* 117, 2938–2948. doi: 10.1161/CIRCULATIONAHA.107.743161

Díaz-Zuccarini, V., Di Tomaso, G., Agu, O., and Pichardo-Almarza, C. (2014). Towards personalised management of atherosclerosis via computational models in vascular clinics: technology based on patient-specific simulation approach. *Healthc. Technol. Lett.* 1, 13–18. doi: 10.1049/htl.2013.0040

Erren, T. C. (2007). The case for a posteriori hypotheses to fuel scientific progress. *Med. Hypotheses* 69, 448–453. doi: 10.1016/j.mehy.2006.12.026

Gallo, D., Steinman, D. A., and Morbiducci, U. (2016). Insights into the co-localization of magnitude-based versus direction-based indicators of disturbed shear at the carotid bifurcation. *J. Biomech.* doi: 10.1016/j.jbiomech.2016.02.010. [Epub ahead of print].

Gasser, T. C., Ogden, R. W., and Holzapfel, G. A. (2006). Hyperelastic modelling of arterial layers with distributed collagen fibre orientations. *J. R. Soc. Interface* 3, 15–35. doi: 10.1098/rsif.2005.0073

Gijsen, F. J., van de Vosse, F. N., and Janssen, J. D. (1999). The influence of the non-newtonian properties of blood on the flow in large arteries: steady flow in a carotid bifurcation model. *J. Biomech.* 32, 601–608. doi: 10.1016/S0021-9290(99)00015-9

Hagan, P. G., Nienaber, C. A., Isselbacher, E. M., Bruckman, D., Karavite, D. J., Russman, P. L., et al. (2000). The International Registry of Acute Aortic Dissection (IRAD). *JAMA* 283, 897–903. doi: 10.1001/jama.283.7.897

Himburg, H. A. (2004). Spatial comparison between wall shear stress measures and porcine arterial endothelial permeability. *Am. J. Physiol. Heart Circ. Physiol.* 286, H1916–H1922. doi: 10.1152/ajpheart.00897.2003

Isgum, I., van Ginneken, B., and Olree, M. (2004). Automatic detection of calcifications in the aorta from CT scans of the abdomen. 3D computer-aided diagnosis. *Acad. Radiol.* 11, 247–257. doi: 10.1016/S1076-6332(03)00673-1

Karmonik, C., Bismuth, J. X., Davies, M. G., and Lumsden, A. B. (2008). Computational hemodynamics in the human aorta: a computational fluid dynamics study of three cases with patient-specific geometries and inflow rates. *Technol. Health Care* 16, 343–354.

Khan, I. A., and Nair, C. K. (2002). Clinical, diagnostic, and management perspectives of aortic dissection. *Chest* 122, 311–28.

Kim, S., and Giddens, D. P. (2015). Mass transport of low density lipoprotein in reconstructed hemodynamic environments of human carotid arteries: the role of volume and solute flux through the endothelium. *J. Biomech. Eng.* 137:041007. doi: 10.1115/1.4028969

Ku, D. N., Giddens, D. P., Zarins, C. K., and Glagov, S. (1985). Pulsatile flow and atherosclerosis in the human carotid bifurcation. Positive correlation between plaque location and low oscillating shear stress. *Arteriosclerosis* 5, 293–302. doi: 10.1161/01.ATV.5.3.293

Lanzer, P., Boehm, M., Sorribas, V., Thiriet, M., Janzen, J., Zeller, T., et al. (2014). Medial vascular calcification revisited: review and perspectives. *Eur. Heart J.* 35, 1515–1525. doi: 10.1093/eurheartj/ehu163

Levesque, M. J., Liepsch, D., Moravec, S., and Nerem, R. M. (1986). Correlation of endothelial cell shape and wall shear stress in a stenosed dog aorta. *Arteriosclerosis* 6, 220–229. doi: 10.1161/01.ATV.6.2.220

Lusis, A. J. (2000). Atherosclerosis. *Nature* 407, 233–241. doi: 10.1038/35025203

Lusis, A. J. (2012). Life After GWAS: functional genomics in vascular biology. *Arterioscler. Thromb. Vasc. Biol.* 32, 169–169. doi: 10.1161/ATVBAHA.111.243543

Malayeri, A. A., Natori, S., Bahrami, H., Bertoni, A. G., Kronmal, R., Lima, J. A., et al. (2008). Relation of aortic wall thickness and distensibility to cardiovascular risk factors (From the Multi-Ethnic Study of Atherosclerosis [MESA]). *Am. J. Cardiol.* 102, 491–496. doi: 10.1016/j.amjcard.2008.04.010

Malek, A. M., Alper, S. L., and Izumo, S. (1999). Hemodynamic shear stress and its role in atherosclerosis. *JAMA* 282, 2035–2042.

Marzo, A., Singh, P., Reymond, P., Stergiopulos, N., Patel, U., and Hose, R. (2009). Influence of inlet boundary conditions on the local haemodynamics of intracranial aneurysms. *Comput. Methods Biomech. Biomed. Engin.* 12, 431–444. doi: 10.1080/10255840802654335

Meng, H., Tutino, V. M., Xiang, J., and Siddiqui, A. (2014). High WSS or Low WSS? Complex interactions of hemodynamics with intracranial aneurysm initiation, growth, and rupture: toward a unifying hypothesis. *AJNR. Am. J. Neuroradiol.* 35, 1254–1262. doi: 10.3174/ajnr.A3558

Meyer, G., Merval, R., and Tedgui, A. (1996). Effects of pressure-induced stretch and convection on low-density lipoprotein and albumin uptake in the rabbit aortic wall. *Circ. Res.* 79, 532–540. doi: 10.1161/01.RES.79.3.532

Morbiducci, U., Gallo, D., Cristofanelli, S., Ponzini, R., Deriu, M. A., Rizzo, G., et al. (2015). A rational approach to defining principal axes of multidirectional wall shear stress in realistic vascular geometries, with application to the study of the influence of helical flow on wall shear stress directionality in aorta. *J. Biomech.* 48, 899–906. doi: 10.1016/j.jbiomech.2015.02.027

Morbiducci, U., Ponzini, R., Gallo, D., Bignardi, C., and Rizzo, G. (2013). Inflow boundary conditions for image-based computational hemodynamics impact of idealized versus measured velocity profiles in the human aorta. *J. Biomech.* 46, 102–109. doi: 10.1016/j.jbiomech.2012.10.012

Nadra, I. (2005). Proinflammatory activation of macrophages by basic calcium phosphate crystals via protein kinase C and MAP kinase pathways: a vicious cycle of inflammation and arterial calcification? *Circ. Res.* 96, 1248–1256. doi: 10.1161/01.RES.0000171451.88616.c2

Olgac, U., Kurtcuoglu, V., and Poulikakos, D. (2008a). Computational modeling of coupled blood-wall mass transport of LDL: effects of local wall shear stress. *Am. J. Physiol. Heart Circul. Physiol.* 294, H909–H919. doi: 10.1152/ajpheart.01082.2007

Olgac, U., Kurtcuoglu, V., and Poulikakos, D. (2008b). Computational modeling of coupled blood-wall mass transport of LDL: effects of local wall shear stress. *Am. J. Physiol. Heart Circul. Physiol.* 294, H909–H919. doi: 10.1152/ajpheart.01082.2007

Peiffer, V., Sherwin, S. J., and Weinberg, P. D. (2013a). Does low and oscillatory wall shear stress correlate spatially with early atherosclerosis? A systematic review. *Cardiovasc. Res.* 99, 242–250. doi: 10.1093/cvr/cvt044

Peiffer, V., Sherwin, S. J., and Weinberg, P. D. (2013b). Computation in the rabbit aorta of a new metric – the transverse wall shear stress – to quantify the multidirectional character of disturbed blood flow. *J. Biomech.* 46, 2651–2658. doi: 10.1016/j.jbiomech.2013.08.003

Raghavan, M. L., and Vorp, D. A. (2000). Toward a biomechanical tool to evaluate rupture potential of abdominal aortic aneurysm: identification of a finite strain constitutive model and evaluation of its applicability. *J. Biomech.* 33, 475–482. doi: 10.1016/S0021-9290(99)00201-8

Reymond, P., Crosetto, P., Deparis, S., Quarteroni, A., and Stergiopulos, N. (2013). Physiological simulation of blood flow in the aorta: comparison of hemodynamic indices as predicted by 3-D FSI, 3-D rigid wall and 1-D models. *Med. Eng. Phys.* 35, 784–791. doi: 10.1016/j.medengphy.2012.08.009

Roy, D., Holzapfel, G. A., Kauffmann, C., and Soulez, G. (2014). Finite element analysis of abdominal aortic aneurysms: geometrical and structural reconstruction with application of an anisotropic material model. *IMA J. Appl. Mathematics* 79, 1011–1026. doi: 10.1093/imamat/hxu037

Sáez, P., Malvè, M., and Martínez, M. A. (2015). A theoretical model of the endothelial cell morphology due to different waveforms. *J. Theor. Biol.* 379, 16–23. doi: 10.1016/j.jtbi.2015.04.038

Sangiorgi, G., Rumberger, J. A., Severson, A., Edwards, W. D., Gregoire, J., Fitzpatrick, L. A., et al. (1998). Arterial calcification and not lumen stenosis is highly correlated with atherosclerotic plaque burden in humans: a histologic study of 723 coronary artery segments using nondecalcifying methodology. *J. Am. Coll. Cardiol.* 31, 126–133. doi: 10.1016/S0735-1097(97)00443-9

Schindelin, J., Arganda-Carreras, I., Frise, E., Kaynig, V., Longair, M., Pietzsch, T., et al. (2012). Fiji: an open-source platform for biological-image analysis. *Nat. Methods* 9, 676–682. doi: 10.1038/nmeth.2019

Suo, J., Ferrara, D. E., Sorescu, D., Guldberg, R. E., Taylor, W. R., and Giddens, D. P. (2006). Hemodynamic shear stresses in mouse aortas: implications for atherogenesis. *Arterioscler. Thromb. Vasc. Biol.* 27, 346–351. doi: 10.1161/01.ATV.0000253492.45717.46

Tedgui, A., and Lever, M. J. (1984a). Filtration through damaged and undamaged rabbit thoracic aorta. *Am. J. Physiol.* 247 (5 Pt 2), H784–H791.

Tedgui, A., and Lever, M. J. (1984b). Filtration through damaged and undamaged rabbit thoracic aorta. *Am. J. Physiol. Heart Circ. Physiol.* 247, H784–91.

Tegner, J., Skogsberg, J., and Bjorkegren, J. (2006). Thematic review series: systems biology approaches to metabolic and cardiovascular disorders. Multi-organ whole-genome measurements and reverse engineering to uncover gene networks underlying complex traits. *J. Lipid Res.* 48, 267–277. doi: 10.1194/jlr.R600030-JLR200

Tsai, T. T., Fattori, R., Trimarchi, S., Isselbacher, E., Myrmel, T., Evangelista, A., et al. (2006). Long-term survival in patients presenting with type B acute aortic dissection: insights from the international registry of acute aortic dissection. *Circulation* 114, 2226–2231. doi: 10.1161/CIRCULATIONAHA.106.622340

Xiang, J., Natarajan, S. K., Tremmel, M., Ma, D., Mocco, J., Hopkins, L. N., et al. (2010). Hemodynamic-morphologic discriminants for intracranial aneurysm rupture. *Stroke* 42, 144–152. doi: 10.1161/STROKEAHA.110.592923

3

Spatiotemporal Dynamics of Insulitis in Human Type 1 Diabetes

Kyle C. A. Wedgwood[1]*, *Sarah J. Richardson*[2], *Noel G. Morgan*[2] *and* *Krasimira Tsaneva-Atanasova*[3, 4]

[1] *Centre for Biomedical Modelling and Analysis, University of Exeter, Exeter, UK,* [2] *University of Exeter Medical School, University of Exeter, Exeter, UK,* [3] *College for Engineering, Mathematics and Physical Sciences, University of Exeter, Exeter, UK,* [4] *Engineering and Physical Sciences Research Council Centre for Predictive Modelling in Healthcare, University of Exeter, Exeter, UK*

Edited by:
Joseph L. Greenstein, Johns Hopkins University, USA

Reviewed by:
Josep Bassaganya-Riera, Virginia Tech, USA
Anmar Khadra, McGill University, Canada

***Correspondence:**
Kyle C. A. Wedgwood
k.c.a.wedgwood@exeter.ac.uk

Type 1 diabetes (T1D) is an auto-immune disease characterized by the selective destruction of the insulin secreting beta cells in the pancreas during an inflammatory phase known as insulitis. Patients with T1D are typically dependent on the administration of externally provided insulin in order to manage blood glucose levels. Whilst technological developments have significantly improved both the life expectancy and quality of life of these patients, an understanding of the mechanisms of the disease remains elusive. Animal models, such as the NOD mouse model, have been widely used to probe the process of insulitis, but there exist very few data from humans studied at disease onset. In this manuscript, we employ data from human pancreases collected close to the onset of T1D and propose a spatio-temporal computational model for the progression of insulitis in human T1D, with particular focus on the mechanisms underlying the development of insulitis in pancreatic islets. This framework allows us to investigate how the time-course of insulitis progression is affected by altering key parameters, such as the number of the CD20+ B cells present in the inflammatory infiltrate, which has recently been proposed to influence the aggressiveness of the disease. Through the analysis of repeated simulations of our stochastic model, which track the number of beta cells within an islet, we find that increased numbers of B cells in the peri-islet space lead to faster destruction of the beta cells. We also find that the balance between the degradation and repair of the basement membrane surrounding the islet is a critical component in governing the overall destruction rate of the beta cells and their remaining number. Our model provides a framework for continued and improved spatio-temporal modeling of human T1D.

Keywords: type 1 diabetes, insulitis, agent-based modeling, spatio-temporal dynamics, peri-islet basement membrane

1. INTRODUCTION

Type 1 diabetes (T1D) is an auto-immune disease characterized by the selective destruction of pancreatic beta cells in the islets of Langerhans by the immune system (Eisenbarth, 1986; Atkinson, 2012; Boitard, 2012; La Torre and Lernmark, 2012; Pugliese, 2014; Richardson et al., 2014; Roep and Tree, 2014). This destruction takes place during an inflammatory phase, known as insulitis, in which various immune cells infiltrate the islets (Lecompte, 1958; Gepts, 1965; Willcox et al., 2009; Morgan et al., 2014). As the beta cell mass decreases over the course of the disease, the ability of the islets to secrete sufficient quantities of insulin to properly regulate blood glucose levels becomes compromised. As a result, patients with type 1 diabetes ultimately become reliant on the lifelong administration of external insulin.

Well-defined genetic components have been identified which predispose individuals to T1D; in particular the HLA-genotype (Itoh et al., 1993; Somoza et al., 1994). However, this alone is not sufficient to predict which individuals will develop the disease with precision. Since monozygotic twin studies have shown limited pairwise concordance for T1D (Barnett et al., 1981; Lo et al., 1991; Redondo et al., 2001), it is clear that environmental factors, such as viral infection, vitamin D status and childhood nutrition may also contribute significantly to the development of T1D (Knip et al., 2005).

Among the different subtypes of immune cells which infiltrate islets, CD8+ T cells are considered as the likely mediators of beta cell destruction (Bottazzo et al., 1985; Itoh et al., 1993; Somoza et al., 1994). It is widely believed that these promote beta cell apoptosis by both direct and indirect mechanisms and that macrophages then clear dying and dead beta cells very quickly. Other T lymphocytes, such as those expressing CD4, are also thought to play a role, though their precise functions are less clear (Willcox et al., 2009; Richardson et al., 2011). In addition to the T cells, B lymphocytes (CD20+) are also present in significant numbers during certain stages of insulitis in some patients. Indeed, recent evidence has suggested that the number of B cells and/or the ratio of B-cells to CD4+ cells present in the infiltrate can be used to classify the disease into two distinct phenotypes. These have been defined as "hyper-immune," characterized by elevated numbers of CD20+ cells and a rapid loss of beta cell mass whereas, by contrast, the "pauci-immune" phenotype, is associated with a lower proportion of B cells and a much slower destruction of beta cells (Morgan et al., 2014; Leete et al., 2016). The mechanisms by which the B cells affect the rate of disease progression are unknown, but it is possible that they collaborate with CD8+ T cells to drive beta cell loss (Huppa and Davis, 2003).

Progress in understanding the cellular and molecular mechanisms underlying insulitis in humans has been hindered by the paucity of available samples from patients who died at, or close to, disease onset. Fewer than 200 such samples are available worldwide (Gepts, 1965; Foulis and Stewart, 1984; Klöppel et al., 1985; Dotta et al., 2007; Walker et al., 2011; Campbell-Thompson et al., 2012; Pugliese, 2014), and inferring the time course of a disease process from histological samples is fraught with difficulty since a range of assumptions and extrapolations about the likely progression are inevitably required to achieve this.

To offset this problem, much work has been performed using animal models such as the non-obese diabetic (NOD) mouse as a proxy for the human condition (Kachapati et al., 2012). The advantages of this are clear—experiments can be performed over relatively short periods of time and analysis of circulating and pancreatic lymphyocyte populations is achieved more readily during the course of disease. Such studies have, therefore, been extremely informative as a means to identify important cellular and molecular factors involved (Lally and Bone, 2003), although considerable effort needs to be made to verify that results are translatable to humans. While a large number of potential therapies have been identified in rodents, a means to prevent the human disease remains elusive (Brehm et al., 2012; In't Veld, 2014; Pugliese et al., 2014; Reed and Herold, 2015).

An alternative approach, which may shed light on human insulitis is offered by mathematical modeling. By constructing environments that mimic the pancreas and immune system, experiments can be performed *in silico*, and the results compared to empirical data arising from studies of relevant human tissue. Through model construction and analysis, candidate mechanisms giving rise to T1D can be interrogated in a systematic way. To this end, a number of authors have developed models describing: beta cell function (Bertram and Sherman, 2004), defective macrophage clearing (Marée et al., 2006), immune cell populations (Mahaffy and Edelstein-Keshet, 2007), multi-clonal populations of immune cells (Khadra et al., 2009, 2011; Jaberi-Douraki et al., 2015), immune cell cycles (Jaberi-Douraki et al., 2014b), and apoptotic stress generated by the loss of beta cell mass (Jaberi-Douraki et al., 2014c).

A common theme among the models outlined above is that they can be regarded as "lumped models," in that they deal with averaged quantities taken over the whole body. For certain applications, this seems appropriate, as it is commensurate with current clinical practice. The biomarkers clinicians have available are typically derived from blood samples, in particular, measures of HbA1c and fasting glucose levels (American Diabetes Association, 2006; Inzucchi, 2012). Other biomarkers, such as measures of C-peptide (NIC, 2015) and islet cell and anti-insulin antibodies (Taplin and Barker, 2008) can be used to identify individuals who may be susceptible to diabetes and, though it is not currently routine clinical practice, they can also be used to aid classification of diabetes (Jones and Hattersley, 2013). Biomarkers obtained from blood samples are, by their nature, whole body measures. However, it is known from histological samples that insulitis displays a pronounced spatial dependence (Willcox et al., 2009; Morgan et al., 2014). This is due in part to the distribution of the islets within the pancreas, but is also a consequence of the presence of a basement membrane around the islets (Korpos et al., 2013), establishing a barrier to immune cell infiltration. Interactions and communication between immune cells also play a role and are likely to contribute to this spatial dependence.

Insulitis is a spatially heterogeneous process, both within an individual islet and across the pancreas as a whole. Islets that are heavily infiltrated can be located near to islets that are free from inflammation. Moreover, within an infiltrated islet, the destruction of beta cells appears not to follow a "wave-like" profile, as might be expected, but seems more random (Willcox et al., 2009; Morgan et al., 2014). These observations lend credence to the notion that immune cell communication is a critical component of insulitis.

To investigate insulitis and its relevance to the progression of T1D, these spatial aspects must thus be taken into account. Moreover, since the number of immune cells infiltrating any given islet is low (Willcox et al., 2009; Morgan et al., 2014), density based approaches, such as ordinary differential equation (ODE) modeling, cannot be applied with precision and alternatives are required.

In this manuscript, we construct an agent-based model of the insulitis process within a single islet. The model is posed on a regular domain that represents the space surrounding the islet. The principal agents in our model are the immune cells and we describe how they locate and target beta cells and how they interact with one another. By using an agent-based approach, we avoid the problems associated with density-based (averaged) descriptions of cell mass. We begin by introducing the relevant biology for our problem. Following this, we describe the development of, and results from, the agent-based model simulations. Finally, we conclude with a discussion of the model and its potential extensions.

2. RELEVANT BIOLOGY

In this section, we highlight the biology corresponding to the components of our model.

2.1. Cell Types

Our model consists of three distinct cell types—two types of lymphocyte and the beta cells. We consider in particular CD8+ T lymphocytes and CD20+ B lymphocytes. Of the various immune cells that have been shown to be involved in human insulitis, these are the ones that are believed to be most relevant to the questions under investigation in our study. The CD8+ cells are implicated in the destruction of the beta cells, while the CD20+ cells are deduced to play an important role, since their absence is associated with a weakly aggressive phenotype (Morgan et al., 2014). In our model, the B cells will essentially act as antigen presenting cells to the T cells (Rodríguez-Pinto, 2005). We assume that macrophages efficiently clear apoptotic beta cells. We do not account for mitosis in any of the cell types; whilst division and selection of relevant lymphocytes has been demonstrated in the lymphatic system, evidence for it in the peri-islet space is minimal (Willcox et al., 2010). Immune cells are assumed to have a finite lifespan and beta cells die following interactions with T cells (Cnop et al., 2010). We do not consider other cell types, such as α and δ cells, that are also found within the islets of Langerhans (Kim et al., 2009).

2.2. Basement Membrane

Individual islets are contained within an encapsulating membrane. This membrane is comprised of various laminins and collagen (Korpos et al., 2013) and acts as barrier to invading immune cells. In mice with T1D, these compounds are lost, suggesting that the basement membrane has been degraded. In our model, we assume that the T cells are responsible for degrading the membrane, but in principle, other cells could be also responsible for this. We also make the assumption that the membrane can be repaired over time.

2.3. Cell Movement

We assume that beta cells have a fixed location, but that immune cells are free to move around. The direction of this movement is mediated by a chemokine gradient, which will be described below (Stein and Nombela-Arrieta, 2005). In addition, immune cells of different types are attracted to one another, instigated by different chemical signaling pathways (Janeway et al., 1985). Cells are not allowed to pass through the basement membrane, nor through each other, but are allowed to remain in contact with one another.

2.4. Chemokine

The beta cells are considered to secrete a chemokine signal that attracts immune cells toward the islet (Christen and Von Herrath, 2004). In our study, we assume that the chemokine molecules are bound to the membrane of the beta cells and may also be cleaved off so that they diffuse freely.

2.5. Cell Interactions

In their role as antigen presenting cells, the B cells can cause the T cells to enter an activated state, in which both their sensitivity to the chemokine gradient and their killing efficiency are enhanced (Friedl et al., 2005). We assume that activated T cells have a shorter lifespan than their non-activated counterparts (Green et al., 2003).

When a T cell encounters a beta cell, it triggers a pro-apoptotic pathway within the beta cell, ultimately resulting in the death of the beta cell (Cnop et al., 2005).

The behavior of the cells and membrane are summarized below.

2.6. Summary of Basic Behavior

Below, we will summarize the behavior of each of the cells types included in our modeling.

T cells
- Move up chemokine gradients
- Degrade basement membrane
- Kill beta cells
- Interact with B cells

When activated by B cells, T cells become more sensitive to the chemokine gradient and become more effective at killing beta cells.

B cells
- Move up chemokine gradients
- Form productive interactions with T cells
- Activate T cells

Beta cells
- Immobile
- Killed by T cells

We do not include in our model the possibility of beta cell replication. Studies have suggested that beta cell proliferation is increased, particularly in the early stages of insulitis (Willcox et al., 2010). Here, we disregard this effect.

Basement membrane
- Degraded by T cells.
- Can self-repair.

3. METHODS

3.1. Building a Cellular Automaton

ODE-based modeling approaches average the dynamics associated with insulitis over space essentially representing the entire pancreas. Such approaches have provided numerous insights into the potential roles of T and B cell involvement in T1D and we refer the reader to Jaberi-Douraki et al. (2014a) for a recent and comprehensive review of this literature. However, we know that insulitis is heterogeneous and that the number of immune cells directly involved in the destruction of the beta cells in humans is low - far lower than would be necessary for a density approximation to be justified (Willcox et al., 2009; Morgan et al., 2014). Instead, we employ a different, agent-based, modeling approach (An et al., 2009; Holcombe et al., 2012). Our agents, the immune and beta cells, occupy finite area within a two dimensional space. In a real islet, cells are free to move in three dimensional space, but for ease of modeling (as a starting point of first approximation) and for a more direct comparison with data, we shall here restrict ourselves to the planar case. The agents are given specific rules to locomote, interact with one another and with the basement membrane. Simulations will then be performed to track emergent behavior arising from the basic rules. In this way, the model will evolve over a spatially extended domain in which cell-cell interactions can be explicitly defined.

Note that the rules that govern our agents' dynamics are not intended to be exhaustive lists for all potential behaviors exhibited by the cells. Instead, they are intended to represent a minimal set of interactions that can explain the observed phenomena. Additionally, the model can be iteratively developed to account for new interactions as required.

3.2. Cell Movement

In this modeling framework, the cells are treated as being discrete circular particles (Levine et al., 2000; Palsson and Othmer, 2000; Tijskens et al., 2003; Maini and Baker, 2011; Bruna and Chapman, 2012; Plank and Simpson, 2012). Cell movement is simulated by modeling the forces that act upon cells from different sources and then by resolving these forces using Newton's second law of motion. In the absence of external forces, intrinsic forces acting upon the immune cells cause them to diffuse randomly, obeying a random walk. In the presence of a chemokine gradient, these same cells will seek to move up the gradient. Cells may be attracted to one another and may be in contact with one another for extended periods of time, that is, we allow cells to overlap, but they may not pass through each other.

3.3. Immune Cell Forces

Before discussing the governing equations, it is useful to define indexing sets for each of the cell types. In the following, we define the sets $\mathcal{T}$, $\mathcal{B}$, and $\boldsymbol{\beta}$ as sets containing the indices for T, B, and beta cells respectively. From Newton's second law of motion, we have:

$$m_i \frac{\mathrm{d}^2 x_i}{\mathrm{d}t^2} = -\eta_i \frac{\mathrm{d}x_i}{\mathrm{d}t} + F_i + \sigma \xi_i(t), \quad i \in \mathcal{T} \cup \mathcal{B}, \tag{1}$$

for each cell where $x_i \in \mathbb{R}^2$ is the cell's location, $m_i \in \mathbb{R}_{\geq 0}$ is the mass of the cell, $\eta_i \in \mathbb{R}_{\geq 0}$ is the cell's viscosity and $F_i \in \mathbb{R}^2$ is the force acting on the cell. The final term in Equation (1) represents a Gaussian white noise process: $\langle \xi(t) \rangle = 0$, $\langle \xi(t)\xi(t+s) \rangle = \delta(\mathrm{t-s})$, with strength σ such that the immune cells perform random walks in the absence of any other forces (Palsson and Othmer, 2000; Middleton et al., 2014). In biological systems, we assume that the cells have low Reynold's numbers such that inertial forces are small compared to viscous ones. Under this approximation, we can replace (Equation 1) with:

$$\frac{\mathrm{d}x_i}{\mathrm{d}t} = \frac{1}{\eta_i} F_i + \sigma \xi_i(t), \quad i \in \mathcal{T} \cup \mathcal{B}$$

For simplicity, we set $\eta_i = 1$ for all cells. Note that, in a general mathematical framework, specific values for η_i can be absorbed into the definition of F_i. For our model definition, we include heterogeneity between immune cells through variations in their sensitivity to chemical gradients, as we shall discuss in Section 3.11. We will also initially assume that the cells within the subgroups are homogeneous with respect to their geometry, that is, they share a common radius r_{imm}.

The force F can be broken up into its constituent parts:

$$F_i = F_i^{\mathrm{chemo}} + F_i^{\mathrm{cell\text{-}cell}} + F_i^{\mathrm{islet}} + F_i^{\beta}. \tag{2}$$

In the above, F_i^{chemo} represents chemotactic force, $F_i^{\mathrm{cell\text{-}cell}}$ represents cell attraction and repulsion, whilst F_i^{islet} represents the interaction with the basement membrane and F_i^{β} represents interaction with the beta cells.

3.4. Chemokine Signaling

Since the chemokine is a chemical signal, we establish a gradient using a reaction-diffusion equation:

$$\frac{\partial C}{\partial t} = D\nabla^2 C + f(C). \tag{3}$$

We assume that the chemokine evolves on a much faster timescale compared to the movement of the cells, so we assume the chemokine to be at steady state by setting the LHS of Equation (3) to zero. The domain on which we simulate our model represents only a small space around an individual islet. We note that this region is small compared to the entire pancreas, and further that the chemokine is free to diffuse out of our prescribed region. Typically, when solving PDEs, boundary conditions are applied at the edge of the domain over which the equation is being solved. However, as we are treating our region as only a small part of a much larger one, we feel it is more appropriate to not apply boundary conditions, leaving them open. The reaction term is given by local decay with point sources given by the locations of the beta cells. The chemokine equation now reads:

$$D\nabla^2 C = \lambda C - \sum_{j \in \beta} \nu_j(t)\delta(x - x_j), \tag{4}$$

where λ is the degradation rate of the chemokine and x_j are the centroids of the beta cells. To reflect the fact that dead beta cells will not secrete chemokine, we assume that the production of chemokine is dependent on the current viability of the cell $\nu_j(t)$. For a viable beta cell, $\nu = 1$, whilst dead beta cells have $\nu = 0$. This equation can itself be recast as:

$$QC = \sum_{j\in\beta} \nu_j(t)\delta(x - x_j), \quad Q = (\lambda - D\nabla^2). \tag{5}$$

The Green's function for the operator Q at a given time is a Gaussian function and so we lump parameters together and choose a form for the chemokine signal given by:

$$C(x,t) = \sum_{j\in\beta} \alpha \nu_j(t) \exp\left(-(x - x_j)^2/(2\sigma_C^2)\right), \tag{6}$$

where α now represents the strength of the chemokine signal and σ_C represents the range over which it decays. In order to represent our assumption that the chemokine exists in forms in which it is bound to the membrane and forms in which it freely diffuses, we replace Equation (6) by:

$$C(x,t) = \sum_{k=1}^{2}\sum_{j\in\beta} \alpha \nu_j(t) C_0^k \exp\left(-(x - x_j)^2/(2\sigma_{C,k}^2)\right), \tag{7}$$

where $\sigma_{C,1}$ is small and reflects the membrane bound chemokine, whilst C_0^k, $k = 1, 2$ control the proportion of chemokine that is membrane bound vs. that which is freely diffusing. The immune cells respond to this gradient via:

$$F_i^{\text{chemo}} = S_i \nabla C,$$

where S_i is the sensitivity of the i'th cell to the chemokine gradient. We note that this represents the notion that immune cells tend to move up a potential gradient in which the locations of the beta cells are given by local peaks of this potential. The use of such potentials to model forces is an integral part of this modeling framework (Middleton et al., 2014).

3.5. Cell-Cell Interactions

We assume that T and B cells are attracted to one another. For simplicity's sake, we assume that this can be modeled via another Gaussian function. It is worth noting that this could represent attraction acting at a distance, or could represent a local effect acting to keep cells in contact with one another (or some combination of the two). We do not allow cells to pass through one another. We can achieve both of these through the use of another potential. Common choices for these potentials include the Lennard–Jones potential and the Morse potential (Middleton et al., 2014). We shall use the following potential:

$$U_i^{\text{cell-cell}} = \sum_{j\neq i, j\in\mathcal{T}\cup\mathcal{B}} A_a^{ij} \exp(-|x_j - x_i|^2/d_a^2) - A_r(|x_j - x_i| - 2r_{\text{imm}})^6 H(|x_j - x_i| - 2r_{\text{imm}}). \tag{8}$$

The first of these terms is the attraction between immune cells of different types—B cells seek out T cells to activate and T cells are also attracted to B cells. This is achieved by setting:

$$A_a^{ij} = \begin{cases} A_a^0 & \chi_i \neq \chi_j \\ 0 & \chi_i = \chi_j \end{cases},$$

where χ_i is an indicator function taking value 1 if cell i is a T cell and 0 if it is B cell. The spatial scale of this attraction is set by d_a.

The second term accounts for the fact that cells cannot pass through one another. If the distance between cells is greater than the sum of their radii, that being $2r_{\text{imm}}$, they exert no repulsive force on one another, achieved through the inclusion of a Heaviside function H. However, if this distance falls below this sum, a large repulsive force is exerted. This is known as a hard-core potential (Levine et al., 2000). Note that some authors choose this potential to be infinite at the point of contact, so that cells cannot occupy or share any part of physical space (Bruna and Chapman, 2012). However, this necessitates choosing a smaller time step and, since cells can remain in contact with one another for significant periods of time, we will not make such a choice. Note also that the parameter A_r is shared amongst all immune cells. Once again, the forces acting on the i'th cell can then be expressed as:

$$F_i^{\text{cell-cell}} = -\nabla U_i^{\text{cell-cell}}. \tag{9}$$

Note that, since immune cells move around the domain, the potential "landscape," together with its peaks and troughs, is constantly evolving.

3.6. Islet Interactions

For simplicity, and to remain consistent with our chosen geometry of the cells, the basement membrane encapsulating the islet is chosen to be a circle centred at the origin with radius R. Cells cannot pass through the membrane, so when it is present, immune cells remain on the inside or outside of the islet. This can be reflected through the use of another hard-core potential. Whilst we have not yet defined the dynamics for the membrane, we shall assume that there is a integrity threshold, h, below which the membrane does not prevent cells from passing through. Since we do not expect the membrane to have the same integrity across its whole length, we also need to account for the location at which the cell interacts with the islet. The relevant hard-core potential is then given by:

$$\begin{aligned} U_i^{\text{islet}} &= A_{\text{islet}} H(m(\theta) - h) H(r_{\text{imm}} - |d_i - R|) \\ &\quad (r_{\text{imm}} - |d_i - R|)^6, \\ d_i &= |x_i - x_{\text{islet}}|, \end{aligned} \tag{10}$$

where $m(\theta) \in [0, 1]$ is the membrane integrity at angle $\theta \in \mathbb{S}^1$ around the membrane, $x_{\text{islet}} \in \mathbb{R}^2$ is the location of the center of the islet and A_{islet} is a constant indicating the strength of repulsion. Once more, the force acting on the cells is then given by:

$$F_i^{\text{islet}} = -\nabla U_i^{\text{islet}}. \tag{11}$$

Finally, we must define the forces acting upon the immune cells exerted by the beta cells. This is functionally the same as the repulsive force between cells (Equation 8) and is given by:

$$U_i^\beta = A_r \sum_{j\in\beta} v_j(t) H(|x_j - x_i| - (r_\beta + r_{\text{imm}})) \\ (|x_j - x_i| - (r_\beta + r_{\text{imm}}))^6, \quad (12)$$

where r_β is the radius of the beta cells. The time-dependence of v_j here reflects that the beta cells may die, after which we no longer need to consider repulsive effects generated by them (assuming the dead cell bodies are cleared by macrophages), and so v_j is set to 0 for that cell. The final force in Equation (2) is defined through

$$F_i^\beta = -\nabla U_i^\beta. \quad (13)$$

3.7. Immune Cell Lifespan

The timescale over which insulitis takes place is long compared to the average lifespan of an immune cell. Thus, we need to incorporate immune cell death into our model. We assume that T and B cells have a lifespan of L_T, and L_B respectively. Each cell then has a counter χ_i, which is incremented by setting $\chi_i \mapsto \chi_i + 1$ at each time step. Activated T cells are expected to have shorter lifespans than unactivated T cells (Green et al., 2003), so for those cells, we instead update the counter via $\chi_i \mapsto \chi_i + \Delta\chi$ where $\Delta\chi > 1$.

When χ_i exceeds L_T (or L_B) for a given T (B) cell, it is considered to be dead and is removed. We keep the number of T and B cells constant throughout the simulation by assuming that each dead cell is replaced by a newly arriving one. We assume that the vasculature is sufficiently dense that immune cells can enter at any point in the extra-islet space. As such, the location of the new cell is drawn randomly from a uniform distribution over the extra-islet space, achieved through the use of polar coordinates. The cell's counter is reset to 0 and, if the cell is a T cell, it is chosen to be in the unactivated state.

3.8. Membrane Dynamics

In our simplified geometry, the encapsulating membrane is a represented by a circle, which can be parameterized by a single variable $\theta \in \mathbb{S}^1$. The equation governing the evolution of the membrane viability, m, is given by:

$$\frac{dm(\theta)}{dt} = \alpha_m(1 - m(\theta)) \\ - \lambda_m \sum_{i\in\mathcal{T}} \exp\left(-(|x_i - x_m(\theta)| - r_{\text{imm}})/d_m\right), \quad (14)$$

where α_m and λ_m are respectively the repair and degradation rates of the membrane and $x_m(\theta)$ is the location in real space of the membrane at position θ. The term in the sum represents the fact that we expect T cells to break down the membrane and this can only occur when T cells are close enough. The range over which the T cells can degrade the membrane is set by d_m.

We note that as beta cells are destroyed, the morphology of the affected islets may change. In particular, some islets in T1D present with small size and irregular outline (Gepts, 1965). The membrane in our study, where it exists, simply follows a circle with fixed radius, and thus does not account for these changes. However, we note that not all degranulated islets have small size and irregular outline (Gepts, 1965). Moreover, the morphology of islets, irrespective of insulitis, is highly variable. It is thus difficult to infer how the morphology of a given islet varies over the course of the insulitic process. Nevertheless, it seems likely that changes to islet morphology will occur, and we discuss approaches to incorporate this into our model in Section 5.1.1.

3.9. Activation and Apoptosis

When T cells and B cells are in contact with one another, we assume that B cells can activate non-activated T cells and that this process has a characteristic time course. We thus describe the activation level of the i'th T cell, a_i, via:

$$\frac{da_i}{dt} = \begin{cases} \sum_{j\in\mathcal{B}} H(|x_j - x_i - 2r_{\text{imm}} + 0.1|) - \lambda_a a_i & a_i < 1 \\ 0 & a_i \geq 1, \end{cases} \quad i \in \mathcal{T}, \quad (15)$$

where λ_a is the decay rate of the activation signal. The constant 0.1 is included inside the Heaviside function to account for the fact that we are not explicitly modeling cell-cell contacts, and the use of hard-core potentials tends to make cells move apart quickly when they are close. Our specific choice for this constant is selected such that cells are allowed to remain in contact with one another. We note that if this value is selected to be too large, cells can essentially occupy the same location, whilst values that are too small will cause cells to "bounce" off one another. We wish to avoid both of these behaviors, and over a range of choices, we found that a value of 0.1 satisfied these criteria. Once the activation signal reaches 1, the cell is activated and remains so until that cell dies, (or until the end of the simulation if that occurs first), reflected by setting the RHS to zero upon the activation reaching 1. Activated cells have an increased sensitivity to the chemokine signal and enhanced killing rate. This is captured in the model by increasing the sensitivity and killing parameters:

$$S_i \mapsto S_i + \Delta S, \quad \kappa_i \mapsto \kappa_i + \Delta\kappa,$$

where ΔS and $\Delta\kappa$ are positive constants.

The dynamics for the apoptosis of beta cells b_i follows a similar prescription:

$$\frac{db_i}{dt} = \begin{cases} \sum_{j\in\mathcal{T}} \kappa_j H(|x_j - x_i - 2r_{\text{imm}} + 0.1|) - \lambda_b b_i & b_i < 1 \\ 0 & b_i \geq 1, \end{cases} , i \in \beta,$$

where λ_b is the decay rate of the apoptotic signal, κ_j is the killing rate of the j'th T cell, and we include the constant, 0.1, for the same reasons as for (Equation 15). If b_i exceeds 1 for a given cell i, that cell is assumed to have been killed and is removed from the simulation. This is achieved by setting its viability, v_i, to be zero.

3.10. Implementation

Since the system we are solving is a Langevin equation, we must use an appropriate numerical method to deal with the stochasticity. For computational efficiency, we shall use the forward Euler–Maruyama scheme. This method only has strong order $dt^{1/2}$ but requires the fewest function evaluations of any of the solvers for stochastic differential equations (SDEs). The one exception to this is the equation governing the membrane dynamics where the explicit Euler scheme is unstable and causes solutions to blow up. For this equation, we use an implicit, backward Euler scheme. This keeps the total order accuracy of solutions consistent with the other equations, but provides the required stability without incurring additional computational cost.

Note that the membrane (Equation 14) treats distinct points along the membrane as being distinct from one another. This is the only system that is grid-based - all of the other equations are grid-free. In order to solve this system, we first discretize the membrane into N_θ points, $\theta_i = -\pi + 2\pi(i-1)/N_\theta$, $i = 1, N$, and subsequently solve (Equation 14) at each of these points. To find the value of the m at a point not on the discrete grid, we use band-limited interpolation for periodic signals as described in Schanze (1995).

3.11. Initial Conditions and Parameters

The immune cells are initially located at random, non-overlapping positions within annular domain with inner radius R and outer radius 1mm. The islet is positioned at the center of the domain, which for simplicity, we set as our origin. To distribute the beta cells, N_β cells are arranged in concentric circles within the islet. Next, for $T = 1000$ timesteps, they are allowed to freely evolve governed by:

$$\frac{dx_i}{dt} = F_i^{\text{islet}} + F_i^{\beta-\beta} + \sigma\xi_i(t), \quad i \in \beta, \tag{16}$$

where x_i is the center of the i'th beta cell, F_i^{islet} is the same as Equation (13) operating on the beta cells rather than the immune cells with:

$$F_i^{\beta-\beta} = -\nabla U_i^{\beta-\beta}, \tag{17}$$

where

$$U_i^{\beta-\beta} = \sum_{j\neq i, j\in\beta} H(|x_j - x_i| - 2r_\beta)(|x_j - x_i| - 2r_\beta)^6, \tag{18}$$

where $x_i \in \mathbb{R}^2$ now represents the location of the beta cell. After the n steps, the beta cells are fixed in location for the remainder of the simulation. Note that during this process, the immune cells are fixed in position. We distribute cells in this manner as it (a) generates a more realistic islet geometry compared with spacing them evenly within the islet and (b) it overcomes the known computational problems with randomly distributing non-overlapping disks in a confined region (Song et al., 2008). In **Figure 1**, we show the initial and final configuration of the beta cells following this approach. All simulations and analyses were performed in Matlab.

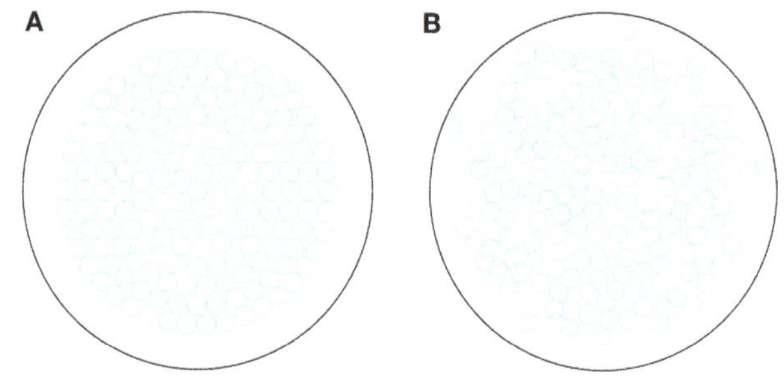

FIGURE 1 | (A) Initial configuration of beta cells within an islet, distributed in concentric circles. **(B)** Final configuration of the beta cells evolved for $T = 1000$ timesteps, according to Equations (16–18).

Other initial conditions are given by $m = 1$ for all θ, $a_i = 0$ for all $i \in \mathcal{T}$, and $b_i = 0$, $\nu_i = 1, \forall i \in \beta$.

The sensitivities of the T and B cells to the chemokine gradient are drawn from distinct Gaussian distribution with means S_T and S_B and variances σ_T and σ_B respectively. Parameters governing the geometry of our domain, such as immune cell and beta cell sizes were chosen to be matched to available data. In particular, beta cell diameters were taken from Saisho et al. (2013), immune cell diameters were based on values found in Wang et al. (2012), numbers of cells in each immune cell population were taken from Willcox et al. (2009) and islet composition (in terms of beta cells) were based on those found for humans in Kim et al. (2009). Data regarding the *in vivo* lifespan of specific effector lymphocytes in humans are rare and these may also be heterogeneous even within the same population. However, effector lymphocytes are unlikely to have lifespans ranging beyond several weeks (Sprent, 1993). To address the uncertainty in lifespans, we performed simulations across a range of values, from 3 to 7 weeks and compared results.

Other parameters in our model were tuned to provide the required behaviour, though we comment that much of the general behavior observed in our simulations is robust to parameter variations, suggesting that the model outcomes are robust. We note that certain processes, such as the degradation and repair of the peri-islet basement membrane, are phenomenological in nature, and as such, it is difficult to ascribe to them meaningful parameter values. Whilst the islet membrane is a physical component of the real system, we represents its viability by a scalar in the range 0–1, and so parameter values should be interpreted with respect to this scaling. Similar comments hold for the chemokine signal. Though a chemoattractant gradient is thought to exist in the peri-islet space, it is not clear what its composition, and subsequent properties, might be. As such, we have chosen a prototypical form to represent our chemokine signal, and investigated how changes to the strength of this signal impacts upon the resulting dynamics. All parameter values are summarized in **Table 1** and referenced in text where they are altered for specific numerical experiments.

4. RESULTS

In **Figure 2**, we show the initial state of a prototypical simulation, and the state of the system at $T = 4000$ days. Here, we

TABLE 1 | Table of parameter values and meanings for the agent-based model.

N_T	Number of T cells	30 Ref: Willcox et al. (2009)
N_B	Number of B cells	{5, 30} Ref: Willcox et al. (2009); Leete et al. (2016)
N_β	Number of beta cells	166 Ref: Kim et al. (2009)
σ	Strength of Wiener process	1.0 μm days^{-1}
r_{imm}	Radius of immune cells	4.0 μm Ref: Wang et al. (2012)
r_β	Radius of beta cells	6.4 μm Ref: Saisho et al. (2013)
R	Radius of islet	120.0 μm Ref: Kim et al. (2009)
S_T	Mean sensitivity of T cells	600 N^{-8} μm mM^{-1}
S_B	Mean sensitivity of B cells	700 N^{-8} μm mM^{-1}
σ_T	Variance of sensitivity of T cells	0.5 N^{-8} μm mM^{-1}
σ_B	Variance of sensitivity of B cells	0.5 N^{-8} μm mM^{-1}
$\bar{\kappa}$	Killing rate	0.2 cell^{-1} day^{-1}
$\alpha_{C,1}$	Chemokine strength	0.5 μM
$\sigma_{C,1}$	Range of chemokine signal	200.0 μm
$\alpha_{C,2}$	Chemokine strength	10.0 μM
$\sigma_{C,2}$	Range of chemokine signal	8.0 μm
A_a^0	Attraction strength between T and B cells	1.0 N^{-8} μm
d_a	Attraction range between T and B cells	4.0 μm
A_r	Repulsion rate for hard-core potential between cells	10.0 N^{-8} μm
A_{islet}	Repulsion rate for hard-core potential of islet	100.0 μm
α_m	Repair rate of membrane	0.01 day^{-1}
λ_m	Degradation rate of membrane	0.1 cell^{-1} day^{-1}
d_m	Membrane degradation range	4.0 μm
λ_a	Decay rate of activation signal	0.5 day^{-1}
λ_b	Decay rate of apoptotic signal	0.5 day^{-1}
ΔS	Activated sensitivity boost	400 N^{-8} μm mM^{-1}
$\Delta\kappa$	Activated killing rate boost	0.9 cell^{-1} day^{-1}
L_T	Lifespan of T cells	{21, 28, 56} days Ref: Sprent (1993)
L_B	Lifespan of B cells	{21, 28, 56} days Ref: Sprent (1993)
$\Delta\kappa$	Activated killing rate boost	0.9 cell^{-1} day^{-1}
$\Delta\chi$	Activated cell lifespan counter	2
dt	Time-step for EM scheme	0.001 day^{-1}

Where indicated, references show the studies upon which specific parameter values were based. Parameters are altered as discussed in the text for specific numerical experiments.

can see that the islet exhibits significant beta cell loss. We also see that immune cells congregate at the islet membrane and that a small number infiltrate the islet itself. Across all parameter sets used in our experiments, this behavior is preserved. In the Supplementary Material, we provide videos showing the full time-course of typical simulations with base parameters as indicated in **Table 1**, lymphocyte lifespan of 21 days and other parameters selected to reflect a variety of conditions, as discussed in the forthcoming sections. These simulations may be compared with the averaged results shown in **Figures 3–7**.

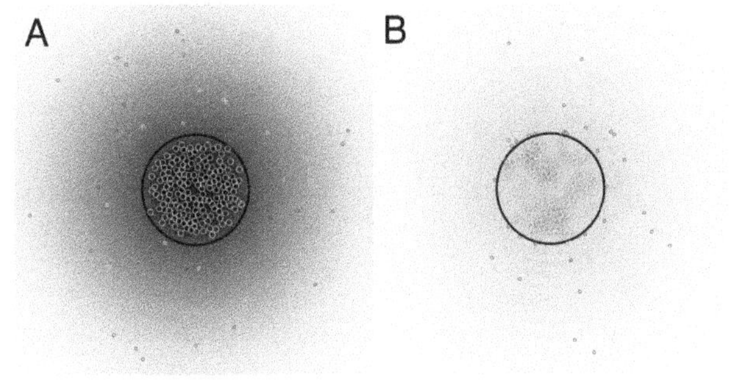

FIGURE 2 | (A) Initial state of system. **(B)** State of the system after $T = 4000$ days. A video of the full time-course of this simulation can be found in the Supplementary Material. In both panels, the blue circles represent beta cells, the red circles represent T cells, the green circles represent B cells, the black circle represents the peri-islet basement membrane and the black gradient represents a chemokine signal. Parameters are as in **Table 1**.

For all parameter sets, we simulate 100 realizations and average at specific time points to develop a time-course. During the simulations, we keep track of the number of immune cells within the islet and the number of viable beta cells. We take the total area encompassing the viable beta cells as being reflective of the beta cell mass. We shall now report both the time-courses of the remaining beta cell mass as a percentage of the initial mass (calculated as the ratio of the total mass of viable beta cells at time t and the initial mass of beta cells, averaged over the 100 realizations), and the number of immune cells in the islet for a number of conditions, along with the standard error of the mean (SEM) at each time point. In each case, we shall use the base set of parameters as indicated in **Table 1**, changing specific values where indicated. In all cases, to assess how the lifespan of the immune cells affects the resulting dynamics, we vary this between 21, 28, and 56 days as indicated in the figures.

4.1. Varying Number of B Cells

We begin by examining a key condition in this study, namely how the number of B cells in the peri-islet space impacts the rate of destruction of beta cells. To address this, we compare a condition with few B cells, setting $N_B = 5$, and one with a higher number of B cells, with $N_B = 30$. These results are summarized in **Figure 3**.

We observe that when the immune cell lifespan is short, neither the low nor the high B cell condition exhibit significant beta cell destruction. As the lifespan is increased, we observe a pronounced increase in this rate. For an immune cell lifespan of 56 days, we see that the high B cell condition reaches an equilibrium value at around 3000 days. Note that the equilibrium value is below the critical threshold of ~20% of remaining beta cell mass (Cnop et al., 2005) (however, it should be noted that this threshold is a whole-body threshold and may not be valid for an individual islet).

4.2. Varying Repair Rate of Membrane

We have assumed in our model that the peri-islet basement membrane is capable of being repaired at a rate α_m. If we decrease α_m, the rate of degradation of the membrane may be such that

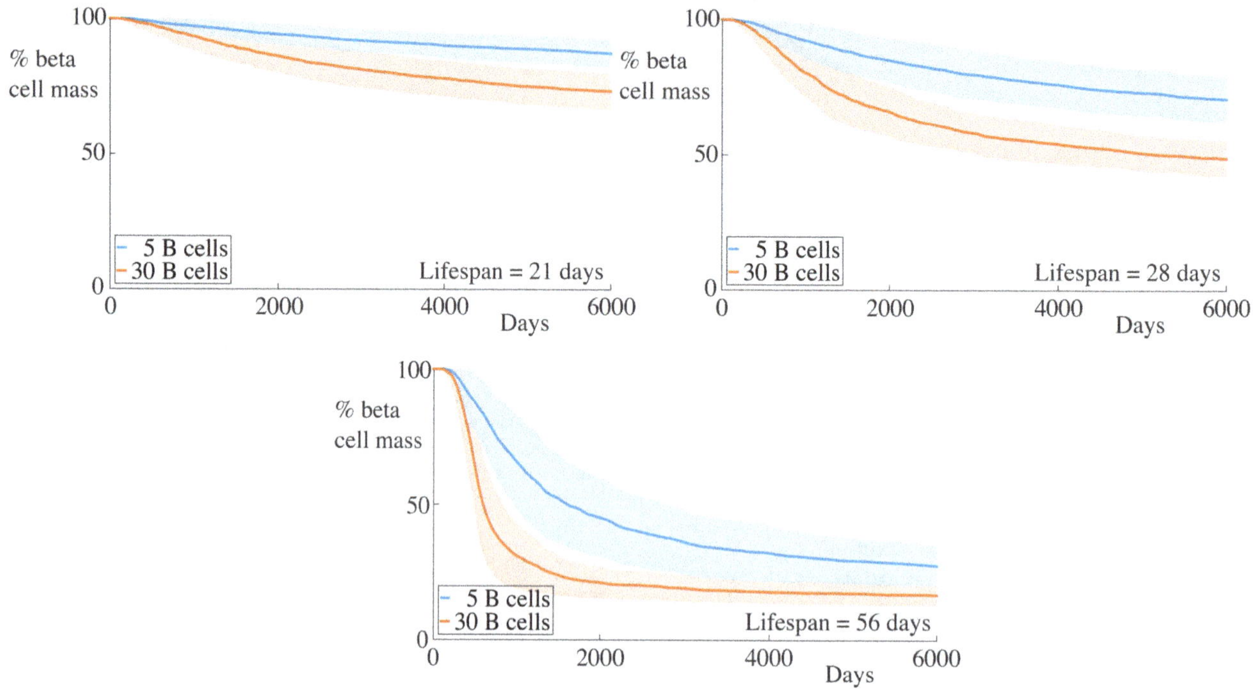

FIGURE 3 | Effect of variation of B cell number on beta cell mass loss in the agent-based model. Each graph shows the mean ± SEM for 100 simulations. Across all immune cell lifespans, we find that increasing the number of B cells significantly increases the destruction rate of the beta cells. Parameters are as in **Table 1**. Videos of typical simulations for the case with immune cell lifespans of 21 days may be found in **Videos S1**, **S2**.

holes caused by infiltrating T cells are not repaired (at least over the timescale of our simulation). In the extreme case, the membrane is incapable of repair, in which case we set $\alpha_m = 0$. In **Figure 4**, we display the time-course of a simulation in such a regime.

It is clear from these graphs that removing the repair mechanism of the membrane dramatically increase the susceptibility of the beta cells to cytotoxicity. In the most extreme case, in the situation with few B cells and a short immune cell lifespan, we now see a huge difference in both the rate of beta cell destruction and the final cell mass at the end of our simulation. In the case with no membrane repair, the final beta cell mass of almost all simulations is below ~20% of the initial mass. These results highlight the importance of the peri-islet basement membrane to the progression of insulitis, and also suggest that ongoing repair of this membrane is critical to slowing the infiltration of the immune cells.

We also observe, in contrast to the results in **Figure 3**, that the immune cell lifespan now has a more muted affect on the resulting dynamics. Whilst there is an increase in the rate of beta cell destruction with increasing lifespan, the more significant contribution appears to be to lower the equilibrium value of remaining beta cell mass. This likely arises due to the fact that, as the beta cell population decreases, so too does the chemokine signal to which the immune cells respond. This means that the cells (on average) move into the islet at a slower rate, since their paths becomes more dominated by noise. If the cell lifespan is too short, the immune cells may die before entering the islet if there are insufficient beta cells generating a chemokine gradient.

4.3. Increasing the Membrane Degradation Rate

Instead of reducing the basement membrane repair rate, we now increase the efficacy of the immune cells of degrading the membrane, setting $\lambda_m = 0.2$ and display the results in **Figure 5**. For comparison with the condition in which the encapsulating membrane is not repaired, we additionally plot results from this case in the same figure, as indicated.

As expected, increasing the rate of membrane degradation increases the rate of beta cell destruction and subsequently decreases the final remaining beta cell mass. Compared with the condition with no membrane repair, we see that the initial rate of beta cell destruction is greater in the case with high membrane degradation. However, as the insulitis process continues, this trend becomes reversed. Remarkably, across all cases varying the number of B cell and the immune cell lifespan, the time at which this occurs appears to be approximately the same. We also see that the final beta cell mass is lower for the condition with no repair than for that with high membrane degradation.

At the beginning of simulations, the ability of the T cells to degrade the basement membrane at a faster rate allows them to infiltrate the islet and subsequently kill the beta cells more quickly. Whilst the initial degradation of the membrane in the no repair condition is slower, the fact that "holes" in the

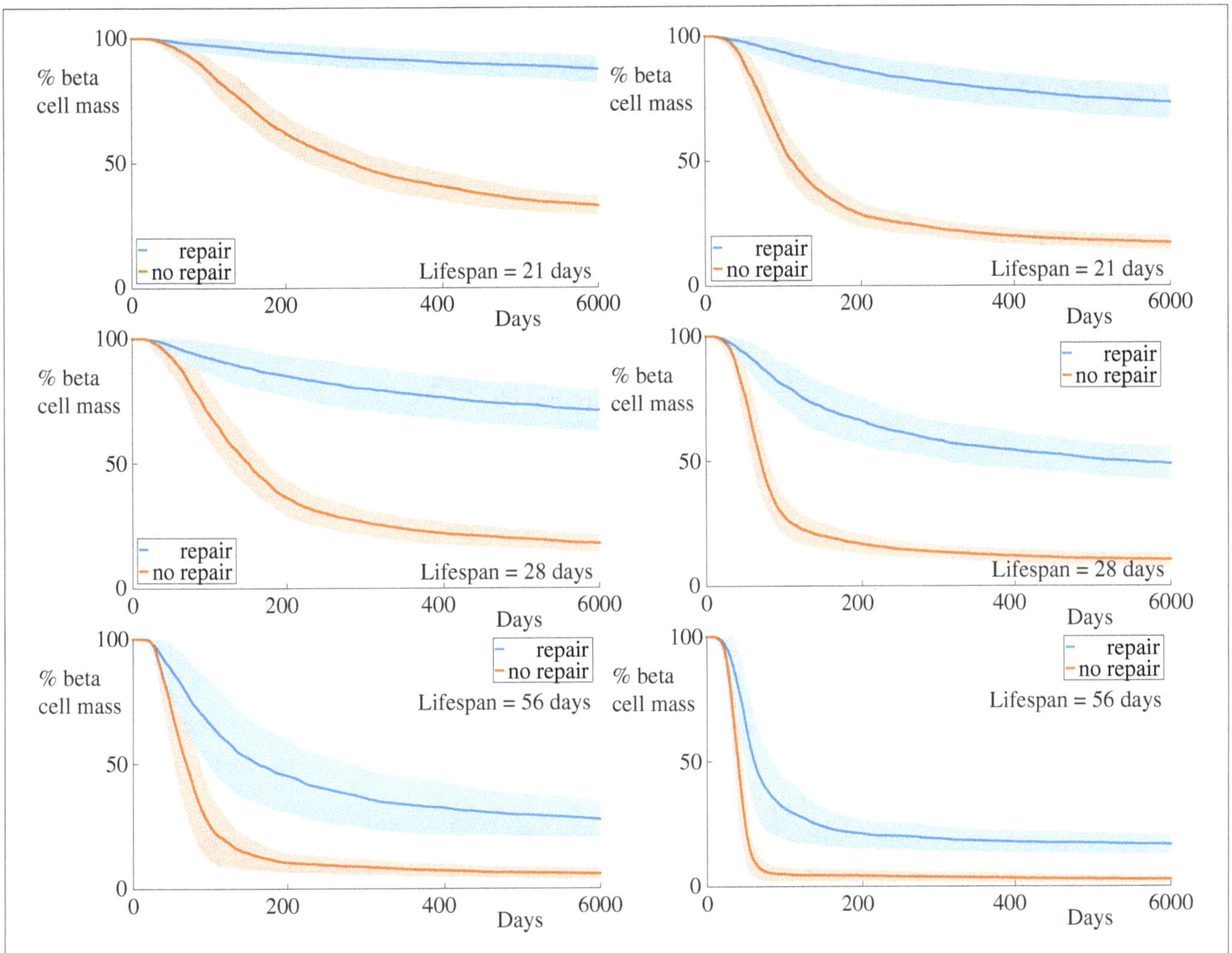

FIGURE 4 | Time-course of beta cell mass comparing conditions with and without peri-islet basement repair. Each graph shows the mean ± SEM for 100 simulations. Data for simulations with 5 B cells are displayed in the left column, whilst those for simulations with 30 B cells are depicted in the right column. In the case without membrane repair, we observe both an increase in the rate of beta cell destruction and pronounced differences in the final beta cell mass across all immune cell lifespans and B cell number. Parameters are as in **Table 1** with $\alpha_m = 0$. Videos of typical simulations for the case with immune cell lifespans of 21 days may be found in **Videos S3**, **S4**.

encapsulating membrane caused by T cells are not repaired means that, at later times, immune cells can simply pass through these holes.

When the membrane can be repaired, the rate of immune cell infiltration is governed by the balance between how quickly it can be degraded and how quickly it can be repaired. As the chemokine signal across the peri-islet space becomes weaker when beta cells die, the average rate that T cells reach the basement membrane falls, since the strength of the signal is a contributory factor in determinining the transit time for a T cell toward the islet. As there are fewer T cells aggregrating at the islet membrane at any given time, the degradation of the membrane becomes slower, and hence the beta cells are destroyed at a slower rate as time progresses. At later times the arrival rate of T cells may be so slow that the membrane is repaired at the same rate as it is degraded, on average, and so islet infiltration by the immune cells will be halted.

These observations account for the slower rate of beta cell destruction and higher remaining beta cell mass for the high degradation condition compared to that with no membrane repair. This further highlights the importance of continual basement membrane repair to the slowing of insulitis.

4.4. Increasing the Killing Efficiency of the T Cells

We now increase the killing rate of the T cells. To reflect the fact that we assume that activated T cells are responsible for most of the beta cell destruction, we achieve this by doubling the killing rate boost, setting $\Delta\kappa = 1.8$. The results for these experiments are displayed in **Figure 6**.

We observe that increasing the killing rate of the immune cells does not significantly affect the death rate of beta cells, either quantitatively or qualitatively. In all cases, we observe an increase

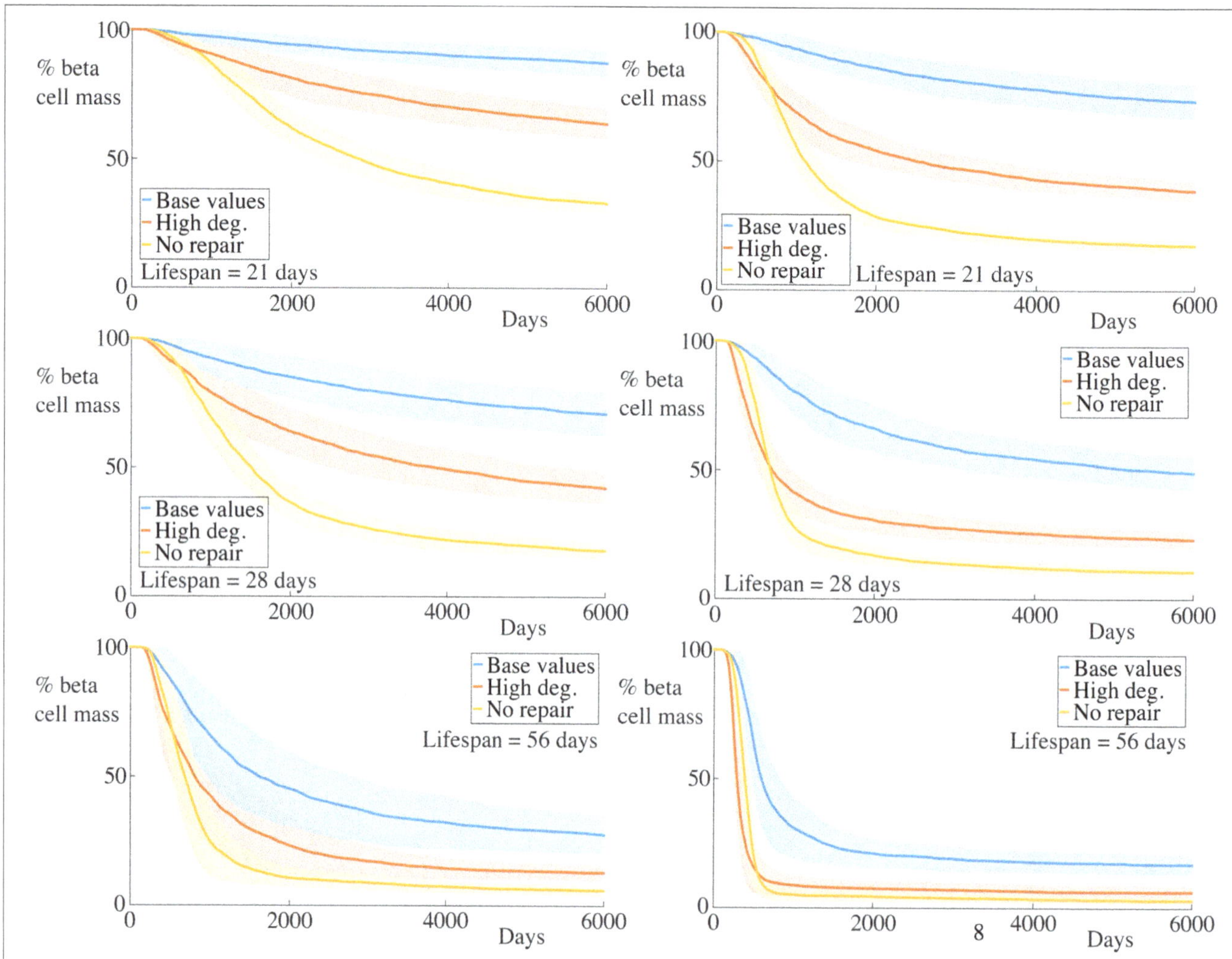

FIGURE 5 | Time-course of beta cell mass comparing with higher membrane degradation rates. Each graph shows the mean ± SEM for 100 simulations. The left column shows results for simulations with 5 B cells whilst the right column shows those with 30 B cells. Data for the condition with the base value for membrane degradation but with $\alpha_m = 0$, as reported in **Figure 4** are also shown. An increase in the beta cell destruction rate, as well as lower final beta cell masses are observed across all immune cell lifespans and number of B cells. Note, however, that whilst the initial rate of beta cell destruction for the condition with high membrane degradation is higher than that for the case with no membrane repair, the final beta cell mass is higher for the former compared with the latter. We refer the reader to the text in Section 4.3 for further interpretation of this observation. Parameters are as in **Table 1** with $\lambda_m = 0.2$ for the high degradation condition and $\alpha_m = 0$ for the no membrane repair condition. Videos of typical simulations for the case with immune cell lifespans of 21 days may be found in **Videos S5**, **S6**.

in the beta cell mass destruction rate, but this is fairly minimal. The limited impact of increased killing rate can be understood by noting that, whilst the killing rate is an important factor in the overall insulitis process, this is dominated by transit time to and within the islet, as well as activation by B cells. Once inside the islet, T cells can interact with and kill beta cells but the total number of beta cells an individual T cell can destroy will be influenced more strongly by its remaining lifespan than on its killing rate. This highlights the need to incorporate the spatial aspects of the immune response and that high T cell killing rates do not necessarily significantly affect the progression of insulitis.

We comment that these observations at first seem at odds with previous results from the NOD mouse where the avidity of a T cell population is well-correlated with its pathogenic potential, and the progression of T1D occurs through avidity maturation (Amrani et al., 2000). Theoretical studies have supported the viewpoint that higher avidity CD8+ T cell populations lead to increased rates of beta cell destruction (Khadra et al., 2011) and that low avidity populations may have a protective effect against T1D development (Khadra et al., 2009). However, recent results in the NOD model have suggested that the highest avidity CD8+ cells do not escape thymocyte negative selection, and accordingly, the cells responsible for beta cell destruction in the murine pancreas may be regarded as those with "intermediate avidity" (Han et al., 2016). Furthermore, the paucity of data on the specificity of lymphocytes involved in human insulitis makes it difficult to propose quantitative statements about the effective killing potential of individual T cells. This is particularly difficult since the majority of the results regarding the avidity of

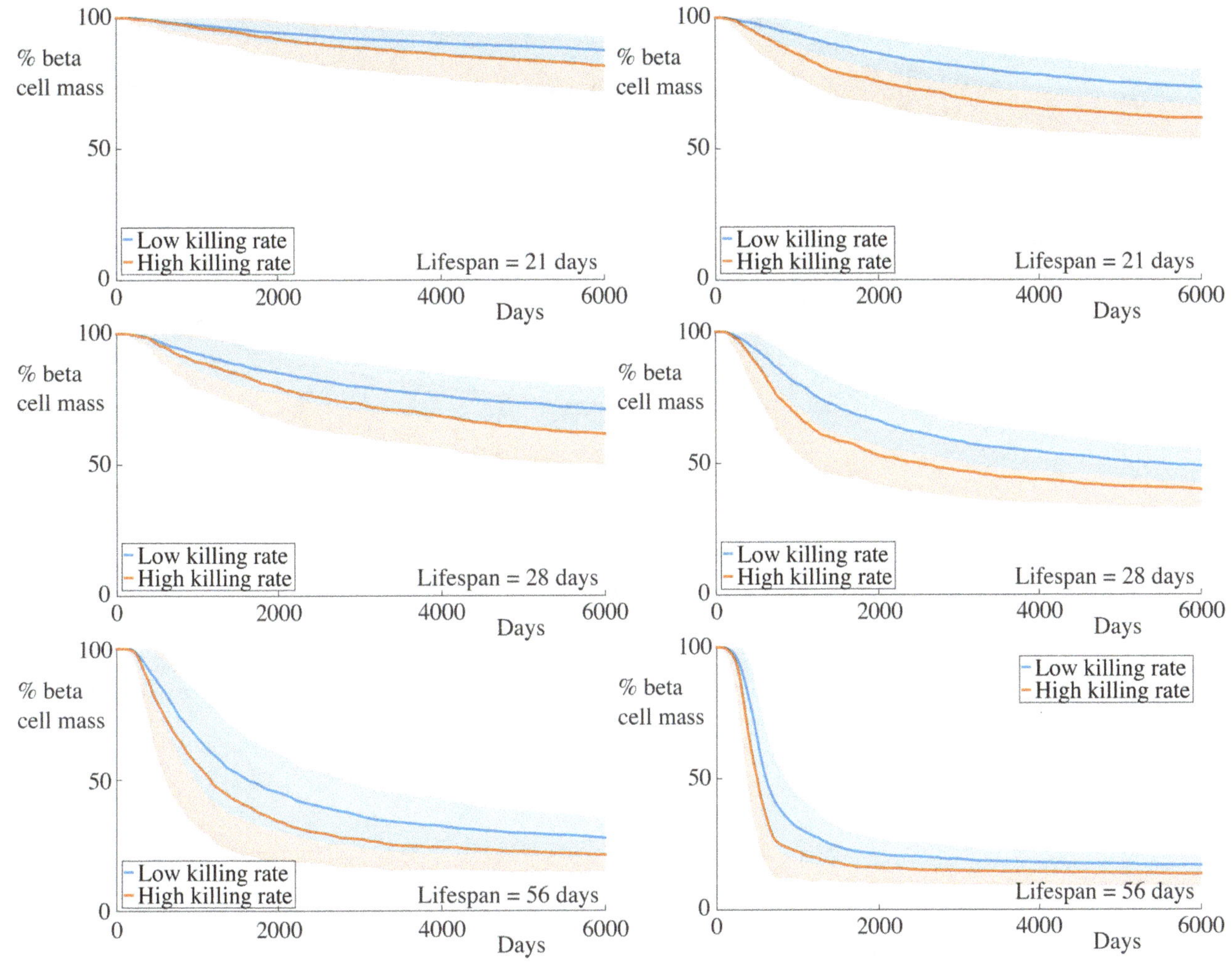

FIGURE 6 | Time-course of beta cell mass comparing with higher activated killing rates. Each graph shows the mean ± SEM for 100 simulations. The left column shows results for simulations with 5 B cells whilst the right column shows those with 30 B cells. Although an increase in the rate of beta cell destruction is observed, this is relatively small in all cases. Parameters are as in **Table 1** with $\Delta\kappa = 1.8$.

human lymphocytes come from peripheral blood measurements and are not necessarily indicative of avidities within the pancreas (Reijonen et al., 2004; Standifer et al., 2009). Finally, the lifespan of a T cell that has successfully infiltrated an islet should still have a significant impact on the number of beta cells it can destroy, assuming its avidity is sufficiently high.

4.5. Increasing the Rate of Chemokine Production

To reflect increases in the rate of chemokine production, we now double α_C^1 and α_C^2 simultaneously, setting $\alpha_C^1 = 1.0$ and $\alpha_C^2 = 20.0$. Note that, since we are assuming a quasi-steady state approximation for the chemokine, this simply adjusts the profile of the chemokine signal across our domain. The results for these simulations are depicted in **Figure 7**.

As for the situation with increased killing rates, we observe similar qualitative behavior to the case with lower chemokine production rates. However, in contrast to the former case, the quantitative differences between the beta cell death rate and final cell mass are much greater. This further highlights the contribution of cell movement in the overall inflammatory response.

4.6. Combining High Chemokine Signals and Reduced Membrane Repair

In our final numerical experiment, whose results are shown in **Figure 8**, we simultaneously increase the strength of the chemokine signal, setting $\alpha_C^1 = 1.0, \alpha_C^2 = 20.0$, as in the previous subsection, and remove the membrane repair, setting $\alpha_m = 0$. As expected, with these combined alterations, we now observe significantly high rates of beta cell destruction for all cases considered. Moreover, the final beta cell mass at the end of our simulations is below 20% of the original cell mass. For the cases with the longest lived immune cells, those having a lifespan of 56 days, we now see that essentially all of the beta cells have been destroyed within 1000 days.

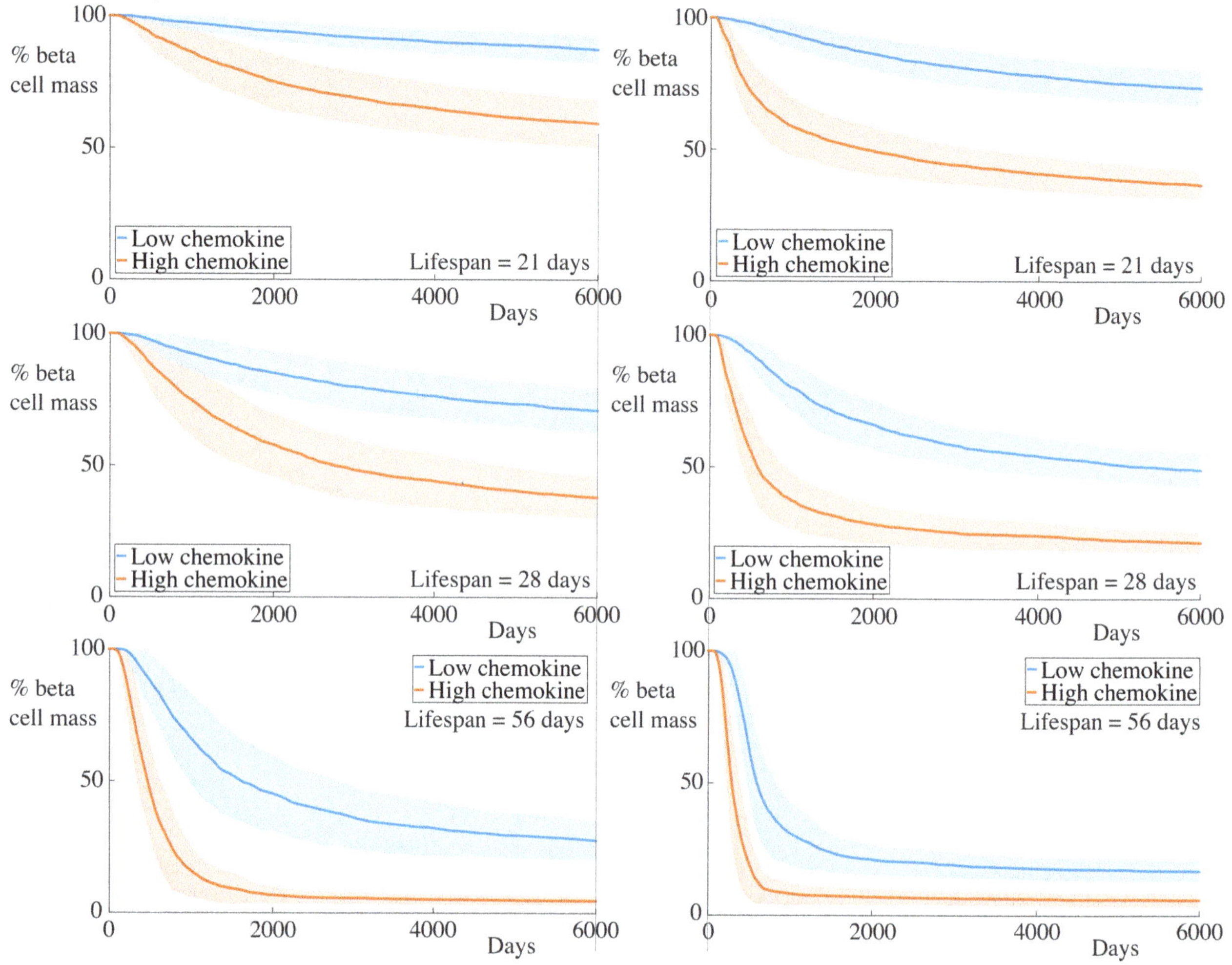

FIGURE 7 | Time-course of beta cell mass comparing conditions with the base and increased strength of chemokine signal. Each graph shows the mean ± SEM for 100 simulations. The left column shows results for simulations with 5 B cells whilst the right column shows those with 30 B cells. In the high chemokine signal parameter regime, we observe both an increase in the rate of beta cell destruction and pronounced differences in the final beta cell mass across all immune cell lifespans and B cell number. These differences are greater than those displayed for higher T cell killing rates in this figure. Parameters are as in **Table 1** with $\alpha_C^1 = 1.0$ and $\alpha_C^2 = 20.0$. Videos of typical simulations for the case with immune cell lifespans of 21 days may be found in **Videos S7**, **S8**.

4.7. Immune Cell Invasion Profiles

We now consider how the invasion of immune cells is dependent on the remaining beta cell mass and report the dependence of the number of immune cells within the islet (i.e., within the basement membrane) on the remaining beta cell mass. These results are shown in **Figure 9** for the case with 5 B cells and in **Figure 10** for the case with 30 B cells. The results for the case with high chemokine signals with no basement membrane repair, and for the case with increased membrane degradation are omitted, but are qualitatively similar to the case with no repair alone. In both figures, we only show results when the immune cell lifespan is 56 days.

For situations with 5 B cells, we observe two qualitative behaviors amongst the range of cases considered. Using the default parameters and those with high killing rates, we observe no infiltration of the islet by the B cells. The immune cell population undergoes an initial increase to a maximum value, followed by an approximately linear decay with the remaining beta cell mass. The behavior with high chemokine signals or no membrane repair does show islet invasion by the B cells. In these cases, the T cell population now both increases and decreases more gradually as a function of the beta cell mass, and attains a higher maximum value than in either of the two previous cases.

If we now increase the number of B cells to 30, we observe, in all cases, invasion of the islets by B cells and a qualitatively similar profile of T cell infiltration. With the default parameters, or those with high killing rates, the maximum value of invading T cells occurs when approximately 50% of the beta cell mass remains. In contrast to the T cell population, the number of invading B cells varies little with the beta cell mass. In the case with no membrane repair, this is no longer true, and we now find that the B cell profile follows the T cell profile and furthermore, that the

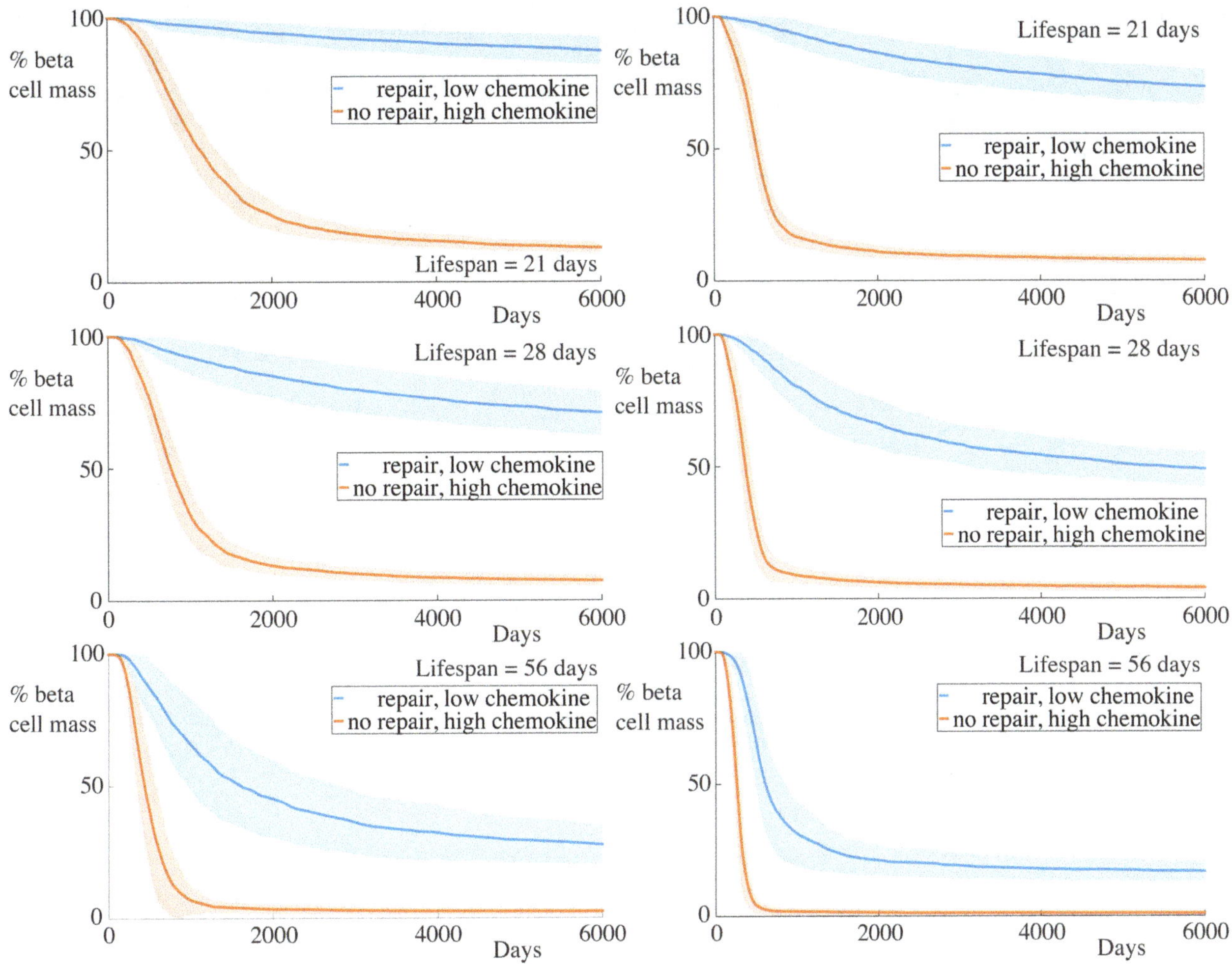

FIGURE 8 | Time-course of beta cell mass with high chemokine signals without basement membrane repair. Each graph shows the mean ± SEM for 100 simulations. The left column shows results for simulations with 5 B cells whilst the right column shows those with 30 B cells. In all cases, the rate of beta cell loss is high and the remaining cell mass at the end of the experiments is low. Parameters are as in **Figure 7** with $\alpha_m = 0$.

maximum value of immune cell infiltration occurs at low values of the remaining beta cell mass.

4.8. Trial-to-Trial Variability and the Role of Space

For each of the scenarios listed, we report the standard error of the mean for all recorded values at each time point. In general, we observe that variability across our simulations for a given scenario is low, suggesting that the model behavior is robust and that noise does not significantly affect the intrinsic model dynamics. In particular, scenarios with either very high rates or very low rates of beta cell destruction exhibit very little variability. Variation is highest in scenarios that lead to intermediate rates of beta cell death, suggesting that the role of noise in such situations is much more significant in these conditions.

The results reported in **Figures 3–10** reflect spatial averages that are taken over the peri-islet space. As such, they do not reflect the spatial component of the dynamics produced by the agent-based model. These effects are, in general, quite subtle. In particular, since we assume that activation of CD8+ cells by CD20+ cells requires them to be co-localized, the spatial arrangement of the immune cells will play an important role in determining the overall rate of beta cell destruction. The geometry in this initial version of the model has been chosen to be simplistic in nature. In future versions, the geometry will be adjusted to better match that of real islets as discussed in Section 5.1.1. One could imagine scenarios in which the routes taken by immune cells when migrating toward the islet could influence the average number of interactions with one another, and further, that these paths would be influenced by prominent vasculature near the islet.

The spatial configuration of the immune cells also plays a significant role in determining how quickly the basement membrane is degraded. Where many CD8+ cells accumulate at specific points along the periphery of the islet, we would expect high rates of membrane degradation. Conversely, if the cells

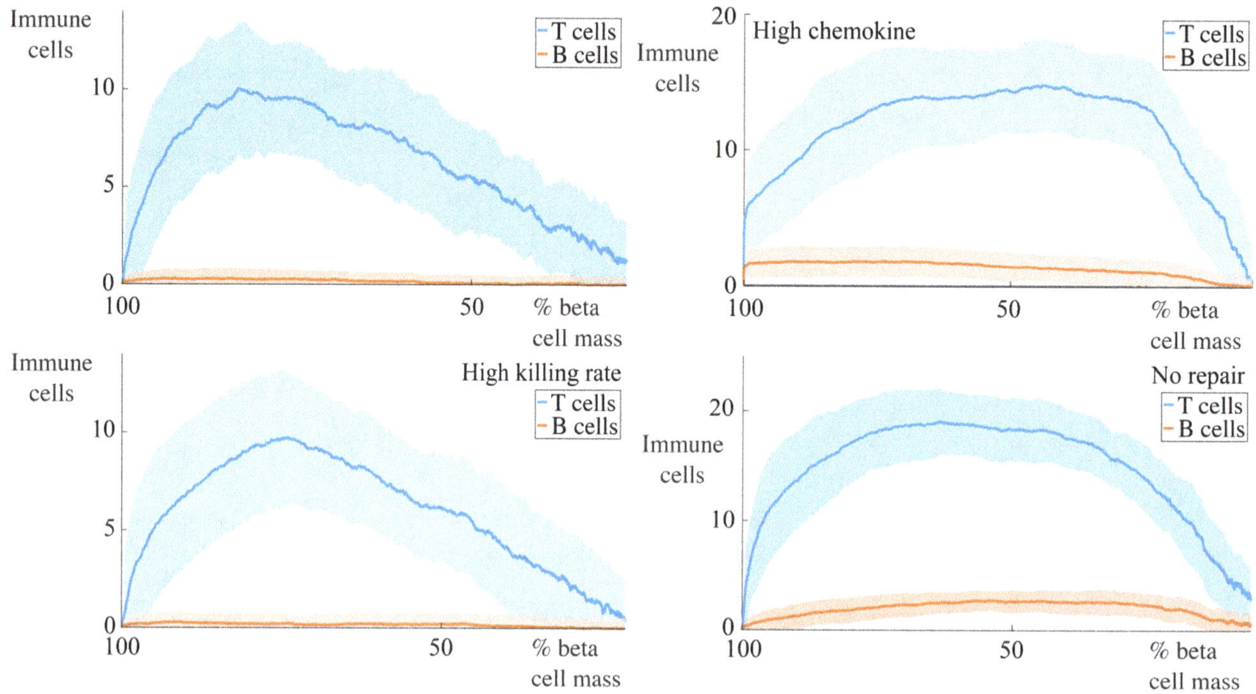

FIGURE 9 | Profiles of immune cell invasion dependent on the remaining beta cell mass in cases with 5 B cells. Each graph shows the mean ± SEM for 100 simulations. We report the number of T and B cells within the islet in a number of conditions. The top-left panel shows results using the parameters in **Table 1**, the bottom-left uses these values with $\Delta\kappa = 1.8$, in the top-right, we set $\alpha_C^1 = 1.0$ and $\alpha_C^2 = 20.0$, and the bottom-right has $\alpha_m = 0$.

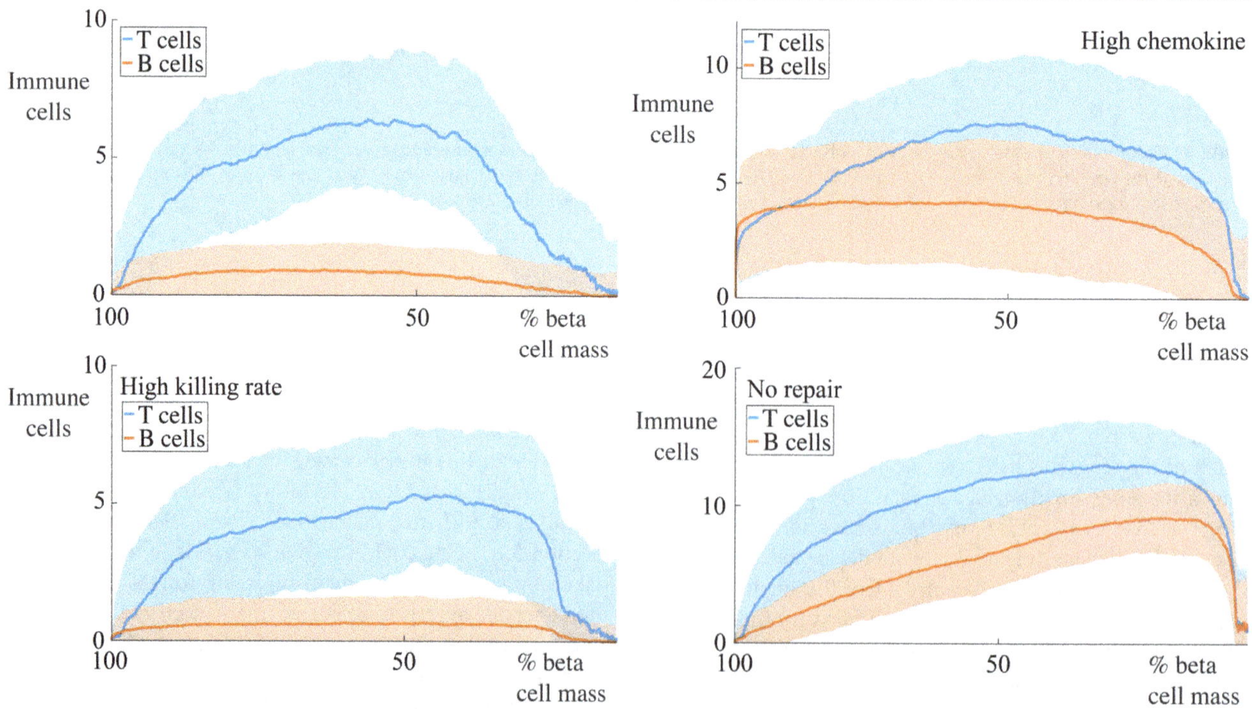

FIGURE 10 | Profiles of immune cell invasion dependent on the remaining beta cell mass in cases with 30 cells. Each graph shows the mean ± SEM for 100 simulations. We report the number of T and B cells in the same conditions as displayed in **Figure 9**.

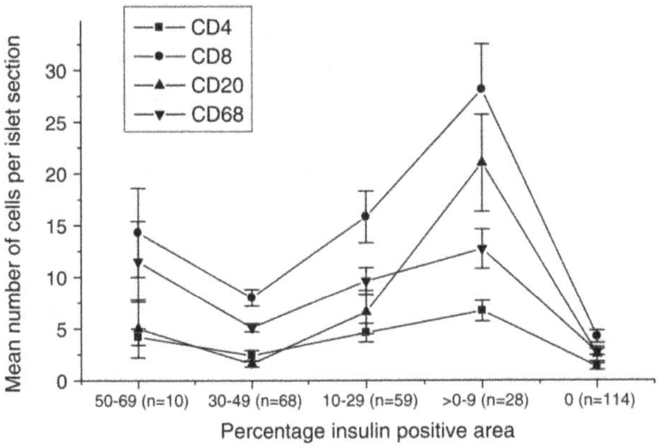

FIGURE 11 | Data showing the immune cell population against the percentage of the islet which is positive for insulin, as calculated via morphometry. This percentage can be thought of as being representative of the residual beta cell mass. Each point is the mean ± SEM for the number of relevant immune cells detected in each islet section, averaged over 279 islets across 29 patients with T1D. The number of islets analyzed in each insulin-positive category is given in parenthesis. Permission to reprint this figure from Willcox et al. (2009) was granted by John Wiley and Sons (Licence number: 3971831129485; www.interscience.wiley.com).

are evenly distributed along the membrane, average degradation rates are likely to be low. Our results suggest that the dynamical processes governing the repair and destruction of the basement membrane play a key role in determining the overall rate of beta cell destruction and so it is clear that the spatial distribution of immune cells along this membrane will also be critical.

These spatial considerations are the key determinant for the degree of variability observed between trials, as reported in each of the figures. In the Supplementary Material, we provide animations of typical simulations in a subset of the scenarios listed to demonstrate the subtle role that these effects have on the insulitis process.

5. DISCUSSION

In this manuscript, we have constructed a spatial, agent-based model of immune cell invasion of a prototypical islet of Langerhans. The agents in the model are immune and beta cells, and each obeys rules that broadly match the general behavior of these cell types. The model attempts to mimic a real islet, allowing for *in silico* experiments to be performed through model simulation. The agent-based framework allows both the spatial aspects of insulitis to be investigated, and avoids issues of applying ODE-based approaches to a system that has low cell numbers. It is intended that this will lay down the framework for more extensive spatio-temporal modeling of insulitis, with an ultimate aim to understand its core mechanisms and to devise strategies to slow or halt its progress.

Overall, the observed phenomena in the model are well matched to real data, which have been collected and analyzed from human patient samples (Willcox et al., 2009; Morgan et al., 2014), and are reported in **Figure 11**. We note that data for the pauci-immune phenotype have recently appeared in Leete et al. (2016), but since our model is primarily focussed on the hyper-immune phenotype, we highlight these results here. Results from the agent-based model have the same qualitative behavior as displayed by these data. This in turn provides evidence that the mechanisms included within our model are sufficient to capture qualitatively the key features of the inflammatory response. In particular, we note that the case with no membrane repair appears to provide the best qualitative match to the existing data, in which we see a rise in the number of infiltrating immune cells during disease progression, with a peak at around 10–20% of remaining beta cell mass. We note that scenarios with low overall rates of beta cell death are associated with low levels of B cell infiltration. This further supports the viewpoint that the CD20+ cells play an important role in determining the rate of beta cell destruction. By adding and removing mechanisms in a systematic way, we can probe what effect their inclusion or exclusion will have on the resulting dynamics and subsequent prognosis. It is important to note that, since we investigate the disease at the level of the islets, we can make predictions about how modifying certain cell properties will affect the inflammation of the islets, rather than considering whole body responses.

Accordingly, in the present study, we have performed several parameter studies to examine how the relative contributions of core processes affects insulitis. We have demonstrated that, under the assumptions used in constructing the model, an increased number of B cells in the peri-islet space gives rise to a faster destruction of the beta cells. This effect can be amplified by factoring in manipulations of other processes, such as the killing rate of the beta cells by T lymphocytes and the repair rate of the islet's basement membrane. Simply increasing the killing rate of the T cells did not have a significant impact of the rate of the beta cell destruction. However, increasing the strength of the chemokine signal to which the immune cells respond did show a pronounced increase in the death rate. These facts highlight the importance of the chemotactic process by which the immune cells enter the islet, find and kill the beta cells.

In scenarios in which the islet basement membrane is unable to repair itself, or is degraded more rapidly, we also see significant increases in the beta cell destruction. Moreover, we also find that the final beta cell mass at the end of our simulations is substantially depleted. This highlights the importance of this membrane as a barrier to the infiltrating immune cells. In particular, when chemokine signals are sufficiently high, if the membrane is incapable of being repaired, all of the beta cells may ultimately be killed, whereas in other situations, we find residual beta cells even after long simulation times (not shown). The importance of the basement membrane during insulitis in our model is in line with a recent study of this membrane in samples from human patients with T1D (Bogdani, 2016).

Upon examining the dependence of the immune cell invasion on the remaining beta cell mass, we find in most cases the profile of B cell infiltration is relatively flat. In cases with low B cell numbers, low chemokine signals and with a membrane that is repaired, there is essentially no B cell invasion. In situations with higher numbers of B cells, the beta cell mass at which the T cell invasion is maximal is shifted to lower values. In the case with no membrane repair, this maximal point is shifted to very low values of the remaining cell mass. Moreover, in this instance, we also see that the B cell invasion profile follows that of the T cells. This condition is most consistent with the human data reported in Willcox et al. (2009), suggesting that, for our model, low membrane repair rates are an important factor in the insulitis process.

The rise and subsequent decrease in T cell populations observed in our simulations is in accordance with experimental results of T1D specific autoimmune responses in the NOD model (Trudeau et al., 2003). These dynamics have been described as "waves" and have been explained mathematically by transitions through Hopf and homoclinic bifurcations in Mahaffy and Edelstein-Keshet (2007) and by transient bistability of the autoimmune state in Jaberi-Douraki et al. (2014b). Interestingly, in the latter of these two theoretical studies, inclusion of multiple clones of T cells with different avidities can give rise to multiple waves in the T cell dynamics, which can significantly impact the rate of overall beta cell destruction. This highlights potential avenues for intervention through the promotion of lower avidity T cell populations.

It should be noted that data on the specificities of T cells within human insulitic islets are rare, but that cells with varying antigen specificities may be present (Coppieters et al., 2012; Babon et al., 2016). Thus, a potential development of our model would be to include multiple clonal populations to explore these dynamics in the spatial context, particularly given the potential insights that may be applicable from the theoretical results in Mahaffy and Edelstein-Keshet (2007) and Jaberi-Douraki et al. (2014b).

Agent-based and lattice-based approaches are becoming increasingly popular in biological modeling. ODE based approaches, which have long been a linchpin of the mathematical biology community become ill fit for purpose when wishing to describe processes in which the number of agents is small. Considering stochastic variants of ODEs and partial differential equations (PDEs) models is one potential approach to account for the variability that arises from low cell numbers, but this too is only a good approximation when the number of cells is sufficiently high. Moreover, these approaches do not typically account for trial-to-trial variability that may be important for describing individual results. Note that, at various levels of description, systems of differential equations can still form part of agent-based models.

One of the attractive properties of ODE and PDE modeling when compared with agent-based simulations is the computational cost associated with the latter, whereas density (continuum) based methods typically are low dimensional in nature. Indeed, a significant barrier to the widespread use of agent- and lattice-based models in the past has been the requirement of significant computational power. In recent years, the technological advances in computing have started to erode this barrier. In addition, domain decomposition methods (Chen et al., 2007; Tapia and D'Souza, 2011) and the ability to parallelise computations on large scales on commercially available graphics processing units (Chen et al., 2007; Tapia and D'Souza, 2011; Harvey et al., 2015) has made agent-based simulations achievable universally.

Here, we have focussed on a specific type of agent-based model. This is based on a discrete particle formulation, where each cell is parameterized by a finite area and location. Broadly, agent-based models can be divided into lattice-based and lattice-free methods (Plank and Simpson, 2012). The discrete particle approach is lattice-free, meaning that cells are free to move anywhere in the domain. One limitation of this approach is that the cells are restricted to be described as circles; that is, their shape cannot be deformed. Immune cell deformation may play an important role when considering how cells pass through degraded parts of the basement membrane. In the current model formulation, cells must "wait" until the gaps in the membrane are sufficiently large before they can pass through.

An alternative to the discrete particle formulation is the lattice-based cellular Potts model. This modeling framework has been growing in popularity and has been applied to cell migration in a number of contexts, including vasculogenesis (Merks et al., 2006; Daub and Merks, 2013), morphogenesis (Chen et al., 2007; Marée et al., 2007) and wound healing (Scianna, 2015). Although the cellular Potts model is lattice-based, the cells within the framework can occupy many lattice sites and thus their shape, as well as their location, evolves over time. On a fine enough lattice, such models can look remarkably similar to the system they are approximating. However, with the increase in spatial resolution of the lattice comes an associated computational cost. It should be noted that there are several actively maintained packages for agent-based modeling, such as CHASTE (Mirams et al., 2013) and CompuCell3D (Swat et al., 2012).

One approach that combines the flexibility of lattice-free approaches together whilst allowing for cells to change shape is to treat points on the cell membrane as particles and resolve forces acting upon them, under the assumption that the cell remains bounded by its membrane (Elliott et al., 2012). However, the mathematical machinery and computing power required to embed such a description into our domain are prohibitive for our purposes. A "halfway" house could, in which the cells are treating as deformable ellipsoids, is another possible way of relaxing our assumption that cells are perfectly circular (Palsson and Othmer, 2000).

5.1. Future Directions

The modeling framework described within this manuscript is not intended to be a fully comprehensive endpoint for the study of the spatio-temporal dynamics associated with insulitis. Instead, it is intended to be the first step to building a general virtual environment in which to simulate the invasion of islets during T1D. We will now highlight some potential areas for model development.

5.1.1. Geometry

We have treated perhaps the simplest possible geometry that is reflective of the islet. The model is planar in nature, whereas real islets and immune cells are three dimensional structures. We have ignored the mechanical effects on the basement membrane: as beta cells are removed from the islet, the morphology of the islet itself may change (Brereton et al., 2014). There is currently no structure imposed upon the extra-islet space. Accurate modeling, for example, in which the vasculature is taken into account may constrain locations at which immune cells are allowed to enter the domain. Through the use of imaging techniques, it should be possible to build a more faithful representation of both the intra- and extra-islet space and include these in the model.

5.1.2. Different Cell Types

The model could be developed to include more cell types, such as the macrophages that clear apoptotic beta cells, or different endocrine and exocrine cells which are not targeted by immune cells, but may slow their movement. There may be other cell-cell interactions that may be important during insulitis, for example the role of CD4+ cells in the activation of CD8+ cells (Castiglioni et al., 2005). Even restricting the model to the two immune cell types described in this manuscript, we could consider different clonal populations of the T cells (Khadra et al., 2011). One hypothesized strategy to slow the progression of insulitis is to promote the replication of a low avidity clone of T cells, that will out compete the higher avidity clones for space and resources, leaving only relatively passive T cells (Amrani et al., 2000; Khadra et al., 2009; Bluestone et al., 2015). Such a mechanism could be tested in our modeling framework.

5.1.3. Cell Proliferation

Currently, none of the cell types included in the model undergo cell division. In general, post-natal beta cell replication in humans is thought to occur at low rates, and we have disregarded it here. Similarly, immune cell replication in and around the islet is also assumed to be negligible. However, there is some evidence to suggest both that beta cell proliferation is enhanced during insulitis, at least during its early phases (Willcox et al., 2010; Dirice et al., 2014) and that immune cell replication also takes place within the islet itself (Graham et al., 2012) in the NOD model, though these findings have not been corroborated in human tissue. In particular, data collected from *ex vivo* human samples suggested that very few lymphocytes in the peri-islet space were positive for proliferative markers (Willcox et al., 2010). In order for the hypothesis of promotion of low avidity clones to be tested, the model would have to include cell division, since the action of long-lived memory cells is an integral part of that theory, however this may require coupling of the agent-based model with one representing the lymphatic system.

5.1.4. Multiple Islets

One of the striking features of insulitis is its heterogeneity, even within an individual patient. Islets near to one another can display different inflammatory profiles—insulitis in one islet thus does not imply insulitis in nearby islets. By extending our domain to include multiple islets, we can use the modeling framework to investigate under what scenarios this can occur. In particular, we can hope to identify what cellular interactions, on a broad scale, must occur for such heterogeneity in disease progression to be observed. One simple way to incorporate the influence of other islets in the current framework would be to include additional point sources of chemokine outside of the working domain. This would have subtle effects on the dynamics of cell movement, particularly when the number of beta cells becomes low.

5.1.5. Membrane Degradation

In the current model prescription, the membrane integrity is assumed to be compromised in a simple way by the action of nearby immune cells. This process is solely dependent on the proximity of the immune cells to the membrane. A more sophisticated approach might be to model the cells secreting proteases which diffuse freely and erode the membrane when they meet it. Whilst this is perhaps better fitted to the true processes that degrade the membrane, it is still not clear which cells are responsible for mediating this process. One potential route would thus be to consider the effects of having different immune cells responsible for this process within the modeling framework.

5.1.6. Waxing and Waning

One of the phenomena not accounted for in our description is the honeymoon period (Akirav et al., 2008). This is a phase lasting, in some cases, up to and beyond a year, typically immediately following diagnosis, in which the requirement for exogenous insulin is diminished. Though understanding of the mechanisms giving rise to this transient asymptomatic period is poor, one notable theoretical study demonstrated that this can occur naturally (in rare cases) when taking into account the role of endoplasmic reticulum stress and subsequent beta cell apoptosis (Jaberi-Douraki et al., 2014c). This study highlighted the possibility that elevations in the maximal unfolded protein response due to metabolic therapies could benefit patients by helping to meet metabolic demand, but despite this, they could not ultimately maintain a sufficient population of healthy beta cells to regulate blood glucose levels.

It is not clear whether the honeymoon period is reflective of events occurring at the level of individual islets, whether it is only present at the systemic level and whether it arises as a result of some innate periodicity present in either the immune system or in the beta cells. Modeling studies involving multiple islets may help to address some of these questions.

5.1.7. Therapeutic Intervention

Once a realistic geometry, motivated by real data, has been constructed, and the model sufficiently developed, an ultimate goal would be to test therapeutic interventions that target the pancreas specifically, such as the transplantation/implantation of encapsulated islets or populations of beta cells (Rickels et al., 2005; Robertson, 2010). We have discussed one hypothesis involving the promotion of low avidity T cell clones (Amrani et al., 2000; Khadra et al., 2009). One potential method for doing this is through the intravenous injection of *ex vivo*-selected and *ex vivo*-expanded autologous regulatory T cells, such as CD4+ cells (Bluestone et al., 2015), another is to incite the same expansion of regulatory T cells *in vivo* through the intravenous injection of nanoparticles coated with pMHC (Sugarman et al., 2013). Our modeling framework could easily be extended to include the effect of infusion of either the cells themselves or the nanoparticles to test the spatial aspects of these interventions.

5.1.8. Cell-Cell Coupling

It is evident from the multiple sources of data that communication between immune cells is an important aspect of insulitis, though the exact processes by which these cells exchange information is unknown. One potential avenue of investigation is to assess specific cell interactions based on known immunological mechanisms (Huppa and Davis, 2003; Friedl et al., 2005). These can be incorporated into an agent-based model to see what the overall effects of these specific types of interaction are, with a view to identifying which signaling pathways are most likely to result in the behavior observed in patient data (McLennan et al., 2012).

5.1.9. Comparison with Animal Models

Animal models, and in particular, the NOD mouse model have been extensively used to unravel potential causes and treatment options for human T1D. That true parallels between the disease in rodents and man may be fewer than is ideal, and the fact that treatments that work in mice are not as efficacious in humans (In't Veld, 2014) has led to questions about the suitability of animal models for human disease. From a modeling perspective, this opens up interesting questions. In particular, is it that the underlying mechanisms of insulitis truly are different between different species, or are some processes universal, albeit with potential different time courses and parameters? The former suggests that in order to find a cure for human T1D, only focus on the human condition is likely to yield fruitful results. However, the latter option suggests that, if relations between mice and men for those preserved pathways can be found, that the models can still provide vital insights. An *in silico* modeling framework seems like a natural place to address such a question as it allows quantitative probing of such relationships.

5.1.10. Experimental Design

One of the critical factors hindering progress in understanding T1D is the lack of human pancreatic samples from which to extract data. Moreover, these provide data only at one specific timepoint, so that very little is known about the time course of insulitis in humans.

One role for mathematical modeling is to identify avenues to explore to aid our understanding of disease mechanisms. The present study has suggested that the balance between repair and degradation of the peri-islet basement membrane surrounding the islets is a critical factor which determines the overall rate of beta cell destruction. Presently, little is known about the composition of this membrane in humans, though recent studies have identified its components in both healthy and infiltrated islets in the NOD mouse (Korpos et al., 2013). We thus feel that gaining an improved understanding of the structure of the human peri-islet basement membrane will significantly our improve understanding of insulitis.

Results from the agent-based model also suggest that the chemoattractant produced by the beta cells, and the lymphocytes' response to it is also important for governing disease progression. As such, *in vitro* experiments to assess and quantify the rate of immune cell migration to candidate chemokines would be useful to explore how immune cells initially reach the islet.

Finally, little is understood about the interaction between CD20+ and CD8+ cells during insulitis. The co-localization of these immune cell types in pancreatic samples suggest an interaction between the two (Huppa and Davis, 2003), and as such, we speculate that this could give rise to the activation of T cells. However, experiments have yet to confirm this for the immune cells involved in human T1D. Moreover, the specific dynamics of this process need to be better understood before the true role of the CD20+ cells in human insulitis can be revealed.

Overall, we believe that the proposed spatio-temporal framework has great potential as a tool to investigate insulitis in humans, and by focussing attention on the pancreas, it maximizes the use of the currently available human tissue data. We have identified a number of potential avenues in which the framework could be developed and a number of questions that could be addressed in doing so and we hope to continue in this line of research.

AUTHOR CONTRIBUTIONS

All authors contributed to the initial design and subsequent development of the research project and model. KW developed code to run simulations and analyzed results from them. All authors drafted, edited and approved the submitted manuscript.

FUNDING

This work was generously supported by the Wellcome Trust Institutional Strategic Support Award (WT105618MA). KT gratefully acknowledges the financial support of the EPSRC via grant EP/N014391/1. We are also pleased to acknowledge financial support from the European Unions Seventh Framework Programme PEVNET [FP7/2007-2013] under grant agreement number 261441 to NM. The participants of the PEVNET consortium are described at http://www.uta.fi/med/pevnet/publications.html. Additional support was from a JDRF Career Development Award (5-CDA-2014-221-A-N) to SR and project grant 15/0005156 from Diabetes UK (to NM and SR).

Video S1 | Typical example of agent-based model simulation using parameters in Table 1, with immune cell lifespans of 28 days and $B = 5$.

Video S2 | Typical example of agent-based model simulation using parameters in Table 1, with immune cell lifespans of 28 days and $B = 30$.

Video S3 | Typical example of agent-based model simulation with no peri-islet membrane repair, as indicated in Section 4.2 with immune cell lifespans of 28 days, and $B = 5$.

Video S4 | Typical example of agent-based model simulation with no peri-islet membrane repair, as indicated in Section 4.2 with immune cell lifespans of 28 days, and $B = 30$.

Video S5 | Typical example of agent-based model simulation with high rates of peri-islet membrane degradation, as indicated in Section 4.3 with immune cell lifespans of 28 days, and $B = 5$.

Video S6 | Typical example of agent-based model simulation with high rates of peri-islet membrane degradation, as indicated in Section 4.3 with immune cell lifespans of 28 days, and $B = 30$.

Video S7 | Typical example of agent-based model simulation using in the high chemokine regime, as indicated in Section 4.5 with immune cell lifespans of 28 days, and $B = 5$.

Video S8 | Typical example of agent-based model simulation using in the high chemokine regime, as indicated in Section 4.5 with immune cell lifespans of 28 days, and $B = 30$.

REFERENCES

(2015). *Type 1 Diabetes in Adults: Diagnosis and Management*. Technical Report August, National Institute for Health and Care Excellence.

Akirav, E., Kushner, J. A., and Herold, K. C. (2008). Beta-cell mass and type 1 diabetes: going, going, gone? *Diabetes* 57, 2883–2888. doi: 10.2337/db07-1817

American Diabetes Association (2006). Diagnosis and classification of diabetes mellitus. *Diabetes Care* 29, S43–S48. doi: 10.2337/dc10-S062

Amrani, A., Verdaguer, J., Serra, P., Tafuro, S., Tan, R., and Santamaria, P. (2000). Progression of autoimmune diabetes driven by avidity maturation of a T-cell population. *Nature* 406, 739–742. doi: 10.1038/35021081

An, G., Mi, Q., Dutta-Moscato, J., and Vodovotz, Y. (2009). Agent-based models in translational systems biology. *Wiley Interdiscip. Rev. Syst. Biol. Med.* 1, 159–171. doi: 10.1002/wsbm.45

Atkinson, M. A. (2012). The pathogenesis and natural history of type 1 diabetes. *Cold Spring Harbor Perspect. Med.* 2, 1–18. doi: 10.1101/cshperspect.a007641

Babon, J. A. B., DeNicola, M. E., Blodgett, D. M., Crèvecoeur, I., Buttrick, T. S., Maehr, R., et al. (2016). Analysis of self-antigen specificity of islet-infiltrating t cells from human donors with type 1 diabetes. *Nat. Med.* 22, 1482–1487. doi: 10.1038/nm.4203

Barnett, A. H., Eff, C., Leslie, R. D., and Pyke, D. A. (1981). Diabetes in identical twins. A study of 200 pairs. *Diabetologia* 20, 87–93. doi: 10.1007/bf00262007

Bertram, R., and Sherman, A. (2004). A calcium-based phantom bursting model for pancreatic islets. *Bullet. Math. Biol.* 66, 1313–1344. doi: 10.1016/j.bulm.2003.12.005

Bluestone, J. A., Buckner, J. H., Fitch, M., Gitelman, S. E., Gupta, S., Hellerstein, M. K., et al. (2015). Type 1 diabetes immunotherapy using polyclonal regulatory T cells. *Sci. Trans. Med.* 7:315ra189. doi: 10.1126/scitranslmed.aad4134

Bogdani, M. (2016). Thinking outside the cell: a key role for hyaluronan in the pathogenesis of human type 1 diabetes. *Diabetes* 65, 2015–2114. doi: 10.2337/db15-1750

Boitard, C. (2012). Pancreatic islet autoimmunity. *Presse Med.* 41, e636–e650. doi: 10.1016/j.lpm.2012.10.003

Bottazzo, G. F., Dean, B. M., McNally, J. M., MacKay, E. H., Swift, P. G., and Gamble, D. R. (1985). *In situ* characterization of autoimmune phenomena and expression of HLA molecules in the pancreas in diabetic insulitis. *New Engl. J. Med.* 313, 353–360. doi: 10.1056/NEJM198508083130604

Brehm, M. A., Powers, A. C., Shultz, L. D., and Greiner, D. L. (2012). Advancing animal models of human type 1 diabetes by engraftment of functional human tissues in immunodeficient mice. *Cold Spring Harb. Perspect. Med.* 2:a007757. doi: 10.1101/cshperspect.a007757

Brereton, M. F., Iberl, M., Shimomura, K., Zhang, Q., Adriaenssens, A. E., Proks, P., et al. (2014). Reversible changes in pancreatic islet structure and function produced by elevated blood glucose. *Nat. Commun.* 5:4639. doi: 10.1038/ncomms5639

Bruna, M., and Chapman, S. J. (2012). Excluded-volume effects in the diffusion of hard spheres. *Phys. Rev. E Stat. Nonline. Soft Matter Phys.* 85, 24–29. doi: 10.1103/physreve.85.011103

Campbell-Thompson, M., Wasserfall, C., Kaddis, J., Albanese-O'Neill, A., Staeva, T., Nierras, C., et al. (2012). Network for Pancreatic Organ Donors with Diabetes (nPOD): developing a tissue biobank for type 1 diabetes. *Diabetes Metab. Res. Rev.* 28, 608–617. doi: 10.1002/dmrr.2316

Castiglioni, P., Gerloni, M., Cortez-Gonzales, X., and Zanetti, M. (2005). CD8 T cell priming by B lymphocytes is CD4 help dependent. *Eur. J. Immunol.* 35, 1360–1370. doi: 10.1002/eji.200425530

Chen, N., Glazier, J. A., Izaguirre, J. A., and Alber, M. S. (2007). A parallel implementation of the Cellular Potts Model for simulation of cell-based morphogenesis. *Comput. Phys. Commun.* 176, 670–681. doi: 10.1016/j.cpc.2007.03.007

Christen, U., and Von Herrath, M. G. (2004). IP-10 and type 1 diabetes: a question of time and location. *Autoimmunity* 37, 273–282. doi: 10.1080/08916930410001713124

Cnop, M., Hughes, S. J., Igoillo-Esteve, M., Hoppa, M. B., Sayyed, F., van de Laar, L., et al. (2010). The long lifespan and low turnover of human islet beta cells estimated by mathematical modelling of lipofuscin accumulation. *Diabetologia* 53, 321–330. doi: 10.1007/s00125-009-1562-x

Cnop, M., Welsh, N., Jonas, J.-C., Jo, A., and Lenzen, S. (2005). Mechanisms of pancreatic beta-cell death in type 1 and type 2 diabetes: many differences, few similarities. *Diabetes* 54, S97–S107. doi: 10.2337/diabetes.54.suppl_2.S97

Coppieters, K., Dotta, F., Amirian, N., Campbell, P. D., Kay, T. W. H., Atkinson, M. A., et al. (2012). Demonstration of islet-autoreactive CD8 t cells in insulitic lesions from recent onset and long-term type 1 diabetes patients. *J. Exp. Med.* 209, 51–60. doi: 10.1084/jem.20111187

Daub, J. T., and Merks, R. M. H. (2013). A cell-based model of extracellular-matrix-guided endothelial cell migration during angiogenesis. *Bullet. Math. Biol.* 75, 1377–1399. doi: 10.1007/s11538-013-9826-5

Dirice, E., Kahraman, S., Jiang, W., El Ouaamari, A., De Jesus, D. F., Teo, A. K. K., et al. (2014). Soluble factors secreted by T cells promote β-cell proliferation. *Diabetes* 63, 188–202. doi: 10.2337/db13-0204

Dotta, F., Censini, S., van Halteren, A. G. S., Marselli, L., Masini, M., Dionisi, S., et al. (2007). Coxsackie B4 virus infection of beta cells and natural killer cell insulitis in recent-onset type 1 diabetic patients. *Proc. Natl. Acad. Sci. U.S.A.* 104, 5115–5120. doi: 10.1073/pnas.0700442104

Eisenbarth, G. S. (1986). Type I diabetes mellitus. A chronic autoimmune disease. *New Engl. J. Med.* 314, 1360–1368. doi: 10.1056/NEJM198605223142106

Elliott, C. M., Stinner, B., and Venkataraman, C. (2012). Modelling cell motility and chemotaxis with evolving surface finite elements. *J. R. Soc. Interface* 9, 3027–3044. doi: 10.1098/rsif.2012.0276

Foulis, A., and Stewart, J. (1984). The pancreas in recent-onset Type 1 (insulin-dependent) diabetes mellitus: insulin content of islets, insulitis and associated changes in the exocrine acinar tissue. *Diabetologia* 26, 456–461. doi: 10.1007/BF00262221

Friedl, P., den Boer, A. T., and Gunzer, M. (2005). Tuning immune responses: diversity and adaptation of the immunological synapse. *Nat. Rev. Immunol.* 5, 532–545. doi: 10.1038/nri1647

Gepts, W. (1965). Pathologic anatomy of the pancreas in juvenile diabetes mellitus. *Diabetes* 14, 619–633. doi: 10.2337/diab.14.10.619

Graham, K. L., Krishnamurthy, B., Fynch, S., Ayala-Perez, R., Slattery, R. M., Santamaria, P., et al. (2012). Intra-islet proliferation of cytotoxic T lymphocytes contributes to insulitis progression. *Eur. J. Immunol.* 42, 1717–1722. doi: 10.1002/eji.201242435

Green, D. R., Droin, N., and Pinkoski, M. (2003). Activation-induced cell death in T cells. *Immunol. Rev.* 193, 70–81. doi: 10.1034/j.1600-065X.2003.00051.x

Han, B., Serra, P., Yamanouchi, J., Amrani, A., Elliott, J. F., Dickie, P., et al. (2016). Developmental control of CD8+ t cell-avidity maturation in autoimmune diabetes. *J. Clin. Invest.* 115, 1879–1887. doi: 10.1172/JCI24219

Harvey, D. G., Fletcher, A. G., Osborne, J. M., and Pitt-Francis, J. (2015). A parallel implementation of an off-lattice individual-based model of multicellular populations. *Comput. Phys. Commun.* 192, 130–137. doi: 10.1016/j.cpc.2015.03.005

Holcombe, M., Adra, S., Bicak, M., Chin, S., Coakley, S., Graham, A. I., et al. (2012). Modelling complex biological systems using an agent-based approach. *Integr. Biol.* 4, 53–64. doi: 10.1039/C1IB00042J

Huppa, J. B., and Davis, M. M. (2003). T-cell-antigen recognition and the immunological synapse. *Nat. Rev. Immunol.* 3, 973–983. . doi: 10.1038/nri1245

In't Veld, P. (2014). Insulitis in human type 1 diabetes: a comparison between patients and animal models. *Semin. Immunopathol.* 36, 569–579. doi: 10.1007/s00281-014-0438-4

Inzucchi, S. E. (2012). Diagnosis of diabetes. *New Engl. J. Med.* 367, 542–550. doi: 10.1056/NEJMcp1103643

Itoh, N., Hanafusa, T., Miyazaki, A., Miyagawa, J., Yamagata, K., Yamamoto, K., et al. (1993). Mononuclear cell infiltration and its relation to the expression of major histocompatibility complex antigens and adhesion molecules in pancreas biopsy specimens from newly diagnosed insulin-dependent diabetes mellitus patients. *J. Clin. Invest.* 92, 2313–2322. doi: 10.1172/JCI116835

Jaberi-Douraki, M., Liu, S. W., Pietropaolo, M., and Khadra, A. (2014a). Autoimmune responses in T1DM: quantitative methods to understand onset, progression and prevention of disease. *Pediatr. Diabetes* 15, 162–174. doi: 10.1111/pedi.12148

Jaberi-Douraki, M., Pietropaolo, M., and Khadra, A. (2014b). Predictive models of type 1 diabetes progression: understanding T-cell cycles and their implications on autoantibody release. *PLoS ONE* 9:e93326. doi: 10.1371/journal.pone.0093326

Jaberi-Douraki, M., Pietropaolo, M., and Khadra, A. (2015). Continuum model of T-cell avidity: understanding autoreactive and regulatory T-cell responses in type 1 diabetes. *J. Theor. Biol.* 383, 93–105. doi: 10.1016/j.jtbi.2015.07.032

Jaberi-Douraki, M., Schnell, S., Pietropaolo, M., and Khadra, A. (2014c). Unraveling the contribution of pancreatic beta-cell suicide in autoimmune type 1 diabetes. *J. Theor. Biol.* 375, 77–87. doi: 10.1016/j.jtbi.2014.05.003

Janeway, C. A. Jr., Bottomly, K. I. M., Horowitz, J. A. Y., Kaye, J., Jones, B., and Tite, J. (1985). Modes of cell: cell communication in the immune system. *J. Immunol.* 135, 739s–742s.

Jones, A. G., and Hattersley, A. T. (2013). The clinical utility of C-peptide measurement in the care of patients with diabetes. *Diabetic Med.* 30, 803–817. doi: 10.1111/dme.12159

Kachapati, K., Adams, D., Bednar, K., and Ridgway, W. M. (2012). The non-obese diabetic (NOD) mouse as a model of human type 1 diabetes. *Methods Mol. Biol.* 933, 3–16. doi: 10.1007/978-1-62703-068-7_1

Khadra, A., Pietropaolo, M., Nepom, G. T., and Sherman, A. (2011). Investigating the role of T-cell avidity and killing efficacy in relation to type 1 diabetes prediction. *PLoS ONE* 6:e14796. doi: 10.1371/journal.pone.0014796

Khadra, A., Santamaria, P., and Edelstein-Keshet, L. (2009). The role of low avidity T cells in the protection against type 1 diabetes: a modeling investigation. *J. Theor. Biol.* 256, 126–141. doi: 10.1016/j.jtbi.2008.09.019

Kim, A., Miller, K., Jo, J., Kilimnik, G., and Wojcik, P. (2009). Islet architecture. *Islets* 1, 129–136. doi: 10.4161/isl.1.2.9480

Klöppel, G., Löhr, M., Habich, K., Oberholzer, M., and Heitz, P. U. (1985). Islet pathology and the pathogenesis of type 1 and type 2 diabetes mellitus revisited. *Surv. Synth. Pathol. Res.* 4, 110–125.

Knip, M., Veijola, R., Virtanen, S. M., Hyo, H., Vaarala, O., and Åkerblom, H. K. (2005). Environmental triggers and determinants of type 1 diabetes. *Diabetes* 54, S125–S136. doi: 10.2337/diabetes.54.suppl_2.S125

Korpos, É., Kadri, N., Kappelhoff, R., Wegner, J., Overall, C. M., Weber, E., et al. (2013). The peri-islet basement membrane, a barrier to infiltrating leukocytes in type 1 diabetes in mouse and human. *Diabetes* 62, 531–542. doi: 10.2337/db12-0432

La Torre, D., and Lernmark, A. (2012). Immunology of beta cell destruction. *Adv. Exp. Med. Biol.* 771, 537–583. doi: 10.1007/978-90-481-3271-3_24

Lally, F. J., and Bone, A. J. (2003). "Animal models of type 1 diabetes," in *Textbook of Diabetes, 3rd Edn.*, eds J. C. Pickup and G. Williams (Oxford: Blackwell Science), 19.1–19.17.

Lecompte, P. (1958). Insulitis in early juvenile diabetes. *AMA Arch. Pathol.* 66, 450–457.

Leete, P., Willcox, A., Krogvold, L., Dahl-Jørgensen, K., Foulis, A. K., Richardson, S. J., et al. (2016). Differential insulitic profiles determine the extent of beta cell destruction and the age at onset of type 1 diabetes. *Diabetes* 65, 1362–1369. doi: 10.2337/db15-1615

Levine, H., Rappel, W.-J., and Cohen, I. (2000). Self-organization in systems of self-propelled particles. *Phys. Rev. E* 63:017101. doi: 10.1103/PhysRevE.63.017101

Lo, S. S., Tun, R. Y., Hawa, M., and Leslie, R. D. (1991). Studies of diabetic twins. *Diabetes Metab. Res. Rev.* 7, 223–238. doi: 10.1002/dmr.5610070403

Mahaffy, J. M., and Edelstein-Keshet, L. (2007). Modeling cyclic waves of circulating t cells in autoimmune diabetes. *SIAM J. Appl. Math.* 67, 915–937. doi: 10.1137/060661144

Maini, P. K., and Baker, R. E. (2011). Modelling collective cell motion in biology. *Adv. Appl. Math.* 47, 688–709. doi: 10.1007/978-3-319-06923-4_1

Marée, A., Grieneisen, V., and Hogeweg, P. (2007). "Single-cell based models in biology and medicine," in *The Cellular Potts Model and Biophysical Properties of Cells, Tissues and Morphogenesis* (Basel: Birkhäuser Verlag), 107–136.

Marée, A. F. M., Kublik, R., Finegood, D. T., and Edelstein-Keshet, L. (2006). Modelling the onset of Type 1 diabetes: can impaired macrophage phagocytosis make the difference between health and disease? *Philos. Trans. Ro. Soc. A* 364 1267–1282. doi: 10.1098/rsta.2006.1769

McLennan, R., Dyson, L., Prather, K. W., Morrison, J. A., Baker, R. E., Maini, P. K., et al. (2012). Multiscale mechanisms of cell migration during development: theory and experiment. *Development* 139, 2935–2944. doi: 10.1242/dev.081471

Merks, R. M. H., Brodsky, S. V., Goligorsky, M. S., Newman, S. A., and Glazier, J. A. (2006). Cell elongation is key to *in silico* replication of *in vitro* vasculogenesis and subsequent remodeling. *Dev. Biol.* 289, 44–54. doi: 10.1016/j.ydbio.2005.10.003

Middleton, A. M., Fleck, C., and Grima, R. (2014). A continuum approximation to an off-lattice individual-cell based model of cell migration and adhesion. *J. Theor. Biol.* 359, 220–232. doi: 10.1016/j.jtbi.2014.06.011

Mirams, G., Arthurs, C., Bernabeu, M., Bordas, R., Cooper, J., Corrias, A., et al. (2013). Chaste: an open source C++ library for computational physiology and biology. *PLOS Comput. Biol.* 9:e1002970. doi: 10.1371/journal.pcbi.1002970

Morgan, N. G., Leete, P., Foulis, A. K., and Richardson, S. J. (2014). Islet inflammation in human type 1 diabetes mellitus. *IUBMB Life* 66, 723–734. doi: 10.1002/iub.1330

Palsson, E., and Othmer, H. G. (2000). A model for individual and collective cell movement in Dictyostelium discoideum. *Proc. Natl. Acad. Sci. U.S.A.* 97, 10448–10453. doi: 10.1073/pnas.97.19.10448

Plank, M. J., and Simpson, M. J. (2012). Models of collective cell behaviour with crowding effects: comparing lattice-based and lattice-free approaches. *J. R. Soc. Interface* 9, 2983–2996. doi: 10.1098/rsif.2012.0319

Pugliese, A. (2014). Advances in the etiology and mechanisms of type 1 diabetes. *Discov. Med.* 18, 141–150.

Pugliese, A., Vendrame, F., Reijonen, H., Atkinson, M. A., Campbell-Thompson, M., and Burke, G. W. (2014). New insight on human type 1 diabetes Biology: nPOD and nPOD-transplantation. *Curr. Diabetes Rep.* 14:530. doi: 10.1007/s11892-014-0530-0

Redondo, M. J., Yu, L., Hawa, M., Mackenzie, T., Pyke, D. A., Eisenbarth, G. S., et al. (2001). Heterogeneity of type I diabetes: analysis of monozygotic twins in Great Britain and the United States. *Diabetologia* 44, 354–62. doi: 10.1007/s001250051626

Reed, J. C., and Herold, K. C. (2015). Thinking bedside at the bench: the NOD mouse model of T1DM. *Nat. Rev. Endocrinol.* 11, 308–314. doi: 10.1038/nrendo.2014.236

Reijonen, H., Mallone, R., Heninger, A.-K., Laughlin, E. M., Kochik, S. A., Falk, B., et al. (2004). GAD65-specific CD4 t-cells with high antigen avidity are prevalent in peripheral blood of patients with type 1 diabetes. *Diabetes* 53, 1987–1994. doi: 10.2337/diabetes.53.8.1987

Richardson, S. J., Morgan, N. G., and Foulis, A. K. (2014). Pancreatic pathology in type 1 diabetes mellitus. *Endocr. Pathol.* 25, 80–92. doi: 10.1007/s12022-014-9297-8

Richardson, S. J., Willcox, A., Bone, A. J., Morgan, N. G., and Foulis, A. K. (2011). Immunopathology of the human pancreas in type-I diabetes. *Semin. Immunopathol.* 33, 9–21. doi: 10.1007/s00281-010-0205-0

Rickels, M. R., Schutta, M. H., Markmann, J. F., Barker, C. F., Naji, A., and Teff, K. L. (2005). β-Cell function following human islet transplantation for type 1 diabetes. *Diabetes* 54, 100–106. doi: 10.2337/diabetes.54.1.100

Robertson, R. P. (2010). Islet transplantation a decade later and strategies for filling a half-full glass. *Diabetes* 59, 1285–1291. doi: 10.2337/db09-1846

Rodríguez-Pinto, D. (2005). B cells as antigen presenting cells. *Cell. Immunol.* 238, 67–75. doi: 10.1016/j.cellimm.2006.02.005

Roep, B. O., and Tree, T. I. M. (2014). Immune modulation in humans: implications for type 1 diabetes mellitus. *Nat. Rev. Endocrinol.* 10, 229–242. doi: 10.1038/nrendo.2014.2

Saisho, Y., Butler, A. E., Manesso, E., Elashoff, D., Rizza, R. A., and Butler, P. C. (2013). Beta-cell mass and turnover in humans. *Diabetes Care* 36, 111–117. doi: 10.2337/dc12-0421

Schanze, T. (1995). Sinc interpolation of discrete periodic signals. *IEEE Trans. Signal Process.* 43, 1502–1503. doi: 10.1109/78.388863

Scianna, M. (2015). An extended cellular potts model analyzing a wound healing assay. *Comput. Biol. Med.* 62, 33–54. doi: 10.1016/j.compbiomed.2015.04.009

Somoza, N., Vargas, F., Roura-Mir, C., Vives-Pi, M., Fernández-Figueras, M. T., Ariza, A., et al. (1994). Pancreas in recent onset insulin-dependent diabetes mellitus. Changes in HLA, adhesion molecules and autoantigens, restricted T cell receptor V beta usage, and cytokine profile. *J. Immunol.* 153, 1360–1377.

Song, C., Wang, P., and Makse, H. A. (2008). A phase diagram for jammed matter. *Nature* 453, 629–632. doi: 10.1038/nature06981

Sprent, J. (1993). Lifespans of naive, memory and effector lymphocytes. *Curr. Opin. Immunol.* 5, 433–438. doi: 10.1016/0952-7915(93)90065-Z

Standifer, N., Burwell, E. A., Gersuk, V. H., Greenbaum, C. J., and Nepom, G. T. (2009). Changes in autoreactive t cell avidity during type 1 diabetes development. *Clin. Immunol.* 132, 312–320. doi: 10.1016/j.clim.2009.04.013

Stein, J. V., and Nombela-Arrieta, C. (2005). Chemokine control of lymphocyte trafficking: a general overview. *Immunology* 116, 1–12. doi: 10.1111/j.1365-2567.2005.02183.x

Sugarman, J., Tsai, S., Santamaria, P., and Khadra, A. (2013). Quantifying the importance of pmhc valency, total pmhc dose and frequency on nanoparticle therapeutic efficacy. *Immunol. Cell Biol.* 91, 350–359. doi: 10.1038/icb.2013.9

Swat, M., Thomas, G. L., Belmonte, J. M., Shirinifard, A., Hmeljak, D., and Glazier, J. A. (2012). Multi-scale modeling of tissues using CompuCell3D. *Comput. Methods Cell Biol. Methods Cell Biol.* 110, 325–366. doi: 10.1016/B978-0-12-388403-9.00013-8

Tapia, J. J., and D'Souza, R. M. (2011). Parallelizing the cellular potts model on graphics processing units. *Comput. Phys. Commun.* 182, 857–865. doi: 10.1016/j.cpc.2010.12.011

Taplin, C. E., and Barker, J. M. (2008). Autoantibodies in type 1 diabetes. *Autoimmunity* 41, 11–18. doi: 10.1080/08916930701619169

Tijskens, E., Ramon, H., and De Baerdemaeker, J. (2003). Discrete element modelling for process simulation in agriculture. *J. Sound Vibr.* 266, 493–514. doi: 10.1016/S0022-460X(03)00581-9

Trudeau, J. D., Kelly-smith, C., Verchere, C. B., Elliott, J. F., Dutz, J. P., Finegood, D. T., et al. (2003). Prediction of spontaneous autoimmune diabetes in nod mice by quantification of autoreactive t cells in peripheral blood. *J. Clin. Invest.* 111, 217–223. doi: 10.1172/JCI200316409

Walker, J. N., Johnson, P. R. V., Shigeto, M., Hughes, S. J., Clark, A., and Rorsman, P. (2011). Glucose-responsive beta cells in islets isolated from a patient with long-standing type 1 diabetes mellitus. *Diabetologia* 54, 200–202. doi: 10.1007/s00125-010-1930-6

Wang, L., Abbasi, F., Ornatsky, O., Cole, K. D., Misakian, M., Gaigalas, A. K., et al. (2012). Human CD4+ lymphocytes for antigen quantification: characterization using conventional flow cytometry and mass cytometry. *Cytometry A* 81A, 567–575. doi: 10.1002/cyto.a.22060

Willcox, A., Richardson, S. J., Bone, A. J., Foulis, A. K., and Morgan, N. G. (2009). Analysis of islet inflammation in human type 1 diabetes. *Clin. Exp. Immunol.* 155, 173–181. doi: 10.1111/j.1365-2249.2008.03860.x

Willcox, A., Richardson, S. J., Bone, A. J., Foulis, A. K., and Morgan, N. G. (2010). Evidence of increased islet cell proliferation in patients with recent-onset type 1 diabetes. *Diabetologia* 53, 2020–2028. doi: 10.1007/s00125-010-1817-6

4

Analysis of a Compartmental Model of Endogenous Immunoglobulin G Metabolism with Application to Multiple Myeloma

*Felicity Kendrick[1], Neil D. Evans[1], Bertrand Arnulf[2], Hervé Avet-Loiseau[3], Olivier Decaux[4], Thomas Dejoie[5], Guillemette Fouquet[6], Stéphanie Guidez[7], Stéphanie Harel[2], Benjamin Hebraud[8], Vincent Javaugue[7], Valentine Richez[9], Susanna Schraen[6], Cyrille Touzeau[5], Philippe Moreau[5], Xavier Leleu[7], Stephen Harding[10] and Michael J. Chappell[1]**

[1] School of Engineering, University of Warwick, Coventry, UK, [2] Hôpital Saint-Louis, Paris, France, [3] Unité de Génomique du Myélome, Institut Universitaire du Cancer de Toulouse Oncopole, Toulouse, France, [4] Centre Hospitalier Universitaire de Rennes, Rennes, France, [5] Centre Hospitalier Universitaire de Nantes, Nantes, France, [6] Centre Hospitalier Régional Universitaire de Lille, Lille, France, [7] Centre Hospitalier Universitaire de Poitiers, Poitiers, France, [8] Centre Hospitalier Universitaire de Toulouse, Toulouse, France, [9] Centre Hospitalier Universitaire de Nice, Nice, France, [10] Department of Research and Development, The Binding Site Group Limited, Birmingham, UK

Edited by:
Krasimira Tsaneva-Atanasova,
University of Exeter, UK

Reviewed by:
Marc Thilo Figge,
Leibniz-Institute for Natural Product Research and Infection Biology, Hans-Knoell-Institute, Germany
Stefano Severi,
University of Bologna, Italy

***Correspondence:**
Michael J. Chappell
m.j.chappell@warwick.ac.uk

Immunoglobulin G (IgG) metabolism has received much attention in the literature for two reasons: (i) IgG homeostasis is regulated by the neonatal Fc receptor (FcRn), by a pH-dependent and saturable recycling process, which presents an interesting biological system; (ii) the IgG-FcRn interaction may be exploitable as a means for extending the plasma half-life of therapeutic monoclonal antibodies, which are primarily IgG-based. A less-studied problem is the importance of endogenous IgG metabolism in IgG multiple myeloma. In multiple myeloma, quantification of serum monoclonal immunoglobulin plays an important role in diagnosis, monitoring and response assessment. In order to investigate the dynamics of IgG in this setting, a mathematical model characterizing the metabolism of endogenous IgG in humans is required. A number of authors have proposed a two-compartment nonlinear model of IgG metabolism in which saturable recycling is described using Michaelis–Menten kinetics; however it may be difficult to estimate the model parameters from the limited experimental data that are available. The purpose of this study is to analyse the model alongside the available data from experiments in humans and estimate the model parameters. In order to achieve this aim we linearize the model and use several methods of model and parameter validation: stability analysis, structural identifiability analysis, and sensitivity analysis based on traditional sensitivity functions and generalized sensitivity functions. We find that all model parameters are identifiable, structurally and taking into account parameter correlations, when several types of model output are used for parameter estimation. Based on these analyses we estimate parameter values from the limited available data and compare them with previously published parameter values. Finally we show how the model can be applied in future studies of treatment effectiveness in IgG multiple myeloma with simulations of serum monoclonal IgG responses during treatment.

Keywords: biomedical systems, lumped-parameter systems, identifiability, parameter identification, sensitivity analysis, immunoglobulin G, metabolism, multiple myeloma

1. INTRODUCTION

Immunoglobulin G (IgG) is protected from degradation by the neonatal Fc receptor (FcRn), resulting in an unusually long metabolic half-life at normal concentrations (~23 days; Rosenthal and Tan, 2010) and a high serum concentration in healthy adults (10–16 $g\,l^{-1}$; Hall and Yates, 2010). The half-life of IgG is not constant, but varies with its serum concentration, due to saturation of recycling receptors. Elevated IgG concentrations saturate receptors such that a greater proportion of circulating IgG is degraded; conversely at low concentrations a greater proportion of IgG is recycled and the half-life is extended. Circulating IgG is internalized into intracellular endosomes in order to be degraded. FcRn expressed within the cells binds IgG inside the acidic environment of endosomes with a pH-dependent affinity. FcRn then sequesters the bound IgG away from the degradation pathway and back to the cell membrane, releasing it once again into the circulation. Those IgG molecules that are not bound to FcRn continue to follow the pathway to be degraded in lysosomes (Junghans and Anderson, 1996).

In multiple myeloma, clonal plasma cells in the bone marrow secrete a unique, monoclonal immunoglobulin (Ig). Half of patients have IgG-producing clones and are said to have IgG myeloma (Anderson, 2003). The monoclonal Ig produced by the cancer offers a convenient opportunity for clinicians to monitor the response of the tumor to therapy via the secreted protein, which is readily quantified in a blood sample. The cancer itself is only accessible by bone marrow biopsy or aspirate, both of which are unpleasant, invasive procedures. The concentration of monoclonal Ig in the blood is therefore the preferred measure by which the tumor is monitored; patient monitoring in clinical trials and the non-trial setting alike is heavily reliant on measurements of monoclonal Ig concentration in the blood (Kumar et al., 2016).

IgG myeloma patients typically present with an elevated concentration of serum monoclonal IgG. During treatment, the malignant plasma cells are killed and the production rate of monoclonal IgG correspondingly decreases, resulting in a fall in serum monoclonal IgG concentration. In this way, the serum monoclonal IgG response is used as a surrogate for the tumor response to treatment. The possible effects of the metabolism of IgG on its application as a cancer marker in multiple myeloma have been little studied, but are acknowledged in the literature. Sullivan and Salmon (1972) first brought the issue of IgG metabolism to the attention of the multiple myeloma community. Serum monoclonal IgG concentration, plasma volume, and IgG synthesis rate per cell were measured in 11 patients with IgG myeloma. Calculating the fractional catabolic rate of IgG using the equation provided by Waldmann and Strober (1969), Sullivan and Salmon (1972) estimated the tumor burden at a number of time points during treatment for each patient, concluding that increases and decreases in the tumor burden were underestimated by increases and decreases in monoclonal IgG. More recently, Bradwell et al. (2013), Koulieris et al. (2012), and Durie et al. (2006) have cited the concentration-dependent metabolism of IgG as a possible explanation for why monoclonal IgG may be seen as an unreliable response marker in multiple myeloma.

In order to investigate the dynamics of IgG in multiple myeloma, a mathematical model characterizing the metabolism of endogenous IgG in humans is required. Many mathematical models of IgG metabolism have been published in the literature (more than 20 at the time of writing), usually with the aim of describing the pharmacokinetics of therapeutic monoclonal antibodies that are similarly regulated by FcRn. Many of the models are therefore pharmacokinetic in nature: their parameter values are obtained from animal experiments and they may be physiologically based, with up to ten organs and the lymphatic system explicitly represented in the model (Hansen and Balthasar, 2003; Ferl et al., 2005; Garg and Balthasar, 2007; Fang and Sun, 2008; Urva et al., 2010; Chen and Balthasar, 2012; Deng et al., 2012; Xiao, 2012; Yan et al., 2012; Fronton et al., 2014; Ng et al., 2014). Physiologically based pharmacokinetic (PBPK) models may be unnecessarily complex for investigating serum IgG dynamics in multiple myeloma, particularly considering the limited human-derived data that are available for parameter estimation. More suitably, several authors have proposed a comparatively simple two-compartment model of IgG metabolism in which saturable recycling by FcRn is described using Michaelis–Menten kinetics (Waldmann and Strober, 1969; Kim et al., 2007; Hattersley et al., 2013). They also provide certain parameter values for humans.

In order to investigate serum IgG responses in IgG multiple myeloma, the parameter values used are highly important in order to have confidence in model-based predictions. Parameter estimation using limited data is an important problem in the mathematical modeling of physiological systems. Methods for parameter identification including structural identifiability analysis and sensitivity analysis should be used in the early stages of the model validation process; specifically, these analyses address whether parameters can be estimated from the available measurements and, where further experiments are possible, inform experiment design. In this paper we analyse the nonlinear two-compartment model of IgG metabolism (Waldmann and Strober, 1969; Kim et al., 2007; Hattersley et al., 2013) and the available measurements in humans for structural identifiability and sensitivity, in order to make optimal use of the limited data available in the literature. Having considered the identifiability problem, we estimate parameter values from the data, with the intention that the model can be used in the future to make generalized predictions for patients.

2. METHODS

2.1. Experimental Data

Data for parameter estimation were obtained from the literature. Studies of protein metabolism involve intravenously injecting a subject with radioisotopically labeled protein, known as a tracer, and then monitoring the proportion of tracer remaining in the blood over a period of time following administration. Radioactive tracers allow for distinction between the injected dose and the endogenously produced protein, enabling direct visualization of the distribution and elimination processes of a protein despite it being homeostatic. The radioactive label (usually iodine) remains bound to the protein until the protein is degraded, at which point the label is released and rapidly excreted in urine. Several tracer studies were performed for IgG in humans in the middle of the last century and the results collated by Waldmann and Strober (1969).

2.1.1. Individual Timecourse Data

The data from a single subject consist of the timecourse of the proportion of an administered dose of radiolabeled IgG remaining in plasma and the proportion remaining in the whole body, calculated by subtracting the radioactivity in urine from the administered dose. The data collected from an individual are shown in **Figure 1A**. The data have been extracted from a plot by Solomon et al. (1963) using OriginPro[1]. Seven plots of this type have been found by the authors in the literature. The data in these plots are assumed to arise from seven individuals to whom we refer as subjects A–G. The data for subjects A–D are taken from Solomon et al. (1963), for subjects E and F from Waldmann and Terry (1990) and for subject G from Waldmann and Strober (1969). Subjects A and C have IgG myeloma, subject D has macroglobulinemia and subject E has familial hypercatabolic hypoproteinemia. These conditions do not preclude the subjects from this study but there may be a correlation between these conditions and individuals' parameter values of IgG metabolism; this is discussed in Section 3.2.

2.1.2. Fractional Catabolic Rate and Half-Life

In compartmental analysis, parameters are often considered as either micro constants or macro (hybrid) constants. The micro constants are dependent upon the assumed structure of the compartmental model, whereas the macro constants can be determined directly from the profile of concentration or radioactivity over time, such as the exponents of a multi-exponential profile, and do not assume a particular model structure (Riviere, 2011).

Waldmann and Strober (1969) have plotted two macro parameters, the fractional catabolic rate (FCR) and the terminal half-life ($T_{1/2}$), that can be calculated directly from an individual subject's timecourse of radioactivity. The FCR is defined as the fraction of the administered IgG in plasma that is catabolized per day and is calculated by dividing the rate at which the administered dose leaves the body at any time $t > 0$ by the amount of the dose remaining in plasma at that time. The rate at which the dose leaves the body is given by the slope of the timecourse of the dose remaining in the whole body. The $T_{1/2}$ is defined as the time taken for half of the administered IgG to be eliminated, after completion of the distribution phase. This is obtained from the terminal slope of the timecourse observations plotted on a logarithmic scale.

The plots of FCR and $T_{1/2}$ provided by Waldmann and Strober (1969) are reproduced in **Figures 1B,C**. Each point in these plots was obtained from the timecourse data of a single subject, an example of which is shown in **Figure 1A**. The parameters have been taken from a large number of subjects (FCR – 41 subjects; $T_{1/2}$ – 44 subjects) with a wide range of plasma concentrations of IgG, in order to capture the concentration-dependent behavior of IgG metabolism. Macro parameters are functions of the micro parameters of the assumed compartmental structure—therefore in this paper the FCR and $T_{1/2}$ data are used in the estimation of the parameters of the underlying compartmental model.

2.2. Model of Endogenous IgG Metabolism

The nonlinear two-compartment model of endogenous IgG metabolism, with Michaelis–Menten kinetics describing the rate of recycling by FcRn receptors (Waldmann and Strober, 1969; Kim et al., 2007; Hattersley et al., 2013), is given by:

$$\begin{aligned}\dot{x}_1(t) &= -\left(k_{21} + k_{31} - \frac{V_{\max}}{K_M + x_1(t)}\right)x_1(t) + k_{12}x_2(t) + I(t)\\ \dot{x}_2(t) &= k_{21}x_1(t) - k_{12}x_2(t)\end{aligned} \tag{1}$$

where $x_1(t)$ and $x_2(t)$ represent the quantities in µmol of IgG in plasma and in a peripheral compartment, respectively. $I(t)$ represents the synthesis of IgG into plasma in µmol day^{-1}. Rate constants k_{ij} represent material flow from compartment j to compartment i. The rate constant of the removal of IgG from the plasma compartment into intracellular endosomes for degradation is given by k_{31}, with the indices denoting the transfer from plasma to a third compartment representing intracellular endosomes, which is omitted from the model. The rate of FcRn-mediated recycling, as a fraction of the quantity of IgG in plasma, is given by $V_{\max}/(K_M + x_1(t))$. The parameters $V_{\max}$ and K_M are the maximum absolute rate of FcRn-mediated recycling in µmol day^{-1} and the Michaelis constant, representing the quantity of IgG in plasma in µmol at which the absolute recycling rate is half $V_{\max}$. Those IgG molecules which are removed from the plasma compartment into intracellular endosomes and which do not get recycled by FcRn are degraded in lysosomes. The amino acid products of lysosomal degradation are reused in the synthesis of new proteins (Appelqvist et al., 2013). A schematic of the system model is shown in **Figure 2**. **Table 1** summarizes the model states and parameters.

All states and parameters can only take non-negative values. The rate at which IgG is recycled cannot exceed the rate at which it leaves the plasma compartment to be degraded in intracellular endosomes; equivalently, the net elimination rate must be positive for all states and input rates: $k_{31} - \frac{V_{\max}}{K_M} > 0$.

When the production rate of IgG is assumed constant, $I(t) = I_0$, in order to determine the model's steady states, solving $\dot{x}_1(t) = 0$ and $\dot{x}_2(t) = 0$ simultaneously gives the equilibrium point:

$$\begin{aligned}\hat{x}_1 &= \frac{-k_{31}K_M + I_0 + V_{\max} + \sqrt{4k_{31}K_M I_0 + (-k_{31}K_M + I_0 + V_{\max})^2}}{2k_{31}}\\ \hat{x}_2 &= \frac{k_{21}}{k_{12}}\hat{x}_1.\end{aligned} \tag{2}$$

The stability of this equilibrium point for all parameter values is demonstrated in Section 2.2.1.

[1]Originpro (2016). Windows. Northampton: OriginLab Corporation.

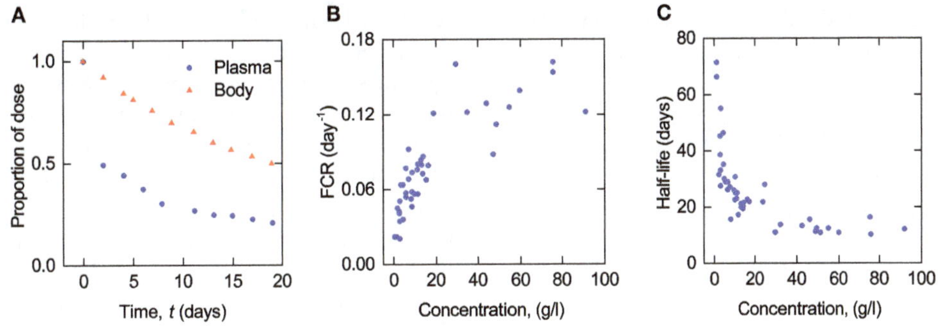

FIGURE 1 | (A) Proportion of administered IgG remaining in plasma (blue circles) and the body (red triangles) in a typical normal subject; data from Solomon et al. (1963). Plasma concentration dependence of **(B)** fractional catabolic rate (FCR) and **(C)** half-life ($T_{1/2}$) of IgG; redrawn from Waldmann and Strober (1969) with permission from S. Karger AG, Basel.

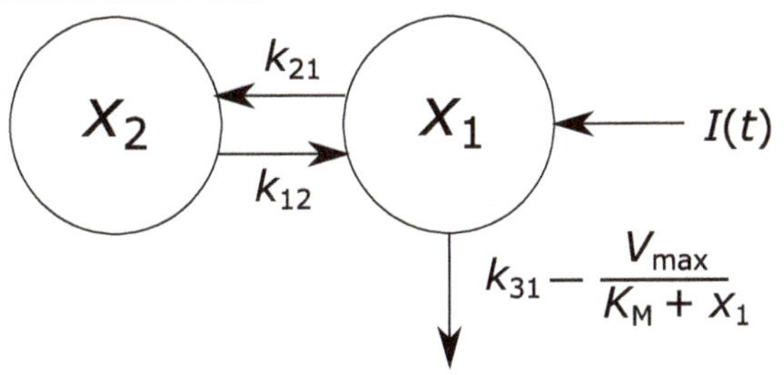

FIGURE 2 | Endogenous IgG metabolism model schematic.

TABLE 1 | States and parameters of IgG metabolism model.

Name	Units	Physiological interpretation
x_1	μmol	Quantity of IgG in the central (plasma) compartment
x_2	μmol	Quantity of IgG in the peripheral (tissue) compartment
k_{21}	day^{-1}	Rate constant of flow of IgG from plasma to peripheral compartment
k_{31}	day^{-1}	Rate constant of flow of IgG from plasma into endosomes by pinocytosis
k_{12}	day^{-1}	Rate constant of flow of IgG from peripheral compartment to plasma
V_{max}	μmol day^{-1}	Maximum absolute recycling rate
K_M	μmol	Michaelis constant; the quantity of IgG in plasma at which the absolute recycling rate is half V_{max}

2.2.1. Stability of Steady States

Linearizing the system described by Equation (1) about the equilibrium point gives:

$$\begin{pmatrix} \dot{x}_1(t) \\ \dot{x}_2(t) \end{pmatrix} = \begin{pmatrix} -k_{21} - k_{31} - \frac{V_{max}\hat{x}_1}{(K_M+\hat{x}_1)^2} & k_{12} \\ k_{21} & -k_{12} \end{pmatrix} \begin{pmatrix} x_1(t) \\ x_2(t) \end{pmatrix}. \quad (3)$$

According to the Routh–Hurwitz stability criterion, the two-state system is stable provided the coefficients of the characteristic polynomial of the linearized system are positive (Routh, 1877). The coefficients of the characteristic polynomial are given by:

$$\begin{aligned} a_2 &= 1 \\ a_1 &= \frac{k_{31}K_M^2 - K_M V_{max} + 2k_{31}K_M\hat{x}_1 + k_{31}\hat{x}_1^2 + k_{12}(K_M+\hat{x}_1)^2 + k_{21}(K_M+\hat{x}_1)^2}{(K_M+\hat{x}_1)^2} \\ a_0 &= \frac{k_{12}\left(k_{31}\left(K_M+\hat{x}_1\right)^2 - K_M V_{max}\right)}{\left(K_M+\hat{x}_1\right)^2}. \end{aligned} \quad (4)$$

The denominators in the expressions for a_0 and a_1 are always positive. All parameters and the steady state $\hat{x}_1$ are positive. The sign of a_0 is thus given by the sign of $\left(k_{31}\left(K_M+\hat{x}_1\right)^2 - K_M V_{max}\right)$. For stability of the equilibrium point it is necessary that $\left(k_{31}\left(K_M+\hat{x}_1\right)^2 - K_M V_{max}\right) > 0$. This condition is met when $k_{31}K_M^2 - K_M V_{max} > 0$, or equivalently $k_{31} - \frac{V_{max}}{K_M} > 0$. The sign of a_1 is given by the sign of its numerator, $k_{31}K_M^2 - K_M V_{max} + 2k_{31}K_M\hat{x}_1 + k_{31}\hat{x}_1^2 + k_{12}(K_M + \hat{x}_1)^2 + k_{21}(K_M + \hat{x}_1)^2$. Once again, the sign of a_1 is positive provided that $k_{31} - \frac{V_{max}}{K_M} > 0$.

Both of the coefficients a_0 and a_1 are positive provided that all parameter values are positive and $k_{31} - \frac{V_{max}}{K_M} > 0$. Referring back to Equation (1), that is the condition which ensures a positive IgG elimination rate for all $x_1 > 0$. A negative elimination rate does not make sense physiologically and as such parameter values are not permitted which violate this condition. The equilibrium point is thus stable for all permitted parameter values.

2.3. Model of Observed Measurements

In this section we consider how the observable measurements (timecourse of radioactivity, FCR and $T_{1/2}$) relate to the system model. Tracer experiments are designed specifically so that the tracer-labeled protein observes linear kinetics, despite the mode of metabolism being in fact nonlinear (Anderson, 1983). A linear model describing the timecourse observations is derived here.

2.3.1. Timecourse Observations

Assuming that the radiolabeled IgG dose and unlabeled endogenous IgG are indistinguishable by the system, both are described by the model in Equation (1). The injected and endogenous IgG can be explicitly represented by letting $x_i(t) =$

$x_{i,\mathrm{T}}(t) + x_{i,\mathrm{E}}(t)$ for $i = 1, 2$, with "T" denoting tracer and "E" denoting endogenous IgG. Then, from Equation (1), the dynamics of labeled and unlabeled IgG are given by:

$$\begin{aligned}\dot{x}_{1,\mathrm{T}}(t) &= -\left(k_{21} + k_{31} - \frac{V_{\max}}{K_{\mathrm{M}} + x_{1,\mathrm{E}}(t) + x_{1,\mathrm{T}}(t)}\right) x_{1,\mathrm{T}}(t) \\ &\quad + k_{12}x_{2,\mathrm{T}}(t) \\ \dot{x}_{2,\mathrm{T}}(t) &= k_{21}x_{1,\mathrm{T}}(t) - k_{12}x_{2,\mathrm{T}}(t) \\ \dot{x}_{1,\mathrm{E}}(t) &= -\left(k_{21} + k_{31} - \frac{V_{\max}}{K_{\mathrm{M}} + x_{1,\mathrm{E}}(t) + x_{1,\mathrm{T}}(t)}\right) x_{1,\mathrm{E}}(t) \\ &\quad + k_{12}x_{2,\mathrm{E}}(t) + I_{\mathrm{E}} \\ \dot{x}_{2,\mathrm{E}}(t) &= k_{21}x_{1,\mathrm{E}}(t) - k_{12}x_{2,\mathrm{E}}(t)\end{aligned} \tag{5}$$

where $x_{i,\mathrm{T}}(t)$ and $x_{i,\mathrm{E}}(t)$ represent the quantities in μmol of radiolabeled and endogenous IgG in compartment i, respectively.

The intravenous bolus injection of tracer can be treated as a non-zero initial condition for $x_{1,\mathrm{T}}(t)$; thus the initial conditions of the tracer are given by:

$$\begin{aligned}x_{1,\mathrm{T}}(0) &= D \\ x_{2,\mathrm{T}}(0) &= 0\end{aligned} \tag{6}$$

where D is the dose of tracer in μmol. The production rate of endogenous IgG, I_{E} μmol day^{-1}, is assumed constant. The initial conditions of the endogenous IgG are given by the equilibrium point in Section 2.2, with $I_0 = I_{\mathrm{E}}$. The experimenter measures the proportion of the initially injected radioactivity in plasma and in the whole body. The observation functions are thus given by:

$$\begin{aligned}y_1(t) &= x_{1,\mathrm{T}}(t)/D \\ y_2(t) &= \left(x_{1,\mathrm{T}}(t) + x_{2,\mathrm{T}}(t)\right)/D.\end{aligned} \tag{7}$$

A sufficiently small quantity of radiolabeled IgG, typically 0.5–1 mg (3.33 x 10^{-3}–6.67 x 10^{-3} μmol) (Solomon et al., 1963), is administered into plasma so as not to perturb the steady state of the endogenous protein. Thus $x_{1,\mathrm{E}}$ and $x_{2,\mathrm{E}}$ can be assumed constant. Then the equations describing the tracer dynamics are no longer coupled with those describing the endogenous IgG dynamics. A second assumption is required in order to derive a linear model: the quantity of tracer, $x_{1,\mathrm{T}}(t)$, is assumed to be much smaller than the quantity of the subject's endogenous IgG, $x_{1,\mathrm{E}}$. Thus the term $\frac{V_{\max}}{K_{\mathrm{M}} + x_{1,\mathrm{E}} + x_{1,\mathrm{T}}(t)}$ can be approximated by $\frac{V_{\max}}{K_{\mathrm{M}} + x_{1,\mathrm{E}}}$. In this way, the elimination rate of the tracer is determined by the quantity of the subject's endogenous plasma IgG only. A further simplification can be made by noticing that for a linear model, the initial conditions and observation gain cancel out (see Equations 6, 7). The equations describing the tracer kinetics are thus given by:

$$\begin{aligned}\dot{x}_{1,\mathrm{P}}(t) &= -\left(k_{21} + k_{31} - \frac{V_{\max}}{K_{\mathrm{M}} + x_{1,\mathrm{E}}}\right) x_{1,\mathrm{P}}(t) + k_{12}x_{2,\mathrm{P}}(t) \\ \dot{x}_{2,\mathrm{P}}(t) &= k_{21}x_{1,\mathrm{P}}(t) - k_{12}x_{2,\mathrm{P}}(t)\end{aligned} \tag{8}$$

where $x_{1,\mathrm{P}}(t)$ and $x_{2,\mathrm{P}}(t)$ represent the proportion of the radiolabeled IgG dose D in the central and peripheral compartments, respectively, at time t. $x_{1,\mathrm{E}}$ represents the quantity of the subject's endogenous IgG in the central compartment, which is assumed to remain in steady state. All other parameters are defined as in Section 2.2.

The initial conditions of the model are now given by:

$$\begin{aligned}x_{1,\mathrm{P}}(0) &= 1 \\ x_{2,\mathrm{P}}(0) &= 0.\end{aligned} \tag{9}$$

The corresponding observation functions are given by:

$$\begin{aligned}y_1(t) &= x_{1,\mathrm{P}}(t) \\ y_2(t) &= x_{1,\mathrm{P}}(t) + x_{2,\mathrm{P}}(t).\end{aligned} \tag{10}$$

The linearized model represented by Equations (8–10) is a valid approximation of the nonlinear model (Equations 5–7) when

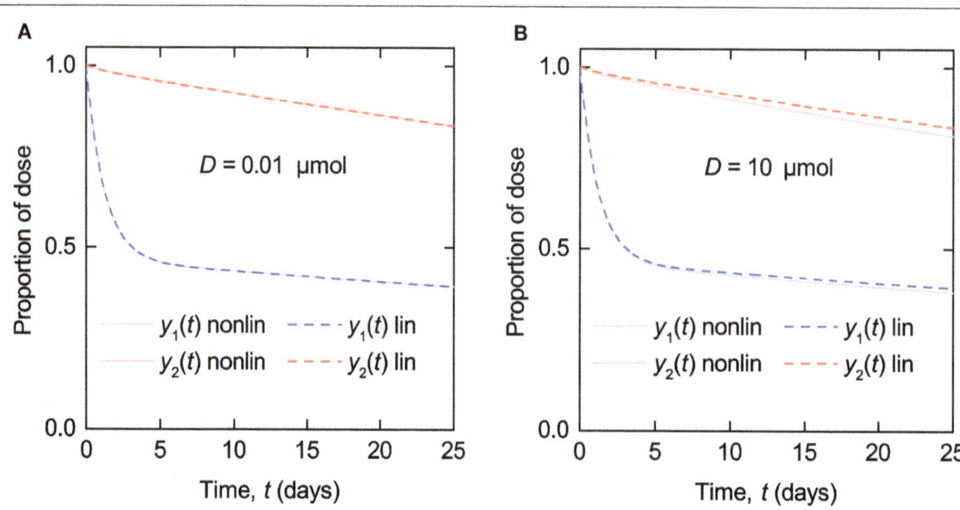

FIGURE 3 | Simulations of timecourse responses $y_1(t)$ and $y_2(t)$ as described by Equations (5–7) (nonlinear model - solid line) and Equations (8–10) (linearized model - dashed line). The quantity of endogenous IgG in plasma at $t = 0$, $x_{1,\mathrm{E}}(0)$, is 5 μmol. The tracer dose D is **(A)** 0.01 μmol and **(B)** 10 μmol.

the administered dose of radiolabeled IgG, D, is sufficiently smaller than the quantity of endogenous IgG in plasma at $t = 0$, $x_{1,E}(0)$. In **Figure 3** simulations of the nonlinear model and the linearized model are compared. The parameter values used are those estimated in this paper and summarized in **Table 6**. The production rate of endogenous IgG is set to $I_E = 0.0727$ μmol day^{-1} to give $x_{1,E}(0) = 5$ μmol for the nonlinear model and $x_{1,E} = 5$ μmol for the linearized model, representing the lower limit of the quantities of endogenous IgG in plasma seen in the data. In **Figure 3A** the tracer dose D is 0.01 μmol, representing the upper limit of administered tracer doses (Solomon et al., 1963). The nonlinear and linearized model responses are indistinguishable, illustrating that for typical tracer doses the linearized model is a valid approximation of the nonlinear model. In **Figure 3B** the tracer dose D is 10 μmol, 1,000 times larger; at this point the assumptions weaken and there is a noticeable difference between the responses of the two models.

2.3.2. Fractional Catabolic Rate and Half-Life

The FCR is defined as the proportion of the radiolabeled IgG in plasma that is catabolized per day. From Equation (8) this is given by:

$$\mathrm{FCR} = k_{31} - \frac{V_{\max}}{K_M + x_{1,E}}. \tag{11}$$

The terminal half-life, $T_{1/2}$, is related to the elimination phase of the kinetics, after the distribution phase is complete. The model described by Equation (8) is a linear two-compartment model with the solutions for $x_{1,P}(t)$ and $x_{2,P}(t)$ given by the bi-exponential functions:

$$\begin{aligned} x_{1,P}(t) &= A_{11}\exp(\lambda_1 t) + A_{12}\exp(\lambda_2 t) \\ x_{2,P}(t) &= A_{21}\exp(\lambda_1 t) + A_{22}\exp(\lambda_2 t) \end{aligned} \tag{12}$$

where A_{ij} and λ_j are macro constants, with $|\lambda_1| > |\lambda_2|$. By definition, $T_{1/2}$ is given by:

$$T_{1/2} = -\frac{\log 2}{\lambda_2}. \tag{13}$$

Solving Equation (8) for λ_2 and substituting into Equation (13) gives the following expression for $T_{1/2}$ in terms of the micro parameters of the model:

$$T_{1/2} = 2\log 2 \Bigg/ \left(k_{12} + k_{21} + k_{31} - \frac{V_{\max}}{K_M + x_{1,E}} - \sqrt{-4k_{12}\left(k_{31} - \frac{V_{\max}}{K_M + x_{1,E}}\right) + \left(k_{12} + k_{21} + k_{31} - \frac{V_{\max}}{K_M + x_{1,E}}\right)^2}\right). \tag{14}$$

From Equations (11, 14), we find that the relationship between $T_{1/2}$ and FCR is given by:

$$T_{1/2} = \frac{2\log 2}{k_{12} + k_{21} + \mathrm{FCR} - \sqrt{-4k_{12}\mathrm{FCR} + (k_{12} + k_{21} + \mathrm{FCR})^2}}. \tag{15}$$

3. RESULTS

3.1. Structural Identifiability of Model Parameters

Structural identifiability addresses the question of whether model parameters can be uniquely identified from available observations, under the assumption of the availability of ideal (i.e., noise-free) and continuous observational data. Structural identifiability of parameters does not imply that they are identifiable in practice, from observations that are inevitably measured with noise; therefore in this paper structural identifiability analysis is used alongside sensitivity analysis.

Here we determine which of the model parameters are structurally uniquely identifiable from the following measurements: an individual subject's timecourse, FCR vs. the quantity of endogenous IgG in plasma, and $T_{1/2}$ vs. the quantity of endogenous IgG in plasma.

3.1.1. Individual Timecourse

Here the transfer function method is used (Bellman and Åström, 1970). To apply this approach the system described by Equations (8–10) is re-written in vector-matrix notation as

$$\begin{aligned} \dot{x}(t,p) &= A(p)x(t,p) + B(p)u(t) \\ x(0,p) &= 0 \\ y(t,p) &= C(p)x(t,p), \end{aligned} \tag{16}$$

where $x(t,p) = (x_{1,P}(t), x_{2,P}(t))$, and $y(t,p) = (y_1(t), y_2(t))$ are column vectors representing the state and the observation, respectively. $u(t)$ represents the single input to the system, an impulse at time $t = 0$, given by $u(t) = \delta(t)$. $A(p)$ and $C(p)$ are 2×2 matrices and $B(p)$ is a column vector. $A(p)$, $B(p)$, and $C(p)$ are given by:

$$A(p) = \begin{pmatrix} -\left(k_{21} + k_{31} - \frac{V_{\max}}{K_M + x_{1,E}}\right) & k_{12} \\ k_{21} & -k_{12} \end{pmatrix},$$

$$B(p) = \begin{pmatrix} 1 \\ 0 \end{pmatrix}, C(p) = \begin{pmatrix} 1 & 0 \\ 1 & 1 \end{pmatrix}. \tag{17}$$

Note that the administration of a bolus dose is now represented as an impulse at time $t = 0$, rather than a non-zero initial condition, such that $x(0,p) = 0$.

Taking Laplace transforms of Equation (16), the input-output relation is described by $Y(s) = G(s)U(s)$, where $G(s)$ is the transfer function matrix, given by $G(s) = C(p)(sI - A(p))^{-1}B(p)$, where I is the 2×2 identity matrix. $G(s)$ has two elements, corresponding to the two measured outputs, which are given by:

$$G_1(s) = \frac{s + k_{12}}{s^2 + \left(k_{31} - \frac{V_{max}}{K_M + x_{1,E}} + k_{12} + k_{21}\right) s + \left(k_{31} - \frac{V_{max}}{K_M + x_{1,E}}\right) k_{12}}$$
$$G_2(s) = \frac{s + k_{12} + k_{21}}{s^2 + \left(k_{31} - \frac{V_{max}}{K_M + x_{1,E}} + k_{12} + k_{21}\right) s + \left(k_{31} - \frac{V_{max}}{K_M + x_{1,E}}\right) k_{12}}. \tag{18}$$

Let $\Phi(p) = (\phi_1(p), ..., \phi_4(p))$, where $p = (k_{12}, k_{21}, k_{31}, V_{max}, K_M, x_{1,E})$, denote the (distinct) coefficients of s in Equation (18). The coefficients, $\Phi(p)$, are uniquely determinable from the input-output relationship of the system and are given by:

$$\begin{aligned}
\phi_1(p) &= k_{12} \\
\phi_2(p) &= k_{12} + k_{21} \\
\phi_3(p) &= k_{12}\left(k_{31} - \frac{V_{max}}{K_M + x_{1,E}}\right) \\
\phi_4(p) &= k_{31} - \frac{V_{max}}{K_M + x_{1,E}} + k_{12} + k_{21}.
\end{aligned} \tag{19}$$

Introducing an alternative parameter vector $\bar{p} = \left(\bar{k}_{12}, \bar{k}_{21}, \bar{k}_{31}, \bar{V}_{max}, \bar{K}_M, \bar{x}_{1,E}\right)$ and equating $\Phi(p) = \Phi(\bar{p})$, it can readily be seen from $\phi_1(p)$ and $\phi_2(p)$ that the parameters k_{12} and k_{21} are uniquely determined (i.e., $k_{12} = \bar{k}_{12}$ and $k_{21} = \bar{k}_{21}$) and therefore structurally globally identifiable from the timecourse of radioactivity remaining in plasma and the body. The parameters k_{31}, V_{max}, K_M and $x_{1,E}$ are not uniquely identifiable; however the FCR (given by Equation 11) is uniquely identifiable.

3.1.2. Fractional Catabolic Rate

The relationship between the FCR and $x_{1,E}$ is given by Equation (11). The SolveAlways function was used in Mathematica[2] to find out whether the parameters k_{31}, V_{max}, and K_M are uniquely determinable by the relationship in Equation (11). Introducing an alternative parameter vector $\left(\bar{k}_{31}, \bar{V}_{max}, \bar{K}_M\right)$ and solving the equation:

$$k_{31} - \frac{V_{max}}{K_M + x_{1,E}} = \bar{k}_{31} - \frac{\bar{V}_{max}}{\bar{K}_M + x_{1,E}}, \tag{20}$$

over all values of $x_{1,E}$, gives $\left(\bar{k}_{31}, \bar{V}_{max}, \bar{K}_M\right) = (k_{31}, V_{max}, K_M)$ as the only solution for the unknown parameters. Therefore, the parameters k_{31}, V_{max}, and K_M are uniquely determinable from the relationship between the FCR and $x_{1,E}$.

3.1.3. Terminal Half-Life

The relationship between $T_{1/2}$ and $x_{1,E}$ is given by Equation (14). We now wish to know whether the parameter vector $p = (k_{12}, k_{21}, k_{31}, V_{max}, K_M)$ is uniquely determinable from the relationship in Equation (14). From Equation (13), this is equivalent to asking whether p is uniquely determinable from the relationship between λ_2 and $x_{1,E}$, given by:

$$\lambda_2 = \frac{1}{2}\left(- k_{12} - k_{21} - k_{31} + \frac{V_{max}}{K_M + x_{1,E}} + \sqrt{-4k_{12}\left(k_{31} - \frac{V_{max}}{K_M + x_{1,E}}\right) + \left(k_{12} + k_{21} + k_{31} - \frac{V_{max}}{K_M + x_{1,E}}\right)^2}\right). \tag{21}$$

The structural identifiability problem amounts to determining whether there exists an alternative parameter vector $\bar{p}$ such that $\lambda_2(x_{1,E}, p) = \lambda_2(x_{1,E}, \bar{p})$ with $p \neq \bar{p}$. λ_2 is one of the roots of the characteristic polynomial equation, given by:

$$\lambda^2 + \left(k_{12} + k_{21} + k_{31} - \frac{V_{max}}{K_M + x_{1,E}}\right)\lambda + k_{12}\left(k_{31} - \frac{V_{max}}{K_M + x_{1,E}}\right) = 0. \tag{22}$$

We wish to know whether there exists an alternative parameter vector $\bar{p} \neq p$, such that:

$$\begin{aligned}
\lambda^2 + \left(k_{12} + k_{21} + k_{31} - \frac{V_{max}}{K_M + x_{1,E}}\right)\lambda + k_{12}\left(k_{31} - \frac{V_{max}}{K_M + x_{1,E}}\right) \\
= \lambda^2 + \left(\bar{k}_{12} + \bar{k}_{21} + \bar{k}_{31} - \frac{\bar{V}_{max}}{\bar{K}_M + x_{1,E}}\right)\lambda \\
+ \bar{k}_{12}\left(\bar{k}_{31} - \frac{\bar{V}_{max}}{\bar{K}_M + x_{1,E}}\right).
\end{aligned} \tag{23}$$

The coefficients of the quadratic are unique, therefore

$$\begin{aligned}
k_{12} + k_{21} + k_{31} - \frac{V_{max}}{K_M + x_{1,E}} &= \bar{k}_{12} + \bar{k}_{21} + \bar{k}_{31} - \frac{\bar{V}_{max}}{\bar{K}_M + x_{1,E}} \\
k_{12}\left(k_{31} - \frac{V_{max}}{K_M + x_{1,E}}\right) &= \bar{k}_{12}\left(\bar{k}_{31} - \frac{\bar{V}_{max}}{\bar{K}_M + x_{1,E}}\right).
\end{aligned} \tag{24}$$

The only solution to Equation (24), over all values of $x_{1,E}$ and for positive parameter values only, is $p = \bar{p}$. Therefore the parameters $k_{12}, k_{21}, k_{31}, V_{max}$, and K_M are uniquely determinable from the relationship between $T_{1/2}$ and $x_{1,E}$.

3.1.4. Summary of Structural Identifiable Parameters

From an individual's timecourse of radioactivity remaining in plasma and the body, described by Equations (8–10), the parameters k_{12} and k_{21} are structurally uniquely identifiable. The parameters k_{31}, V_{max}, K_M, and $x_{1,E}$ are not uniquely identifiable; however the FCR is uniquely identifiable. The parameter $x_{1,E}$ may be measured independently; however the parameters k_{31}, V_{max}, and K_M remain unidentifiable even when $x_{1,E}$ is known. This result is intuitive, as the unidentifiable parameters describe the nonlinear behavior which is not demonstrated when the quantity of endogenous IgG in plasma, $x_{1,E}$, is constant.

[2]Mathematica 10.4 (2016). Windows. Champaign: Wolfram Research Inc.

In order to show the nonlinear behavior of the model, observations need to be made over a range of steady state quantities of endogenous IgG in plasma, $x_{1,E}$. Given a set of timecourses, each described by Equations (8–10), for a range of different values of $x_{1,E}$, it is possible to measure the FCR, $T_{1/2}$ and $x_{1,E}$ for each timecourse. From the relationship between the FCR and $x_{1,E}$, the parameters k_{31}, V_{max} and K_M are structurally globally identifiable. From the relationship between $T_{1/2}$ and $x_{1,E}$, the parameters $k_{12}, k_{21}, k_{31}, V_{max}$, and K_M are structurally globally identifiable. The structurally identifiable parameters are summarized in **Table 2**.

3.2. Estimation of Parameters from Individual Timecourse Data

The model parameters k_{12} and k_{21} along with the FCR are structurally globally identifiable from the individual timecourse measurements $y_1(t)$ and $y_2(t)$, as described by the linearized model in Equations (8–10). These three parameters were estimated from timecourse data from seven individuals whom we refer to as subjects A–G. The data are described in Section 2.1.

Parameter values were estimated for each subject by analytically solving the linear ODE system and minimizing the sum of squared residual errors between the model output and the data, using the function NonlinearModelFit in Mathematica[2]. For each subject, the model outputs $y_1(t)$ and $y_2(t)$ were simultaneously fitted to the plasma and whole body timecourse data, respectively. Three examples of the fits (subjects A–C) are shown in **Figure 4**. The corresponding plots for all subjects are provided in the Supplementary Material.

The parameter estimates and their standard errors are given in **Table 3**. For all three parameters across all subjects, the standard errors are small relative to their respective parameter estimates. The distribution of parameter estimates among the seven subjects is illustrated in **Figure 5**. The mean and median of each parameter are summarized in **Table 3**. The root mean squared error (RMSE), as a measure of the goodness-of-fit, for each fitted timecourse is also provided in **Table 3**.

As stated in Section 2.1, several subjects have IgG myeloma, macroglobulinemia or familial hypercatabolic hypoproteinemia. Patients with familial hypercatabolic hypoproteinemia do not express FcRn, explaining the large value of the FCR (0.247 day^{-1}) for subject E. The parameter V_{max} (not estimated here) for subject E should be equal to zero, reflecting the absence of recycling receptors. Subjects A and C have IgG myeloma and subject D has macroglobulinemia. The high or low values of the FCR in these patients should be explained by the concentration-dependent catabolism of IgG, as described by Equation (11), with abnormally high or low plasma IgG concentrations likely occurring as symptoms of the respective disease.

TABLE 2 | Structurally identifiable parameters.

Observation	Structurally globally identifiable parameters
Individual subject's timecourse	k_{12}, k_{21}
FCR vs. $x_{1,E}$	k_{31}, V_{max}, and K_M
$T_{1/2}$ vs. $x_{1,E}$	$k_{12}, k_{21}, k_{31}, V_{max}$, and K_M

3.2.1. Sensitivity to Model Parameters

Along with structural identifiability, parameter identification requires sensitivity of the model output to the parameters. There are two types of sensitivity function: traditional sensitivity functions (TSFs) and generalized sensitivity functions (GSFs) (Thomaseth and Cobelli, 1999). Used together, TSFs and GSFs can provide insight in terms of the information about individual parameters provided by a measured output over the time duration of the experiment.

In order to estimate a model parameter from measurements of a model output it is necessary that the measured output is sensitive to the parameter over the time interval of the experiment (Banks et al., 2007). The TSF of a measured model output with respect to one component of the model parameter vector is given by the partial derivative of the output with respect to the parameter, for example:

$$s_{TSF,y_1,k_{12}}(t) = \frac{\partial y_1}{\partial k_{12}}(t) \tag{25}$$

is the TSF of the model output $y_1(t)$ with respect to the parameter k_{12}. The TSF is locally defined for the "true" parameter vector of the system. For the inverse problem in this section the true model

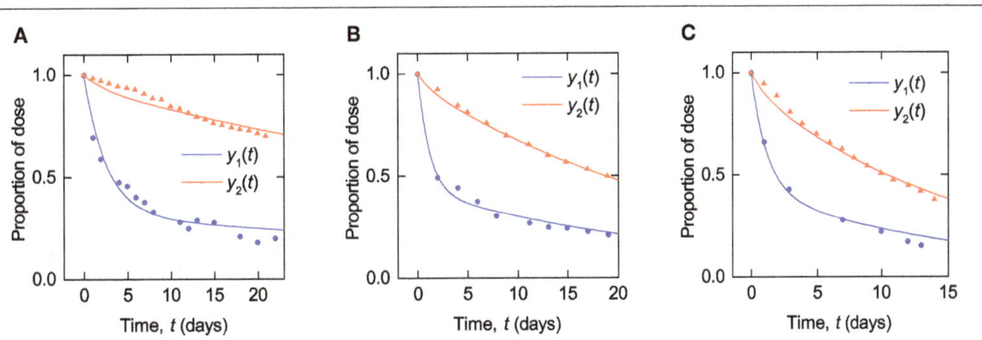

FIGURE 4 | Timecourse fits: model described by Equations (8–10) fitted to timecourse data extracted from plots in Solomon et al. (1963) **for (A–C)** subjects A, B, and C.

TABLE 3 | Parameter estimates and their standard errors (SE) and the root mean squared error (RMSE) for each fitted timecourse.

Subject	FCR (day^{-1})		k_{12} (day^{-1})		k_{21} (day^{-1})		RMSE
	Estimate	SE	Estimate	SE	Estimate	SE	
A	0.0359	0.00169	0.130	0.0182	0.231	0.0218	0.0336
B	0.0761	0.00190	0.381	0.0539	0.426	0.0546	0.0182
C	0.125	0.00397	0.382	0.0643	0.378	0.0516	0.0235
D	0.0311	0.000863	0.432	0.0746	0.347	0.0559	0.0136
E	0.247	0.00632	0.341	0.125	0.140	0.0333	0.0197
F	0.0728	0.00108	0.358	0.0233	0.476	0.0268	0.0134
G	0.0766	0.00149	0.656	0.0538	0.965	0.0716	0.0222
Mean	0.0950		0.383		0.423		
Median	0.0761		0.381		0.378		

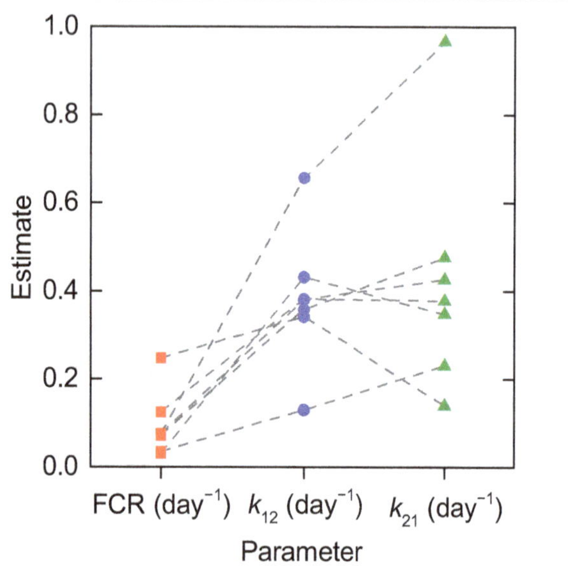

FIGURE 5 | Parameter estimates for individual timecourses. Dashed lines connect the estimates obtained for an individual subject.

and parameter vector for each subject is unknown; therefore the TSFs are calculated for the estimated parameter vectors for each subject to investigate parameter sensitivity over a range of parameter vectors that are likely to be seen in individuals.

The TSFs for the timecourse outputs $y_1(t)$ and $y_2(t)$ are plotted in **Figure 6**, evaluated at the parameter estimates for subjects A, B, and C, respectively, provided in **Table 3**. The TSFs were calculated in Mathematica[2]. The TSFs show that the model outputs are sensitive to all three parameters over the time interval of observation for each of the parameter vectors. The corresponding plots for all subjects are provided in the Supplementary Material. A similar pattern is observed for the remaining subjects D–G.

A shortcoming of the TSF is that it does not account for correlation between parameters. An alternative function, the GSF, takes account of parameter correlations and quantifies the information content of a model output on an individual parameter over the time duration of observation. The GSFs for a general model output function $y(t) = f(t, \theta_0)$ with parameter vector θ_0 are defined as:

$$s_{GSF,y,\theta}(t_l) = \sum_{i=1}^{l} \frac{1}{\sigma^2(t_i)} [F^{-1} \times \nabla_\theta f(t_i, \theta_0)] \bullet \nabla_\theta f(t_i, \theta_0), \quad (26)$$

where

$$F = \sum_{j=1}^{n} \frac{1}{\sigma^2(t_j)} \nabla_\theta f(t_j, \theta_0) \nabla_\theta f(t_j, \theta_0)^{\mathrm{T}} \quad (27)$$

is the Fisher information matrix and the model output $y(t) = f(t, \theta_0)$ is observed with error at discrete times $t_l, l = 1, ..., n$. The measured values of the output are given by $Y_l = f(t_l, \theta_0) + \epsilon_l$ with $\sigma^2(t_l)$ the variance of the error on the observation at time t_l, ϵ_l (Thomaseth and Cobelli, 1999). In the definition of the GSF the true parameter vector θ_0 is assumed known. Here the GSFs are calculated for the estimated parameter vectors for each subject, in order to investigate the inverse problem for the different dynamics seen in individuals.

The GSFs for the timecourse outputs $y_1(t)$ and $y_2(t)$ are plotted in **Figure 6**, for the parameter estimates of subjects A, B, and C, respectively, provided in **Table 3**. The GSFs were calculated in Mathematica[2]. Unlike the TSF, the GSF is defined only at discrete measurement times. In order to obtain an approximation of the smooth function for continuous measurement data, as in Thomaseth and Cobelli (1999) and Banks et al. (2007), we assume a high rate of sampling, calculating the GSF as though a measurement is taken every 0.1 days for each subject. We assume that the variances of the errors are equal at all measurement times, such that the variance terms cancel out in the definition of the GSF (see Equations 26, 27).

According to the interpretation of GSFs given by Thomaseth and Cobelli (1999), a steep increase in the GSF of a particular parameter between 0 and 1 indicates the interval on which the information on that parameter, provided by the measured output, is concentrated. **Figures 6H,I** show the ideal pattern of three distinct intervals of steep increase between 0 and 1 for

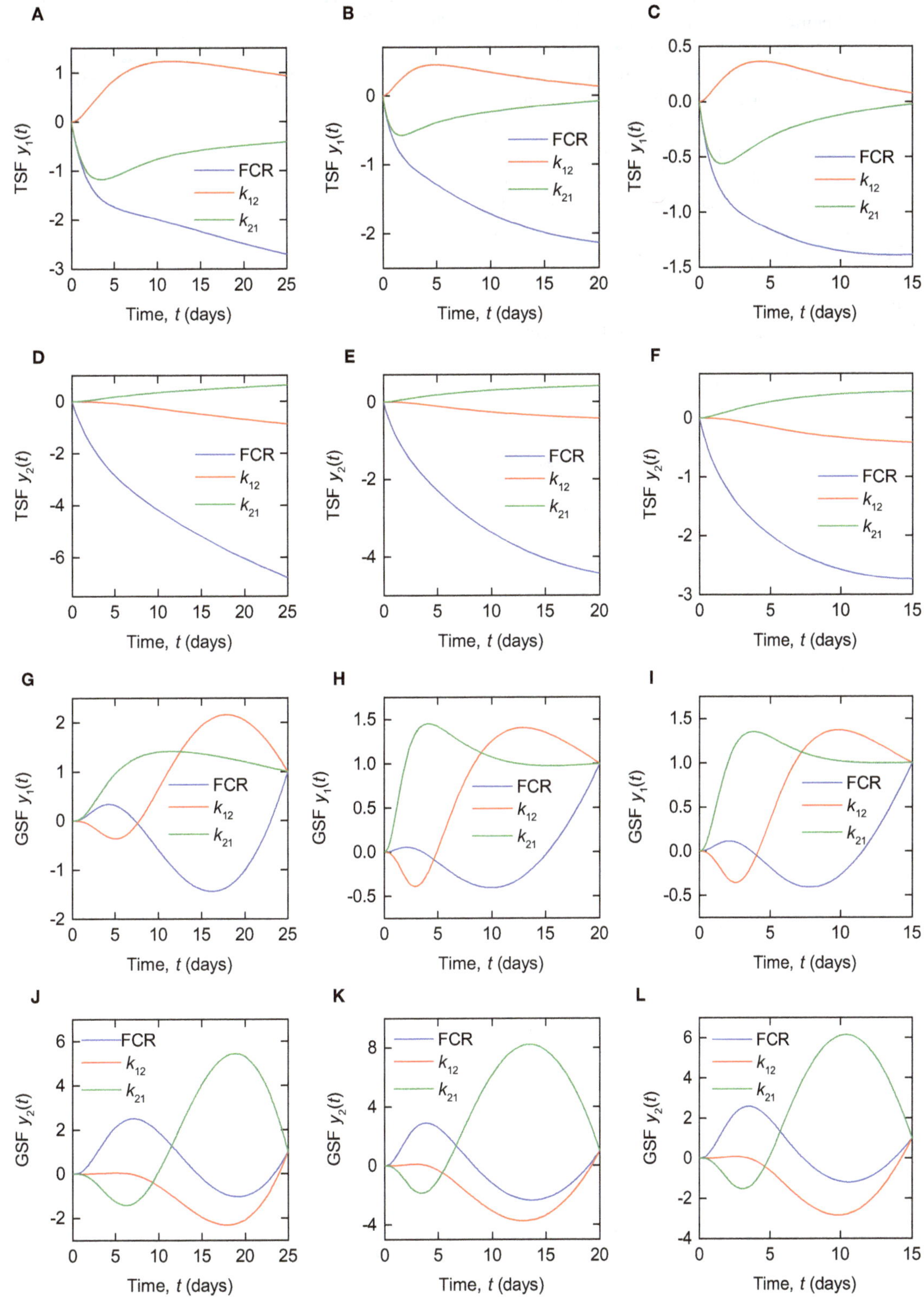

FIGURE 6 | Traditional sensitivity functions (TSFs) of timecourse outputs $y_1(t)$, for (A–C) subjects A, B and C, and $y_2(t)$, for **(D–F)** subjects A, B, and C. Generalized sensitivity functions (GSFs) of timecourse outputs $y_1(t)$, for **(G–I)** subjects A, B, and C, and $y_2(t)$, for **(J–L)** subjects A, B, and C.

the three parameters, respectively, with the information on k_{21} concentrated at the beginning of the experiment, followed by k_{12} and then the FCR. Intuitively this makes sense, because the dose is administered to the first compartment, from where it transfers to the second compartment in the initial days of the experiment. The pattern shown by the GSFs of $y_1(t)$ for subject A, depicted in **Figure 6G**, appears slightly different to that shown by subjects B and C; however, if we extend the experiment time to 50 days for subject A, the GSFs then appear extremely similar to those of subjects B and C. This is due to the slower dynamics exhibited by this subject—see **Table 3** and **Figure 4**. The pattern exhibited by the GSFs of $y_2(t)$ for each patient indicates a high correlation between the parameters, producing large oscillations in the GSFs. The corresponding plots for all subjects are provided in the Supplementary Material.

The GSF is clearly a useful tool for understanding the behavior of estimators of correlated parameters; however, it is still important to use the TSF alongside the GSF when analysing parameter sensitivity. As mentioned, the GSFs of $y_1(t)$ for subject A appear more similar to those for subjects B and C when the duration of the experiment is extended. The trajectory of the GSF is dependent on the times at which data are collected, and is forced to equal one at the final measurement time. This can result in misleading GSFs when the observation interval is defined over a period of low sensitivity (as defined by the TSF) of the output to the parameters. For this reason we use the TSF and GSF alongside one another. This is discussed in more detail by Banks et al. (2007).

3.3. Estimation of Parameters from Fractional Catabolic Rate and Half-Life

As shown in Section 3.1, the parameters k_{31}, V_{max}, and K_M are structurally unidentifiable from an individual subject's timecourse data, assuming the linearized model given by Equations (8–10). It is therefore necessary to make use of the relationships between FCR and $T_{½}$, respectively, and the quantity of endogenous IgG in plasma, $x_{1,E}$. Unfortunately, these relationships are not known for an individual subject; obtaining them would require performing the tracer experiment over a range of different plasma concentrations of endogenous IgG within an individual subject, which is not practically feasible. We therefore estimate parameters from the FCR and $T_{½}$ measurements taken from a sample of patients with a wide range of endogenous IgG plasma concentrations, as though the data arose from an individual subject, in what may be described as a naive pooled approach (Wright, 1998). It is therefore not possible to gain a sense of the distribution of the parameters k_{31}, V_{max}, and K_M within the population as for those parameters estimated from the individual timecourse data, k_{12} and k_{21}, as illustrated in **Figure 5**.

The parameters k_{12}, k_{21}, k_{31}, V_{max}, and K_M were estimated from FCR vs. $x_{1,E}$ and $T_{½}$ vs. $x_{1,E}$ data, simultaneously. Waldmann and Strober (1969) provide plots of FCR and $T_{½}$ vs. plasma endogenous IgG concentration in $g\,l^{-1}$. The data were extracted using the Digitizer tool in OriginPro[1]. The plasma IgG concentration in $g\,l^{-1}$ was converted to $\mu mol\,l^{-1}$ by dividing by the molar mass of IgG, 0.15 $g\,\mu mol^{-1}$. The concentration was then multiplied by an average plasma volume of 3 l (Solomon et al., 1963) to obtain the quantity $x_{1,E}$ in μmol.

TABLE 4 | Parameter estimates and standard errors estimated from FCR and T½ data.

Parameter	Units	Estimate	Standard error (SE)	95% confidence interval
k_{31}	day^{-1}	0.159	0.0111	(0.137, 0.181)
V_{max}	$\mu mol\,day^{-1}$	40.0	10.5	(19.1, 60.9)
K_M	μmol	272	55.4	(162, 382)
k_{12}	day^{-1}	0.158	0.155	(−0.150, 0.467)
k_{21}	day^{-1}	0.187	0.231	(−0.273, 0.647)

The FCR and $T_{½}$ data were fitted simultaneously by the model outputs described in Equations (11, 14). The parameter values were estimated by minimizing the sum of squared residual errors between the model outputs and the measured values, using the function NonlinearModelFit in Mathematica[2]. Due to the different scales of the parameters ($0.02 < \text{FCR} < 0.17\ day^{-1}$; $10 < T_{½} < 72$ days) the $T_{½}$ data points were assigned different weights to the FCR data points. It was assumed that the standard deviation of the residual errors is of the order of the size of the measured values, for both FCR and $T_{½}$, respectively. Therefore, the weights given to the $T_{½}$ data points were equal to the squared mean of the measured $T_{½}$ values divided by the squared mean of the measured FCR values, and the weights given to the FCR data points were set to 1. Using this approach, the variance of the $T_{½}$ residuals was assumed to be 1.08×10^5 times the variance of the FCR residuals.

The data and model fits are shown in **Figure 7**. The parameter estimates and their standard errors are given in **Table 4**. The standard errors are almost as large as, or larger than, the estimates themselves for the parameters k_{12} and k_{21}, indicating that these parameters cannot be estimated with a reasonable level of precision. This may be because the measured output $T_{½}$ is insensitive to variations in the parameters k_{12} and k_{21} or due to correlations between the parameters.

3.3.1. Sensitivity to Model Parameters and Parameter Correlations

The sensitivity of the outputs FCR and $T_{½}$ to the model parameters is illustrated by the TSFs shown in **Figure 8**. In **Figure 8** the TSFs are evaluated for the parameter estimates given in **Table 4**. Due to the wide range in parameter estimates (0.158–272), each TSF is multiplied by the value of the corresponding parameter estimate; thus the normalized TSF can be seen as representing the sensitivity of the output to variation in a parameter proportional to its value (for the particular parameter values used here). The GSFs of the FCR with respect to the parameters k_{31}, V_{max}, and K_M were calculated for the parameter estimates in **Table 4**. We assume measurements taken in intervals of 10 μmol between 0 μmol and 2000 μmol plasma endogenous IgG, $x_{1,E}$, in order to get an approximately smooth function. Again we assume that the variances of the errors are equal over

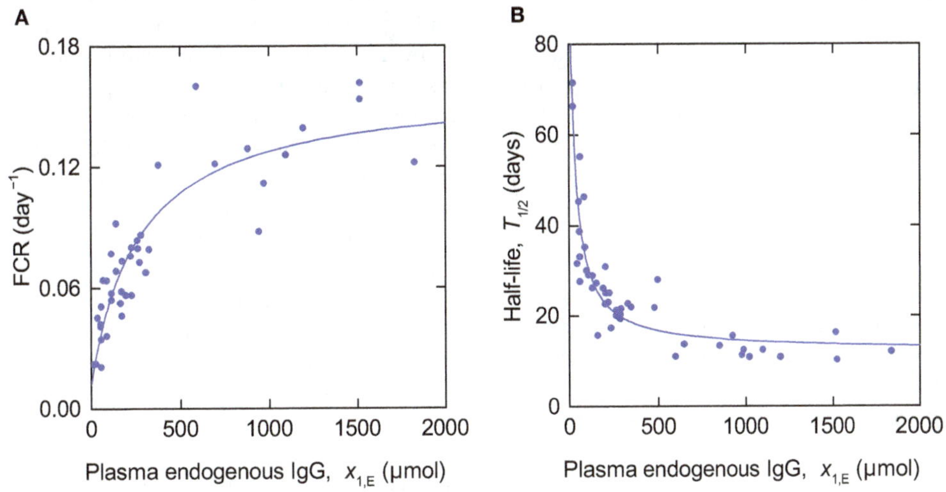

FIGURE 7 | Expressions for (A) FCR (Equation 11) and **(B)** $T_{1/2}$ (Equation 14) fitted to data from Waldmann and Strober (1969).

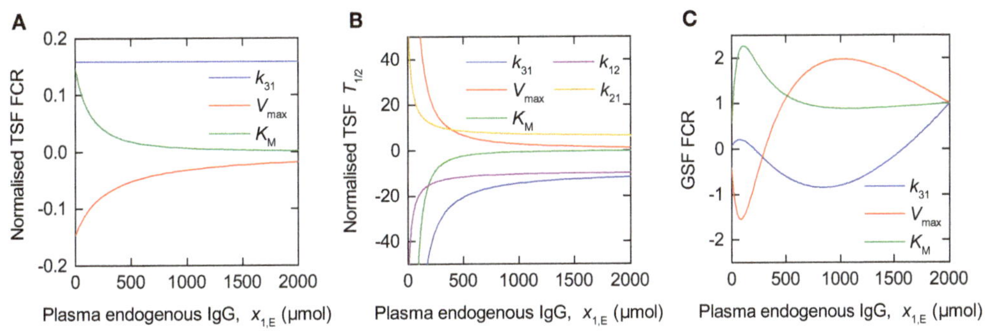

FIGURE 8 | Traditional sensitivity functions (TSFs) of (A) FCR and **(B)** $T_{1/2}$ and generalized sensitivity functions (GSFs) of **(C)** FCR with respect to model parameters.

all concentrations, such that the variance terms cancel out in the definition of GSF (see Equations 26, 27).

Figure 8A shows the TSFs of the FCR with respect to the parameters k_{31}, V_{max}, and K_M, evaluated for the parameter estimates in **Table 4**. The plot shows that, for the parameter vector used, the FCR is sensitive to all three parameters over the range of plasma endogenous IgG concentrations measured, with greater sensitivity to K_M and V_{max} at smaller concentrations. The similarity between the TSFs of the FCR with respect to K_M and V_{max}, respectively, may cause high correlation between these two parameters.

Figure 8C shows the GSFs of the FCR with respect to k_{31}, V_{max}, and K_M. The GSFs show a similar pattern to those of the timecourse observations shown in **Figure 6**, however the larger magnitude of oscillation indicates a strong correlation between the parameters. The GSFs indicate that, for the parameter vector used, measurements at very small quantities of plasma endogenous IgG, below around 100 μmol, have the greatest influence on the estimate of K_M, then the region between around 100 and 1.000 μmol has the greatest influence over the estimate of V_{max}, with, finally, the information on k_{31} available at the remaining larger quantities. This interpretation is consistent with the TSFs given in **Figure 8A**: the sensitivity to K_M decreases rapidly at small quantities, followed by the sensitivity to V_{max}, with the sensitivity to k_{31} constant, that is insensitive to the quantity $x_{1,E}$ itself.

Figure 8B shows the TSFs of $T_{1/2}$ with respect to the parameters k_{12}, k_{21}, k_{31}, V_{max}, and K_M. The plot shows that $T_{1/2}$ is much more sensitive to all of the parameters at smaller quantities of endogenous IgG. The similarity between the trajectories of the TSFs with respect to all parameters suggests that they are highly correlated. It was not possible to calculate the GSFs of $T_{1/2}$ due to the Fisher information matrix being ill-conditioned, such that the inverse could not be computed. If we attempt to estimate all five parameters from the $T_{1/2}$ data alone, we obtain standard errors of the order of 10×10^6 and higher, and correlation coefficients of 1 or -1 between the parameters.

When the parameters are estimated from FCR vs. $x_{1,E}$ and $T_{1/2}$ vs. $x_{1,E}$ simultaneously, we find high correlation coefficients between the parameters, with the correlation matrix given by:

$$\begin{array}{c} \\ k_{31} \\ V_{max} \\ K_M \\ k_{12} \\ k_{21} \end{array} \begin{array}{c} \begin{array}{ccccc} k_{31} & V_{max} & K_M & k_{12} & k_{21} \end{array} \\ \begin{pmatrix} 1 & 0.928 & 0.839 & -0.290 & -0.300 \\ 0.928 & 1 & 0.976 & -0.228 & -0.253 \\ 0.839 & 0.976 & 1 & -0.0890 & -0.115 \\ -0.290 & -0.228 & -0.0890 & 1 & 0.996 \\ -0.300 & -0.253 & -0.115 & 0.996 & 1 \end{pmatrix} \end{array}.$$

Parameters k_{31}, V_{max}, and K_M are highly correlated pairwise, with the strongest correlation between K_M and V_{max}. k_{12} and k_{21} are highly correlated with one another (correlation coefficient of 0.996) and not with the other parameters, explaining why they cannot be estimated with a reasonable level of precision from these data.

3.4. Simulations of IgG Responses in IgG Multiple Myeloma

In order to simulate monoclonal IgG responses in IgG multiple myeloma, the model of endogenous IgG metabolism given by Equation (1) needs to explicitly account for monoclonal IgG produced by the malignant plasma cells, and polyclonal IgG produced by healthy plasma cells, since both types of IgG undergo the same processes of recycling and elimination and therefore one is influenced by the other. The dynamics of monoclonal and polyclonal IgG in an IgG myeloma patient may be described by:

$$\begin{aligned} \dot{x}_{1,m}(t) &= -\left(k_{21} + k_{31} - \frac{V_{max}}{K_M + x_{1,m}(t) + x_{1,p}(t)}\right) x_{1,m}(t) \\ &\quad + k_{12}x_{2,m}(t) + I_m(t) \\ \dot{x}_{2,m}(t) &= k_{21}x_{1,m}(t) - k_{12}x_{2,m}(t) \\ \dot{x}_{1,p}(t) &= -\left(k_{21} + k_{31} - \frac{V_{max}}{K_M + x_{1,m}(t) + x_{1,p}(t)}\right) x_{1,p}(t) \\ &\quad + k_{12}x_{2,p}(t) + I_p(t) \\ \dot{x}_{2,p}(t) &= k_{21}x_{1,p}(t) - k_{12}x_{2,p}(t), \end{aligned} \tag{28}$$

where $x_{1,m}(t)$ and $x_{2,m}(t)$ are the quantities of monoclonal IgG in plasma and in the peripheral space, respectively, $x_{1,p}(t)$ and $x_{2,p}(t)$ are the quantities of polyclonal IgG in plasma and in the extravascular space, respectively, $I_m(t)$ is the production rate of monoclonal IgG in μmol day^{-1}, $I_p(t)$ is the production rate of polyclonal IgG in μmol day^{-1}, and all other parameters are as previously defined.

It is assumed that the production rate of monoclonal IgG, $I_m(t)$, is determined by the number of myeloma cells, or tumor burden. Modeling the response of the myeloma cell population under therapy is in itself a significant problem. Here we assume a highly simplified, phenomenological model which nevertheless shows good qualitative agreement with responses seen in real patients. In the following simulations the production rate of monoclonal IgG, $I_m(t)$, is given by:

$$I_m(t) = \left(I_{m,0} - I_{m,\infty}\right) \exp\left(-k_{kill}t\right) + I_{m,\infty}, \tag{29}$$

where $I_m(0) = I_{m,0}$ μmol day^{-1}, $I_m(t)$ tends to $I_{m,\infty}$ μmol day^{-1} for large t, and k_{kill} day^{-1} is the rate constant of tumor kill.

In **Figure 9** we present simulations of monoclonal IgG responses in IgG myeloma patients during treatment. Simulations of the plasma concentration of monoclonal IgG are shown alongside plasma IgG concentrations from six IgG myeloma patients, taken from the Intergroupe Francophone du Myélome (IFM) 2009-02 clinical trial (Leleu et al., 2013). The plasma concentration of monoclonal IgG was measured by serum protein electrophoresis at regular intervals during treatment. The simulated quantities of monoclonal and polyclonal IgG are defined by Equation (28), with the monoclonal IgG production rate $I_m(t)$ given by Equation (29). The polyclonal IgG production rate $I_p(t)$ is assumed to remain constant at $I_{p,0} = 15$ μmol day^{-1}, as in normal individuals (Waldmann and Strober, 1969). At time $t = 0$ the system is assumed to be in steady state, such that the initial conditions of monoclonal and polyclonal IgG are given by:

$$\begin{aligned} x_{1,m}(0) &= \frac{I_{m,0}}{I_0} \frac{-k_{31}K_M + I_0 + V_{max} + \sqrt{4k_{31}K_M I_0 + (-k_{31}K_M + I_0 + V_{max})^2}}{2k_{31}} \\ x_{2,m}(0) &= \frac{k_{21}}{k_{12}} x_{1,m}(0) \\ x_{1,p}(0) &= \frac{I_{p,0}}{I_0} \frac{-k_{31}K_M + I_0 + V_{max} + \sqrt{4k_{31}K_M I_0 + (-k_{31}K_M + I_0 + V_{max})^2}}{2k_{31}} \\ x_{2,p}(0) &= \frac{k_{21}}{k_{12}} x_{1,p}(0) \end{aligned} \tag{30}$$

where $I_0 = I_{m,0} + I_{p,0}$.

In order to convert the quantity of plasma monoclonal IgG in μmol into concentration in g l^{-1} the simulated quantity was multiplied by the molecular weight of IgG, 0.15 g μmol^{-1}, and divided by an average plasma volume of 3 l. For the parameters k_{12} and k_{21} the mean values of the respective parameter estimates from the individual fits in Section 3.2 are used; for the parameters k_{31}, V_{max}, and K_M the values estimated from FCR and $T_{1/2}$ vs. $x_{1,E}$ in Section 3.3 are used. The parameters of the model describing the monoclonal IgG production rate, given by Equation (29), namely $I_{m,0}$, $I_{m,\infty}$ and k_{kill}, were manually adjusted in order to produce simulations that approximately replicate the responses seen in patients. The parameter values used are provided in **Table 5**.

The purpose of these simulations is to demonstrate how the model analyzed in this paper may be used in the future to investigate monoclonal IgG responses in IgG myeloma. Patient data are presented alongside the simulations to show that they provide good qualitative agreement with in vivo responses, supporting the suitability of the model for future investigations. In these simulations, the initial and final monoclonal IgG production rates, $I_{m,0}$ and $I_{m,\infty}$, and the rate at which the monoclonal IgG production rate falls during treatment, k_{kill}, have been varied, whilst the parameters of the metabolic model have been fixed. In reality there will be inter-patient variability in both the parameters of the tumor response model and the metabolic model. Here we have not explicitly modeled the response in

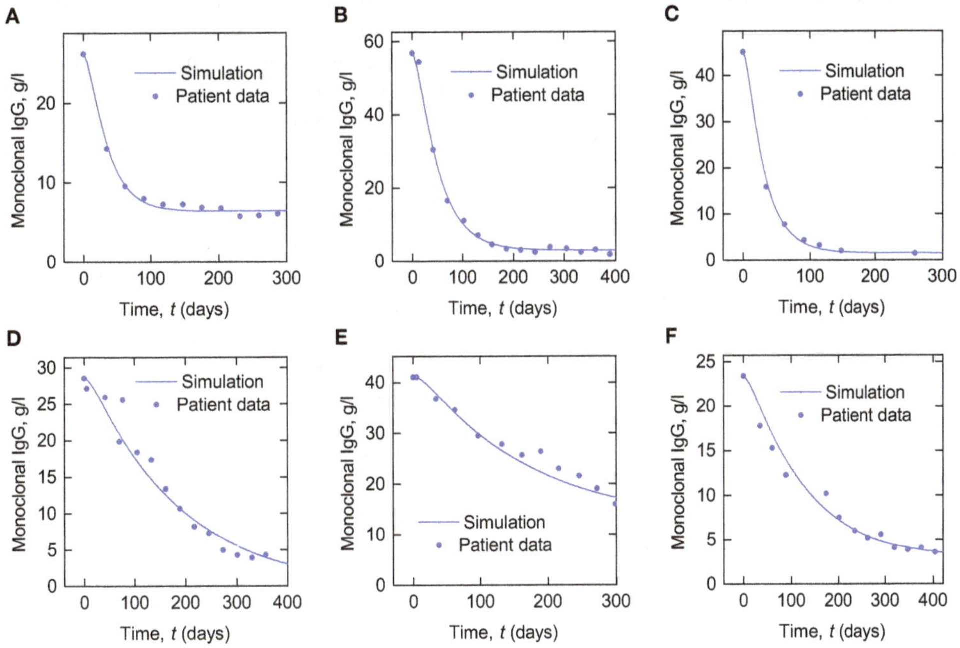

FIGURE 9 | Simulations of plasma monoclonal IgG responses in IgG myeloma alongside data from six IgG myeloma patients (A–F).

the myeloma cell population itself—only the production rate of monoclonal IgG. Additional assumptions are required in order to investigate the relationship between the tumor response and the serum monoclonal IgG response. The simplest approach is to assume that the rate of synthesis per myeloma cell is constant and as such the total body synthesis rate $I_m(t)$ is directly proportional to the total myeloma cell population. It is possible that the cellular synthesis rate may vary over the course of treatment; however studies of in vitro IgG synthesis in plasma cells taken from IgG myeloma patients have shown that, whilst the cellular synthesis rate varies between patients, for an individual patient it remains constant over a period of months (Salmon and Smith, 1970). In the present study we have also assumed constant polyclonal IgG production, however it is known that normal plasma cells in the bone marrow are frequently suppressed by the clonal cell presence. If we assume that the cellular rate of IgG synthesis remains constant in normal plasma cells then suppression of these cells likely contributes to a decrease in overall polyclonal IgG synthesis. There are several complex mechanisms involved, but fundamentally the suppression of polyclonal cells is believed to be due to competition between monoclonal and polyclonal cells for survival niches in the bone marrow microenvironment (Paiva et al., 2011). Mathematically modeling normal plasma cell suppression could be the subject of future research. It has also been assumed that the system is in steady state at the commencement of treatment. Future studies will be required to validate this assumption or employ alternative models in which the tumor is growing initially, and to assess the impact of the chosen assumptions on any conclusions drawn from the simulations.

4. DISCUSSION

The aim of this study was to analyse a previously published model of endogenous IgG metabolism and available measurements in humans with respect to parameter identifiability. The model was linearized to replicate experimental conditions in which small doses of administered IgG exhibit linear timecourse responses. The linearized model was then analyzed in terms of parameter structural identifiability and parameter sensitivity. The analyses show that certain important parameters are not structurally identifiable from an individual's timecourse response; however they are structurally identifiable using the relationships between the FCR and $T_{1/2}$, respectively, and the quantity of endogenous IgG in plasma, which is assumed to remain in steady state.

A limitation of the linear model of an individual's timecourse response, given by Equations (8–10), is that the parameters k_{31}, V_{max}, and K_M are structurally unidentifiable. It is not known whether these parameters are structurally identifiable in the nonlinear model of coupled radiolabeled and endogenous IgG responses given by Equations (5–7). There are two reasons this approach was not pursued here: firstly, structural identifiability analysis of a four-state nonlinear model with rational terms presents a more challenging problem; secondly, if the nonlinear model were found to be structurally identifiable, the responses available nonetheless do not demonstrate nonlinear behavior due to the small doses of radiolabeled IgG administered (see **Figure 3**)—therefore the parameters representing nonlinear behavior may not be practically identifiable by fitting the nonlinear model, and there is an increased risk of fitting the noise in the data with the increased model complexity.

TABLE 5 | Parameter values used to produce the simulations in Figure 9.

Parameter	Units	Panel					
		A	B	C	D	E	F
$I_{m,0}$	μmol day^{-1}	61	152	116	68	105	53
$I_{m,\infty}$	μmol day^{-1}	11.5	5	2.5	0	24	5
k_{kill}	day^{-1}	0.055	0.03	0.07	0.007	0.0065	0.01
$I_{p,0}$	μmol day^{-1}	15	15	15	15	15	15
k_{12}	day^{-1}	0.38	0.38	0.38	0.38	0.38	0.38
k_{21}	day^{-1}	0.42	0.42	0.42	0.42	0.42	0.42
k_{31}	day^{-1}	0.16	0.16	0.16	0.16	0.16	0.16
V_{max}	μmol day^{-1}	40	40	40	40	40	40
K_M	μmol	270	270	270	270	270	270

Structural identifiability analysis alone does not imply that parameters can necessarily be estimated in practice. In this paper two sensitivity functions were investigated: the traditional sensitivity function (TSF) and generalized sensitivity function (GSF). TSFs and GSFs were computed and plotted for the timecourse outputs $y_1(t)$ and $y_2(t)$, using the parameter values estimated from the timecourse data of seven individuals. The TSFs show that the measured timecourse outputs are sensitive to the model parameters over the duration of the experiment, which is different for each individual. In addition, the GSFs show the influence of the duration of observation taking into account correlation between the parameters; for example, subject A is observed over a relatively short time period compared to subjects B and C, considering its slower dynamics. The GSF curves of $y_1(t)$ for subject A show a larger magnitude of oscillation than those of subjects B and C; however when the duration of observation is increased for subject A, the GSFs of $y_1(t)$ are remarkably similar for the three subjects. This would suggest that the estimation of the parameters for subject A would benefit from a longer duration of observation. If the tracer experiments were to be repeated, the insights obtained from the GSFs in particular could be used to inform the sampling grid of measurements. With the data that are currently available, k_{12}, k_{21}, and the FCR are estimated with a good level of precision for all subjects.

TSFs and GSFs were also calculated for the relationship between the FCR and $x_{1,E}$. The TSFs indicate that the FCR is most sensitive to the parameters V_{max} and K_M at small concentrations of plasma endogenous IgG; this is also indicated in the GSFs, which show that the information on K_M is concentrated very close to zero, followed by V_{max} and finally k_{31}. The TSFs for the $T_{1/2}$ imply high correlation between all the parameters. From the simultaneous estimation of the parameters from both FCR and $T_{1/2}$ data, the parameters k_{31}, V_{max}, and K_M are estimated with a reasonable level of precision. We notice that the estimated values of k_{12} and k_{21} are quite different from the averages of the estimates of the same parameters obtained from fitting individual timecourses. From the FCR and $T_{1/2}$ data, the 95% confidence interval estimates of k_{12} and k_{21} are given by $(-0.150, 0.467)$ and $(-0.273, 0.647)$, respectively. These large intervals containing zero imply that the parameters k_{12} and k_{21} are not well estimated from these data. The estimates of k_{12} and k_{21} from the FCR and $T_{1/2}$ data are very highly correlated with one another, but not with the remaining three parameters; this offers an explanation for why they are not well estimated from these data. Nevertheless, the 95% confidence intervals for k_{12} and k_{21} contain the averages of the individual estimates from timecourse data, 0.38 and 0.42, respectively. If we fix k_{12} to 0.38 and k_{21} to 0.42 in the estimation from the FCR and $T_{1/2}$ data, the newly obtained parameter estimates of k_{31}, V_{max}, and K_M are 0.161, 45.6, and 307, respectively. These values fall well within the confidence intervals of the previous estimation where k_{12} and k_{21} are unconstrained.

Table 6 compares previously published parameter values alongside the parameter values estimated in this paper. For the parameters k_{12} and k_{21} the mean value of the parameter among the seven subjects is chosen to represent the average; for the parameters k_{31}, V_{max}, and K_M the values estimated from FCR and $T_{1/2}$ vs. $x_{1,E}$ data from many subjects are taken to represent the population average. At first glance the newly estimated parameter values are not wildly dissimilar to the previously published values. Waldmann and Strober (1969) give the values of $k_{31} = 0.18$ day^{-1} and $V_{max}/w = 147$ mg day^{-1} kg^{-1}, where w is body weight in kg. Assuming a 70 kg human, this is equivalent to $V_{max} = 68.6$ μmol day^{-1}. The value of k_{31} was obtained 'from the asymptotic value of the plot of the IgG fractional catabolic rate versus its concentration'. These are the same data that were utilized in this paper as described in Section 2.1. The authors subtracted the value of the FCR for each individual from 0.18 to find the fractional recycling rate. The fractional recycling rate (FRR) is thus given by:

$$\text{FRR} = k_{31} - \text{FCR} = k_{31} - \left(k_{31} - \frac{V_{max}}{K_M + x_{1,E}}\right) = \frac{V_{max}}{K_M + x_{1,E}}. \tag{31}$$

They then multiplied the plasma concentration of endogenous IgG by the plasma volume per kg of body weight for each individual to get the quantity of endogenous IgG in plasma per kg of body weight, $x_{1,E}/w$. Finally the authors multiplied the FRR by $x_{1,E}/w$ to obtain the absolute recycling rate per kg of body weight, which we will call ARR. They then plotted the reciprocal of the ARR against the reciprocal of $x_{1,E}/w$ to obtain a straight line relationship, given by:

$$\frac{1}{\text{ARR}} = \frac{K_M}{V_{max}}\frac{w}{x_{1,E}} + \frac{w}{V_{max}}. \tag{32}$$

From the intercept the authors obtained the value of $V_{max}/w = 147$ mg kg^{-1} day^{-1}.

Kim et al. (2007) provide values for all model parameters. Again using the FCR vs. endogenous IgG concentration data first published by Waldmann and Strober (1969) and described here in Section 2.1, Kim et al. (2007) estimate K_M/w using a least-squares fitting. The equation fitted to the data is

$$\text{FCR} = k_{31} - \frac{V_{max}/w}{\frac{v_1}{w}\left(\frac{K_M}{v_1} + \frac{x_{1,E}}{v_1}\right)}, \tag{33}$$

where v_1 is the plasma volume. The authors fit Equation (33) to FCR vs. $x_{1,E}/v_1$ to obtain K_M/v_1 whilst fixing the remaining

TABLE 6 | Comparison with published parameter values.

Parameter	Units	Previously published values			This paper
		Waldmann and Strober (1969)	Kim et al. (2007)	Hattersley et al. (2013)	
k_{12}	day^{-1}	–	0.158	0.41	0.38
k_{21}	day^{-1}	–	0.156	0.51	0.42
k_{31}	day^{-1}	0.18	0.18†	0.13	0.16
V_{max}	µmol day^{-1}	68.6*	68.6*†	103	40
K_M	µmol	–	420**	530	270

*Assuming 70 kg human. **Assuming 3 l plasma volume. †Taken from Waldmann and Strober (1969).

parameters as follows: $k_{31} = 0.18$ day^{-1} (Waldmann and Strober, 1969), $V_{max}/w = 0.98$ µmol kg^{-1} (Waldmann and Strober, 1969), and $v_1/w = 0.042$ l kg^{-1} (Waldmann and Terry, 1990). By this approach the authors obtain a value of $K_M/v_1 = 140$ µmol l^{-1}. Assuming a plasma volume of 3 l this is equates to $K_M = 420$ µmol. The authors also estimate the parameters k_{12} and k_{21} by curve fitting to tracer experiment data from Waldmann and Terry (1990).

The parameter values provided by Hattersley et al. (2013) were obtained by methods described in Hattersley (2009). Hattersley (2009) estimates parameters k_{12} and k_{21} by fitting the model to tracer experiment data in Waldmann and Strober (1969). For the remaining model parameters, k_{31}, V_{max}, and K_M, the author takes a completely different approach, fitting the model directly to serum IgG data from an IgG myeloma patient, assuming a delayed logistic function to describe the production of tumor-produced IgG. For this approach, the parameters k_{12}, k_{21}, and v_1 were fixed.

In this paper data from a number of subjects have been used for parameter estimation: parameters k_{12} and k_{21} were estimated individually for seven subjects and parameters k_{31}, V_{max}, and K_M were estimated from the pooled data of around 40 subjects. In order to make predictions of IgG responses in IgG multiple myeloma that can be generalized across patients, a model which characterizes average, or expected, behavior is advantageous. One of the limitations of this study is that a full population approach has not been taken. Fitting the timecourse data individually and summarizing the parameter estimates by the sample mean or median can be seen as a two-stage approach, and fitting the FCR and $T_{1/2}$ data as though they arose from an individual is essentially a naive pooled approach; these approaches have been shown to be inferior to a full population approach (Wright, 1998). A population approach has not been taken here due to the limited data that are available in the literature. In future studies it may be possible with further experiments to apply a full population approach and gain information on the distribution of all parameters within the population. Furthermore, the work presented here could be enhanced with a simulation study in which parameters are estimated from synthetic data, in order to provide additional understanding of the identification problem and inform the design of future experiments.

In Section 3.4 we have shown how the model can be extended to simulate monoclonal IgG responses in IgG myeloma. The assumptions behind these simulations are discussed in detail in that section. In IgG myeloma patients the serum monoclonal IgG response is used as a surrogate for the tumor response to treatment; however the relationship between the tumor response and the monoclonal IgG response is inevitably influenced by the natural elimination of IgG from the body, which is known to have a non-constant relationship with serum IgG concentration. It is our intention that the model analyzed in this paper can be used in the future as the basis of more detailed investigations into the dynamics of IgG responses in IgG multiple myeloma; for example, is the prediction of long-term outcomes by monoclonal IgG responses influenced by the concentration-dependent and comparatively long half-life of the protein? Such future studies could impact upon how responses to treatment are assessed in patients.

In addition, the concentration-dependent elimination of IgG may be implicated in the pharmacokinetics of monoclonal antibody (mAb) agents for multiple myeloma, for example daratumumab, which is currently undergoing evaluation in patients with relapsed or refractory disease (Costello, 2017). Due to FcRn-mediated recycling, the longevity of daratumumab is predicted to be shorter in patients with high monoclonal IgG loads whereas low monoclonal IgG concentrations may favorably affect the pharmacokinetic profile of the agent, such that doses could be administered at less frequent intervals. The pharmacokinetics of various mAbs have been well studied; however the use of mAbs in multiple myeloma is recent and the dynamic response of the tumor-produced IgG in IgG myeloma patients adds an additional level of complexity. With the appropriate data it would be highly interesting from a pharmacological point of view to couple mathematical models of the mAb disposition and the tumor-produced IgG response, which is in turn directly affected by the efficacy of the agent.

IgG metabolism is implicated in other medical applications beyond patient monitoring in multiple myeloma, including antibody mediated rejection of transplants, infection and intravenous IgG (IVIG) therapy. In medical applications, mathematical models can be used to investigate biomedical systems *in silico*, allowing many scenarios and interventions to be simulated whilst avoiding the costs associated with human and animal experimentation. For biomedical applications, the parameter values used are of high importance as they can greatly affect conclusions drawn from simulations. In this paper structural identifiability analysis and sensitivity analysis are presented as a first stage in the model validation process, showing which model parameters are identifiable from which measured outputs. Together structural identifiability analysis and sensitivity analysis can be used to inform parameter estimation and improve confidence in the methodology used. In order for the model to applied in other biomedical scenarios, validation against patient data would be necessary. A limitation of the model analyzed here is that it may not contain a sufficient level of detail for all applications. Future work may involve comparing this simple model with more complex models that are based closely on the biological mechanisms, such as the model presented by Ferl et al. (2005).

5. CONCLUSIONS

In this research a previously published model of endogenous IgG metabolism in humans has been analyzed and parameter values estimated using limited data from the literature. The analyses herein provide an understanding of how parameter estimates have been obtained and sources of uncertainty; for future applications in which the parameter values themselves are of key importance, an understanding of how they were obtained and why is crucial. The parameterized model can have a wide-ranging impact in studies of endogenous IgG metabolism in biomedical applications, not limited to investigations of IgG as a response marker in IgG multiple myeloma, supporting therapeutic interventions and patient monitoring.

AUTHOR CONTRIBUTIONS

FK performed model analyses and parameter estimation. FK, MC, and NE wrote the manuscript. MC, NE, and SHarding initiated, coordinated, and supervised the work. XL provided discussion on the clinical application of the work. XL, BA, HA, OD, TD, GF, SG, SHarel, BH, VJ, PM, VR, SS, and CT provided data from the Intergroupe Francophone du Myélome (IFM) 2009-02 clinical trial. All authors reviewed and approved the final manuscript.

FUNDING

This research was supported by a Biotechnology and Biological Sciences Research Council (BBSRC) studentship, through the Midlands Integrative Biosciences Training Partnership (MIBTP), and an Engineering and Physical Sciences Research Council (EPSRC) Impact Acceleration Account (IAA) award.

REFERENCES

Anderson, D. H. (1983). *Compartmental Modeling and Tracer Kinetics, Volume 50 of Lecture Notes in Biomathematics*. Berlin; Heidelberg: Springer-Verlag.

Anderson, K. (2003). Multiple myeloma: how far have we come? *Mayo Clin. Proc.* 78, 15–17. doi: 10.4065/78.1.15

Appelqvist, H., Wäster, P., Kågedal, K., and Öllinger, K. (2013). The lysosome: from waste bag to potential therapeutic target. *J. Mol. Cell Biol.* 5, 214–226. doi: 10.1093/jmcb/mjt022

Banks, H. T., Dediu, S., and Ernstberger, S. L. (2007). Sensitivity functions and their use in inverse problems. *J. Inv. Ill-Posed Problems* 15, 683–708. doi: 10.1515/jiip.2007.038

Bellman, R., and Åström, K. (1970). On structural identifiability. *Math. Biosci.* 7, 329–339. doi: 10.1016/0025-5564(70)90132-X

Bradwell, A., Harding, S., Fourrier, N., Mathiot, C., Attal, M., Moreau, P., et al. (2013). Prognostic utility of intact immunoglobulin Ig'κ/Ig'λ ratios in multiple myeloma patients. *Leukemia* 27, 202–207. doi: 10.1038/leu.2012.159

Chen, Y., and Balthasar, J. P. (2012). Evaluation of a catenary PBPK model for predicting the *in vivo* disposition of mAbs engineered for high-affinity binding to FcRn. *AAPS J.* 14, 850–859. doi: 10.1208/s12248-012-9395-9

Costello, C. (2017). An update on the role of daratumumab in the treatment of multiple myeloma. *Ther. Adv. Hematol.* 8, 28–37. doi: 10.1177/2040620716677523

Deng, R., Meng, Y. G., Hoyte, K., Lutman, J., Lu, Y., Iyer, S., et al. (2012). Subcutaneous bioavailability of therapeutic antibodies as a function of FcRn binding affinity in mice. *MAbs* 4, 101–109. doi: 10.4161/mabs.4.1.18543

Durie, B. G., Harousseau, J.-L., Miguel, J. S., Bladé, J., Barlogie, B., Anderson, K., et al. (2006). International uniform response criteria for multiple myeloma. *Leukemia* 20, 1467–1473. doi: 10.1038/sj.leu.2404284

Fang, L., and Sun, D. (2008). Predictive physiologically based pharmacokinetic model for antibody-directed enzyme prodrug therapy. *Drug Metab. Dispos.* 36, 1153–1165. doi: 10.1124/dmd.107.019182

Ferl, G. Z., Wu, A. M., and DiStefano, J. J. (2005). A predictive model of therapeutic monoclonal antibody dynamics and regulation by the neonatal Fc receptor (FcRn). *Ann. Biomed. Eng.* 33, 1640–1652. doi: 10.1007/s10439-005-7410-3

Fronton, L., Pilari, S., and Huisinga, W. (2014). Monoclonal antibody disposition: a simplified PBPK model and its implications for the derivation and interpretation of classical compartment models. *J. Pharmacokinet. Pharmacodyn.* 41, 87–107. doi: 10.1007/s10928-014-9349-1

Garg, A., and Balthasar, J. P. (2007). Physiologically-based pharmacokinetic (PBPK) model to predict IgG tissue kinetics in wild-type and FcRn-knockout mice. *J. Pharmacokinet. Pharmacodyn.* 34, 687–709. doi: 10.1007/s10928-007-9065-1

Hall, A., and Yates, C. (2010). *Immunology (Fundamentals of Biomedical Science), 1st Edition*. Oxford: Oxford University Press.

Hansen, R. J., and Balthasar, J. P. (2003). Pharmacokinetic/pharmacodynamic modeling of the effects of intravenous immunoglobulin on the disposition of antiplatelet antibodies in a rat model of immune thrombocytopenia. *J. Pharm. Sci.* 92, 1206–1215. doi: 10.1002/jps.10364

Hattersley, J. G. (2009). *Mathematical Modeling of Immune Condition Dynamics: A Clinical Perspective*. Ph.D. thesis, University of Warwick.

Hattersley, J. G., Chappell, M. J., Zehnder, D., Higgins, R. M., and Evans, N. D. (2013). Describing the effectiveness of immunosuppression drugs and apheresis in the treatment of transplant patients. *Comput. Methods Programs Biomed.* 109, 126–133. doi: 10.1016/j.cmpb.2011.12.013.

Junghans, R. P., and Anderson, C. L. (1996). The protection receptor for IgG catabolism is the β2-microglobulin-containing neonatal intestinal transport receptor. *Proc. Natl. Acad. Sci. U.S.A.* 93, 5512–5516.

Kim, J., Hayton, W. L., Robinson, J. M., and Anderson, C. L. (2007). Kinetics of FcRn-mediated recycling of IgG and albumin in human: pathophysiology and therapeutic implications using a simplified mechanism-based model. *Clin. Immunol.* 122, 146–155. doi: 10.1016/j.clim.2006.09.001

Koulieris, E., Panayiotidis, P., Harding, S. J., Kafasi, N., Maltezas, D., Bartzis, V., et al. (2012). Ratio of involved/uninvolved immunoglobulin quantification by

Hevylite[TM] assay: clinical and prognostic impact in multiple myeloma. *Exp. Hematol. Oncol.* 1:9. doi: 10.1186/2162-3619-1-9

Kumar, S., Paiva, B., Anderson, K. C., Durie, B., Landgren, O., Moreau, P., et al. (2016). International Myeloma Working Group consensus criteria for response and minimal residual disease assessment in multiple myeloma. *Lancet Oncol.* 17, 328–346. doi: 10.1016/S1470-2045(16)30206-6

Leleu, X., Attal, M., Arnulf, B., Moreau, P., Traulle, C., Marit, G., et al. (2013). Pomalidomide plus low-dose dexamethasone is active and well tolerated in bortezomib and lenalidomide-refractory multiple myeloma: intergroupe francophone du myélome 2009-02. *Blood* 121, 1968–1975. doi: 10.1182/blood-2012-09-452375

Ng, C. M., Loyet, K. M., Iyer, S., Fielder, P. J., and Deng, R. (2014). Modeling approach to investigate the effect of neonatal Fc receptor binding affinity and anti-therapeutic antibody on the pharmacokinetic of humanized monoclonal anti-tumor necrosis factor-α IgG antibody in cynomolgus monkey. *Eur. J. Pharm. Sci.* 51, 51–58. doi: 10.1016/j.ejps.2013.08.033

Paiva, B., Pérez-Andrés, M., Vídriales, M.-B., Almeida, J., de las Heras, N., Mateos, M.-V., et al. (2011). Competition between clonal plasma cells and normal cells for potentially overlapping bone marrow niches is associated with a progressively altered cellular distribution in MGUS vs myeloma. *Leukemia* 25, 697–706. doi: 10.1038/leu.2010.320

Riviere, J. E. (2011). *Comparative Pharmacokinetics: Principles, Techniques, and Applications, 2nd Edition.* Hoboken, NJ: Wiley-Blackwell.

Rosenthal, K. S., and Tan, M. J. (2010). *Microbiology and Immunology, 3rd Edition.* Maryland Heights, MO: Mosby.

Routh, E. J. (1877). *A Treatise on the Stability of a Given State of Motion: Particularly Steady Motion.* London: MacMillan and Company.

Salmon, S. E., and Smith, B. A. (1970). Immunoglobulin synthesis and total body tumor cell number in IgG multiple myeloma. *J. Clin. Invest.* 49, 1114–1121. doi: 10.1172/JCI106327

Solomon, A., Waldmann, T., and Fahey, J. (1963). Clinical and experimental metabolism of normal 6.6 s γ-globulin in normal subjects and in patients with macroglobulinemia and multiple myeloma. *J. Lab. Clin. Med.* 62, 1–17.

Sullivan, P. W., and Salmon, S. E. (1972). Kinetics of tumor growth and regression in IgG multiple myeloma. *J. Clin. Invest.* 51, 1697–1708. doi: 10.1172/JCI106971

Thomaseth, K., and Cobelli, C. (1999). Generalized sensitivity functions in physiological system identification. *Ann. Biomed. Eng.* 27, 607–616.

Urva, S., Yang, V., and Balthasar, J. (2010). Physiologically based pharmacokinetic model for T84. 66: a monoclonal anti-CEA antibody. *J. Pharm. Sci.* 99, 1582–1600. doi: 10.1002/jps.21918

Waldmann, T. A., and Strober, W. (1969). Metabolism of immunoglobulins. *Prog. Allergy* 13, 1–110.

Waldmann, T. A., and Terry, W. D. (1990). Familial hypercatabolic hypoproteinemia. A disorder of endogenous catabolism of albumin and immunoglobulin. *J. Clin. Invest.* 86, 2093–2098. doi: 10.1172/JCI114947

Wright, P. (1998). Population based pharmacokinetic analysis: why do we need it; what is it; and what has it told us about anaesthetics? *Br. J. Anaesth.* 80, 488–501. doi: 10.1093/bja/80.4.488

Xiao, J. J. (2012). Pharmacokinetic models for FcRn-mediated IgG disposition. *J. Biomed. Biotechnol.* 2012:282989. doi: 10.1155/2012/282989

Yan, X., Chen, Y., and Krzyzanski, W. (2012). Methods of solving rapid binding target-mediated drug disposition model for two drugs competing for the same receptor. *J. Pharmacokinet. Pharmacodyn.* 39, 543–560. doi: 10.1007/s10928-012-9267-z

5

Quantifying the Frictional Forces between Skin and Nonwoven Fabrics

Kavinda Jayawardana [1], Nicholas C. Ovenden [1] and Alan Cottenden [2]*

[1] Department of Mathematics, University College London (UCL), London, UK, [2] Department of Bioengineering and Medical Physics, University College London (UCL), London, UK

When a compliant sheet of material is dragged over a curved surface of a body, the frictional forces generated can be many times greater than they would be for a planar interface. This phenomenon is known to contribute to the abrasion damage to skin often suffered by wearers of incontinence pads and bed/chairbound people susceptible to pressure sores. Experiments that attempt to quantify these forces often use a simple capstan-type equation to obtain a characteristic coefficient of friction. In general, the capstan approach assumes the ratio of applied tensions depends only on the arc of contact and the coefficient of friction, and ignores other geometric and physical considerations; this approach makes it straightforward to obtain explicitly a coefficient of friction from the tensions measured. In this paper, two mathematical models are presented that compute the material displacements and surface forces generated by, firstly, a membrane under tension in moving contact with a rigid obstacle and, secondly, a shell-membrane under tension in contact with a deformable substrate. The results show that, while the use of a capstan equation remains fairly robust in some cases, effects such as the curvature and flaccidity of the underlying body, and the mass density of the fabric can lead to significant variations in stresses generated in the contact region. Thus, the coefficient of friction determined by a capstan model may not be an accurate reflection of the true frictional behavior of the contact region.

Edited by:
Krasimira Tsaneva-Atanasova, University of Exeter, UK

Reviewed by:
Adam Jan Gadomski, University of Science and Technology, Poland
Luis Vilhena, University of Coimbra, Portugal

***Correspondence:**
Nicholas C. Ovenden
n.ovenden@ucl.ac.uk

Keywords: friction, nonwoven fabrics, shell theory, capstan equation, abrasion dermatitis, incontinence, free boundary

1. INTRODUCTION

Pressure ulcers are an area of localized cutaneous damage typically associated with pressure from bony protuberances on aged skin. They can develop when a large amount of pressure is applied to an area of skin over a short period of time or occur when less pressure is applied over a prolong period of time. When pressure is applied to soft tissue, it may result in completely or partially obstructed blood flow to the soft tissue, starving the tissue of oxygen and nutrients, which eventually leads to necrosis in the affected area, and thus an ulcer. Shear (i.e., constant and prolonged static friction) is also a cause, as it can pull on blood vessels that feed the skin, consequently restricting the blood flow. Pressure ulcers often occur in very sedentary individuals, such as those with impaired mobility (Maklebust and Sieggreen, 2001).

It is assumed that friction contributes to skin damage via stripping of the epidermal layer of the skin, creating an environment conducive to further skin damage due to friction. An alteration in the coefficient of friction increases the skin's adherence to the outside surface, which can eventually lead to wounds and infections. A publication by Murray et al. (2001) highlights many preventive measures for pressure ulcers. For example, to eliminate shear and friction, it is recommended that the exposed skin is covered by protective dressings, padding or sheepskin. For those who are bedridden, elevating the foot of bed to 20° is advised when sitting to prevent sliding as well as maintaining the head of the bed at the lowest possible elevation, consistent with the individual's medical condition and comfort.

Athletes, due to excessive movement, are also subject to repeated mechanical trauma to the skin, which is often painful (Bergfeld and Taylor, 1985). Conditions such as fissure of the nipple (Powell, 1983; Conklin, 1990) and friction blisters (Herring and Richie, 1990) are well documented traumas to the skin caused by friction. It is also established that friction plays a role in the development of dermatitis in other settings. In many cases, it is observed that friction damages the stratum corneum and the stratum basale to varying degrees (Wilkinson, 1985). Friction is also believed to play a significant role in incontinence-associated dermatitis with published literature on the coefficient of friction (measuring techniques and actual measurements) between human skin and fabrics in incontinence related publications (Berg, 1987; Cottenden et al., 2008a,b; Cottenden and Cottenden, 2009). The NHS estimates that between 3 and 6 million adults in the UK have some degree of urinary incontinence (Irwin et al., 2006), and the prevalence is set to increase due to an aging population. It is documented that the wearing of incontinence pads over prolong periods of time is a major cause of incontinence associated dermatitis. Although the pads absorb moisture, they can also act as a barrier that prevents water from escaping. This leads to over-hydration of the stratum corneum in the epidermis (i.e., the upper most part of the skin). Scheuplein and Blank (1971) found that an increase in skin hydration leads to an increase in the thickness of the stratum corneum, resulting in a weakening of the cell structure. Moreover, tests conducted on adults and on infants showed that over-hydration of the stratum corneum is responsible for a threefold increase in the coefficient of friction.

It is estimated that over 400, 000 individuals develop a new pressure ulcer annually in the UK (mainly the elderly) and approximately 51,000 of them will be admitted to hospital (Farage et al., 2007). A study conducted in 1993 showed that the cost to the NHS of treating pressure ulcers was around £180 – £321 million, approximately 0.4–0.8% of total health spending (Touche, 1993). However, more recently this figure was considered to be a substantial underestimate, even allowing for inflation (Bennett et al., 2004). Bennett et al. (2004) found that the cost of treating pressure ulcers in UK (excluding methicillin-resistant Staphylococcus aureus (MRSA), surgical interventions and litigation costs) ranges between £1.4 to £2.1 billion annually, which is over 4% of gross NHS expenditure. The costs were deduced by estimating the daily cost of the resources required to deliver protocols of care reflecting good clinical practice. In the USA, the Omnibus Budget Reconciliation Act in 1987 made it easier for claimants to prove that a provider had been negligent following the development of pressure ulcers. Between 1992 and 1996, the median settlement value following successful litigation for negligence regarding a pressure ulcer was $279,000 (Thomson and Brooks, 1999).

The above evidence strongly indicates that preventive measures are immensely important. Lowering the prevalence of pressure ulcers should lead to fewer resources being spent their treatment, thus reducing overall spending in the NHS. To design new products and protocols that reduce skin damage, there has been a great deal of experimental work aiming to accurately quantify the mechanical forces generated by fabrics and other materials in contact with human skin and its response (Dowson, 1996; Sivamani et al., 2003; Silver et al., 2003; Cottenden et al., 2008b). In these experiments it is vital to quantify the relationship between normal and tangential forces across the contact region and this is often achieved by determining a coefficient of friction (Gwosdow et al., 1986; Zhang and Mak, 1999; Cottenden et al., 2008b).

In recent years, the UCL Continence & Skin Technology group has developed and validated a novel method for measuring friction between fabrics and skin (Cottenden et al., 2008a,b), as well as developing mathematical models (Cottenden and Cottenden, 2009) to determine coefficients of friction. One aim of this research is to predict the magnitude of potentially damaging frictional forces generated between skin in contact with certain fabrics, such as the nonwoven materials used in incontinence pads. **Figure 1** shows the basic setup of the experiments and a typical force trace obtained by the tensometer measurements. The tensometer is used to measure the force needed to drag a strip of nonwoven fabric (30 mm wide in the work referenced here) over the volar forearm of a volunteer while a weight (an applied mass, m) is attached to the other end of the strip. The arm is supported such that its upper surface, in the vicinity of the fabric strip, is held horizontal and at the same height as the tensometer grips, while the other end of the strip hangs vertically in a plane perpendicular to the pull direction of the tensometer. The angle of the arc of contact between the arm and strip is typically $\pi/2$ although, in the group's work using arm phantoms (Cottenden et al., 2008a), angles other than $\pi/2$ were achieved by adjusting the height of the tensometer grips relative to the arm phantoms.

One remarkable aspect of the experimental results obtained is that their rather simple capstan-type model (Rao et al.,

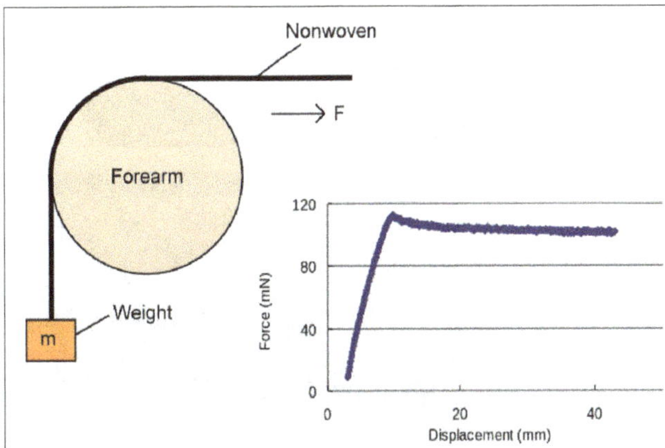

FIGURE 1 | Basic setup of experiments to determine the coefficients of friction and a plot of a typical tensometer trace from Cottenden et al. (2008a,b). Image courtesy of Dr. Sabrina Falloon, UCL.

2003; Jung et al., 2008), based only on the ratio of tensions at either end of the fabric strip and the arc of contact, appears able to produce reliable values for the coefficient of friction from experiments with various applied weights across different human subjects. This is despite the fact that during experiments significant skin deformation, rucking and wrinkling is often observed (**Figure 2**).

The capstan equation is the most perfect example of a belt-friction model, which describes behavior of a belt-like object moving over a rigid obstacle subjected to friction (Rao et al., 2003). Consider a membrane (i.e., a two-dimensional elastic body) with zero-Poisson's ratio or a *string* (i.e., a one-dimensional elastic body with arbitrary Poisson's ratio) over a *rough* rigid cylinder subject to appropriate boundary conditions such that the body in question is at *limiting equilibrium*, in other words it is at the point of slipping. This is a simple belt-friction problem and its properties are well-described by the capstan equation. However, we ask: what if our rigid contact body is no longer a cylinder, but some arbitrary geometry? Should a simple capstan equation apply to these geometries? Would a capstan-type equation still apply in the case where the underlying body is not rigid but *deformable*. Are the coefficients of friction obtained by a capstan equation reliable in such cases? These questions are the main focus of this paper and we examine the aspects of geometry and deformability of the underlying body separately in Sections 2 and 3 respectively.

In Section 2, we look at a thin membrane pulled dynamically at a constant speed over a rigid body. The section begins by introducing the capstan equation and highlighting some experimental data on arm phantoms that does not fit the predictions of the capstan model. We then extend an established model for Coulomb's law of static friction (Kikuchi and Oden, 1988) to curvilinear coordinates and use it in conjunction with a numerical model of a thin membrane pulled dynamically over a rigid body to investigate how the calculated coefficient of friction varies with (i) membrane parameters such as Poisson's ratio, Young's modulus and mass density; (ii) the speed and the applied tensions at the membrane edges; and (iii) the underlying geometry of the body, specifically, the Gaussian curvature. In Section 3, we examine the behavior of shells supported by elastic foundations when subjected to a friction condition. Faced with a free-boundary problem at the contact region, we use the same model for Coulomb's law of static friction (Kikuchi and Oden, 1988) to derive a more computationally tractable displacement-based static friction condition. We then take the overlying shell theory and use the displacement-based friction condition to transform the model into a constrained elastic two-body contact problem and explicitly derive the governing equations and the boundary conditions for the static friction problem of a thin shell-membrane on an elastic foundation. Finally, we present some numerical results to examine how the thickness

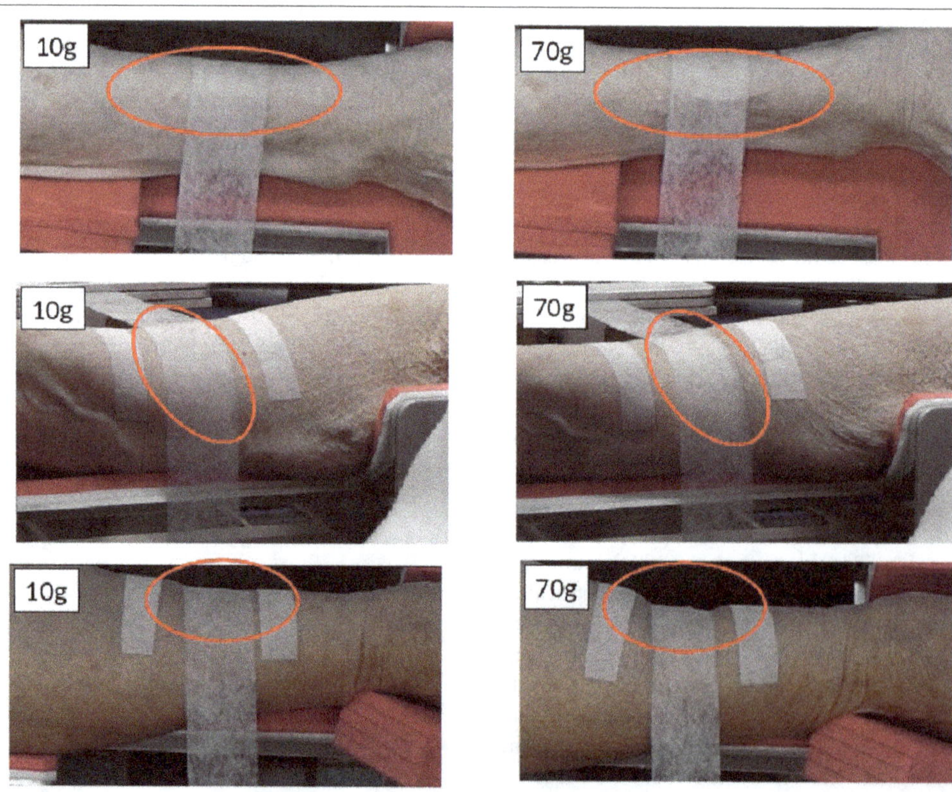

FIGURE 2 | Pictures of deformation in the skin and underlying soft tissue generated during measurements of friction between strips of nonwoven fabric and the volar forearms of female volunteers. Forearms were held horizontal while a tensometer pulled nonwoven strips over the skin surface with an applied mass secured to the opposite (hanging) end of the fabric (10 and 70g). Images courtesy of Dr. Sabrina Falloon, UCL. The work was conducted with the approval of London Stanmore Research Ethics Committee and The Whittington Hospital NHS R&D office, September 2011.

and elasticity of the foundation affect the displacement and shear stress across the contact region. Discussion and conclusions follow in Section 4.

2. MODELING A NON-WOVEN FABRIC AS A MEMBRANE SUPPORTED BY A RIGID FOUNDATION WITH FRICTION

2.1. Capstan Equation and Applications in Friction Modeling

The capstan equation or, as otherwise known, Euler's equation of tension transmission, is the relationship governing the maximum applied tension T_{max} with respect to the minimum applied tension T_0 of an elastic string wound around a rough cylinder. The governing equation is given by

$$T_{max} = T_0 \exp(\mu_F \theta)\,, \quad (1)$$

where θ is the contact angle and μ_F is the coefficient of friction. By *string* we mean a one-dimensional elastic body and *rough* is an engineering term implying that the contact area exhibits friction. Note that the coefficient of friction is the physical ratio of the shear force and the normal force between two contacting bodies. In engineering, the capstan equation describes a body under a load, in equilibrium, involving friction between rope and a wheel-like circular object, and thus it is widely used to analyse the tension transmission behavior of cable-like bodies in contact with circular profiled surfaces (Jung et al., 2008; Baser and Konukseven, 2010) as well as in the field of robotics (Behzadipour and Khajepour, 2006).

As an example of the application of a capstan-type equation to experimental results, consider the results of Cottenden et al. (2008a) where the coefficient of friction is determined from experiments as that shown in **Figure 1** of a non-woven fabric strip in contact with various arm phantoms made from Plaster of Paris and covered in Neoprene. Note, in particular, Figure 11 of Cottenden et al. (2008a) which shows the coefficients of friction obtained with different geometries, applied weights, and contact angles. While the capstan equation proves quite successful in obtaining coefficients of friction, the authors observe a steady increase in the mean coefficient of friction as the applied weight increases. Such dependence contradicts the assumptions of Equation (1). There is also variation in the coefficients of friction measured across different geometries and contact angles which seems to get wider as the applied mass is reduced. The authors acknowledge the apparent dependence of the coefficient of friction on the applied weight in Cottenden et al. (2008a) but highlight that the mean variation is small compared to the scatter of the data. The authors suggest that the departure from the capstan equation is likely to be caused by an interaction between the Neoprene on the underlying body and the moving nonwoven fabric at large tension.

The raw data of the case of a fabric strip in contact with an arm phantom of cylindrical cross section with a $\frac{127}{360}\pi$ contact angle can be found in Karavokiros' masters thesis (see table 2a of Karavokiros, 2007); for convenience, this raw data is reproduced here in **Table 1**. The table shows five repeated measurements involving dragging the fabric strip over the arm phantom with different applied masses (m in **Figure 1**). **Figure 3** is a plot using these data of the measured tension ratio $\delta\tau = T_{max}/T_0$, vs. the applied mass, where the mean ratio obtained for each applied mass is also shown. Note that Equation (1) implies that the tension ratio is constant for all applied masses, i.e., $\delta\tau = \exp(\mu_d\theta_0)$, where μ_d and θ_0 are constants. However, **Figure 3** suggests as the applied mass increases, the tension ratio increases also. Of course, such an effect could be attributed to a number of factors, including experimental errors, but one possibility is that the standard capstan equation is simply not valid in these cases. To test this, we now develop two, more sophisticated, three-dimensional numerical models of a thin compliant sheet over an underlying body and vary the geometry and material properties of both the thin sheet and underlying body, as well as the applied tensions.

TABLE 1 | Tensometer readings: Plaster of Paris cylinder covered in Neoprene with $\frac{127}{360}\pi$ contact angle, where g is the acceleration due to gravity.

Applied Mass	Recorded Tension (F) 10^{-3}N					
(m) in grams	Repeat Measurements					Mean
10	16.0g	15.0g	15.0g	15.0g	16.0g	15.6g
30	51.0g	54.0g	51.0g	50.0g	51.0g	51.4g
50	88.0g	87.0g	89.0g	87.0g	90.0g	88.2g
70	125g	124g	128g	122g	124g	125g

Applied mass m and tension F refer to the experimental schematic in ***Figure 1****. Data courtesy of Karavokiros (2007).*

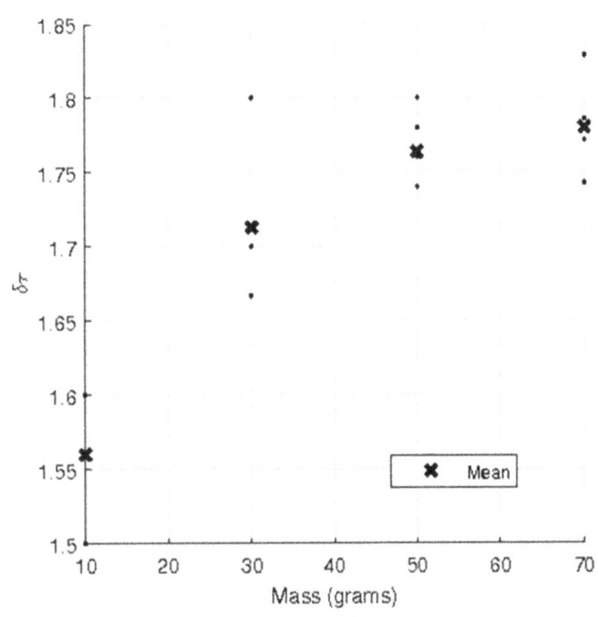

FIGURE 3 | Tension ratio against applied mass from the raw experimental data in Table 1 obtained from Karavokiros (2007).

2.2. Belt-Friction Model

In this section, we derive a pure-traction belt-friction model to describe the behavior of dynamic membranes supported by static rigid foundations.

Let $\omega \subset \mathbb{R}^2$ be a simply connected open bounded domain with the sufficiently smooth boundary $\partial\omega$ and let $\boldsymbol{\sigma} \in C^2(\omega; \mathbf{E}^3)$ be an injective immersion. Now assume that an isotropic elastic membrane is in contact with a rigid surface with positive mean curvature initially such that, in the stress-free configuration of the membrane, the contact area can be parameterized by the immersion $\boldsymbol{\sigma}(x^1, x^2)$ in curvilinear coordinates. Now we assert that the membrane is dynamic but the contact area remains static and constant as ω. Invoking a similar terminology to that of Cottenden and Cottenden (2009), the governing Cauchy momentum equations for the membrane stress $\tau^{\alpha\beta}$ $(\alpha, \beta = 1, 2)$ in tangential and normal directions can be derived thus

$$\nabla_\alpha \tau^{\alpha\beta}(\boldsymbol{u}) + f_r^\beta(\boldsymbol{u}) + g_r^\beta = \varrho\left(\partial_{tt}u^\beta + \Gamma^\beta_{\alpha\gamma}\partial_t u^\alpha \partial_t u^\gamma\right), \quad (2)$$

$$F_{[\mathrm{II}]\alpha\gamma}\tau^{\alpha\gamma}(\boldsymbol{u}) + f_r^3(\boldsymbol{u}) + g_r^3 = \varrho F_{[\mathrm{II}]\alpha\gamma}\partial_t u^\alpha \partial_t u^\gamma, \quad (3)$$

with the boundary conditions

$$n_\alpha \tau^{\alpha\beta}(\boldsymbol{u})|_{\partial\omega} = \tau_0^\beta|_{\partial\omega}, \quad (4)$$

where $\boldsymbol{u} \in C^2(\omega; \mathbb{R}^2)$ is the displacement field and ϱ is the mass density of the membrane, $\boldsymbol{g}_r \in C^0(\omega; \mathbb{R}^3)$ is an external loading field, $f_r^\beta(\boldsymbol{u})$ are the shear force densities, $f_r^3(\boldsymbol{u})$ is the normal reaction density, $\boldsymbol{n} \in C^0(\omega; \mathbb{R}^2)$ is the unit outward normal to the boundary and $\boldsymbol{\tau}_0 \in C^0(\omega; \mathbb{R}^2)$ is the traction field applied to the boundaries of the membrane. The covariant *second fundamental form tensor* of $\boldsymbol{\sigma}$ with respect to the curvilinear coordinates is defined as

$$F_{[\mathrm{II}]\alpha\beta} = N_i \partial_{\alpha\beta}\sigma^i, \ \forall\, \alpha, \beta \in \{1, 2\}.$$

where $\boldsymbol{N} = ||\partial_1\boldsymbol{\sigma} \times \partial_2\boldsymbol{\sigma}||^{-1}(\partial_1\boldsymbol{\sigma} \times \partial_2\boldsymbol{\sigma})$ is the unit normal to the surface $\boldsymbol{\sigma}$, $\times$ is the Euclidean cross product and $||\cdot||$ is the Euclidean norm.

Notice that this is a pure-traction problem. This implies that the boundary traction field $\boldsymbol{\tau}_0$ cannot be arbitrary chosen. To proceed, we assume that the velocity, the acceleration and the force density fields are known and fixed prior to the problem. Typically here, in line with the tensometer experiments, we assume that the membrane moves at a given constant speed. The time-dependent translational part of the displacement is thus easily subtracted out, leaving the residual displacement u^β. We now invoke the compatibility condition for pure-traction problems (see Section 1.3.4 of Necas et al., 2011 or Section 1.8 of Ciarlet, 2000) to find

$$\begin{aligned}&\int_{\partial\omega} \tau_0^\beta w_\beta \, d(\partial\omega)\\ &+ \int_\omega \left(f_r^\beta(\boldsymbol{u}) + g_r^\beta - \varrho\left(\partial_{tt}u^\beta + \Gamma^\beta_{\alpha\gamma}\partial_t u^\alpha\partial_t u^\gamma\right)\right) w_\beta \, d\omega = 0,\\ &\forall\, \boldsymbol{w} \in \{\boldsymbol{v} \in \boldsymbol{H}^1(\omega) \mid \boldsymbol{\epsilon}(\boldsymbol{v}) = \boldsymbol{0}\},\end{aligned} \quad (5)$$

where $\boldsymbol{\epsilon}(\cdot)$ is the strain tensor of a true-membrane. The compatibility condition implies that the internal forces are balanced by all the applied external forces.

Suppose now that the contact area is rough and the friction law governing this region is given by the model in chapter of Kikuchi and Oden (1988) for Coulomb's law of static friction for the slip case. For a flat contact surface with normal (compressive) and tangential stress fields given by σ_n and $\boldsymbol{\sigma}_T$ respectively, Coulomb's law of static friction can be expressed by the following relationship between these surface stresses and the tangential displacement field $\boldsymbol{u}_T$:

$$\begin{aligned}|\boldsymbol{\sigma}_T| &< \nu_F|\sigma_n| \Rightarrow \boldsymbol{u}_T = 0,\\ |\boldsymbol{\sigma}_T| &= \nu_F|\sigma_n| \Rightarrow \boldsymbol{u}_T = -\lambda\boldsymbol{\sigma}_T \quad \text{for some } \lambda \geqslant 0.\end{aligned}$$

Here, ν_F is the coefficient of friction in respect to Coulomb's law of friction. Kikuchi and Oden point out that it is not possible to mathematically analyse a variational problem for Coulomb's law using conventional mathematical methods. As a result, the question of existence of solutions to such friction problems remains an open one. To circumvent this difficulty, Kikuchi and Oden propose the following regularized version of the static friction law involving a small parameter ε:

$$-\boldsymbol{\sigma}_T = \nu_F|\sigma_n|\frac{\boldsymbol{u}_T}{|\boldsymbol{u}_T|} \quad \text{if } |\boldsymbol{u}_T| \geqslant \varepsilon, \quad (6)$$

$$-\boldsymbol{\sigma}_T = \nu_F|\sigma_n|\frac{\boldsymbol{u}_T}{\varepsilon} \quad \text{if } |\boldsymbol{u}_T| < \varepsilon. \quad (7)$$

In the limit $\varepsilon \to 0$, Coulomb's original law of static friction is recovered. For the slip case, we assume the first condition holds for large tangential displacements which, on converting this to a membrane in contact with a rigid surface in curvilinear coordinates, yields

$$f_r^\beta(\boldsymbol{u}) + \nu_F\left(u_\alpha u^\alpha\right)^{-\frac{1}{2}} u^\beta f_r^3(\boldsymbol{u}) = 0. \quad (8)$$

We now rearrange the compatibility condition Equation (5) and use Coulomb's law of friction Equation (8) to find a relationship between the coefficient of friction and the external loadings, given by

$$\begin{aligned}&\nu_F \int_\omega \frac{\left(F_{[\mathrm{II}]\beta\gamma}\tau^{\beta\gamma}(\boldsymbol{u}) + g_r^3 - \varrho F_{[\mathrm{II}]\beta\gamma}\partial_t u^\beta\partial_t u^\gamma\right)}{\left(u_\delta u^\delta\right)^{\frac{1}{2}}} u^\alpha w_\alpha \, d\omega\\ &+ \int_{\partial\omega} \tau_0^\alpha w_\alpha \, d(\partial\omega)\\ &+ \int_\omega g_r^\alpha w_\alpha - \varrho\left(\partial_{tt}u^\alpha + \Gamma^\alpha_{\gamma\delta}\partial_t u^\gamma\partial_t u^\delta\right) w_\alpha \, d\omega = 0,\\ &\forall\, \boldsymbol{w} \in \{\boldsymbol{v} \in \boldsymbol{H}^1(\omega) \mid \boldsymbol{\epsilon}(\boldsymbol{v}) = \boldsymbol{0}\}.\end{aligned} \quad (9)$$

The residual displacement u^β and $f_r^j(\boldsymbol{u})$ give us five unknowns and Equations (2), (3), and (8) provide us with five equations. Thus, the system is fully determined with boundary conditions Equation (4). Furthermore, Equation (9) provides us with a stronger system as if the coefficient of friction is unknown then a known traction can close the system and vice versa.

While it is a closed system, it is not straightforward to prove that a solution exists for this model. The problem of proving the existence of solutions arises from the function $f_r^\beta(\boldsymbol{u})$, as it is a function of both $\boldsymbol{u}$ and ∇u. If $f_r^\beta(\boldsymbol{u})$ was purely a function of $\boldsymbol{u}$, then the existence of solutions may be proved by variational methods for semi-linear elliptic equations (see Badiale and Serra, 2010) but with the ∇u dependence, our model is not even a variational problem. Nevertheless, a numerical finite-difference solution is pursued in the next section.

2.3. Numerical Analysis

To conduct numerical experiments, assume that we are dealing with a surface of revolution case where both the contact surface and the unstressed membrane are parameterized by the same immersion. Let this immersion be $\boldsymbol{\sigma}(x^1,x^2) = (x^1,\varphi(x^1)\sin(x^2),\varphi(x^1)\cos(x^2))_E$, where $x^1 \in (0,l)$ and $x^2 \in (-\frac{1}{2}\pi,0)$. To permit changes in lateral curvature we assert that $\varphi(x^1) = r_0 - 16c(l^{-1}x^1 - \frac{1}{2})^4$, where, initially, to keep the contact area as a surface of positive mean curvature, c is a positive parameter with $c < r_0$. Note that l, r_0 are some positive real constants that are specified later. With some calculations, we find the first fundamental form tensor to be $F_{[I]} = \text{diag}((\psi_1)^2,(\psi_2)^2)$, where $\psi_1 = (1+(\varphi'(x^1))^2)^{\frac{1}{2}}$ and $\psi_2 = \varphi(x^1)$. With a few more calculations one can find that

$$\Gamma^1_{11} = (\psi_1)^{-1}\partial_1\psi_1\,,\quad F_{[II]1}{}^{1} = (\psi_1)^{-1}\varphi''(x^1)\left(1+(\varphi'(x^1))^2\right)^{-1}\,,$$
$$\Gamma^2_{21} = (\psi_2)^{-1}\partial_1\psi_2\,,\quad F_{[II]2}{}^{2} = -(\psi_2)^{-1}\left(1+(\varphi'(x^1))^2\right)^{-\frac{1}{2}}\,,$$

where $\Gamma^\gamma_{\alpha\beta}$ are the Christoffel symbols of the second kind and $F_{[II]}$ is the second fundamental form tensor. Now given that $\boldsymbol{u} = (u^1(x^1,x^2),u^2(x^1,x^2))$ is the displacement field, one can derive the following:

$$\begin{aligned}
\nabla_1 u^1 &= \partial_1 u^1 + \Gamma^1_{11}u^1\,,\\
\nabla_1 u^2 &= \partial_1 u^2 + \Gamma^2_{21}u^2\,,\\
\nabla_2 u^1 &= \partial_2 u^1 - (\psi_1)^{-2}(\psi_2)^2\Gamma^2_{21}u^2\,,\\
\nabla_2 u^2 &= \partial_2 u^2 + \Gamma^2_{22}u^2\,.
\end{aligned}$$

Now assume that our membrane is subjected to the acceleration of gravity i.e., subject to $(0,0,-g)_E$ in Cartesian coordinates. With coordinate transforms, from Euclidean to curvilinear, one may re-express acceleration due to gravity in the curvilinear coordinates as gJ, where

$$J = \left(-\varphi'(x^1)(\psi_1)^{-2}\cos(x^2),\varphi^{-1}(x^1)\sin(x^2),-(\psi_1)^{-1}\cos(x^2)\right)\,,$$

and x^2 is the angle that the vector $(\psi_1,\psi_2,0)$ makes with the vector $(0,0,1)_E$.

Now given that ϱ is the mass density, $(0,0,0)$ is the acceleration field and $(0,(\psi_2)^{-1}V,0)$ is the velocity field of the membrane, then we can express the governing equations of the membrane as

$$(\Lambda+\mu)\partial^1\left(\nabla_\alpha u^\alpha\right) + \mu\Delta u^1 + \varrho gJ^1 + f_r^1(\boldsymbol{u}) = -\varrho(\psi_1)^{-2}\Gamma^2_{21}V^2\,,$$
$$(\Lambda+\mu)\partial^2\left(\nabla_\alpha u^\alpha\right) + \mu\Delta u^2 + \varrho gJ^2 + f_r^2(\boldsymbol{u}) = 0\,,$$

and

$$\begin{aligned}
&\left((\Lambda+2\mu)\left(\partial_1 u^1 + \Gamma^1_{11}u^1\right) + \Lambda\left(\partial_2 u^2 + \Gamma^2_{21}u^1\right)\right)F_{[II]1}{}^{1}\\
&+\left(\Lambda\left(\partial_1 u^1 + \Gamma^1_{11}u^1\right) + (\Lambda+2\mu)\left(\partial_2 u^2 + \Gamma^2_{21}u^1\right)\right)\\
&F_{[II]2}{}^{2} + \varrho gJ^3 + f_r^3(\boldsymbol{u}) = \varrho F_{[II]2}{}^{2}V^2\,,
\end{aligned}$$

where $\Lambda = 2\lambda\mu(\lambda+2\mu)^{-1}$ and λ and μ are the first and second Lamé parameters respectively. Assuming that our contact area is rough, the final governing equation required is the friction law Equation (8) where the coefficient of friction ν_F is considered to be an unknown.

Now divide the boundary into sub-boundaries, so that

$$\begin{aligned}
\partial\omega_f &= \left\{\{0\}\times\left(-\frac{1}{2}\pi,0\right)\right\}\cup\left\{\{l\}\times\left(-\frac{1}{2}\pi,0\right)\right\}\\
\partial\omega_{T_0} &= \left\{[0,l]\times\left\{-\frac{1}{2}\pi\right\}\right\}\,,\\
\partial\omega_{T_{\max}} &= \{\{[0,l]\times\{0\}\}\,,
\end{aligned}$$

and we assert that the boundary conditions are

$$\begin{aligned}
&(\psi_1)^2\partial_2 u^1 + (\psi_2)^2\partial_1 u^2|_{\partial\omega} = 0\ (\text{zero-Robin})\,,\\
&(\Lambda+2\mu)\left(\partial_1 u^1 + \Gamma^1_{11}u^1\right) + \Lambda\left(\partial_2 u^2 + \Gamma^2_{21}u^1\right)|_{\partial\omega_f}\\
&\qquad = 0\ (\text{zero-Robin})\,,\\
&\Lambda\left(\partial_1 u^1 + \Gamma^1_{11}u^1\right) + (\Lambda+2\mu)\left(\partial_2 u^2 + \Gamma^2_{21}u^1\right)|_{\partial\omega_{T_0}}\\
&\qquad = \tau_0\ (\text{traction})\,,\\
&\Lambda\left(\partial_1 u^1 + \Gamma^1_{11}u^1\right) + (\Lambda+2\mu)\left(\partial_2 u^2 + \Gamma^2_{21}u^1\right)|_{\partial\omega_{T_{\max}}}\\
&\qquad = \tau_{\max}\ (\text{traction})\,,
\end{aligned}$$

where $\tau_{\max} > \tau_0$ are positive real constants.

Finally, we take Equation (9) and modify it ever so slightly to make it easier to numerically model, thus obtaining the following relation,

$$\begin{aligned}
&(\tau_{\max}-\tau_0)\int_0^l \psi_1\psi_2\,dx^1\\
&+\varrho\int_\omega\left(gJ^\alpha\psi_\alpha + (\psi_1)^{-1}\Gamma^2_{21}V^2\right)\psi_1\psi_2\,dx^1dx^2\\
&-\sqrt{2}\,\nu_F\int_\omega f_r^3(\boldsymbol{u})\psi_1\psi_2\,dx^1dx^2 = 0\,.
\end{aligned}\tag{10}$$

Now we are ready to conduct some numerical experiments. Our goal in this section is to investigate how variables such as the Gaussian curvature, Young's modulus, Poisson's ratio, speed and the mass density of the membrane and the traction may affect the value of the coefficient of friction obtained. Note that for our experiments, we keep the values $\tau_0 = 1$, $l = 1$, $r_0 = 1$ and $g = 9.81$ fixed.

To conduct numerical experiments, we employ a second-order accurate finite-difference method in conjunction with Newton's method for nonlinear systems. On discretising the domain, as we are dealing with curvilinear coordinates, we find that $\Delta x^2 \leq \psi_0 \Delta x^1$, for all $\psi_0 \in \{\psi_1/\psi_2 \mid x^1 \in [0, l]\}$ where Δx^β is a small increment in the x^β direction. For our purposes we use $\Delta x^2 = \frac{1}{N-1}$ and $\psi_0 = \psi_1/\psi_2|_{x^1=\frac{1}{2}l}$, where $N = 250$. We also choose to terminate our iterating process once $|1 - \nu_{Fm+1}/\nu_{Fm}|$ falls below a certain specified tolerance, where ν_{Fm} is the mth iterative solution for the coefficient of friction. Note that to numerically model Equation (10), we use the prismoidal formula (Meserve and Pingry, 1952). Unfortunately, as this is a pure-traction problem iterative schemes can be highly unstable and so, to ensure convergence, the condition $u^2|_{\partial\omega_{T_0}} \leq 0$ is strictly enforced.

We initially ran the numerical code using the following values: $\tau_{\max} = \frac{3}{2}$, $c = 0$, $E = 10^3$, $\nu = \frac{1}{4}$, $V = 0.01$, $\varrho = 0.01$ and with a grid of 160×250 points. The grid size N was varied to confirm accuracy of the converged solution. The coefficient of friction calculated in this case is $\nu_F = 0.195$ to three significant figures. We now proceed to investigate how the variation in certain parameters may change this value and thus see whether this differs from the value predicted by the classical capstan model Equation (1).

We begin by varying the tension applied to the membrane. **Figure 4** is calculated with the values of $\tau_{\max} \in \{1.25, 1.30, 1.35, \ldots 2.00\}$, $c = 0$, $E = 10^3$, $\nu = \frac{1}{4}$, $V = 0.01$ and $\varrho = 0.01$. It shows that as the tension ratio increases, the coefficient of friction also increases. This is intuitive because, as the maximum applied tension increases, the coefficient of friction must increase to maintain a constant speed. Equation (1) predicts a similar trend and so agreement with the classical capstan model seems quite good although the figure shows that the numerical code produces consistently lower values. The principal reason for the discrepancy is the non-zero mass density of the fabric which is investigated below.

Varying Poisson's ratio ν, the Young's modulus E and the speed of the membrane V do not lead to any significant changes in the coefficient of friction. But when we examine varying the mass density of the fabric we find significant alterations to the coefficient of friction determined by the model. **Figure 5** is calculated for three different applied tensions: $\tau_{\max} \in \{1.50, 1.75, 2.00\}$, with values $c = 0$, $E = 10^3$, $\nu = \frac{1}{4}$, $V = 0.01$ and $\varrho \in \{0, 0.003, 0.006, \ldots, 0.03\}$. This shows as the mass density (with respect to the volume) of the membrane increases, the coefficient of friction decreases markedly in all cases. As $\rho \to 0$ the coefficient of friction obtained from Equation (1), e.g., $(2/\pi)\ln(3/2) = 0.258\ldots$, is attained, but even a small mass density, $\rho = O(10^{-3})$ for instance, can lead to the accuracy being significantly reduced to only one significant figure. This feature is ignored in typical capstan equation calculations.

Finally, let us examine what effect varying the Gaussian curvature K has on the coefficient of friction. Note that, for our given geometry, K is a function of x^1, and thus it varies across the contact surface. As our curvature parameter c changes the Gaussian curvature changes across the contact surface. For $c = 0$, $K = 0$ everywhere. For $c > 0$ the contact surface is a barrel-shape with non-negative (i.e., positive or zero) Gaussian curvature whereas for $c < 0$ the contact surface is a saddle-shape with non-positive (i.e., negative or zero) Gaussian curvature across the contact area. This implies that there exists a positive correlation between c and K, i.e., as c increases so does K and so we investigate how c relates to ν_F. **Figure 6** is calculated for non-negative Gaussian curvatures with the values of $\tau_{\max} \in \{1.50, 1.75, 2.00\}$, and $c \in \{0, \frac{1}{40}, \frac{2}{40}, \frac{3}{40}, \frac{4}{40}, \frac{5}{40}, \frac{6}{40}, \frac{7}{40}, \frac{8}{40}, \frac{9}{40}\}$, $E = 10^3$, $\nu = \frac{1}{4}$, $V = 0.01$, and $\varrho = 0.01$. The three plots in the figure for different tension ratios demonstrate that as c increases (i.e., Gaussian curvature increases) the coefficient of friction decreases and there exists an optimum value of c, where the coefficient of friction is at a minimum, given that all other variables are constant. For our simulations, this value is observed around $c = \frac{1}{8}$. The initial decrease in ν_F as c increases is intuitive. To illustrate this, consider a membrane pulled over a surface with high curvature. The higher curvature would leads to higher normal reaction force which, in turn, results in a higher friction force and finally a higher maximum applied tension. Also inclusion of a nonzero lateral curvature (note that we are considering a case with two positive principal curvatures) means that a higher maximum applied tension is required to support the strains in the lateral direction. Thus, for even a relatively small coefficient of friction, a higher maximum applied tension can be observed. Hence, if one kept every variable fixed, except for the Gaussian curvature and the coefficient of friction, then one would expect to see a low coefficient of friction for a high Gaussian curvature. However, the existence of a minimum ν_F for positive c is a surprising outcome suggesting a possible optimum Gaussian curvature that can lead to a minimum coefficient of friction. Conversely, it is possible numerically to extend our range of c values to include $\{-\frac{1}{40}, -\frac{2}{40}, -\frac{3}{40}, -\frac{4}{40}\}$ that represent saddle-type contact regions. These simulations suggest that such saddle-type contact regions lead to significantly higher values of ν_F than for the zero-curvature case (**Figure 7**). Extending that range yet further to $c = -0.02$ (not shown) suggests that no *maximum* coefficient of friction is attained but that the coefficient of friction continues to increase as c becomes more negative. Such intriguing variations in ν_F captured by our model here cannot be simulated by the classical capstan model Equation (1).

3. SHELL-MEMBRANES SUPPORTED BY ELASTIC FOUNDATIONS WITH STATIC FRICTION

In this section, we examine the effect that the deformation of the underlying substrate has on the frictional forces generated in the region of contact. To do this, we derive a shell-membrane model to describe the behavior of an overlying compliant sheet on an elastic foundation subjected to static friction. However, such a computation requires our frictional law (e.g., Coulomb's law of static friction) to be imposed on a *free boundary* as the displacement across the contact region is unknown.

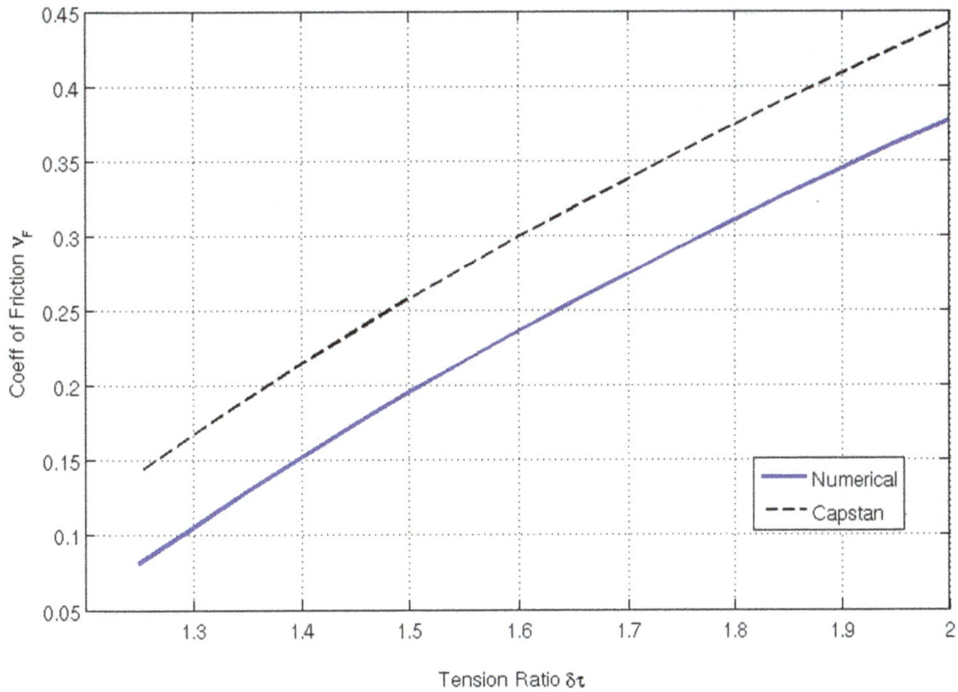

FIGURE 4 | Coefficient of friction relative to $\delta\tau = \tau_{max}/\tau_0$ calculated using our numerical model and compared to the standard capstan result: $\nu_F = \frac{2}{\pi}\ln\delta\tau$.

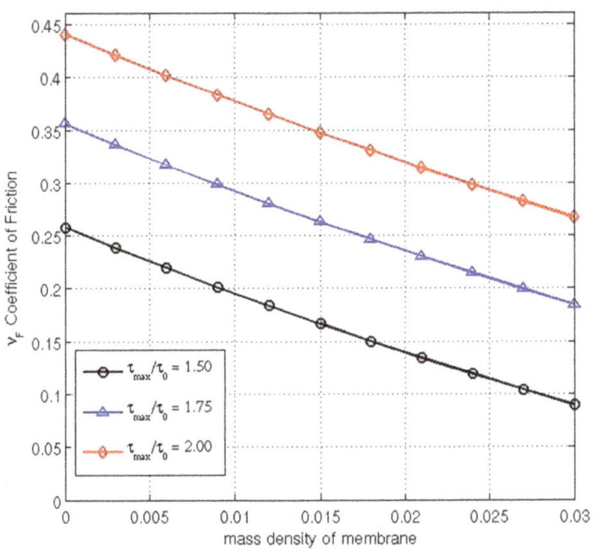

FIGURE 5 | Coefficient of friction relative to ρ for three different values of τ_{max}/τ_0 = [1.50,1.75,2.00].

3.1. Deriving a Displacement-Based Frictional Law

Recall Kikuchi and Oden's model (chapter of Kikuchi and Oden, 1988) for Coulomb's law of friction for slip that we extended to curvilinear coordinates Equation (8). For the static case, by eliminating the regularization parameter ε from their original Equations, (6) and (7), the friction law can be expressed in the following form

$$T_3^{\beta}(\boldsymbol{u}) + \nu_F\,(g_{33})^{\frac{1}{2}}\,(u_\alpha u^\alpha)^{-\frac{1}{2}}\,T_3^3(\boldsymbol{u})u^\beta \leq 0.$$

Unlike the rigid substrate case, this equation is extremely difficult to impose on a free contact boundary between shell-membrane and substrate in its present form. Therefore, a more computationally tractable displacement-based approximation is sought. Assume that $u^\beta \geq 0$ and contract the above equation with u^β. Noting that in our framework $g_{33} = 1$, we thus find

$$\mu\left(u^\alpha\bar{\nabla}_\alpha u^3 + \frac{1}{2}\bar{\nabla}_3(u_\alpha u^\alpha)\right) + \nu_F(u_\alpha u^\alpha)^{\frac{1}{2}}\left(\lambda\bar{\nabla}_\alpha u^\alpha + (\lambda + 2\mu)\bar{\nabla}_3 u^3\right) \leq 0\,.$$

Now assume that this body is in contact with an elastic foundation, thus it permits normal displacements at the boundary, so we assert that only normal derivatives are of any consequence. Hence, we may approximate the above relation as

$$\mu\bar{\nabla}_3(u_\alpha u^\alpha)^{\frac{1}{2}} + \nu_F(\lambda + 2\mu)\bar{\nabla}_3 u^3 \leq 0\,.$$

To simplify matters further, we assert that the condition is independent of any elastic properties of the overlying body. This may be achieved by assuming $\lambda = 0$, and thus we find

$$\bar{\nabla}_3\left((u_\alpha u^\alpha)^{\frac{1}{2}} + 2\nu_F u^3\right) \leq 0\,.$$

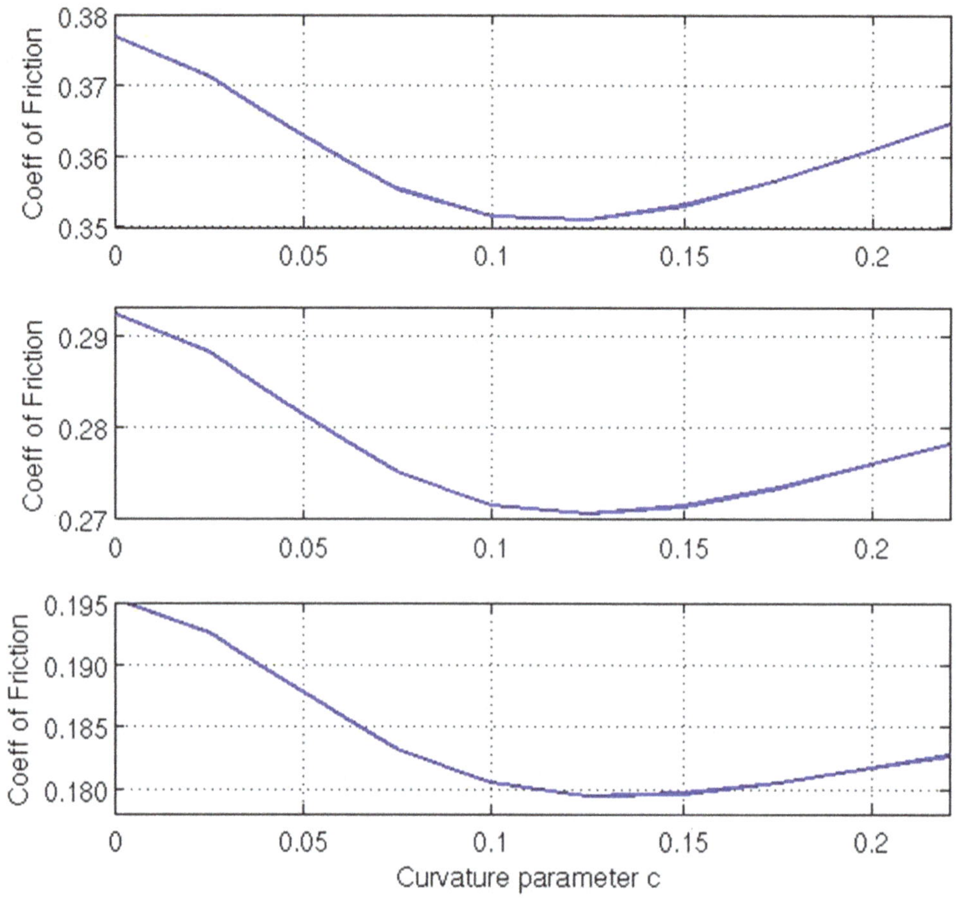

FIGURE 6 | Coefficient of friction relative to lateral curvature parameter *c* where the geometry has non-negative Gaussian curvative. Top: $\tau_{\max}/\tau_0 = 2.00$, **Middle**: $\tau_{\max}/\tau_0 = 1.75$, **Bottom**: $\tau_{\max}/\tau_0 = 1.50$.

By approximating the above condition even further, we arrive at the following hypothesis:

Hypothesis 1. *A shell supported by an elastic foundation with a rough contact area, that has thickness small relative to the radius of mean curvature, satisfies the following displacement-based friction condition*

$$u^3 \leq -\frac{1}{2\nu_F}(u_\alpha u^\alpha)^{\frac{1}{2}}, \tag{11}$$

*where ν_F is the coefficient of friction between shell and the foundation, and **u** is the displacement field of the shell. If $2\nu_F u^3 + (u_\alpha u^\alpha)^{\frac{1}{2}} < 0$, then we say that the the shell is bonded to the foundation, and if $2\nu_F u^3 + (u_\alpha u^\alpha)^{\frac{1}{2}} = 0$, then we say that the shell is at limiting equilibrium.*

The justifications for introducing the hypothesis are as follows. Kikuchi and Oden (1988) assert that the variational form of a linear elastic body subjected to Coulomb's law of static friction, is non-convex and non-differentiable. Thus, the existence of a (unique or otherwise) solution is an open question that cannot be proven with conventional means. But we can show that the variational form, i.e., the energy functional of a linear elastic body, subjected to the displacement-based friction condition from this hypothesis is convex, coercive and differentiable, and thus proving the existence of solutions is perfectly possible. Also, unlike the model of Kikuchi and Oden (1988), our displacement-based condition is independent of the regularization parameter ε and it is well defined for all finite values of $\boldsymbol{u}$. Furthermore, we show that our problem can be numerically modeled without an initial guess of the purely normal stress, which is something Kikuchi and Oden's model is incapable of. Note, however, that we can guarantee that the condition from our hypothesis will hold as we have already asserted that the lower-surface of the shell is not hyperbolic and its mean curvature is positive. Thus, for sensible boundary conditions, we can always expect the normal displacement to be non-positive.

Does our hypothesized displacement-based friction law make physical sense? Well, consider two elastic blocks in a zero-gravity scenario, block-*A* at the top and block-*B* at the bottom, where the base of block-*B* satisfies the zero-Dirichlet boundary condition. Now assume that the contact area of both blocks is rough and that one is applying forces to both the top and to one

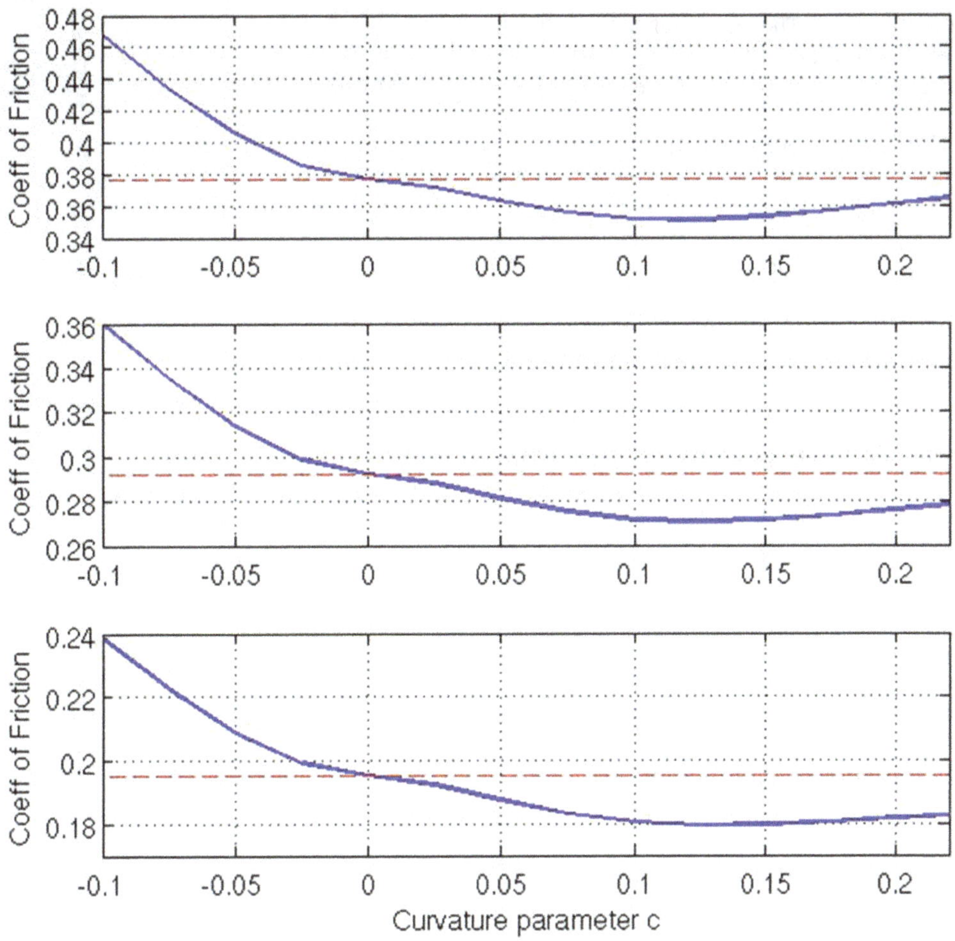

FIGURE 7 | Coefficient of friction relative to c including geometries with positive and negative Gaussian curvature. Top: $\tau_{\max}/\tau_0 = 2.00$, **Middle**: $\tau_{\max}/\tau_0 = 1.75$, **Bottom**: $\tau_{\max}/\tau_0 = 1.50$.

side of block-*A* to mimic respectively compression and shear in the contact region. The higher the compression, the higher the normal displacement is toward the bottom, i.e., $u^3 < 0$, and, the higher the shear, the higher the tangential displacement is in the direction of the applied tangential force. Just as for the case of Coulomb's friction, where the bodies are in respective equilibrium given that the magnitude of the normal stress is above a certain factor of the magnitude of the tangential stress, i.e., $|T^3_3(\boldsymbol{u})| > \nu_F^{-1}|T^\alpha_3(\boldsymbol{u})T^3_\alpha(\boldsymbol{u})|^{\frac{1}{2}}$, we assert that the bodies are in equilibrium given that the normal displacement is below a certain factor of the magnitude of the tangential displacement, i.e., $u^3 < -\frac{1}{2}\nu_F^{-1}(u_\alpha u^\alpha)^{\frac{1}{2}}$, if $u^3 < 0$. Note that this factor may or may not be $\frac{1}{2}\nu_F^{-1}$, but this is the most mathematically logical factor we have derived.

3.2. The Governing Equations

Let $\Omega \subset \mathbb{R}^3$ be a connected open bounded domain that satisfies the segment condition with a uniform $C^1(\mathbb{R}^3; \mathbb{R}^2)$ boundary $\partial\Omega$ such that $\omega, \partial\Omega_0 \subset \partial\Omega$ with $\bar{\omega} \cap \partial\bar{\Omega}_0 = \emptyset$ and $\text{meas}(\partial\Omega_0; \mathbb{R}^2) > 0$, and let $\omega \subset \mathbb{R}^2$ be a connected open bounded plane that satisfies the segment condition with a uniform $C^1(\mathbb{R}^2; \mathbb{R})$ boundary $\partial\omega$. Let $\bar{\boldsymbol{X}} \in C^2(\bar{\Omega}; \mathbf{E}^3)$ be a diffeomorphism and $\boldsymbol{\sigma} \in C^3(\bar{\omega}; \mathbf{E}^3)$ be an injective immersion. Let $\boldsymbol{f} \in \boldsymbol{L}^2(\Omega), \boldsymbol{f}_0 \in \boldsymbol{L}^2(\omega)$ and $\boldsymbol{\tau}_0 \in \boldsymbol{L}^2(\partial\omega)$.

For this section, we assume that $\boldsymbol{u} \in C^2(\Omega; \mathbb{R}^3)$, $u^\alpha|_\omega \in C^3(\omega)$, $u^3|_\omega \in C^4(\omega)$ and $2\nu_F u^3 + (u_\alpha u^\alpha)^{\frac{1}{2}} \leq 0$ everywhere in ω. For the elastic foundation we define $T^{ij}(\boldsymbol{u}) = A^{ijkl}E_{kl}(\boldsymbol{u})$ to be the second Piola-Kirchhoff stress tensor, $E_{ij}(\boldsymbol{u}) = \frac{1}{2}(g_{ik}\bar{\nabla}_j u^k + g_{jk}\bar{\nabla}_i u^k)$ to be the linearized Green-St Venant stress tensor, $A^{ijkl} = \bar{\lambda} g^{ij}g^{kl} + \bar{\mu}(g^{ik}g^{jl} + g^{il}g^{jk})$ to be the elasticity tensor, $\bar{\lambda} = (1 - \bar{\nu} - 2\bar{\nu}^2)^{-1}\bar{\nu}\bar{E}$ as the first Lamé parameter, $\bar{\mu} = \frac{1}{2}(1 + \bar{\nu})^{-1}\bar{E}$ as the second Lamé parameter, $\bar{E}$ as the Young's modulus and $\bar{\nu}$ as Poisson's ratio of the elastic foundation. Furthermore, $\boldsymbol{f}$ is some external force density field acting on the elastic foundation. The covariant *first fundamental form tensor* of $\boldsymbol{\sigma}$ with respect to the curvilinear coordinates is defined as

$$F_{[\mathrm{I}]\alpha\beta} = \partial_\alpha \sigma_i \partial_\beta \sigma^i\,,\ \forall\, \alpha, \beta \in \{1, 2\}\,.$$

Also we regard the indices $\alpha,\beta,\gamma,\delta \in \{1,2\}$. Furthermore, $F_{[\mathrm{I}]\alpha\gamma}F_{[\mathrm{I}]}^{\gamma\beta} = \delta_\alpha^\beta$, $\forall\ \alpha,\beta \in \{1,2\}$. The second fundamental form tensor is as defined above in Section 2.2.

The governing equations of the elastic foundation are given by

$$\bar{\nabla}_i T_j^i(\boldsymbol{u}) + f_j = 0\ ,\ \forall\, j \in \{1,2,3\}\ ,$$

with the following boundary conditions:

$$\boldsymbol{u}|_{\partial\Omega_0} = \boldsymbol{0}\ ,$$
$$\bar{n}_i T_j^i(\boldsymbol{u})|_{\{\partial\Omega\setminus\{\omega\cup\partial\Omega_0\}\}} = 0\ ,\ \forall\, j \in \{1,2,3\}\ ,$$

where $\bar{\boldsymbol{n}}$ is the unit outward normal to the boundary $\partial\Omega$ in curvilinear coordinates.

Turning now to the overlying body, we assume that the body is so thin, and the bending moments are so small, that it can be approximated by a shell-membrane. For the shell-membrane we define $\tau^{\alpha\beta}(\boldsymbol{u}) = B^{\alpha\beta\gamma\delta}\epsilon_{\gamma\delta}(\boldsymbol{u})$ to be the stress tensor and $\eta^{\alpha\beta}(\boldsymbol{u}) = B^{\alpha\beta\gamma\delta}\rho_{\gamma\delta}(\boldsymbol{u})$ to be the negative of the change in the moments density tensor.

$$\epsilon_{\alpha\beta}(\boldsymbol{u}) = \frac{1}{2}\left(\nabla_\alpha(u_\beta|_\omega) + \nabla_\beta(u_\alpha|_\omega)\right) - F_{[\mathrm{II}]\alpha\beta}(u^3|_\omega)$$

is then half the change in the first fundamental form tensor,

$$\begin{aligned}\rho_{\alpha\beta}(\boldsymbol{u}) &= \nabla_\alpha\nabla_\beta(u^3|_\omega) - F_{[\mathrm{II}]\alpha\gamma}F_{[\mathrm{II}]\beta}^{\gamma}(u^3|_\omega) + F_{[\mathrm{II}]\beta\gamma}\nabla_\alpha(u^\gamma|_\omega)\\ &\quad + F_{[\mathrm{II}]\alpha\gamma}\nabla_\beta(u^\gamma|_\omega) + \left(\nabla_\alpha F_{[\mathrm{II}]\beta\gamma}\right)(u^\gamma|_\omega)\end{aligned}$$

is the change in the second fundamental form tensor,

$$B^{\alpha\beta\gamma\delta} = \frac{2\lambda\mu}{\lambda+2\mu}F_{[\mathrm{I}]}^{\alpha\beta}F_{[\mathrm{I}]}^{\gamma\delta} + \mu(F_{[\mathrm{I}]}^{\alpha\gamma}F_{[\mathrm{I}]}^{\beta\delta} + F_{[\mathrm{I}]}^{\alpha\delta}F_{[\mathrm{I}]}^{\beta\gamma})$$

is the elasticity tensor, $\lambda = (1-\nu-2\nu^2)^{-1}\nu E$ is the first Lamé parameter, $\mu = \frac{1}{2}(1+\nu)^{-1}E$ is the second Lamé parameter, E is the Young's modulus and ν is Poisson's ratio of the frictionally coupled shell. $\mathrm{Tr}(T_j^3(\boldsymbol{u})) = T_j^3(\boldsymbol{u})|_\omega$ is the normal stress of the foundation in the contact region, and $\boldsymbol{f}_0$ is some external force density field acting on the overlying shell.

The governing equations for the shell-membrane are determined by considering a variational problem based on the following energy functional:

$$\begin{aligned}J(\boldsymbol{u}) &= \int_\Omega \frac{1}{2}A^{ijkl}E_{ij}(\boldsymbol{u})E_{kl}(\boldsymbol{u}) - f^i u_i\, d\Omega\\ &\quad + \int_\omega \frac{1}{2}B^{\alpha\beta\gamma\delta}\left(h\epsilon_{\alpha\beta}(\boldsymbol{u})\epsilon_{\gamma\delta}(\boldsymbol{u}) + \frac{1}{3}h^3\rho_{\alpha\beta}(\boldsymbol{u})\rho_{\gamma\delta}(\boldsymbol{u})\right)\\ &\quad - hf_0^i u_i\, d\omega - \int_{\partial\omega} h\tau_0^i u_i\, d(\partial\omega)\ ,\end{aligned}$$

where $\boldsymbol{u}$ is subject to the displacement-based friction condition Equation (11) and the region in which the shell is at limiting equilibrium is unknown prior to solving the problem. Following the detailed analysis in chapter of Jayawardana (2016), it is possible to obtain the following governing equations for the shell-membrane:

if $\left[2\nu_F u^3 + (u_\alpha u^\alpha)^{\frac{1}{2}}\right]|_\omega < 0$ (the *bonded* case), **then**

$$\begin{aligned}&\nabla_\alpha\tau_\beta^\alpha(\boldsymbol{u}) + \frac{2}{3}h^2 F_{[\mathrm{II}]\beta}^{\ \alpha}\nabla_\gamma\eta_\alpha^\gamma(\boldsymbol{u}) + \frac{1}{3}h^2\left(\nabla_\gamma F_{[\mathrm{II}]\beta}^{\ \alpha}\right)\eta_\alpha^\gamma(\boldsymbol{u})\\ &\qquad - \frac{1}{h}\mathrm{Tr}(T_\beta^3(\boldsymbol{u})) + f_{0\beta} = 0\ ,\ \forall\,\beta \in \{1,2\}\ ,\\ &F_{[\mathrm{II}]\alpha}^{\ \gamma}\tau_\gamma^\alpha(\boldsymbol{u}) - \frac{1}{3}h^2\nabla_\alpha\left(\nabla_\gamma\eta^{\alpha\gamma}(\boldsymbol{u})\right) + \frac{1}{3}h^2 F_{[\mathrm{II}]\alpha}^{\ \delta}F_{[\mathrm{II}]\gamma}^{\ \alpha}\eta_\delta^\gamma(\boldsymbol{u})\\ &\qquad - \frac{1}{h}\mathrm{Tr}(T_3^3(\boldsymbol{u})) + f_{03} = 0\ ;\end{aligned}$$

if $\left[2\nu_F u^3 + (u_\alpha u^\alpha)^{\frac{1}{2}}\right]|_\omega = 0$ (the *limiting equilibrium* case), **then**

$$\begin{aligned}&\nu_F\nabla_\alpha\tau_\beta^\alpha(\bar{\boldsymbol{u}}) - \frac{1}{2}\frac{u_\beta}{(u_\alpha u^\alpha)^{\frac{1}{2}}}F_{[\mathrm{II}]\alpha}^{\ \gamma}\tau_\gamma^\alpha(\bar{\boldsymbol{u}})\\ &\quad + \frac{2}{3}\nu_F h^2 F_{[\mathrm{II}]\beta}^{\ \alpha}\nabla_\gamma\eta_\alpha^\gamma(\bar{\boldsymbol{u}}) + \frac{1}{6}h^2\frac{u_\beta}{(u_\alpha u^\alpha)^{\frac{1}{2}}}\nabla_\alpha\nabla_\gamma\eta^{\alpha\gamma}(\bar{\boldsymbol{u}})\\ &\quad + \frac{1}{3}\nu_F h^2\left(\nabla_\gamma F_{[\mathrm{II}]\beta}^{\ \alpha}\right)\eta_\alpha^\gamma(\bar{\boldsymbol{u}})\\ &\quad - \frac{1}{6}h^2\frac{u_\beta}{(u_\alpha u^\alpha)^{\frac{1}{2}}}F_{[\mathrm{II}]\alpha}^{\ \delta}F_{[\mathrm{II}]\gamma}^{\ \alpha}\eta_\delta^\gamma(\bar{\boldsymbol{u}})\\ &\quad - \frac{\nu_F}{h}\mathrm{Tr}(T_\beta^3(\bar{\boldsymbol{u}})) + \frac{1}{2h}\frac{u_\beta}{(u_\alpha u^\alpha)^{\frac{1}{2}}}\mathrm{Tr}(T_3^3(\bar{\boldsymbol{u}}))\\ &\quad + \nu_F f_{0\beta} - \frac{1}{2}\frac{u_\beta}{(u_\alpha u^\alpha)^{\frac{1}{2}}}f_{03} = 0\ ,\qquad \forall\,\beta \in \{1,2\}\ ,\end{aligned}$$

where $\bar{\boldsymbol{u}}|_\omega = (u^1, u^2, -\frac{1}{2}\nu_F^{-1}(u_\alpha u^\alpha)^{\frac{1}{2}})|_\omega$ and $(\partial_3\bar{u}^1, \partial_3\bar{u}^2, \partial_3\bar{u}^3)|_\omega = (\partial_3 u^1, \partial_3 u^2, \partial_3 u^3)|_\omega$. Finally, the boundary conditions of the overlying shell are

$$\begin{aligned}\left[n_\alpha\tau_\beta^\alpha(\boldsymbol{u}) + \frac{2}{3}h^2 n_\gamma F_{[\mathrm{II}]\beta}^{\ \alpha}\eta_\alpha^\gamma(\boldsymbol{u})\right]|_{\partial\omega} &= \tau_{0\beta}\ ,\ \forall\,\beta \in \{1,2\}\ ,\\ -\frac{1}{3}h^2 n_\gamma\nabla_\alpha\eta^{\alpha\gamma}(\boldsymbol{u})|_{\partial\omega} &= \tau_{03}\ ,\\ \partial_\beta(u^3|_\omega)|_{\partial\omega} &= 0\ ,\ \forall\,\beta \in \{1,2\}\ ,\end{aligned}$$

where $\boldsymbol{n}$ is the unit outward normal vector to the boundary $\partial\omega$ in curvilinear coordinates and $\boldsymbol{\tau}_0$ is the external traction field acting on the boundary of the overlying shell.

3.3. Numerical Experiments

To conduct numerical experiments, assume that we are dealing with a shell-membrane of thickness h, supported by an elastic foundation, where the unstrained configuration of the foundation has an annular cross-section, characterized by the diffeomorphism $\bar{X}(x,\theta,r) = (x, r\sin(\theta), r\cos(\theta))_{\mathrm{E}}$, where $(x^1,x^2,x^3) = (x,\theta,r)$, $x \in (-L,L)$, $\theta \in (-\pi,\pi]$, and $r \in (a_0,a)$, and assume that the contact region lies within $x \in (-\ell,\ell)$, $\theta \in (-\frac{1}{2}\pi, 0)$, where $0 < \ell < L$. Let the sufficiently smooth field $\boldsymbol{u} = (u^1(x,\theta,r), u^2(x,\theta,r), u^3(x,\theta,r))$ be the displacement field of the foundation. With some calculations one can find the metric tensor to be $\boldsymbol{g} = \mathrm{diag}(1, r^2, 1)$ and the covariant derivatives to be

$$\bar{\nabla}_1 u^1 = \partial_1 u^1\,,\quad \bar{\nabla}_1 u^2 = \partial_1 u^2\,,\quad \bar{\nabla}_1 u^3 = \partial_1 u^3\,,$$
$$\bar{\nabla}_2 u^1 = \partial_2 u^1\,,\quad \bar{\nabla}_2 u^2 = \partial_2 u^2 + r^{-1}u^3\,,\quad \bar{\nabla}_2 u^3 = \partial_2 u^3 - ru^2\,,$$
$$\bar{\nabla}_3 u^1 = \partial_3 u^1\,,\quad \bar{\nabla}_3 u^2 = \partial_3 u^2 + r^{-1}u^2\,,\quad \bar{\nabla}_3 u^3 = \partial_3 u^3\,.$$

With further calculations, one can express the governing equations of the foundation as

$$(\bar{\lambda} + \bar{\mu})\partial^1(\bar{\nabla}_i u^i) + \bar{\mu}\bar{\Delta}u^1 = 0\,,$$
$$(\bar{\lambda} + \bar{\mu})\partial^2(\bar{\nabla}_i u^i) + \bar{\mu}\bar{\Delta}u^2 = 0\,,$$
$$(\bar{\lambda} + \bar{\mu})\partial^3(\bar{\nabla}_i u^i) + \bar{\mu}\bar{\Delta}u^3 = 0\,.$$

The boundary of the foundation can be decomposed into sub-boundaries as

$$\partial\Omega = \bar{\omega} \cup \partial\Omega_0 \cup \partial\Omega_f\,,$$
$$\omega = \{a\} \times (-\tfrac{1}{2}\pi, 0) \times (-\ell, \ell)\,,$$
$$\partial\Omega_0 = \{\{a_0\} \times (-\pi, \pi] \times [-L, L]\} \cup \{(a_0, a] \times (-\pi, \pi] \times \{\{-L\} \cup \{L\}\}\}\,,$$
$$\partial\Omega_f = \{\{a\} \times (-\pi, \pi] \times (-L, L)\} \setminus \bar{\omega}\,.$$

Thus, we can express the boundary conditions of the foundation as

$$\boldsymbol{u}|_{\partial\Omega_0} = 0 \text{ (zero-Dirichlet)},$$
$$\left[\partial_3 u^1 + \partial_1 u^3\right]|_{\partial\Omega_f} = 0 \text{ (zero-Robin)},$$
$$\left[r^2\partial_3 u^2 + \partial_2 u^3\right]|_{\partial\Omega_f} = 0 \text{ (zero-Robin)},$$
$$\left[\bar{\lambda}(\partial_1 u^1 + \partial_2 u^2 + r^{-1}u^3) + (\bar{\lambda} + 2\bar{\mu})\partial_3 u^3\right]|_{\partial\Omega_f} = 0 \text{(zero-Robin)}.$$

Let $\boldsymbol{u}|_\omega = (u^1(x,\theta,a), u^2(x,\theta,a), u^3(x,\theta,a))$ be the displacement field of the shell-membrane. With some calculations, one can find the first fundamental form tensor to be $F_{[I]} = \mathrm{diag}(1, a^2)$ and the covariant derivatives to be

$$\nabla_1 u^1 = \partial_1 u^1\,,\quad \nabla_1 u^2 = \partial_1 u^2\,,$$
$$\nabla_2 u^1 = \partial_2 u^1\,,\quad \nabla_2 u^2 = \partial_2 u^2\,.$$

Considering the case $h^2 \rho_\alpha^\gamma(\boldsymbol{u})\rho_\gamma^\alpha(\boldsymbol{u}) \ll \epsilon_\alpha^\gamma(\boldsymbol{u})\epsilon_\gamma^\alpha(\boldsymbol{u})$, with some further calculations one can express the governing equations of the shell-membrane as:
If $[2\nu_F u^3 + (u^1u^1 + r^2u^2u^2)^{\frac{1}{2}}]|_\omega < 0$, then

$$(\Lambda + \mu)\partial^1(\nabla_\alpha u^\alpha) + \mu\Delta u^1 + \frac{1}{a}\Lambda\partial^1 u^3 - \frac{1}{h}\mathrm{Tr}(T_3^1(\boldsymbol{u})) = 0\,,$$
$$(\Lambda + \mu)\partial^2(\nabla_\alpha u^\alpha) + \mu\Delta u^2 + \frac{1}{a}(\Lambda + 2\mu)\partial^2 u^3 - \frac{1}{h}\mathrm{Tr}(T_3^2(\boldsymbol{u})) = 0\,,$$
$$\Lambda\partial_1 u^1 + (\Lambda + 2\mu)\left(\partial_2 u^2 + \frac{1}{a}u^3\right) + \frac{a}{h}\mathrm{Tr}(T_3^3(\boldsymbol{u})) = 0\,.$$

If $[2\nu_F u^3 + (u^1u^1 + r^2u^2u^2)^{\frac{1}{2}}]|_\omega = 0$, then

$$\begin{aligned}(\Lambda + \mu)\partial^1(\nabla_\alpha u^\alpha) &+ \mu\Delta u^1 - \frac{\Lambda}{2a\nu_F}\partial^1(u^1u^1 + a^2u^2u^2)^{\frac{1}{2}} - \frac{1}{h}\mathrm{Tr}(T_3^1(\boldsymbol{u})) - \frac{(\Lambda + \mu)}{4a\nu_F^2}u^1 - \frac{(\bar{\lambda} + \bar{\mu})}{4h\nu_F^2}u^1\\ &+ \frac{1}{2\nu_F}\frac{u^1}{(u^1u^1 + a^2u^2u^2)^{\frac{1}{2}}}\Big(\Lambda\partial_1 u^1 + (\Lambda + 2\mu)\partial_2 u^2 + \frac{a}{h}\left(\bar{\lambda}(\partial_1 u^1 + \partial_2 u^2) + (\bar{\lambda} + 2\bar{\mu})\partial_3 u^3\right)\Big) = 0\,,\end{aligned}$$

$$\begin{aligned}(\Lambda + \mu)\partial^2(\nabla_\alpha u^\alpha) &+ \mu\Delta u^2 - \frac{(\Lambda + 2\mu)}{2a\nu_F}\partial^2(u^1u^1 + a^2u^2u^2)^{\frac{1}{2}} - \frac{1}{h}\mathrm{Tr}(T_3^2(\boldsymbol{u})) - \frac{(\Lambda + \mu)}{4a\nu_F^2}u^2 - \frac{(\bar{\lambda} + \bar{\mu})}{4h\nu_F^2}u^2\\ &+ \frac{1}{2\nu_F}\frac{u^2}{(u^1u^1 + a^2u^2u^2)^{\frac{1}{2}}}\Big(\Lambda\partial_1 u^1 + (\Lambda + 2\mu)\partial_2 u^2 + \frac{a}{h}\left(\bar{\lambda}(\partial_1 u^1 + \partial_2 u^2) + (\bar{\lambda} + 2\bar{\mu})\partial_3 u^3\right)\Big) = 0\,.\end{aligned}$$

The boundary of the shell-membrane can be decomposed into sub-boundaries as

$$\partial\omega = \partial\omega_{T_0} \cup \partial\omega_{T_{\max}} \cup \partial\omega_f\,,$$
$$\partial\omega_{T_0} = [-\ell, \ell] \times \left\{-\frac{1}{2}\pi\right\}\,,$$
$$\partial\omega_{T_{\max}} = [-\ell, \ell] \times \{0\}\,,$$
$$\partial\omega_f = \{\{-\ell\} \cup \{\ell\}\} \cup \left(-\frac{1}{2}\pi, 0\right)\,.$$

Thus, we can express the boundary conditions of the shell-membranes as

$$\left[\Lambda\partial_1 u^1 + (\Lambda + 2\mu)(\partial_2 u^2 + a^{-1}u^3)\right]|_{\partial\omega_{T_0}} = \tau_0 \text{ (traction)}\,,$$
$$\left[\Lambda\partial_1 u^1 + (\Lambda + 2\mu)(\partial_2 u^2 + a^{-1}u^3)\right]|_{\partial\omega_{T_{\max}}} = \tau_{\max} \text{ (traction)}\,,$$
$$\left[\partial_2 u^1 + a^2\partial_1 u^2\right]|_{\partial\omega_f} = 0 \text{ (zero-Robin)}\,,$$
$$\left[(\Lambda + 2\mu)\partial_1 u^1 + \Lambda(\partial_2 u^2 + a^{-1}u^3)\right]|_{\partial\omega_f} = 0 \text{ (zero-Robin)}\,.$$

A second-order accurate finite-difference method is again employed in conjunction with Newton's method for nonlinear systems. A modestly fine grid is chosen and the iterative process is terminated once $|1 - ||\boldsymbol{u}_{m+1}||_{\ell^2}/||\boldsymbol{u}_m||_{\ell^2}|$ falls below a certain value (10^{-10} in the calculations shown here). We attempt to model a stiff shell-membrane on a relatively flaccid foundation with a large coefficient of friction. To do so, we keep the values $\nu_F = 1$, $h = 0.001$, $a = 1$, $\ell = \frac{1}{4}$, $L = \frac{1}{2}$, $E = 10^3$, $\nu = \frac{1}{4}$, $\bar{\nu} = 0$, $\tau_0 = 1$ and $\tau_{\max} = 2$ fixed for all experiments. Note that some preliminary work in chapter of Jayawardana (2016) found that if (i) the overlying body is relatively thin, (ii) it has a relatively high Young's modulus, (iii) the coefficient of friction is high, and (iv) the mean curvature is a constant, then the solution for the shell model with friction is in relatively good agreement with a numerical model using Kikuchi and Oden's original friction law in chapter of Kikuchi and Oden (1988), thus justifying the choice of our parameters.

Figure 8 shows the total surface displacement and the total surface shear at the outer boundary of the foundation for varying $\bar{E}$ (the Young's modulus of the foundation). The total displacement is calculated by $|\boldsymbol{u}||_{r=a} = (\sum_{\{\Delta x^1,\Delta x^2\}} u_i u^i)^{\frac{1}{2}}|_{r=a}$, for $i \in \{1,2,3\}$ and the total shear is calculated by $|\sum_{\{\Delta x^1,\Delta x^2\}} T^3_\alpha(\boldsymbol{u})T^\alpha_3(\boldsymbol{u})|^{\frac{1}{2}}|_{r=a}$, for $\alpha \in \{1,2\}$. For these simulations, we asserted that $a_0 = \frac{1}{4}$ and $\bar{E} = \{10, 20, 30, 40, 50, 60, 70, 80, 90, 100\}$. From **Figure 8** one can see that as the Young's modulus of the foundation increases, the shear stress experienced on the underlying body surface increases, but the total displacement of the body decreases. Only the decrease in displacement, as the Young's modulus of the foundation increases, seems intuitive as when the Young's modulus increases, one would need to increase the amount of force applied to achieve the same amount of displacement. However, the increase in total surface shear is a more interesting result suggesting higher shear forces occur on the underlying surface as the body becomes stiffer despite the coefficient of friction remaining fixed.

Figure 9 shows the total surface displacement and the total surface shear at the outer boundary of the foundation for varying a_0. For these experiments, we asserted that $a_0 \in \{\frac{5}{20}, \frac{6}{20}, \frac{7}{20}, \frac{8}{20}, \frac{9}{20}, \frac{10}{20}, \frac{11}{20}, \frac{12}{20}, \frac{13}{20}, \frac{14}{20}, \frac{15}{20}\}$ and $\bar{E} = 10$. From **Figure 9**, one can see that as the thickness of the foundation increases, both the total displacement and the shearing stress on the surface of the body increases. The increase in displacement as the thickness of the foundation increases seems intuitive as there is more deformable matter available to be displaced but this appears again to lead to higher shear stress in the contact region. To explore this further, our final figure (**Figure 10**) shows the total surface stress (shear *and* normal stresses) at the outer boundary of the foundation. The total surface stress is calculated by $|\sum_{\{\Delta x^1,\Delta x^2\}} T^3_i(\boldsymbol{u})T^i_3(\boldsymbol{u})|^{\frac{1}{2}}|_{r=a}$, where $i \in \{1,2,3\}$. These plots indicate that not only does a thicker foundation experience higher shear stresses but increasingly higher normal stresses with a consequent increase in the likelyhood of potential damage to the underlying body.

4. CONCLUSIONS

Two numerical models have been presented here that can provide insight into the forces generated by a compliant sheet of material under tension in contact with a rigid or deformable object. The aspects of geometry and deformability of the underlying body were examined separately by the mathematical models presented in Sections 2 and 3 respectively.

The model in Section 2 determines the coefficient of friction in the contact region between a thin membrane pulled dynamically at a constant speed and a rigid underlying body. The model of Kikuchi and Oden (1988) for Coulomb's law of static friction was extended to curvilinear coordinates, and a numerical model was used to investigate how the calculated coefficient of friction varies with different material and physical parameters. For parameters such as Poisson's ratio of the membrane, Young's modulus of

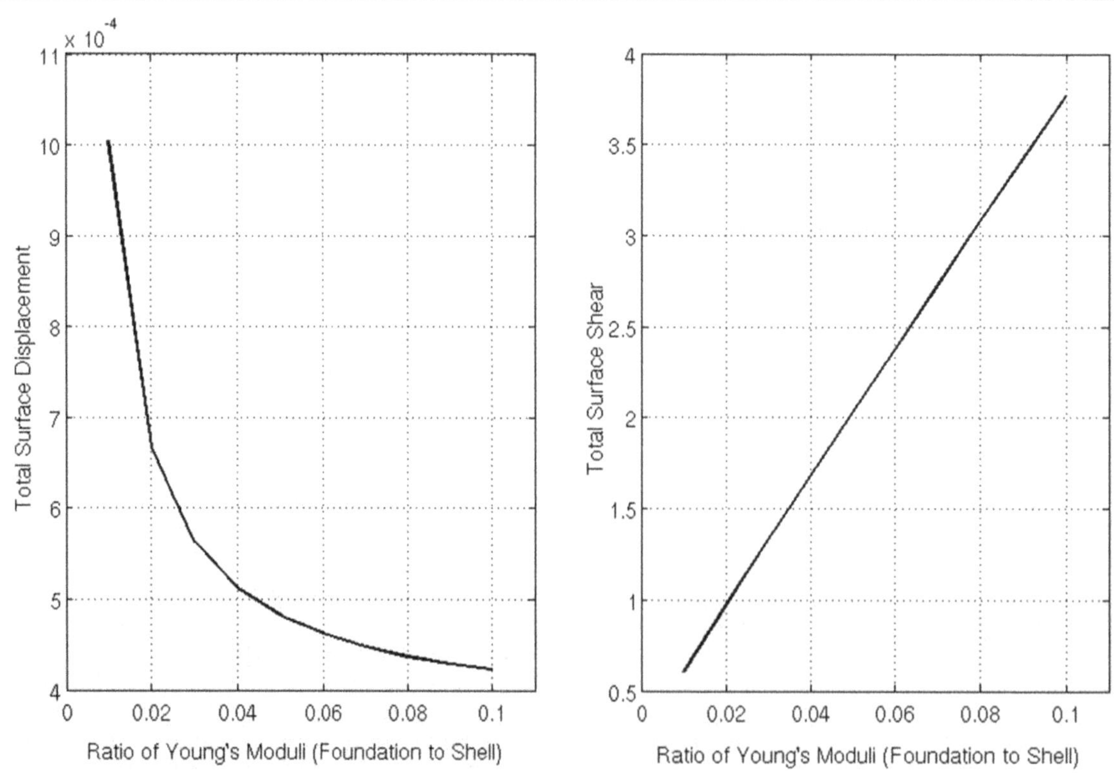

FIGURE 8 | Displacement and the shear in the contact region for varying $\bar{E}/E$.

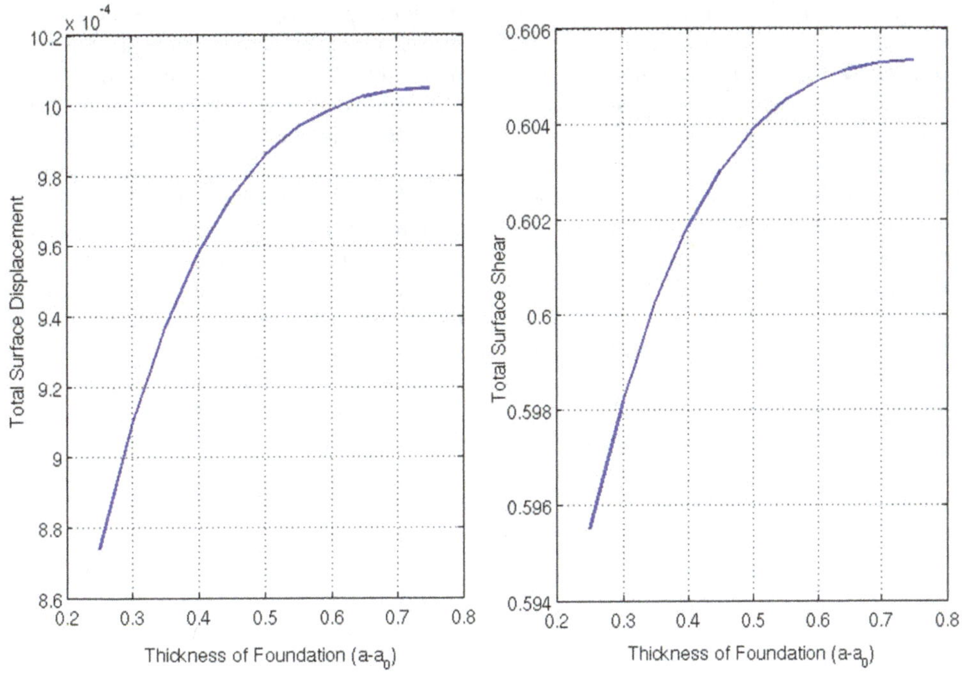

FIGURE 9 | Displacement and the shear in the contact region for varying foundation thickness.

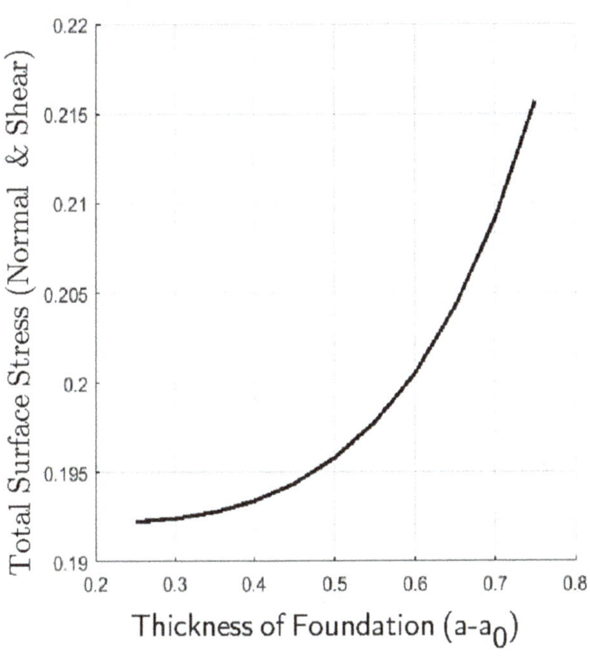

FIGURE 10 | Total surface stress (shear and normal) for varying foundation thickness.

the membrane and the speed of the membrane there was no significant variation in the determined coefficient of friction; this indicates that a capstan-type approach lacking dependence on these parameters should still produce accurate results. However, changes to the mass density of the fabric and the lateral (and thus Gaussian) curvature of the underlying body appear to lead to significant variations in the determined coefficient of friction which would not be captured by a capstan model. Even a remarkably small mass density of the fabric (which is assumed to be negligible in the capstan model) leads to the capstan model overestimating the actual coefficient of friction. This effect can be, to a certain extent, attributed to the fabric weight contributing to the overall tension and how this impacts via the logarithmic relation between the coefficient of friction and the tension ratio.

In varying the Gaussian curvature of the underlying body, the numerical model suggests that for a saddle-type geometry (as seen in real experiments such as **Figure 2**) the capstan approach may lead to a significant underestimate of the coefficient of friction. On the other hand, for a barrel-type geometry with positive Gaussian curvature the converse is true with the capstan equation potentially overestimating the coefficient of friction. The numerical model also indicates the intriguing possibility of an optimal barrel-type geometry where the coefficient of friction is minimized—a surprising result that certainly requires further investigation. We note, in passing, that the effect curvature appears to have on the coefficient of friction here possesses some similarities to how Kelvin's equation governs equilibrium vapor pressure over a curved surface (Skinner and Sambles, 1972; Fisher and Israelachvili, 1981) and how grain shape and size modify the rate of complex matter agglomeration (Gadomski and Rubı, 2003).

The model developed in Section 3 is for a thin shell-membrane under tension in frictional contact with an elastic foundation where static friction is imposed in the region of contact. The fact that our frictional law (e.g., Coulomb's law of static friction) must now be imposed on a *free boundary* because we no longer know *a priori* the location of the contact region significantly increases the computational complexity. To combat this, a modified, more computationally tractable, displacement-based static friction condition is derived from the model of Kikuchi and Oden (1988) for Coulomb's law of static friction in curvilinear coordinates. We can show that a set of governing equations for a two-body contact problem that incorporates this displacement-based static friction condition yields a unique solution. A numerical scheme for the two-body static friction contact problem is then developed where, this time, the coefficient of friction is specified. Using this model, we examine how the normal and tangential stresses and displacements computed by the model vary as we vary the stiffness and thickness of the underlying body. For a shell-membrane supported by an elastic foundation subjected to static friction we observe the following: (i) that as the Young's modulus of the elastic body increases, the magnitude of the displacement of the body surface decreases; (ii) that as the Young's modulus of the elastic body increases, the magnitude of the tangential shear stresses acting on the surface of body increases; (iii) that as the thickness of the elastic body increases, the magnitude of the surface displacements of the body increases; and (iv) that as the thickness of the elastic body increases, the magnitude of the tangential shear stresses acting on the body surface increases.

The shell-membrane in contact with a deformable elastic foundation model appears to indicate that both elastic and geometrical properties of the elastic foundation significantly affect the stress and deformation of the underlying tissue. This is reflected in the numerical results via a strong positive correlation between the thickness of the foundation and the amount of stress transferred from the shell-membrane to deforming the foundation. Indeed, with the applied tension and coefficient of friction fixed in the numerical model, the amount of stress experienced by the underlying elastic body appears to depend rather significantly on its geometry (thickness and curvature) and elastic properties, which are features that are neglected by capstan-type model approaches.

The models presented in this paper highlight some very interesting results and lead to a number of questions which should be pursued in terms of experimental design and to improve quantification of the frictional forces generated by nonwoven fabrics. One important issue is to examine how to measure the curvature of the contact region and designing experiments that can see how curvature affects the relationship between normal and tangential forces in the contact region. It would also be wise to test if the experimental results are indeed sensitive to variations in the mass density of the fabric. An extended capstan model that incorporates mass density is proposed in Jayawardana (2016) and other modifications to the capstan model could be considered. Experiments involving deformable underlying bodies with well-known material parameters would be useful to validate our second model and further our understanding about how experiments on quantifying frictional forces involving human subjects should deal with significant skin deformation, such as in the examples shown in **Figure 2**.

Some future work is planned for the models presented, including proving the existence of a unique solution in the model of Section 2 and validating the implementation of the model of Coulomb's law of friction in Kikuchi and Oden (1988) for dynamic friction. For the elastic foundation model (Section 3), further theoretical research on rigorously applying static and dynamic friction at a contact region which is a free boundary, on introducing finite deformations, and on how such a numerical model can be effectively used to provide insight from experimental data are under investigation. Finally, in terms of accurately modeling the response of the skin and underlying tissue, some consideration should be given to exploring beyond Coulomb's law of friction to incorporate the adhesive and repulsive intermolecular forces (such as via hydration) occurring across a variety of spatial scales, leading to Derjaguin-type frictional laws (Gadomski, 2015).

ETHICS STATEMENT

This paper is fundamentally a mathematical paper but there is one figure (**Figure 2**) containing images of experiments on human subjects from a previous project that is shown as motivation for the mathematical modeling work. The images show deformation in the skin and underlying soft tissue generated during measurements of friction between strips of nonwoven fabric and the volar forearms of female volunteers. The work was conducted with the approval of London Stanmore Research Ethics Committee and The Whittington Hospital NHS R&D office, September 2011. It was carried out in accordance with the recommendations of IRAS (Integrated Research Application System) Help guidance notes for HRA approval (Health Research Authority) with written informed consent from all subjects. All subjects gave written informed consent in accordance with the Declaration of Helsinki. Investigators interacting with elderly and disabled participants underwent DBS checks prior to REC submission and as a requirement for study approval.

AUTHOR CONTRIBUTIONS

Research Design: KJ, NO, and AC. Mathematical Modeling and Analysis: KJ and NO. Clincial data analysis and guidance: AC. KJ, NO, and AC all contributed to the writing of the manuscript.

FUNDING

This work was supported by The Dunhill Medical Trust [grant number R204/0511]; UCL Impact Studentship.

ACKNOWLEDGMENTS

We are most grateful to Dr. Sabrina Falloon and Skevos Karavokiros for providing their insight regarding the experiments on measuring friction as well as for the data and images provided. We are also very grateful to The Dunhill Medical Trust and UCL for funding KJ's Ph.D. studentship.

REFERENCES

Badiale, M., and Serra, E. (2010). *Semilinear Elliptic Equations for Beginners: Existence Results via the Variational Approach*. London: Springer Science & Business Media.

Baser, O., and Konukseven, E. I. (2010). Theoretical and experimental determination of capstan drive slip error. *Mech. Mach. Theory* 45, 815–827. doi: 10.1016/j.mechmachtheory.2009.10.013

Behzadipour, S., and Khajepour, A. (2006). "Cable-based robot manipulators with translational degrees of freedom," in *Industrial Robotics: Theory, Modelling and Control*, ed S. Cubero (Rijeka: InTech Open Access Publisher). doi: 10.5772/5035

Bennett, G., Dealey, C., and Posnett, J. (2004). The cost of pressure ulcers in the U.K. *Age Ageing* 33, 230–235. doi: 10.1093/ageing/afh086

Berg, R. (1987). Etiologic factors in diaper dermatitis: a model for development of improved diapers. *Pediatrician* 14:27.

Bergfeld, W. F., and Taylor, J. S. (1985). Trauma, sports, and the skin. *Am. J. Indus. Med.* 8, 403–413. doi: 10.1002/ajim.4700080422

Ciarlet, P. (2000). *Mathematical Elasticity, Vol. III: Theory of Shells*. Amsterdam: Elsevier Science.

Conklin, R. J. (1990). Common cutaneous disorders in athletes. *Sports Med.* 9, 100–119. doi: 10.2165/00007256-199009020-00004

Cottenden, A. M., Cottenden, D. J., Karavokiros, S., and Wong, W. K. (2008a). Development and experimental validation of a mathematical model for friction between fabrics and a volar forearm phantom. *Proc. Instit. Mech. Eng. H J. Eng. Med.* 222, 1097–1106. doi: 10.1243/09544119JEIM406

Cottenden, A. M., Wong, W. K., Cottenden, D. J., and Farbrot, A. (2008b). Development and validation of a new method for measuring friction between skin and nonwoven materials. *Proc. Instit. Mech. Eng. H J. Eng. Med.* 222, 791–803. doi: 10.1243/09544119JEIM313

Cottenden, D., and Cottenden, A. (2009). An analytical model of the motion of a conformable sheet over a general convex surface in the presence of frictional coupling. *Quart. J. Mech. Appl. Math.* 62, 345–364. doi: 10.1093/qjmam/hbp012

Dowson, D. (1996). "Tribology and the skin surface," in *Bioengineering of the Skin: Skin Surface Imaging and Analysis*, eds K.-P. Wilhelm, P. Elsner, E. Berardesca and H. I. Maibach (Boca Raton, FL: CRC Press) 159–179.

Farage, M. A., Miller, K. W., Berardesca, E., and Maibach, H. I. (2007). Incontinence in the aged: contact dermatitis and other cutaneous consequences. *Contact Derm.* 57, 211–217. doi: 10.1111/j.1600-0536.2007.01199.x

Fisher, L. R., and Israelachvili, J. N. (1981). Experimental studies on the applicability of the Kelvin equation to highly curved concave menisci. *J. Colloid Interface Sci.* 80, 528–541. doi: 10.1016/0021-9797(81)90212-5

Gadomski, A. (2015). Multilevel-interaction friction procedure applicable in case of two opposing surfaces competing with one another - a gedanken experiment. *Phys. Essays* 28, 650–653. doi: 10.4006/0836-1398-28.4.650

Gadomski, A., and Rubı, J. (2003). On the two principal curvatures as potential barriers in a model of complex matter agglomeration. *Chem. Phys.* 293, 169–177. doi: 10.1016/S0301-0104(03)00295-7

Gwosdow, A., Stevens, J., Berglund, L., and Stolwijk, J. (1986). Skin friction and fabric sensations in neutral and warm environments. *Text. Res. J.* 56, 574–580. doi: 10.1177/004051758605600909

Herring, K. M., and Richie D. H. Jr. (1990). Friction blisters and sock fiber composition. a double-blind study. *J. Am. Podiatr. Med. Assoc.* 80, 63–71. doi: 10.7547/87507315-80-2-63

Irwin, D. E., Milsom, I., Kopp, Z., Abrams, P., and Cardozo, L. (2006). Impact of overactive bladder symptoms on employment, social interactions and emotional well-being in six european countries. *BJU Int.* 97, 96–100. doi: 10.1111/j.1464-410X.2005.05889.x

Jayawardana, K. (2016). *Mathematical Theory of Shells on Elastic Foundations: An Analysis of Boundary Forms, Constraints and Applications to Friction and Skin Abrasion*. Ph.D. thesis, University College London, Department of Mathematics.

Jung, J. H., Pan, N., and Kang, T. J. (2008). Capstan equation including bending rigidity and non-linear frictional behavior. *Mech. Mach. Theory* 43, 661–675. doi: 10.1016/j.mechmachtheory.2007.06.002

Karavokiros, S. (2007). *Skin Friction: Validating a Mathematical Model with a Simple Laboratory Model*. M.Sc. Biomedical Engineering and Medical Imaging (BEMI) Dissertation, University College London, Department of Medical Physics and Bioengineering.

Kikuchi, N., and Oden, J. (1988). *Contact Problems in Elasticity: A Study of Variational Inequalities and Finite Element Methods*. Studies in Applied Mathematics. Society for Industrial and Applied Mathematics.

Maklebust, J., and Sieggreen, M. (2001). *Pressure Ulcers: Guidelines for Prevention and Management, 3rd Edn*. Springhouse, PA: Springhouse Corporation.

Meserve, B., and Pingry, R. (1952). Some notes on the prismoidal formula. *Math. Teach.* 45, 257–263.

Murray, L. D., Magazinovic, N., and Stacey, M. C. (2001). Clinical practice guidelines for the prediction and prevention of pressure ulcers. *Primary Intent. Aust. J. Wound Manage.* 9, 88–97.

Necas, J., Simader, C., Necasová, Š., Tronel, G., and Kufner, A. (2011). *Direct Methods in the Theory of Elliptic Equations*. Springer Monographs in Mathematics, Springer, Berlin; Heidelberg.

Powell, B. (1983). Bicyclist's nipples. *JAMA* 249, 2457. doi: 10.1001/jama.1983.03330420019014

Rao, C. L., Lakshinarashiman, J., Sethuraman, R., and Sivakumar, S. M. (2003). *Engineering Mechanics: Statics and Dynamics*. Delhi: PHI Learning Pvt. Ltd.

Scheuplein, R. J., and Blank, I. H. (1971). Permeability of the skin. *Physiol. Rev.* 51, 702–747.

Silver, F. H., Siperko, L. M., and Seehra, G. P. (2003). Mechanobiology of force transduction in dermal tissue. *Skin Res. Technol.* 9, 3–23. doi: 10.1034/j.1600-0846.2003.00358.x

Sivamani, R. K., Goodman, J., Gitis, N. V., and Maibach, H. I. (2003). Coefficient of friction: tribological studies in man–an overview. *Skin Res. Technol.* 9, 227–234. doi: 10.1034/j.1600-0846.2003.02366.x

Skinner, L., and Sambles, J. (1972). The Kelvin equation - a review. *J. Aerosol. Sci.* 3, 199–210. doi: 10.1016/0021-8502(72)90158-9

Thomson, J. S., and Brooks, R. G. (1999). The economics of preventing and treating pressure ulcers: a pilot study. *J. Wound Care* 8, 312–316. doi: 10.12968/jowc.1999.8.6.25879

Touche, R. (1993). *The Cost of Pressure Sores*. Reports to the Department of Health, Department of Health, London.

Wilkinson, D. S. (1985). Dermatitis from repeated trauma to the skin. *Am. J. Indus. Med.* 8, 307–317. doi: 10.1002/ajim.4700080409

Zhang, M., and Mak, A. (1999). *In vivo* friction properties of human skin. *Prosthet. Orthot. Int.* 23, 135–141.

6

Waning Immunity is Associated with Periodic Large Outbreaks of Mumps: A Mathematical Modeling Study of Scottish Data

Dalila Hamami[1*], *Ross Cameron*[2], *Kevin G. Pollock*[2] *and Carron Shankland*[3]

[1] Department of Computing Science, University of Oran1 Ahmed BenBella, Oran, Algeria, [2] Health Protection Scotland, Glasgow, UK, [3] Department of Computing Science and Mathematics, University of Stirling, Stirling, UK

Edited by:
Krasimira Tsaneva-Atanasova, University of Exeter, UK

Reviewed by:
Maria Vittoria Barbarossa, Heidelberg University, Germany
Mauricio Lima, Pontifical Catholic University of Chile, Chile

***Correspondence:**
Dalila Hamami
dalila.hamami@univ-mosta.dz;
dhamami8@gmail.com

Vaccination programs for childhood diseases, such as measles, mumps and rubella have greatly contributed to decreasing the incidence and impact of those diseases. Nonetheless, despite long vaccination programmes across the world, mumps has not yet been eradicated in those countries: indeed, large outbreaks continue. For example, in Scotland large outbreaks occurred in 2004, 2005, and 2015, despite introducing the MMR (Measles-Mumps-Rubella) vaccine more than 20 years ago. There are indications that this vaccine-preventable disease is re-emerging in highly vaccinated populations. Here we investigate whether the resurgence of mumps is due to waning immunity, and further, could a booster dose be the solution to eradicate mumps or would it just extend the period of waning immunity? Using mathematical modeling we enhance a seasonally-structured disease model with four scenarios: no vaccination, vaccinated individuals protected for life, vaccinated individuals at risk of waning immunity, and introduction of measures to increase immunity (a third dose, or a better vaccine). The model is parameterised from observed clinical data in Scotland 2004–2015 and the literature. The results of the four scenarios are compared with observed clinical data 2004–2016. While the force of infection is relatively sensitive to the duration of immunity and the number of boosters undertaken, we conclude that periodic large outbreaks of mumps will be sustained for all except the second scenario. This suggests that the current protocol of two vaccinations is optimal in the sense that while there are periodic large outbreaks, the severity of cases in vaccinated individuals is less than in unvaccinated individuals, and the size of the outbreaks does not decrease sufficiently with a third booster to make economic sense. This recommendation relies on continuous efforts to maintain high levels of vaccination uptake.

Keywords: mumps, vaccination, waning immunity, mathematical and computational modeling and simulation, Bio-PEPA

INTRODUCTION

To prevent, control and eradicate childhood diseases, vaccination programs have been adopted throughout the world. For example the trivalent measles-mumps-rubella vaccine (MMR) (Harling et al., 2005; Le Menach et al., 2014; Cordeiro et al., 2015) has been highly successful for both measles and rubella reduction in many countries.

Despite near eradication of both measles and rubella (Isaacs and Menser, 1990; Glass and Grenfell, 2004; Ueda, 2016), elimination of mumps has not been achieved and could be considered to be re-emerging, despite initial early success in reducing mumps cases. In the last decade, many countries, such as Belgium (Abrams et al., 2014), Korea (Park, 2015), the Netherlands (Snijders et al., 2012), and the US (Dayan et al., 2008) have reported a dramatic increase in the incidence of mumps. In Scotland, 2004/2005 saw a sudden high resurgence in mumps with approximately 4500 cases, 8 years after the second dose of MMR was included in the vaccination program (which was predicted to substantially reduce mumps outbreaks Anderson and May, 1992). One hypothesis is that the resurgence was related to declining vaccine coverage (Nardone et al., 2003; van Boven et al., 2013), in particular, a widespread scare related to autism which led to some parents refusing to vaccinate their children. This can be easily debunked: the herd immunity threshold is estimated at 75–86% (Donaghy et al., 2006) and mumps vaccination levels have stayed above that level (e.g., in Scotland, ranging from 87 to 94% pre-2004). In addition Donaghy et al. (2006) argues that those infected during the 2004/2005 epidemics are characterized by low uptake of a single dose of MMR (catch-up campaign) and being of school age at time when the mumps virus had greatly reduced circulation in that group, delaying infection. The study undertaken by DeStefano et al. (2013) analyzing the number of antigens in both children with and without autism, shows that there is no association between receiving vaccine and developing autism.

A second hypothesis is to link vaccination status and age, e.g., proposing that outbreaks continue in the older population but die out in the increasingly vaccinated population. However, while age structure has shown to be informative in many models of traditionally childhood diseases (Andreasen, 1993; Ferguson et al., 1996; Hethcote, 2000; Brisson et al., 2010), current studies suggest that age is not the key determinant in mumps. Snijders et al. (2012) do not find any significant interaction between these two features. In addition, several studies of different outbreaks occurring at different times and locations in the US and Canada (Centers for Disease Control and Prevention, 2009) indicate that there is no evidence that age is the main factor leading to mumps spread. For instance, the outbreaks occurring in New York (Sulivan, Brooklyn, Rockland county and Orange county), New Jersey and Canada show variable average of infected age groups (Sullivan: 12 years, Brooklyn: 14 years, Rockland county: 12 years, Orange county: 18 years, New Jersey: 19.5 years and Canada: 27.5). However, it was confirmed that all cited cases were related to religious events or camping in Sullivan, with the majority fully vaccinated. It was also reported that the series of outbreaks were due to one fully vaccinated child aged 11 years who had been infected during his travel to UK. Snijders et al. (2012) analyzed a group of infected whose ages ranged in 3–13 years. The authors find out that no significant difference between the attack rate of the group aged 10–13 years and 3–5 years. Considering Scotland specifically, Donaghy et al. (2006) argued that the shift of ages observed in the epidemic in Scotland suggests that the propagation of mumps is becoming more widespread and diverse as the targeted population becomes more dynamic and mobile.

Having rejected the first two hypotheses, the arguments used lead to the third and more plausible hypothesis: MMR vaccine efficacy against mumps reduces over time (van Boven et al., 2013). In 2015 67% of those infected in Scotland were fully vaccinated individuals (1 and 2 doses confounded). Moreover, most primary cases occurred in adolescent and young adults, in contrast to the pre-vaccine era where outbreaks were among children of primary school age. Similar patterns can be found for Belgium in 2012 (Abrams et al., 2014) and in the US in 2006 (Dayan et al., 2008). Serological studies (Heffernan and Keeling, 2009; Park, 2015) show that susceptibility level increases (immunity wanes) as time from vaccination increases; however, the antibody threshold defining the protective level is not well specified for mumps (LeBaron et al., 2009). Even using two doses of the MMR vaccine, existing analyses (Cameron and Smith-Palmer, 2015; Park, 2015) stress that some of the population will remain at risk of disease unless additional control strategies are adopted.

We investigate the hypothesis of waning immunity using mathematically-based computational modeling. The basic model is a seasonal compartmental SEIR model (Anderson et al., 1984, 1987; Keeling and Grenfell, 2002), to which vaccination and immunity is added. We first show that the model produces comparable results to observed mumps data in Scotland[1], matching endemic levels of mumps with occasional larger epidemics, as in 2005 and 2015. Having established the accuracy of the model with historical data, we use it predictively to better understand the relationship between immunity and transmission, to illuminate long-term patterns of resurgent outbreaks, and to determine whether these can be controlled by extending immunity duration (e.g., by using another booster). While modeling has been previously used to investigate mumps and vaccination (Anderson et al., 1987; Abrams et al., 2014; Edmunds et al., 2000), the novelty of our approach lies in consideration of waning immunity and associated optimal control strategies. Our model shows clearly that waning immunity is a driver for a long period of oscillating outbreaks. Moreover, by working with epidemiologists to use mathematics to understand the observed clinical data, we illustrate the power of mathematics to inform public health policy through multi-disciplinary collaboration.

MUMPS EPIDEMIOLOGY IN SCOTLAND

During the period 1988–2015, Health Protection Scotland (HPS), the national surveillance center for Scotland, reported 10943 mumps cases. 10486 of these cases were between 2004 and 2015. Vaccination was introduced in 1988, with a second dose introduced in 1996. **Figure 1** shows the epidemic curve of mumps, and the vaccination uptake curves for both vaccines (MMR1 and MMR2). Observe the initial success of the vaccine (1988–2003) contrasted with a long potential cycle from 2004 to 2015, possible with sub-cycles (2005–2009, 2009–2012, 2012–2015). The 2004/2005 outbreak was related only partly to the

[1] Department of Health Protection Scotland, National surveillance center.

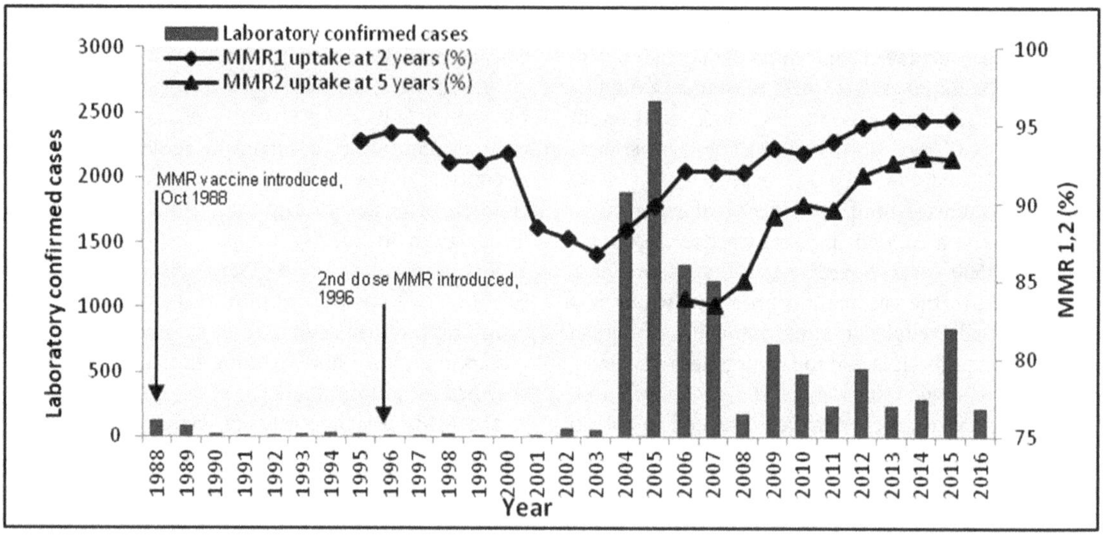

FIGURE 1 | Confirmed mumps cases, Scotland 1988-2016 and MMR vaccine coverage.

decrease in vaccination coverage shown in **Figure 1** (Donaghy et al., 2006). The majority of cases (94%) were born before 1990 (aged 15+ years), with only a few of them receiving only one dose of MMR (around 1%) or none at all. Similarly for the outbreaks in 2009 and 2012. In 2015 the highest incidence of mumps (63%) was related to the group born 1991–2000 (aged 15–24 years). Cameron and Smith-Palmer (2015) argue that the 2015 outbreak was the first where the majority of cases were fully vaccinated. Transmission is a complex feature to model as it can be influenced by many factors (vaccination history, current immunity status, age, opportunity for social mixing, geography, and so on). Moreover, some of these factors are confounded (e.g., age and vaccination history). We propose in this model that vaccination history is used as a proxy for these combined effects. Therefore, the main question arising is: why are vaccinated individuals being infected? Here we focus on the long curve (2005–2015) relating to the long inter-epidemic period. We explore these features within the model presented in Methods, using the Bio-PEPA plugin tool (Ciocchetta and Hillston, 2009) and deterministic simulation to provide time series prediction of the number of infected individuals. The model is parameterised and validated on data up to 2015, and then to further validate its predictive performance it is shown to match 2016 data provided by HPS. The advantages for using the Bio-PEPA formalism (a mathematically-defined computational modeling approach called process algebra) have been fully argued in many works (Ciocchetta and Hillston, 2009; Benkirane et al., 2012; Hamami and Atmani, 2013). Here, the advantages are: formal structuring of interactions between components, a compositional approach to building the epidemiological model, and a range of analysis techniques to support the modeler in understanding the system. The underlying semantics of Bio-PEPA is a continuous time Markov chain.

METHODS

Model Structure, Epidemiological Assumptions, and Parameter Estimates

We consider a compartmental structure for a model of mumps formulated as an extended SEIR (Anderson and May, 1992) model including seasonality and waning immunity: natively susceptible (S1), vaccinated individuals with MMR1 only (V1), vaccinated individuals with both MMR1 and MMR2 (V2), modified susceptible who are vaccinated individuals who have become susceptible (S2), exposed individuals (E), infected individuals (I) and recovered individuals who are regarded as immune for life (R) (Anderson and May, 1992; Greenhalgh and Sfikas, 2003). **Figure 2** shows how these compartments interact.

Our goal is to provide as simple a model as is necessary to demonstrate the impact of waning immunity, therefore we have ignored features which others have chosen to include. For example, the models of Glass and Grenfell (2003) and Barbarossa and Röst (2015) include immunity levels and immune-boosting through vaccination and interactions with infected. Since we have no data on antibody levels as individuals interact we choose not to include this, choosing the simpler scenario which can be parameterised through observed data. Neither do we include age-structure, as mumps has ceased to be a mainly childhood disease. As shown in several works (Donaghy et al., 2006; Centers for Disease Control and Prevention, 2009; Brockhoff et al., 2010; Fanoy et al., 2011), the range of those infected with mumps has become more diverse due to a more mobile susceptible population. Therefore, rather than stratifying the population by age, we assume a more homogenously-mixed population, with routine vaccination, and transmission based on seasonality and immunity status.

This model is general and could be parameterised for any seasonal disease with up to two vaccinations. We use data from

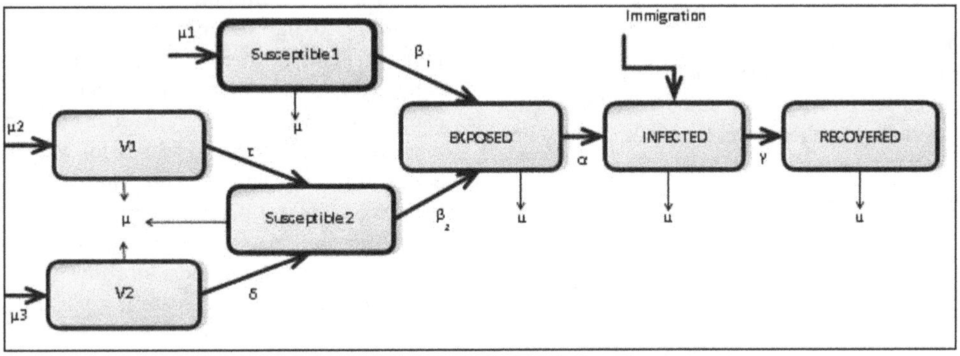

FIGURE 2 | Mumps structure.

Health Protection Scotland (HPS) from 2004 to 2016[2] and some parameters from the literature (Anderson and May, 1992; Keeling and Rohani, 2008). These are detailed in Table 1 Appendix 3 in Supplementary, with some explanatory text.

- Demographic estimation

Birth and death rate (μ) estimated from Scottish demographic data[3].

- Immigration rate estimation (λ)

As the net migration to Scotland is insignificant (typically 15,000 per year), the model has been simplified by having neither mass emigration nor immigration of susceptible individuals. A small constant rate of immigration of infected individuals is required to prevent the disease dying out entirely. This is justified by the knowledge that there is immigration, and there are many populations in the world where mumps is more prevalent and the global population is more mobile, transmitting disease between countries. A small rate of immigration of infectious individuals is estimated as in Finkenstädt et al. (1998) and Benkirane et al. (2012).

- Vaccination rates estimation (μ_1, μ_2, μ_3)

According to vaccination data[4], our basic assumption is an average of 94% MMR1 vaccination coverage (1988–2016) for children aged 0 to 2 years and 90% MMR2 vaccination coverage (1996–2016) for children aged 3 to 5 years. According to past vaccination history (Morgan-Capner et al., 1988; Public Health England, 2013), we estimate the susceptible portion of the remaining unvaccinated population at 20%. Within that proportion of susceptible we consider 11% of those to be aged 10 years or over according to current demographics. It would be more realistic to consider a varying vaccination rate each year; however, we did not want this to confound the patterns obtained through simply waning immunity. We do investigate scenarios in which these average vaccination rates are varied across the simulation period, to show how this affects the pattern of outbreaks.

- Waning immunity estimation (τ, δ)

Our basic assumption is individuals vaccinated with MMR1 and MMR2 (resp. only MMR1) are temporarily protected and that immunity wanes toward susceptibility at constant rate δ (resp. τ). LeBaron et al. (2009) report low antibody levels 4–9 years after MMR1 only, and 7–12 years after MMR2 administration. We also investigate scenarios in which these rates are varied.

- Transmission rate estimation ($\beta1, \beta2, \beta3$)

In our model, the transmission rate depends on two features: seasonality (High, Low) and type of susceptible (native susceptible, modified susceptible) giving four rates: $\beta1$ (High season and native susceptible), $\beta2$ (high season and modified susceptible, $\beta3$ (low season and native susceptible), $\beta4$ (low season and modified susceptible). For seasonality, data report higher number of cases October to May, and fewer between June and September[5]. As most cases occurs in 17–24 year-olds this seasonality is further supported through an assumption that many of that group are likely to be in full-time education, and mixing more in semester-time than in the holiday. As the total number of infected at low season is small we assume $\beta3 = \beta4$. In addition, we assume $\beta2 > \beta1$ (transmission in modified susceptible is higher than in native susceptible). This follows from the model of Scherer and McLean (2002), and is supported by the report of Cameron[6] that within 205 confirmed cases related to two health boards, 137 (67%) individuals were fully vaccinated. As transmission rate is based on the basic reproduction number R_0 (see Table 1 Appendix 3 in Supplementary), a range of proposed values were collected from literature (Anderson et al., 1987; Anderson and May, 1992; van Boven et al., 2013), where R_0 is ranged [4–11]. See Sensitivity Analysis for sensitivity analysis of the particular choices of these rates.

[2]Department of Health Protection Scotland, National surveillance center.
[3]Department of Health Protection Scotland, National surveillance center.
[4]*Information Service Division. Childhood Immunisation Statistics*, Available online at: http://www.isdscotland.org/Health-Topics/Child-Health/Immunisation/

[5]Department of Health Protection Scotland, National surveillance center.
[6]*R. Cameron. Health Protection Scotland, Immunization Team. Personal communication.* (Accessed Jan 21, 2016).

- Incubation rate α and recovery rate γ

Established empirical studies (Anderson et al., 1987; Anderson and May, 1992) estimate the incubation period between 12 and 25 days and the infectious period between 7 and 9 days (Anderson et al., 1987). For modeling convenience, we assume the same period of infection and incubation (Public Health England, 2013) for both natively susceptible and modified susceptible.

- Initial conditions

The initial mix of susceptible, vaccinated, exposed, infected and recovered is calculated for 1996 according to the above assumptions about population based on vaccination beginning in 1988. See Appendix 1 in Supplementary (model component).

The description of the model and parameters above can be summarized by seven ordinary differential equations:

$$\frac{dS1}{dt} = \mu 1N - \frac{\beta(t)\, S1\, I}{N} - \mu\, S1$$

$$\frac{dV1}{dt} = \mu 2N - \tau V1 - \mu\, V1$$

$$\frac{dV2}{dt} = \mu 3N - \delta V2 - \mu\, V2$$

$$\frac{dS2}{dt} = \delta V2 + \tau V1 - \frac{\beta'(t) S2\, I}{N} - \mu\, S2$$

$$\frac{dE}{dt} = \frac{\beta(t) S1\, I}{N} + \frac{\beta'(t) S2\, I}{N} - \alpha E - \mu\, E$$

$$\frac{dI}{dt} = \alpha E - \gamma I - \mu I + \lambda$$

$$\frac{dR}{dt} = \gamma I - \mu R$$

Where:

$$\beta(t)\left(resp.\ \beta'(t)\right) = \begin{cases} \beta 1\left(resp.\ \beta 2(t)\right)\ if\ Time \in [October - May] \\ \beta 3\ if\ Time \in [June - September] \end{cases}$$

This model is coded in Bio-PEPA (see Appendix 1 in Supplementary). Analysis of the model is performed through deterministic simulation. Stochastic simulation was used to guide model development but does not provide additional information when identifying long term trends.

Model Scenarios

To capture the impact of vaccination efficacy and the effect of waning immunity on the population of Scotland for future projection of epidemics, the history of mumps epidemics (from pre-vaccine to post-vaccine era) are reproduced where four strategies are considered:

- *Scenario one.* No vaccination. This is equivalent to the pre-vaccine era and useful for model validation where the whole population is considered susceptible.
- *Scenario two.* Immunity does not wane: τ and δ are zero. This case reflects the introduction of a vaccination protocol to case one, where immunity is assumed to be for life. This is consistent with the period immediately following the introduction of vaccination.
- *Scenario three.* Immunity wanes in vaccinated individuals according to the assumptions above. This scenario reflects modern reality, where mumps is resurgent. Our model is extended to two separate but correlated models: the first model expresses unvaccinated individuals and the second model expresses vaccinated individuals for whom immunity wanes. Scenario three is an extension to case two by introducing the terminology of waning immunity.
- *Scenario four.* An additional medical intervention increases immunity duration. We explore immunity duration across a range (10–80 years). This case is a particular variation of case three, where the immunity duration is specified in the defined range. This scenario is to predictively investigate possible future interventions.

RESULTS

According to observed mumps data in Scotland in **Figure 1**, and in conjunction with observed mumps data in England and Wales in Figures A1, A2 (see Appendix 3 in Supplementary), three different periods of an epidemiological shift in incidence are observed: pre-vaccine, successful post-vaccine and waning immunity period. **Figure 3** depicts time series results for infected cases under scenarios 1–3. Overall, it is clear that mumps occurs every year, regardless of vaccination or waning immunity; however, those factors control the amplitude of the epidemic and the frequency of the highest peaks driving a long term damping oscillation of large outbreaks. After 100 years the difference between the high and low of the cycle is around 25 cases.

Scenario One (No Vaccination = Pre-Vaccine Era)

We begin by checking model performance without vaccine. **Figure 3A** shows an inter-epidemic period of 3 years within an oscillatory pattern of mumps cases. This matches parameter values of incubation period of 13 days, infectious period of 7 days and a mean age of infection of 5 years (all within the ranges of Table 1 Appendix 3 in Supplementary). This is supported by the incidence of mumps in England and Wales (Anderson et al., 1987) and observations in the literature reporting cycles of 2–5 years (Galazka et al., 1999).

We point out that predicted cycles do not damp out during 100 years of simulations. By varying seasonality parameter of the model, including removing seasonality altogether, we observed that after a long period the model reaches an endemic state. To further reinforce the suitability of the model we considered R_0 ranging from [7 to 14]. **Figure 4** (see Appendix 3 in Supplementary) shows that increasing R_0 leads to decreasing the inter-epidemic period from 5 to 3 years.

Scenario Two (up to two vaccinations and immunity is permanent = immediate post-vaccine era)

Turning to the successful post-vaccine era (and assuming life-long immunity), **Figures 3B,D** show a massive decrease of mumps infections consistent with observed data 1988–2003, where waning immunity was not yet an important factor and the number of cases overall dramatically decreased due to the decreased pool of susceptibles, in turn due to vaccination.

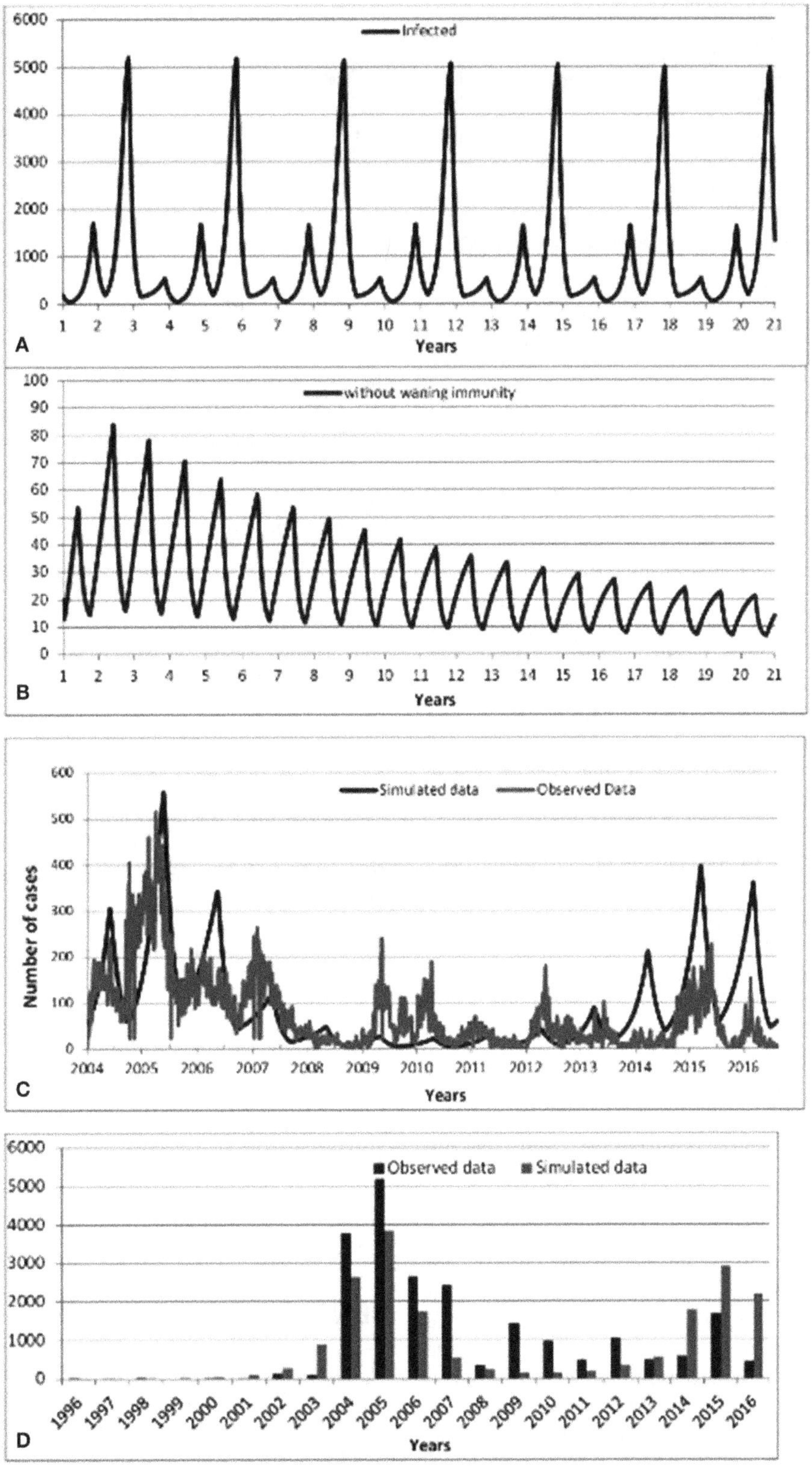

FIGURE 3 | Predicted incidence of mumps from 2004 to 2016:: (A) Scenario 1-No vaccination, **(B)** Scenario 2-Vaccination without waning immunity, **(C)** Scenario 3-Vaccination with waning immunity, **(D)** Predicted-Observed data for mumps from 1996 to 2016.

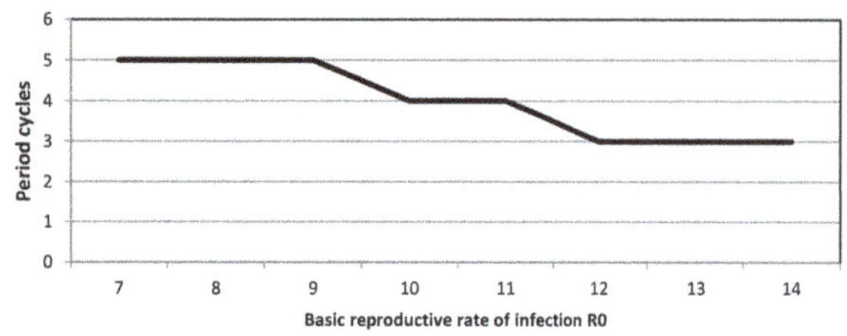

FIGURE 4 | Inter-epidemic period against basic reproductive rate R0 for pre-vaccine era.

Again, this helps to confirm that the model successfully models historical data.

Scenario Three (up to two vaccinations and immunity wanes)

Figure 3C (resp. **Figure 3D**) shows model prediction against observations from Scotland in the post-vaccine era (2004–2016, resp. 1996–2016). **Figure 3C** shows pattern of mumps outbreaks from 2004 to 2016 as waning immunity begins to be more relevant. The simulated data (black solid line) displayed in **Figure 3C** depicts patterns of mumps dynamics qualitatively similar to observed data (gray solid line). Mumps is notoriously under-reported (Takla et al., 2013) as, especially for those in whom immunity has waned, the disease is often milder (and infected do not seek medical attention). Our model has no notion of "level" of infection, therefore sub-clinical, mild, and serious infections are all counted and contribute to disease transmission. Observed data is scaled by two to compensate for under-reporting of mumps. This is a conservative estimate, based on higher uptake of vaccine in Scotland than in Germany (Takla et al., 2013). This is discussed further in the Discussion.

Figure 3D shows that 2005/2015 years were the dominant period reflecting the highest peaks of mumps infection. Some notable gaps are observed (2009, 2010 and 2012); the observed mumps dynamics are inherently stochastic and noisy. **Figures 3C,D** depicts that the simulated data for the year 2016 follows the same patterns as observed data, where the number of infected start to decrease. Qualitatively, the simulation results show that even if vaccination is applied, mumps is occurring each year, where the seasonal patterns of our model depict that the infection increases rapidly over the last few months of the year and the high peak is reached early at the start of the year. This is broadly in agreement with observed data.

Vaccination coverage dips in this period, but this is not the main factor leading to the resurgence and sustainability of mumps, nor is seasonality on its own (as above). We investigate the variability of vaccination coverage by ranging its value from [75 to 95], where 75% is the minimum value related to the threshold level and 95% is the maximum value of applied vaccine coverage in Scotland. **Figure 5** (see Appendix 3 in Supplementary) shows that increasing vaccine coverage leads to a decrease in the peak of infected[7] (from 1694 to 1413). This is 16%, and still produces a large number of cases. Therefore, increasing

[7] Average number of infected corresponds to the average of the highest peaks during 100 years of simulations.

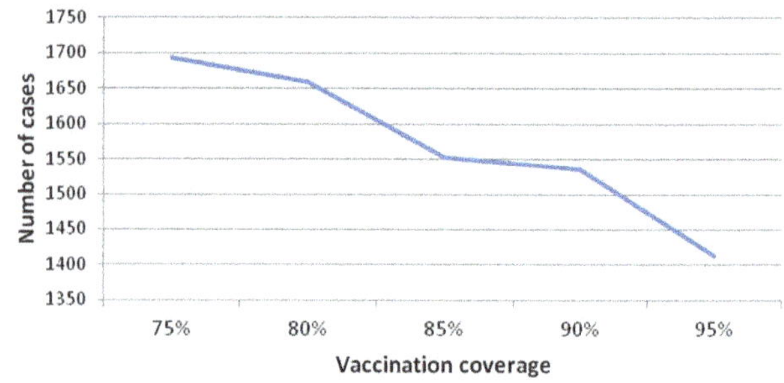

FIGURE 5 | Infected against vaccination coverage.

the vaccination coverage does not prevent disease occurrence. In addition, we note that all experiments (vaccination coverage ranging from [80 to 95]) settle into a 10 year pattern of gently damping oscillations (100 years of simulation), where the large oscillations are up to 2045, and thereafter the outbreaks become more and more regular in height.

To further investigate the impact of waning immunity **Figure 6** depicts separately those infected-unvaccinated and those infected-vaccinated against natively susceptible and modified susceptible over 100 years of temporal prediction. As expected, due to increasing levels of vaccinated individuals in the population, the number of natively susceptibles and infected-unvaccinated decreases over time, reaching a steady state of infection of around 200 individuals. Conversely, waning immunity leads to an increase in the number of modified susceptible and infected-vaccinated, settling into a 10 year pattern with peaks of between 800 and 1200. Therefore, waning immunity and its effects are the dominant portion of any epidemic.

Scenario Four (additional booster-up to three vaccinations and immunity wanes)

Further, we consider scenario 4: the impact of increasing the period of immunity by applying an additional dose of MMR (R and . Cameron , 2016). This could be similarly done by increasing immunity by increasing the efficacy of the vaccination (Public Health England, 2013). We investigate increasing immunity duration in steps from 10 to 80 years (broadly, life expectancy). **Figure 7** compares these scenarios and shows that the average of the number of infected individuals at the peak of each outbreak decreases with increasing duration of immunity, as expected.

SENSITIVITY ANALYSIS

The results above depend on precise parameter values, therefore we used sensitivity analysis to show that the qualitative results of periodic large outbreaks hold across the range. We identify significant parameters reproducing first the observed data, and second leading to the low level endemic state. Table 2 Appendix 3 in Supplementary shows the impact on epidemic amplitude and the periodicity of damping cycles of a series of experiments during 100 years of simulation varying model parameter values for: transmission rates ($\beta 1$, $\beta 2$, $\beta 3$), infectious period (γ),

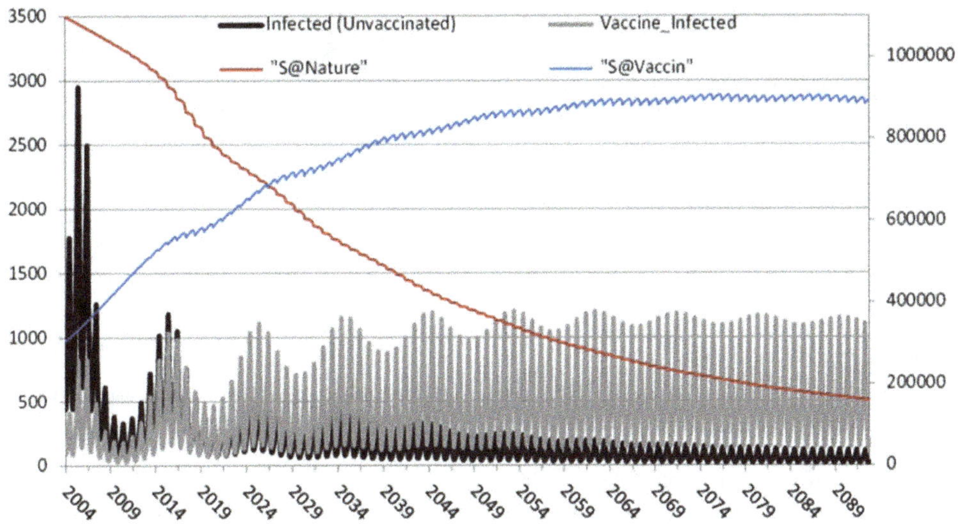

FIGURE 6 | The effect of waning immunity: Left axis: Infected-unvaccinated, Infected-vaccinated. Right axis: natively susceptible and modified susceptible.

incubation period (α), immunity duration (τ, δ), and vaccination rate ($\mu 1$, $\mu 2$, $\mu 3$). The values of the remaining parameters (birth rate, death rate and immigration rate) are fixed.

For all analysis we used ANOVA as implemented in Minitab (Minitab 17 Statistical Software, 2010). The full details of the analysis are in Appendix 2 in Supplementary: as expected, only varying transmission rates and immunity duration impact on results. Increasing R_0 leads to a decrease in period between large outbreaks and therefore an increase in the number of oscillations (see **Figure 8**, Appendix 3 in Supplementary). Smaller immunity durations increase the pool of susceptibles faster and therefore lead to larger and earlier epidemics.

DISCUSSION

Our analysis shows that mumps epidemics will continue, with larger outbreaks of ~1200 every 10 years as shown in **Figure 6**, eventually settling into an endemic state. This is despite high vaccination coverage against mumps (87–95%) since 1988 in Scotland[8] (well above the estimated herd immunity threshold of 75–86% Donaghy et al., 2006).

In this paper, we have presented the results of mathematical modeling using Bio-PEPA, identifying the impact of vaccination and waning immunity in the mumps component of the MMR vaccine. Even though vaccination has been ongoing since 1988, thus largely preventing mumps in children, our results show that waning immunity is the main factor in a repeated pattern of outbreaks. Simulations and analysis undertaken showed that waning immunity over 10 years leads to the highest number of infected and to the longest inter-epidemic period for larger outbreaks.

The first part of this study was to build a seasonal model which reproduces the patterns of the observed data in three scenarios: no vaccination, initial post-vaccine period with immunity for life, and with waning of vaccine-induced immunity as suggested by several sources (Dayan et al., 2008; Snijders et al., 2012; Abrams et al., 2014; Park, 2015). Those show that mumps is present in previously vaccinated individuals with the majority of those affected being university students. While based on Scottish data this is not a peculiarly Scottish phenomenon: for example, in the US (Harling et al., 2005), Korea (Park, 2015) and the Netherlands (Snijders et al., 2012) adolescent individuals were notified as infected despite high vaccine coverage. In these countries, it was observed that the majority of cases were in young adult (18–25 years) who have been fully vaccinated. In the US, where the first dose of MMR was introduced in 1977 and the second dose in 1990, the outbreak occurring in 2006 reached 6584 cases, 63% of whom received two doses of vaccine. For this country it was reported that in 1982 the incidence rate was reduced to 97% and the 3 year cycles observed in the pre-vaccine era disappeared. Moreover, in 2005, 1 year before the resurgence of the outbreak occurred in 2006, the incidence rate was damped to up to 99% where the vaccine coverage reached 91.5%. In the Netherlands, the large epidemic which occurred in 2004 led to the reintroduction of mumps as a notifiable disease. This followed its removal from the notifiable disease register in 1999 as a consequence of low outbreaks and vaccination coverage of at least one dose of MMR of at least 93% since the introduction of routine vaccine in 1987. In Korea, the epidemic of 2013–2014 showed that 99% of infected individuals aged from 13 to 18 years have been fully vaccinated. It is worth noting that Korea is not that different from other countries as in the pre-vaccine era the epidemic cycles were identified at 4 to 5 years and the mean age of infection at 4 to 6 years which shifted to teenagers in the recent outbreaks (2007 and 2013) in time when vaccination coverage rose to 90%.

Waning immunity is expressed in our model by including an additional compartment of modified susceptible, which is

[8]Department of Health Protection Scotland, National surveillance center.

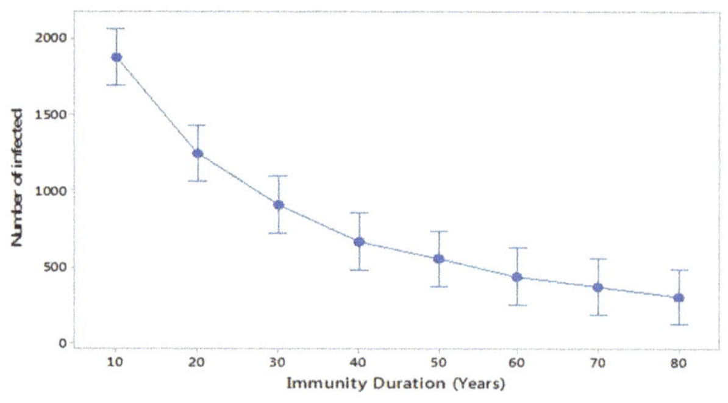

FIGURE 7 | Infected against duration of immunity.

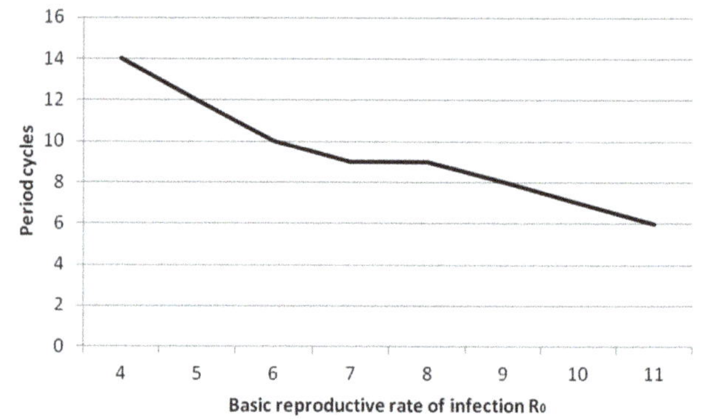

FIGURE 8 | Inter-epidemic period against basic reproductive rate R0 for post-vaccine era.

increased by vaccinated individuals (MMR1 and MMR2) losing their immunity. We find that assuming 5 years of MMR1 vaccine-induced immunity (resp. 10 years of MMR2 vaccine-induced immunity) generates simulation results consistent with more recent mumps post-vaccine data from Scotland (2004–2015). In addition, as our model suggests a 10-year-long gradually damping oscillation, the following trajectory of mumps disease would show a decrease in 2016 and so on, building back up from 2020 to another high peak in the year 2025. The most recent data provided by HPS has confirmed this prediction, where the year 2016 depicts 215 cases compared to 2015 which defines 836 cases. Although our estimates of the amplitude of mumps epidemics are higher than observed data, we conjecture that this can be explained by a low level of reporting. Anecdotally, cases of mumps in vaccinated individuals have much milder symptoms and therefore may be undetected (Public Health England, 2013; Takla et al., 2013; Cordeiro et al., 2015; Gouma et al., 2016).

By considering different values of immunity duration (scenario 4) we can estimate the time needed to reverse the epidemic trend and eliminate mumps. This models the situation that, for example, a new, more effective, vaccine is introduced, or a third vaccine dose is introduced into the national programme. This is shown in **Figure 7**. Even extending immunity to 80 years, a reasonable lifespan, mumps outbreaks still occur. Only by further increasing immunity duration to 150 years eliminates mumps outbreaks, assuming no perturbations occur, such as a new vaccine or new strain of mumps.

It is worth noting that the basic reproductive number R_0 for the pre-vaccine era is estimated at 10.5 which falls in the range [7–14] as cited in literature (Anderson and May, 1992; Keeling and Rohani, 2008) and for the post-vaccine era R_0 is estimated at 6 where in the literature it is quoted at [4–7] (Anderson and May, 1982). Recall that R_0 indicates the number of secondary infections, clearly showing that the number of doses of vaccination and immunity duration has a great impact on decreasing infectious contacts.

Cumulatively, our findings suggest that the more "unprotected" individuals (who were either never vaccinated or lost their immunity), the shorter the period between two high peaks of epidemic outbreak (note the number of cycles in Table 2 for varying values of R_0). In addition, in both cases related to scenarios 1 and 3 (No vaccination and waning immunity), an earlier high peak of mumps is expected. This occurs because the pool of susceptibles is increasing faster as those vaccinated lose their immunity and move to the susceptible state (scenario 3), or the pool of susceptibles is decreasing faster when no vaccination is applied and R_0 is higher (scenario 1). Clearly, controlling the number of susceptible individuals has a great impact on controlling disease. As argued by Gay (1998): to achieve elimination of an epidemic, low levels of susceptible individuals should be maintained, leading the basic reproductive number (R_0) to be <1. We do this here by adjusting immunity duration.

These conclusions illustrate an enhanced understanding of mumps disease in response to mass immunization gained through mathematical modeling. Further, our multi-disciplinary team could explore the potential impact of further vaccination on cyclic outbreaks. Our conclusion for public health services is that they should urge vaccine uptake in those eligible since a high degree of protection is offered by the vaccine overall for those under 18. Considering the possible economic cost/benefit of a third vaccine dose, it seems that while there would be an increased period of immunity, the cyclic outbreaks would continue at about 2/3 the current level, therefore this would not offer significant advantages over the present situation. The Joint Committee on Vaccination and Immunization[9] do not consider these large outbreaks of particular concern, since there has been no formal discussion to introduce a 3rd vaccine dose into the national programme.

We suggest further study with this model could include vaccination programmes targeted to those subject to waning immunity or at higher risk due to social mixing in a diverse population (as in higher education). Such a model might also include economic factors to allow the effect of targeted programmes to be more precisely evaluated. Another interesting facet would be to bring more attention to the level of immunity by analyzing the vaccine/virus content and detect eventual

[9] UK body advising government health policy on vaccination and immunisation.

discrepancy between vaccine strain and mumps outbreak. This might also be linked with a data science approach to analyzing serology of confirmed cases. There are further opportunities to use data science to analyse other features, such as geographic distribution. These developments would allow an enhanced version of **Figure 6** showing waves of outbreaks related to waning immunity, evolution of strains of mumps, and locality.

AUTHOR CONTRIBUTIONS

The Conception or design of the work: CS, DH, and KP. Data collection: RC, KP. Data analysis and interpretation: CS, DH, and KP. Drafting the article: DH, CS. Critical revision of the article: CS, KP, DH, and RC. Final approval of the version to be published: CS, KP, DH, and RC.

FUNDING

DH is grateful to the Algerian government, the University of Mostaganem and to the University of Oran for supporting her series of research visits to the University of Stirling.

ACKNOWLEDGMENTS

We thank the reviewers for their many helpful comments to improve this paper.

REFERENCES

Abrams, S., Beutels, P., and Hens, N. (2014). Assessing mumps outbreak risk in highly vaccinated populations using spatial seroprevalence data. *Am. J. Epidemiol.* 179, 1006–1017. doi: 10.1093/aje/kwu014

Anderson, R. M., Crombie, J. A., and Grenfell, B. T. (1987). The epidemiology of mumps in the UK: a preliminary study of virus transmission, herd immunity and the potential impact of immunization. *Epidemiol. Infect.* 99, 65–84. doi: 10.1017/S0950268800066875

Anderson, R. M., Grenfell, B. T., and May, R. M. (1984). Oscillatory fluctuations in the incidence of infectious disease and the impact of vaccination: time series analysis. *J. Hygiene* 93, 587–608. doi: 10.1017/S0022172400065177

Anderson, R. M., and May, R. M. (1992). *Infectious Diseases of Humans: Dynamics and Control, Vol. 28.* Oxford: Oxford University Press.

Anderson, R., and May, R. M. (1982). Directly transmitted infectious diseases: control by vaccination. *Science* 215:26.

Andreasen, V. (1993). The effect of age-dependent host mortality on the dynamics of an endemic disease. *Math. Biosci.* 114, 29–58. doi: 10.1016/0025-5564(93)90041-8

Barbarossa, M. V., and Röst, G. (2015). Immuno-epidemiology of a population structured by immune status: a mathematical study of waning immunity and immune system boosting. *J. Math. Biol.* 71, 1737–1770. doi: 10.1007/s00285-015-0880-5

Benkirane, S., Norman, R., Scott, E., and Shankland, C. (2012). "Measles epidemics and PEPA: an exploration of historic disease dynamics using process algebra," in *International Symposium on Formal Methods*, eds D. Giannakopoulou and D. Méry (Berlin; Heidelberg: Springer), 101–115. doi: 10.1007/978-3-642-32759-9

Brisson, M., Melkonyan, G., Drolet, M., De Serres, G., Thibeault, R., and De Wals, P. (2010). Modeling the impact of one-and two-dose varicella vaccination on the epidemiology of varicella and zoster. *Vaccine* 28, 3385–3397. doi: 10.1016/j.vaccine.2010.02.079

Brockhoff, H. J., Mollema, L., Sonder, G. J., Postema, C. A., van Binnendijk, R. S., Kohl, R. H., et al. (2010). Mumps outbreak in a highly vaccinated student population, The Netherlands, 2004. *Vaccine* 28, 2932–2936. doi: 10.1016/j.vaccine.2010.02.020

Cameron, R. L., and Smith-Palmer, A. (2015). *Measles, Mumps, Rubella and Whooping Cough Illness, Routine Childhood Vaccine Uptake.* Immunisation Team, HPS Weekly Report, 49, 251–259.

Centers for Disease Control and Prevention (2009). *Mumps outbreak-New York, New Jersey, Quebec, 2009.* MMWR. Morbidity and Mortality Weekly Report, 58:1270.

Ciocchetta, F., and Hillston, J. (2009). Bio-PEPA: a framework for the modelling and analysis of biological systems. *Theor. Comput. Sci.* 410, 3065–3084. doi: 10.1016/j.tcs.2009.02.037

Cordeiro, E., Ferreira, M., Rodrigues, F., Palminha, P., Vinagre, E., and Pimentel, J. P. (2015). Mumps outbreak among highly vaccinated teenagers and children in the Central Region of Portugal, 2012-2013. *Acta Med. Port.* 28, 435–441. doi: 10.20344/amp.5756

Dayan, G. H., Quinlisk, M. P., Parker, A. A., Barskey, A. E., Harris, M. L., Schwartz, J. M. H., et al. (2008). Recent resurgence of mumps in the United States. *N. Eng. J. Med.* 358, 1580–1589. doi: 10.1056/NEJMoa0706589

DeStefano, F., Price, C. S., and Weintraub, E. S. (2013). Increasing exposure to antibody-stimulating proteins and polysaccharides in vaccines is not associated with risk of autism. *J. Pediatr.* 163, 561–567. doi: 10.1016/j.jpeds.2013.02.001

Donaghy, M., Cameron, J. C., and Friederichs, V. (2006). Increasing incidence of mumps in Scotland: options for reducing transmission. *J. Clin. Virol.* 35, 121–129. doi: 10.1016/j.jcv.2005.09.009

Edmunds, W. J., Gay, N. J., Kretzschmar, M., Pebody, R. G., and Wachmann, H. (2000). The pre-vaccination epidemiology of measles, mumps and rubella in Europe: implications for modelling studies. *Epidemiol. Infect.* 125, 635–650. doi: 10.1017/S0950268800004672

Fanoy, E. B., Cremer, J., Ferreira, J. A., Dittrich, S., van Lier, A., Hahné, S. J., et al. (2011). Transmission of mumps virus from mumps-vaccinated individuals to close contacts. *Vaccine* 29, 9551–9556. doi: 10.1016/j.vaccine.2011.09.100

Ferguson, N. M., Nokes, D. J., and Anderson, R. M. (1996). Dynamical complexity in age-structured models of the transmission of the measles virus: epidemiological implications at high levels of vaccine uptake. *Math. Biosci.* 138, 101–130. doi: 10.1016/S0025-5564(96)00127-7

Finkenstädt, B., Keeling, M., and Grenfell, B. (1998). Patterns of density dependence in measles dynamics. *Proc. R. Soc. Lond. B Biol. Sci.* 265, 753–762. doi: 10.1098/rspb.1998.0357

Galazka, A. M., Robertson, S. E., and Kraigher, A. (1999). Mumps and mumps vaccine: a global review. *Bull. World Health Organ.* 77:3.

Galbraith, N. S., Pusey, J., Young, S. J., Crombie, D. L., and Sparks, J. P. (1984). Mumps surveillance in England and Wales 1962-81. *Lancet* 323, 91–94. doi: 10.1016/S0140-6736(84)90015-1

Gay, N. J. (1998). Modeling measles, mumps, and rubella: implications for the design of vaccination programs. *Infect. Control Hosp. Epidemiol.* 19, 570–573. doi: 10.2307/30141782

Glass, K., and Grenfell, B. T. (2003). Antibody dynamics in childhood diseases: waning and boosting of immunity and the impact of vaccination. *J. Theor. Biol.* 221, 121–131. doi: 10.1006/jtbi.2003.3181

Glass, K., and Grenfell, B. T. (2004). Waning immunity and subclinical measles infections in England. *Vaccine* 22, 4110–4116. doi: 10.1016/j.vaccine.2004.02.047

Gouma, S., Hahné, S. J., Gijselaar, D. B., Koopmans, M. P., and van Binnendijk, R. S. (2016). Severity of mumps disease is related to MMR vaccination status and viral shedding. *Vaccine* 34, 1868–1873. doi: 10.1016/j.vaccine.2016.02.070

Greenhalgh, D., and Sfikas, N. (2003). Vaccination programs against mumps in the United Kingdom. *J. Med. Inform. Technol.* 5, MI13–MI22.

Hamami, D., and Atmani, B. (2013). Tuberculosis modelling using Bio-PEPA approach. *Int. J. Med.* 7, 183-190. Available online at: scholar.waset.org/1999.9/9996654

Harling, R., White, J. M., Ramsay, M. E., Macsween, K. F., and van den Bosch, C. (2005). The effectiveness of the mumps component of the MMR vaccine: a case control study. *Vaccine* 23, 4070–4074. doi: 10.1016/j.vaccine.2004.10.020

Heffernan, J. M., and Keeling, M. J. (2009). Implications of vaccination and waning immunity. *Proc. R. Soc. Lond. B Biol. Sci.* 276, 2071–2080. doi: 10.1098/rspb.2009.0057

Hethcote, H. W. (2000). The mathematics of infectious diseases. *SIAM Rev.* 42, 599–653. doi: 10.1137/S0036144500371907

Isaacs, D., and Menser, M. (1990). Measles, mumps, rubella, and varicella. *Lancet* 335, 1384–1387. doi: 10.1016/0140-6736(90)91255-9

Keeling, M. J., and Grenfell, B. T. (2002). Understanding the persistence of measles: reconciling theory, simulation and observation. *Proc. R. Soc. Lond. B Biol. Sci.* 269, 335–343. doi: 10.1098/rspb.2001.1898

Keeling, M. J., and Rohani, P. (2008). *Modeling Infectious Diseases in Humans and Animals.* Princeton, NJ: Princeton University Press.

LeBaron, C. W., Forghani, B., Beck, C., Brown, C., Bi, D., Cossen, C., et al. (2009). Persistence of mumps antibodies after 2 doses of measles-mumps-rubella vaccine. *J. Infect. Dis.* 199, 552–560. doi: 10.1086/596207

Le Menach, A., Boxall, N., Amirthalingam, G., Maddock, L., Balasegaram, S., and Mindlin, M. (2014). Increased measles–mumps–rubella (MMR) vaccine uptake in the context of a targeted immunisation campaign during a measles outbreak in a vaccine-reluctant community in England. *Vaccine* 32, 1147–1152. doi: 10.1016/j.vaccine.2014.01.002

Marco, D., Scott, E., Cairns, D., Graham, A., Allen, J., Mahajan, S., et al. (2012). "Investigating co-infection dynamics through evolution of Bio-PEPA model parameters: a combined process algebra and evolutionary computing approach," in *Computational Methods in Systems Biology* (Berlin; Heidelberg: Springer), 227–246.

Minitab 17 Statistical Software (2010). *[Computer software]*. State College, PA: Minitab, Inc. Available online at: www.minitab.com

Morgan-Capner, P., Wright, J., Miller, C. L., and Miller, E. (1988). Surveillance of antibody to measles, mumps, and rubella by age. *BMJ* 297, 770–772. doi: 10.1136/bmj.297.6651.770

Nardone, A., Pebody, R. G., van Den Hof, S., Levy-Bruhl, D., Plesner, A. M., Rota, M. C., et al. (2003). Sero-epidemiology of mumps in western Europe. *Epidemiol. Infect.* 131, 691–701. doi: 10.1017/S0950268803008768

Park, S. H. (2015). Resurgence of mumps in Korea. *Infect. Chemother.* 47, 1–11. doi: 10.3947/ic.2015.47.1.1

Public Health England (2013). "Mumps: the green book, chapter 23. Mumps immunisation information for public health professionals," in *Immunisation Against Infectious Disease Mumps: Guidance, Data and Analysis 2013.* London: Public Health England.

Savage, E., Ramsay, M., and Crowcroft, N. (2004). *Increases in Mumps Cases in England and Wales in 2004.* Eurosurveillance Weekly, 8, 3-4. Available online at: http://www.eurosurveillance.org/ViewArticle.aspx?ArticleId=2591.

Scherer, A., and McLean, A. (2002). Mathematical models of vaccination. *Br. Med. Bull.* 62, 187–199. doi: 10.1093/bmb/62.1.187

Snijders, B. E., van Lier, A., van de Kassteele, J., Fanoy, E. B., Ruijs, W. L., Hulshof, F., et al. (2012). Mumps vaccine effectiveness in primary schools and households, the Netherlands, 2008. *Vaccine* 30, 2999–3002. doi: 10.1016/j.vaccine.2012.02.035

Takla, A., Wichmann, O., Klinc, C., Hautmann, H., Rieck, T., and Koch, J. (2013). Mumps epidemiology in Germany 2007-11. *Euro Surveill.* 18:20557. doi: 10.2807/1560-7917.ES2013.18.33.20557

Ueda, K. (2016). Epidemiology of rubella and congenital rubella syndrome in Japan before 1989. *Vaccine* 34, 1971–1974. doi: 10.1016/j.vaccine.2015.10.010

van Boven, M., Ruijs, W. L., Wallinga, J., O'Neill, P. D., and Hahne, S. (2013). Estimation of vaccine efficacy and critical vaccination coverage in partially observed outbreaks. *PLoS Comput. Biol.* 9:e1003061. doi: 10.1371/journal.pcbi.1003061

7

Systematic Characterization of Dynamic Parameters of Intracellular Calcium Signals

Laurent Mackay[1†], *Nicholas Mikolajewicz*[2,3†], *Svetlana V. Komarova*[2,3] and *Anmar Khadra*[1*]

[1] Department of Physiology, McGill University, Montreal, QC, Canada, [2] Faculty of Dentistry, McGill University, Montreal, QC, Canada, [3] Shriners Hospital for Children-Canada, Montreal, QC, Canada

Edited by:
Krasimira Tsaneva-Atanasova,
University of Exeter, UK

Reviewed by:
Silvina Ponce Dawson,
University of Buenos Aires, Argentina
James Sneyd,
University of Auckland, New Zealand
Martin Falcke,
Max Delbrück Center for Molecular Medicine (HZ), Germany

***Correspondence:**
Anmar Khadra
anmar.khadra@mcgill.ca

† These authors have contributed equally to this work.

Dynamic processes, such as intracellular calcium signaling, are hallmark of cellular biology. As real-time imaging modalities become widespread, a need for analytical tools to reliably characterize time-series data without prior knowledge of the nature of the recordings becomes more pressing. The goal of this study is to develop a signal-processing algorithm for MATLAB that autonomously computes the parameters characterizing prominent single transient responses (TR) and/or multi-peaks responses (MPR). The algorithm corrects for signal contamination and decomposes experimental recordings into contributions from drift, TRs, and MPRs. It subsequently provides numerical estimates for the following parameters: time of onset after stimulus application, activation time (time for signal to increase from 10 to 90% of peak), and amplitude of response. It also provides characterization of the (i) TRs by quantifying their area under the curve (AUC), response duration (time between ½ amplitude on ascent and descent of the transient), and decay constant of the exponential decay region of the deactivation phase of the response, and (ii) MPRs by quantifying the number of peaks, mean peak magnitude, mean periodicity, standard deviation of periodicity, oscillatory persistence (time between first and last discernable peak), and duty cycle (fraction of period during which system is active) for all the peaks in the signal, as well as coherent oscillations (i.e., deterministic spikes). We demonstrate that the signal detection performance of this algorithm is in agreement with user-mediated detection and that parameter estimates obtained manually and algorithmically are correlated. We then apply this algorithm to study how metabolic acidosis affects purinergic (P2) receptor-mediated calcium signaling in osteoclast precursor cells. Our results reveal that acidosis significantly attenuates the amplitude and AUC calcium responses at high ATP concentrations. Collectively, our data validated this algorithm as a general framework for comprehensively analyzing dynamic time-series.

Keywords: algorithm, calcium imaging, kinetics, osteoclast pathophysiology, parameter characterization, purinergic/P2 receptors, real-time imaging

INTRODUCTION

Cellular biology is vastly populated with dynamic processes, which can be altered dramatically or subtly by pathological causes. Calcium signals, characterized by fast and transient increases in cytosolic free calcium concentration ($[Ca^{2+}]_i$), which vary in amplitude and duration and can exhibit oscillatory dynamics with frequency-dependent downstream effects (Clapham, 2007), represent a prominent example of such dynamic processes (**Figure 1A**). To fully understand the data of such dynamic complexity, a robust methodology to analyse, and characterize these responses is necessary. Numerous studies have investigated $[Ca^{2+}]_i$ dynamics, but the analysis have in many cases been limited to qualitative assessments (Cao et al., 1997; Frame and de Feijter, 1997; Jorgensen et al., 1997; Jørgensen et al., 2003; Isakson et al., 2001; Romanello and D'Andrea, 2001). Studies that have pursued quantitative analysis of calcium time-series reported a number of different, often non-overlapping characteristics of the response (**Table 1**). In cases where experiments were conducted on a smaller-scale, manual analysis was achievable. However, to achieve larger-scale analyses for experiments with hundreds of individual recordings, open-source signal-processing algorithms are required and becoming increasingly relied on to overcome these bottlenecks in productivity. None of 11 published algorithms we examined provided a comprehensive analysis of the entire response observed within a recording (**Table 2**). As a direct consequence of the lack of a standardized methodology to quantify such data-sets, findings from various studies are challenging to compare, relate, and generalize. Hence, the motivation of this study was to achieve faster analysis while standardizing the methodology involved, thereby minimizing user-bias, and ensuring consistency in the analysis of complex physiological signals. While such a tool may or may not change the conclusions of individual studies, it would improve comparability between different studies, and potentially enable meta-analysis of different experiments. We have therefore developed an algorithm that addresses these concerns and focused on the dynamic signals generated by purinergic (P2) receptors to demonstrate its utility.

Purinergic receptors that evoke intracellular responses upon extracellular stimulation with nucleotides, such as ATP and ADP, are known to induce complex $[Ca^{2+}]_i$ signals. P2 receptors are subdivided into two families, P2X and P2Y receptors, which are omnipresent in virtually all mammalian tissue (Burnstock and Verkhratsky, 2009). The mammalian P2X receptor family, consisting of seven subtypes ($P2X_{1-7}$), are ionotropic ligand-gated cation channels that can permit the influx of extracellular calcium upon stimulation (Kaczmarek-Hajek et al., 2012). The mammalian P2Y receptor family, consisting of eight subtypes ($P2Y_{1-2,4,6,11-14}$), are metabotropic G-protein coupled receptors that can indirectly modulate the release of calcium from intracellular calcium stores through inositol triphosphate (von Kugelgen and Hoffmann, 2016). P2 receptors have been demonstrated to play an important role on bone physiology (Lenertz et al., 2015). Since individual bone cells commonly express multiple active P2 receptors (Gallagher and Buckley, 2002), responses to purinergic stimulation result in complex, concentration-dependent $[Ca^{2+}]_i$ transients (Xing et al., 2016). While it remains difficult to experimentally isolate the contribution of individual receptors, a number of studies have demonstrated that various P2 receptor subtypes have distinct calcium response kinetics and signatures. For instance, various P2X receptors desensitize at distinct rates under sustained agonist stimulation (Koshimizu et al., 1999). $P2X_7$-mediated responses in particular are biphasic (Yan et al., 2010) and characterized by sustained $[Ca^{2+}]_i$ elevation (Nobile et al., 2003). It is becoming increasingly clear, however, that P2 receptors cannot be studied and manipulated as individual components, but rather must be regarded as building blocks of a far more "dynamic architecture" that permits diverse functionality and flexibility (Volonte et al., 2006).

The goal of this study is to develop a universal signal-processing algorithm for MATLAB (MathWorks, Natick, MA) that would facilitate and standardize the parameter characterization of time series calcium imaging recordings containing prominent single transient responses (TR) and/or multi-peaked responses (MPR). All signals, no matter their complexity, can be reduced to a set of defined characteristics that describe the magnitude and kinetics of a given response. Based on our expertise and literature review (**Tables 1, 2**), we have selected the following parameters: time of onset after stimulus application (t_{onset}), activation time (time for signal to increases from 10 to 90% of peak; $t_{10-90\%}$), and amplitude of response. Additionally, TRs are specifically described by their area under the curve (AUC), response duration (time between ½ amplitude on ascent and descent of the transient; FWHM), and decay constant of the exponential decay region of the deactivation phase of the response (τ_{decay}, **Figure 1B**) while MPRs are described by their number of peaks (N_{osc}), mean peak magnitude (E), mean periodicity (T), standard deviation of periodicity (σ_T), oscillatory persistence (time between first and last discernable peak; l_{osc}), and duty cycle (fraction of period during which system is active; ξ^{peak}/T, where ξ^{peak} is the width of the oscillatory peaks, **Figure 1C**). Since MPRs can be either stochastic or deterministic (Skupin et al., 2008; Dupont and Combettes, 2009; Dupont et al., 2011), the algorithm reports two sets of MPR parameters. The first describes MPR parameters for all the peaks present, while the second set reports the MPR parameters describing the subset of coherent oscillations, to omit the influence of stochastic processes, and to focus on the deterministic properties of the signal.

Live cell recordings will inevitably contain signal contaminations arising from experimental conditions and instrumentation, including (a) photochemical effects induced by the measurement process and (b) unrelated biological processes. While these imperfections are inherent to the experimental

---

Abbreviations: ADP, Adenosine diphosphate; ATP, Adenosine triphosphate; AUC, Area under curve; $[Ca^{2+}]_i$, Cytosolic free calcium concentration; D(t), Global drift model; E, Mean peak magnitude of MPR; FWHM, Full-width half-max; F(t), Measured fluorescent signal; l_{osc}, Oscillatory persistence; MPR, Multi-peak response; N_{osc}, Number of peaks in MPR; ξ^{peak}, Width of oscillatory peak; P2, Purinergic; σ_T, Standard deviation of periodicity; T, Period of oscillation; $t_{10-90\%}$, Activation time; τ_{decay}, Decay constant; t_{onset}, Time of response onset; TR, Transient response; TV, Total-variational.

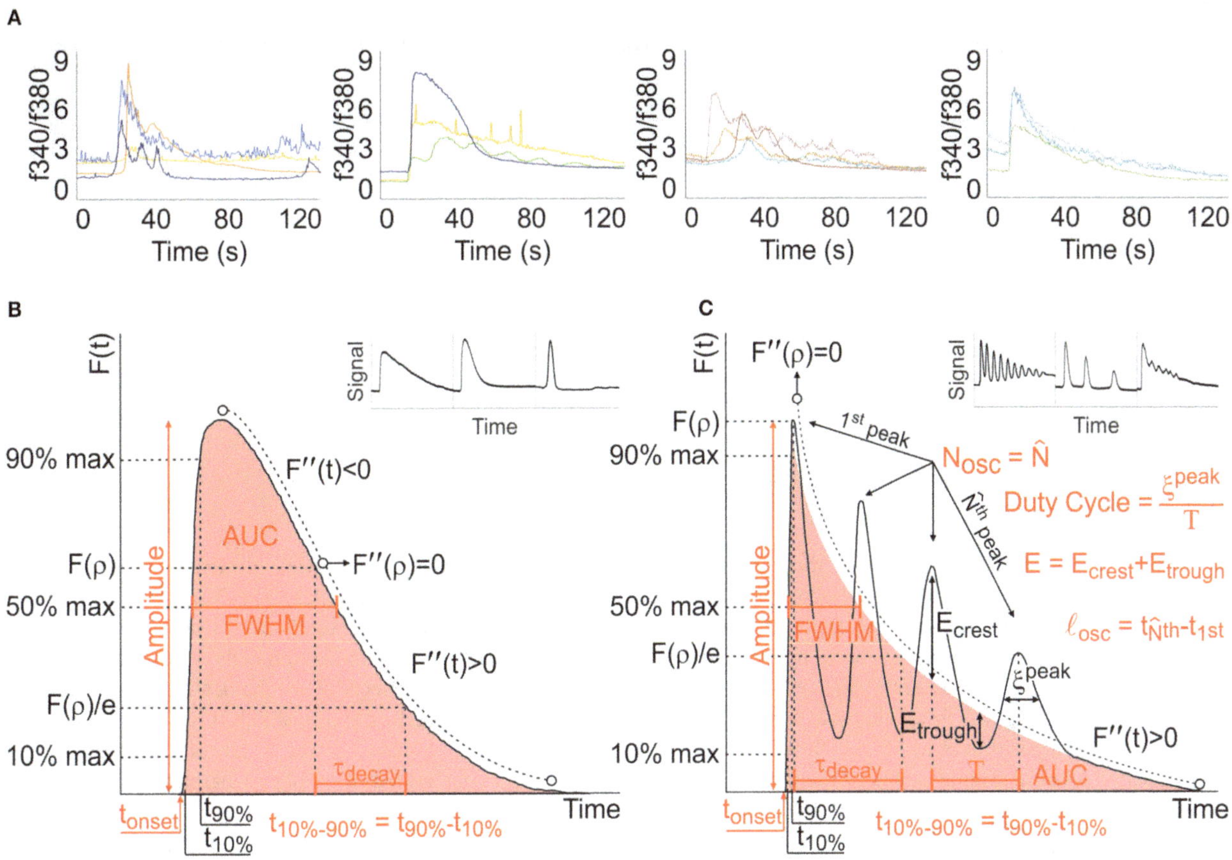

FIGURE 1 | Characterization of dynamic calcium signals. (A) Examples of heterogeneity of $[Ca^{2+}]_i$ responses observed in various live-cell recordings; note many signature forms that TRs and MPRs can exhibit. **(B)** Analysis of parameters for single-peak TRs. Inset: Representative single-peak TRs. **(C)** Analysis of parameters for MPRs. Inset: Representative MPRs. Time of onset, t_{onset}; area under curve, AUC; full-width half-max, FWHM; activation time, $t_{10-90\%}$; decay constant, τ_{decay}; periodicity, T; number of peaks, N_{osc}; oscillatory peak magnitude, E; width of oscillatory peaks, ξ^{peak}; point of inflection in deactivation phase of TR, $F''(\rho)$. Arrows/text in red illustrate how parameters of interest are obtained.

TABLE 1 | Commonly reported parameters in studies investigating calcium dynamics.

Amplitude	t_{onset}	$t_{10-90\%}$	FWHM	AUC	τ_{decay}	Period	References
X	X						Abu Khamidakh et al., 2013
X			X			X	Appleby et al., 2015
X							Churchill et al., 1996
X	X					X	Dickinson and Parker, 2013
X			X				Francis et al., 2016
X	X						Hansen et al., 1993
X	X	X		X		X	James et al., 2011
X							Rast et al., 2015
X							Shabir and Southgate, 2008
X		X	X		X		Smith et al., 2009
X			X				Sun et al., 1997
X							Zhao et al., 2008

process, dynamic processes of interest can still be extracted from these recordings. This process in itself can be complicated and highly subjective depending on the extent to which the raw data are corrupted by noise and drift. Therefore, to reliably evaluate the magnitude and kinetics of these signals, we have developed a systematic way of first identifying unwanted

TABLE 2 | Published signal-processing algorithms.

Amplitude	t_{onset}	$t_{10-90\,\%}$	FWHM	AUC	τ_{decay}	$t_{90-10\%}$	Period	Classifier*	References
X		X	X						Bray et al., 2007
X		X				X			Ellefsen et al., 2014
	X					X			Fritzsche et al., 2015
								X	Juhola et al., 2015
X		X	X			X			Lock et al., 2015
X		X				X	X		Patel et al., 2015
X		X	X		X				Picht et al., 2007
							X		Ruffinatti et al., 2011
X		X				X			Stoehr et al., 2014
X			X		X		X		Steele and Steele, 2014
X									Wong et al., 2010

**Classifier: grouping of time-series by characteristic signature of response.*

signal contaminations, and then removing their effects when determining the parameters of interest. Following algorithm validation, we have also investigated the effect of acidosis on ATP-mediated $[Ca^{2+}]_i$ responses in bone-marrow derived osteoclast precursors to demonstrate the efficacy of this algorithm in characterizing real-time cellular dynamics.

To implement the algorithm in MATLAB, the user is required to store the set of discrete sample points of the measured signal $F(t)$, $\{F_i\}$, $i = 1, 2, \cdots, N$, along with the corresponding discrete time points $\{t_i\}$ in an Excel file (filename.xlsx). The algorithm can then be run using the MATLAB command

```
≫characterizeDocument("filename.xlsx")
```

The MATLAB code required to execute this command along with examples of data and scripts are provided in Supplmentary Materials (Data Sheet 1).

MATERIALS AND METHODS

Cell Culture

All procedures were approved by McGill University's Animal Care Committee and complied with the ethical guidelines of the Canadian Council on Animal Care. Bone marrow precursor cells were isolated from the femur and tibia of 6 week old FVB mice (Charles River), plated at a density of 7.5×10^3 on 48-well glass-bottom plates (No. 1.5 Coverslip, 6 mm glass diameter, uncoated, MatTek Corp.) and cultured for 3 days in αMEM (12,000-022, GIBCO) supplemented with 10% FBS (080152, Wisent), 1% sodium pyruvate (600-110-UL, Wisent), 1% penicillin streptomycin (450-201-EL, Wisent), 50 ng/mL MCSF (300-25, Peprotech), and 50 ng/mL RANKL according to the protocol previously described (Boraschi-Diaz and Komarova, 2016).

Intracellular Calcium Measurements

After 3 days of culture, osteoclast precursors were loaded with fura2-AM, a ratiometric fluorescent calcium dye (F1221, Invitrogen), incubated at room temperature for 30 min and washed twice with physiological solution (130 mM NaCl; 5 mM KCl; 1 mM $MgCl_2$; 1 mM $CaCl_2$; 10 mM glucose; 20 mM HEPES, pH 7.6). The final volume of 270 μL of physiological solution at pH 7.6 or pH 7.0 was added and cells were acclimatized for 10 min to reduce the effects of mechanical agitation that resulted from fura2-AM loading and washing. 10X ATP (Sigma) solutions were prepared in physiological solution at corresponding pH and 30 μL was added after 10 s of baseline $[Ca^{2+}]_i$ recording to obtain a 1X dilution (i.e., final concentrations ranging from 1 μM to 10 mM ATP). $[Ca^{2+}]_i$ was imaged for an additional 110 s at a sampling rate of 2 images per second using a fluorescent inverted microscope (T2000, Nikon). The excitation wavelength was alternated between 340 and 380 nm using an ultra-high-speed wavelength switching illumination system (Lambda DG-4, Quorum Technologies). Regions of interest (ROI) were manually defined and the ratio of the fluorescent emission at 510 nm, following 340 and 380 nm excitation (f340/f380), was calculated and exported using the imaging software (Velocity, Improvision). All data were imported into an excel spreadsheet for subsequent analysis.

Validation and Statistical Analysis

Algorithm performance was evaluated using the algorithm generated figures for 450 individual signal fitting that enabled retrospective visual examination of both response-detection and quality of parameter fitting. Manual and automated estimates were compared using a correlation plot and Bland Altman analysis (Bland and Altman, 1986) to assess the degree of correlation and agreement, respectively. For correlation analysis, the line of exact linear correlation (i.e., $y = x$) is plotted as a reference to assess deviation of the linear regression curve from the desired 1:1 relationship between the manual and automated estimates. For the Bland Altman analysis, we compared the automated (a) and manual (m) parameter estimates of the ith recording to obtain a Z-score, given by

$$Z_i = \frac{x_i - \bar{x}}{\sigma_x}$$

where $x_i = (m_i + a_i)/2$ is the average value of the estimated parameter, $\bar{x}$ is the mean value of x averaged over all recordings, and σ_x is the standard deviation of x overall recordings. Furthermore, the percent difference, Ξ_i, of the ith recording is defined by

$$\Xi_i = \frac{\text{error}}{\text{average value}} \times 100\% = \frac{a_i - m_i}{x_i} \times 100\%.$$

Ξ_i vs. Z_i were plotted to illustrate systematic biases. Negative values of Ξ_i were interpreted as manual estimates being greater than automated estimates, and vice versa. A quantitative estimate of the interval of agreement, within which 95% of differences lie, is defined by

$$95\% \text{ interval of agreement} = \bar{\Xi} \pm 1.96\sigma_{\bar{\Xi}}$$

where $\bar{\Xi}$ is the mean percent difference over all recordings and $\sigma_{\bar{\Xi}}$ is the standard deviation of $\bar{\Xi}$.

Experimental data were expressed as means ± S.E.M. Effect of ATP treatment under control conditions was evaluated using one-way ANOVA followed with Bonferroni *post hoc* test. The effect of acidosis was evaluated using two-way ANOVA with Bonferroni *post hoc* test. Results were accepted as significant at $p < 0.05$. Statistical analysis was performed in MATLAB.

RESULTS AND DISCUSSION

Although the notation used throughout the text implies that the recorded signals are fluorescence, the methodology remains the same for any other type of signals. For fluorescent recordings, the measured fluorescence F may consist of multiple parts: the drift, TR including the activation and deactivation phases, and the superimposed MPR. It can be expressed as the sum of the actual signal, F_{true}, and the normally distributed noise with a standard deviation σ.

$$F(t) = F_{true}(t) + \mathcal{N}(0, \sigma). \tag{1}$$

To characterize parameters that reliably reflect $F_{true}(t)$, $F(t)$ is first preprocessed to remove the effects of noise (Section Noise Characterization) and to estimate the contributions of drift to $F_{true}(t)$ (Section Baseline Drift). Next, the activation phase of TR is fit while simultaneously refining the estimated contribution of the drift. This approach allows us to determine if the recording is consistent with the expected model of a TR superimposed on a drifting baseline (i.e., whether activation phase is followed by a deactivation phase, Section Activation Fitting). If a TR is detected, we proceed by fitting the full set of TR model parameters simultaneously with the drift parameters [Section Transient Response (TR) Model]. In the case where there remain multiple significant deviations in the data from the TR model, we investigate and characterize the presence of oscillatory MPRs (Section Multi-Peaked Responses). The fitting of the TR is refined to remove the effects of the multiple peaks on the initial fit (Section Identifying Coherent Oscillations), in order to provide the best estimate of the baseline around which the MPRs oscillate. The deviations resulting from this secondary fitting of the TR are then analyzed to determine those resulting from coherent oscillatory processes (Section Characterizing Oscillatory Parameters). At each step throughout the fitting procedure, an updated estimate of the optimal set of parameters (e.g., the drift parameters) is obtained. These parameters are then used in a feed-forward manner, where the optimal set of parameters of the preceding fit is used as an initial guess for the subsequent step. This ensures that the algorithm produces high-fidelity fittings. Finally, the algorithm performance and utility is demonstrated with a new data set describing the effect of acidosis on ATP-induced calcium signaling in osteoclast precursors (Section Application to Pathophysiology).

Noise Characterization

The first step in the processing of data is to evaluate four values that will be used for the remainder of the text: the derivative ($\hat{u}$) of the noisy signal, the standard deviation of noise (σ_{-}), indices at which this noise is not prevalent (j), and the noise-to-signal ratio (ϕ). $\hat{u}$ will be used in Section Drift Delimitation to separate the TR from the underlying drift. Data points excluded by j (i.e., noise) will be omitted. The methodology detailed throughout the following section is an iterative procedure. In the instances where those quantities are used, we are referring to the value determined by the final iteration of the procedure.

Euler-Lagrange Formalism

The presence of noise in a recording renders naïve methods of derivative estimation inadequate (Chartrand, 2011). This is particularly exacerbated by the intermittent presence of large amplitude noise (spikes) related to the use of high gain settings on instrumentation. To reliably delimit (i.e., define the boundaries of) the drift in a recording of a noisy transient signal (Section Drift Delimitation), we have adapted the total-variational (TV) technique commonly used to estimate the first derivative of a signal contaminated with various types of noise (Chartrand and Staneva, 2008; Chartrand and Wohlberg, 2010; Chartrand, 2011). This technique performs better than the low-pass filter in distinguishing the drift from the transient response, as it does not indiscriminately remove high frequency components that affect the overall trend of the signal.

Our TV-based methodology seeks a function, $\hat{u}(t)$, which represents the derivative of $F_{true}(t)$, that solves the optimization problem.

$$\min_u \alpha \int_0^L \left|\frac{du}{dx}\right| dx + \frac{1}{2}\int_0^L |(Au(x) - F(x))|^2 dx \tag{2}$$

The first term in Equation (2) is a regularization term which penalizes sudden changes in the derivative (to make the fitting smooth), the second term is an L^2 data fidelity term, where A is the anti-differentiation operator ($Au \approx F_{true}$), and α is a parameter dictating the balance between the two terms. In order to solve this minimization problem, we have to find the stationary solution to the following equation

$$u_t(x) = \alpha \frac{d}{dx}\frac{u'(x)}{|u'(x)|} - A^T(Au(x) - F) \tag{3}$$

derived from the Euler-Lagrange equation associated with Equation (2), where $A^T v(x) = \int_x^L v dx$ is the L^2 -adjoint of A. Within the context of ratiometric fluorescent dyes (such as Fura2 AM used for $[Ca^{2+}]_i$ recordings in this study) the recorded signal is the ratio of two Poisson random variables. The variance and the mean of such a signal follow a complex, and seemingly non-linear, function of the photon count rates at each wavelength. Since these rates are assumed to be unknown *a posteriori* in a recording, we cannot accurately determine how the noise is distributed. However, we will assume that instrumentation and experimental settings contribute to a noise distribution that is approximately Gaussian. Moreover, in the specific case of ratiometric dyes (Section Intracellular Calcium Measurements), we have found that true noise distribution is a complex function of time, but can be represented reliably using a time-dependent Gaussian noise model. Therefore, to reduce data-fidelity and conversely increase regularity in the regions of highest amplitude noise while accurately reproducing data in regions of lowest noise, we find instead a stationary solution to the following equation

$$u_t(x) = \alpha \frac{d}{dx} \frac{u'(x)}{|u'(x) + \varepsilon|} - \frac{A^T (Au(x) - F)}{\psi(x, u) + \eta}, \tag{4}$$

where $\psi(x, u)$ is an iteratively determined weighting function (as described below), and ε and η, are parameters introduced to avoid dividing by zero.

At the nth iteration of the algorithm, we solve for u^{n+1} by setting the left-hand-side of Equation (4) to zero, and linearizing the problem through substituting every u appearing in the denominators by the value of u^n obtained from the previous iterate. For a more detailed descritption of the means used to solve this type of problem (see Vogel, 2002; Chartrand and Staneva, 2008; Chartrand and Wohlberg, 2010). It is known that an appropriate choice of the denominator offsets, ε and η, is necessary to produce acceptable minimizations (Chartrand, 2007), yet this choice is rarely considered beyond their status as parameters that must be tweaked to obtain acceptable results (Li et al., 2007; Chambolle et al., 2009; Oh et al., 2013). In what follows, we detail a methodology on how to determine the parameters α, ε, η, and the function $\psi(t)$, based on the data $F(t)$, and the derivative estimate $u(t)$.

Dynamic Determination of Total-Variational Parameters

Given a set of fluorescence recordings $\{F_i\}$ of length N, at the nth iterate of the regularization algorithm, we identify the set of indices $j = \{i = 1, 2, \cdots, N \,|\, \psi^n(t_i) \neq \infty\}$ (as explained in Section Removal of Noise Spikes) whose data are not likely dominated by noise and thus should contribute to the fidelity term of Equation (4). Letting $\Delta_i = F_i - F_{i-1}$, we can define the weighting sequence $g_i = 1 - |\Delta_i| / \max_i |\Delta_i|$ in order to provide an upper bound on the noise of the signal, given by

$$\sigma_+^n = \frac{\sum_j |\Delta_j| g_j}{\sum_j g_j}.$$

The weights g_i will tend to zero as Δ_i approach their maximum, and converge to a positive number (< 1) as $|\Delta_i|$ approach their minimum. We can thus conclude that the weighted average of $|\Delta_j|$ will identify the smallest differences as being the most informative of the magnitude of noise. Large discrete differenecs, whether they result from transient increases in the noise level or from the fact that the signal is non-stationary, contribute only modestly to the estimate σ_+^n. On the other hand, we can also estimate a lower bound on the noise using

$$\sigma_-^n = mean\left(\left|\zeta_j^n\right|\right), \tag{5}$$

where $\zeta_i^n = F_i - (Au^n)_i$, i.e., by taking the difference between the data and the cumulative integral of u. Because we use the discrete differences Δ_i as our initial solution: $u_i^0 = \Delta_i$, the integral of u^n ($n > 0$) will likely diverge away from the data with each succesive iteration of the algortithm due to the action of the regularization term in Equation (4). This tends to result in σ_-^n being smaller than σ_+^n, although this is not always strictly true. The use of the two different estimators for the noise allows for a more robust performance of the methodology, as both estimators are prone to becoming inaccurate in different scenarios. With an estimate of the noise, we can also estimate the noise-to-signal ratio by

$$\phi^n = \frac{\sigma_+^n}{\max(Au^n)}.$$

The value of ϕ is a critical parameter in our algorithm as it discreminates between small and large values of various quantities. For example, it is employed to calculate an appropriate value of ε, defined in Equation (4), based on the scale of variations of small values of u', as follows

$$\varepsilon^n = std\left(\left\{ |u'_r| : |u'_r| \leq (\phi^n)^2 \max_i \left(|u'_i|\right), r \in i \right\}\right).$$

The calculation of η requires defining another weighting sequence $h_i = |u'_i| / \max_i |u'_i|$, which tends to zero when u is the most regular, as well as an estimate for the upper bound on the total error, χ, between the integral of u and the data in the least regular regions of the solution, given by

$$\chi^n = \frac{\sum_i h_i |\zeta_i^n|}{\sigma_-^n}.$$

When χ is small, η must be made large enough to improve the smoothness of the fitting (at the expense of data-fidelity). This can be acheived by making η a decreasing function of χ. However, if a recording does not contain any rapid jumps or noise spikes (but is nonetheless noisy), such a relation between χ and η will not be sufficient to infer a proper choice of η given χ. Thus, we must include another term independent of χ which will produce modest data-fidelity for signals dominated by drift. We therefore define η to be

$$\eta^n = \frac{\exp}{\sigma_+^n}\left[-\left(\frac{\chi^n+1}{2\sigma_-^n}\max\left(\{|F_i - F_{i-1}| - std(\{|F_r - F_{r-1}| : r = 2,\ldots,N\}) : i = 2,\ldots,N\}\right)\right)^{\frac{\sqrt{\chi^n}}{2}}\right] + mean\left(\underset{d\subset\{\delta^n\}}{arg\,max}\{\,|d|_{\ell^1} : \text{mean}(d) \geq 3\text{std}(d)\,\}\right).$$

which includes the smallest non-zero scale of the weighted differences $\delta_i^n = (1 - h_i^n)\,\Delta_i^n$.

TV methods tend to smooth the fit when there are large amplitude jumps in the data, or where u is large. If a single large jump dominates the derivative, this can lead to excessive local smoothing, which can be resolved by having enhanced data-fidelity at that point. On the other hand, in the presence of large jumps in the data, small noise-driven fluctuations may be under-regularized. In this case, data-fidelity at these points must be reduced. The function ψ allows for local enhancement or reduction of data-fidelity. Unlike ε, η does not depend on ϕ, but data-fidelity must through the weighting function ψ (by making ψ proportional to ϕ). Therefore, we define ψ to be

$$\psi_i^n = \left(\phi^{\sqrt{\chi^n+2}}\right)^{J_1} + \phi^{J_2}, \tag{6}$$

where

$$J_1 = \left[\frac{1}{(\sigma_+^n)^{1+\sqrt{\phi^n}}}\sqrt{\frac{|u_{i+1}^{n-1}|}{\min\left(\{|u_r^{n-1}| : u_r^{n-1} \neq 0\}\right)}}\right],$$

$$J_2 = \left[\sqrt{\frac{\left(\frac{1}{\tau_{\max}} + \frac{1}{3}\right)\sigma_+^n}{(\chi^n + 1)(\omega_i + \delta_i)}} + \frac{mean\left(|u^{n-1}|\right)}{\left|\max(|\upsilon^n|) - |\upsilon_i^n| + \phi^n\right|}\right],$$

$\tau_{\max}^n = \left(mean\left(|\zeta_j^n|\right) + std\left(|\zeta_j^n|\right)\right)/\sigma_+^n$ is a iterative error scale parameter, $\upsilon_i^n = u_{i+1}^{n-1} + (1 - h_{i+1})^2 u_{i-1}^{n-1}$ is an estimator of u_i based on the adjacent values of u, and ω_i is a three-point moving average of δ_i. The exponent J_1 emphasizes data-fidelity (regularity) when the derivative of the following point is large (small), whereas the exponent J_2 emphasizes data-fidelity when Δ_i are small or when the derivative is near its maximum. When Δ_i is large and u_{i-1}, u_i are regular, on the other hand, J_2 is small allowing regularity to propagate forward into regions of signal possessing large ampltidue noise (i.e., where the data is not informative). The balance between the two effects of J_2 along with the χ -dependent terms of Equation (6) produce an acceptable compromise between the regularity of the fit and data-fidelity for recordings across a wide range of signal-to-noise ratio and extremely varied dynamics. Finally, once the maximum relative change between two iterations of the procedure becomes less than $\sqrt{\phi}$, we consider the solution to have achieved quasi-stationarity and terminate the procedure.

Removal of Noise Spikes

When a recording exhibits intermittent periods of high amplitude noise (noise spikes), the data contaminated by these noise spikes is minimally informative. Detecting them allows for determining the indices j (**Figure 2A**). Within our regularization algorithm, this is done by (a) setting ψ to infinity at those time points in such a way that Equation (4) only penalizes irregularity at these time points, and (b) determining the fit at these points based on the surrounding (reliable) data. After each iteration, n, of the regularization algorithm, a smoother fit, Au^n, of the data is obtained. We also obtain a criterion that determines whether or not each point represents a noise spike based on a comparison between the residual differences, ζ^n, and a chosen threshold value. This threshold is specified using the parameter τ_{rm}^n, given by

$$\tau_{rm}^n = (1 - \xi) + \xi\,\tau_{\max}^n,$$

where $\xi = \exp\left(-\left(\left(\sigma_+^n - \sigma_-^n\right)/\max\{\sigma_+^n, \sigma_-^n\}\right) - (\chi^n)^2/N\right)$ is a convergence parameter for the noise rejection method. Positive ζ_i^n are rejected if they are greater than $\tau_{rm}^n\sigma_+^n$, while negative ζ_i^n are rejected if they are less than $\left(\tau_{rm}^n\right)^2\sigma_+^n$. The use of two thresholds is due to the asymmetry of the Poisson statistics underlying data collection using photodetectors. Rejection is achieved by setting $\psi_i = \infty$, which serves as the basis for defining the set of indices j (i.e., $j = \{i = 1, 2, \cdots, N \mid \psi^n(t_i) \neq \infty\}$ as stated before).

Baseline Drift

When fitting data to specific functional forms, it is important to take into account temporal drifting of the baseline in a signal. The specific nature of the processes underlying this drift are not of particular interest here. Rather, we consider them as nuisance trends and aim to remove their effects from the data.

Drift Model

Some drifts are quite slow compared to the timeframe of the experimental recording, and thus can be fairly well represented by linear functions, whereas others are fast and better represented by an exponential decay function. Since, in a given recording, a number of processes will result in the observed drift, we postulate that, the drift throughout the signal can be well fit to a combination of linear and exponential functions

$$d_\iota(t; a_\iota, \tau_\iota, m_\iota) = a_\iota\left[1 - \exp\left(-t/\tau_\iota\right)\right] + m_\iota t \tag{7}$$

where, a_ι, τ_ι, and m_ι are the exponential amplitude, exponential time constant and the slope of the linear component of the ιth drift within the signal, respectively (examples of fitted drifts shown in **Figures 2B,E,F**). With an appropriate choice of data, these are determined using the least squares fitting as discussed in Section Drift Fitting. We have found numerous cases where either the linear or exponential components were not justifed. However, this knowledge is unavailable to us prior to conducting manual or automated analysis of the data, and we cannot assume *a priori* a less general form than Equation (7). Thus, we have

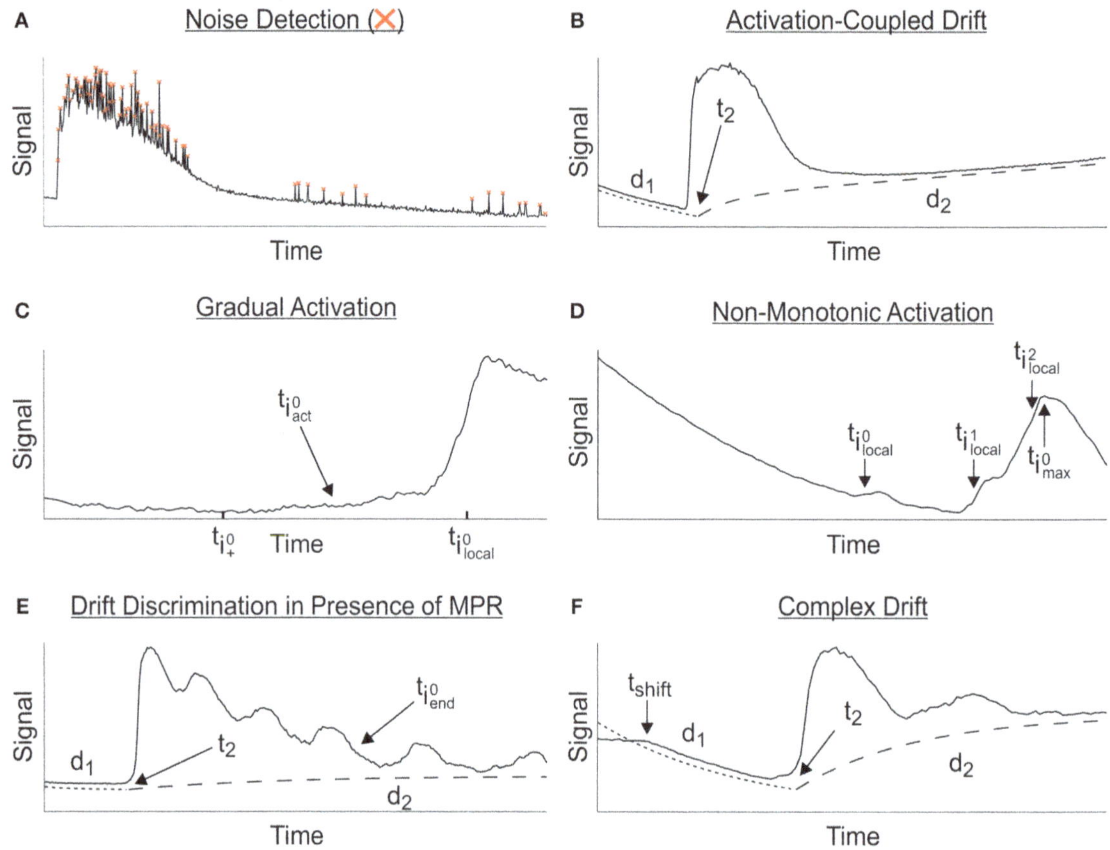

FIGURE 2 | Multitude of features that must be considered during parameter characterization. (A) Experimental recordings are often contaminated with noise; red crosses represent noise detected and corrected. **(B)** Activation-coupled drift; Trajectory of baseline drift, d_1, can shift to secondary drift, d_2 at time t_2. **(C)** Determination of activation in the presence of gradual activation. Region of activation is restricted to $\left[t_{i^0_+}, t_{i^0_{local}}\right]$, within which the most likely point of activation, $t_{i^0_{act}}$, is statistically resolved. **(D)** Non-monotonic activation. Spurts of negligible activity, represented by the local maxima at $t_{i^0_{local}}$ and $t_{i^1_{local}}$, are disregarded as points of activation, in favor of $t_{i^2_{local}}$ which is characterized by the max peak, $t_{i^0_{max}}$. **(E)** Drift discrimination in cases where MPRs are superimposed on TRs. Underlying TR serves as a non-stationary baseline around which the MPR will oscillate within $\left[t_2, t_{i^0_{end}}\right]$ until its contribution becomes negligible and the baseline drift, d_2, dominates the MPR baseline. **(F)** Complex drift; Change in the baseline drift, t_{shift}, prior to onset of TR is ignored in favor of d_1 after the reweighting procedure outlined in Section Drift Fitting.

to rely on the fitting to optimize the contribution of the two compoenets in a data-dependent manner.

It is possible for the signal to exhibit (multiple) drifts with different trends separated perhaps by a TR. In order to capture this effect in a signal, we use a global drift model that combines the intial and secondary drifts (**Figure 2B**) in a semi-piecewise manner in which the initial drift, d_1, continues to contribute to the overall observed drift and succesfully captures the global behavior. This can be written as

$$D(t) = \begin{cases} d_1(t) + z & \text{if } t < t_2 \\ d_1(t) + d_2(t - t_2) + z & \text{if } t \geq t_2 \end{cases}, \tag{8}$$

where t_2 is the time at which the secondary drift, d_2, begins and z is the offset at $t = 0$. Rather than assuming that the drift is similar for all recordings and attempting to construct a standard curve, we assume that a few points in each recording are highly informative of the drift in the baseline.

Drift Delimitation

To fit the drift model to the corresponding portion of the recorded signal, it is necessary to delimit the boundaries of TR by identifying the start of activation and end of deactivation. Firstly, we will aim to estimate the point in time at which the TR of a recording begins. Experimental TRs rarely activate abruptly and simultaneously for a field of imaged cells. Many factors play a role in the heterogeneity of responses observed such as variable diffusion fronts of applied agonist or heterogenous receptor expression among cells. These effects can manifest as very gradual rises or additional small amplitudes prior to a certain activation threshold being surpassed and a rapid activation phase being observed (**Figures 2C,D**). Therefore, analyzing the activation times of all components of the biological unit manually can be subjective. While a simple threshold value in the signal can be effective in detecting activation, the choice of the threshold requires some knowledge of the amplitude of the noise σ and is complicated by the presence of drift in the signal. Instead, we determine the end points of the time intervals dominated by the drift through statistical analysis of an estimate of the first derivative of the signal $\hat{u}(t)$. In other words, to distinguish the drift from TR, the derivative of the latter must change in a way that is more statisically significant than that of the former. The methodology used to obtain the estimate for the derivative is detailed in Section Noise Characterization.

Using the estimated time derivative, we aim to determine (i) the earliest possible time of activation $t_{i^0_{act}}$ (defined as the last time point exhibitng a significant increase in the first derivative before it reaches its maximum value), (ii) the time at which the activation reaches its maximum value $t_{i_{max}}$, and (iii) the time at which the deactivation ends and the signal is once again dominated by the drift $t_{i^0_{end}}$.

i Assuming that the estimated first derivative during activation reaches a local maximum, we can find the most significant local maxima of $\hat{u}$ at the location

$$\mathrm{i}_{local} = \left\{ i : \hat{u}_\mathrm{i} > std\left(\hat{u}\right), \hat{u}_\mathrm{i} \geq \hat{u}_{\mathrm{i}-1} \text{ and } \hat{u}_\mathrm{i} \geq \hat{u}_{\mathrm{i}+1} \right\},$$

where we denote the first significant local maximum by $i^0_{local} = \min\{\mathrm{i}_{local}\}$. By focusing on the portion of the signal containing the first drift and the activation phase of the TR, we restrict our attention to the time interval in which the first derivative $\hat{u}(t)$ is non-negative to disentangle the effects of activation and drift in the data. In other words, we restrict our analysis to $\left[t_{i^0_+}, t_{i^0_{local}}\right]$, where the derivative is non-negative and

$$i^0_+ = \min_{i \in 0,\ldots,i^0_{local}} \left\{ i : \hat{u}_r \geq 0 \quad \forall r = i, \ldots, i^0_{local} \right\}.$$

Without prior knowledge about the sign of the derivative of the baseline drift, we cannot conclude that i^0_+ corresponds to the beginning of activation. To resolve this issue, we employ a statistical test to determine the likely time at which the activation occurs, located at the index

$$i^0_{act} = \max_{i \in i^0_+,\ldots,i^0_{local}} \left\{ i : \hat{u}_\mathrm{i} < std\left(\left\{ \hat{u}_{i^0_+}, \ldots, \hat{u}_{i^0_{local}} \right\}\right) \right\}.$$

This methodology, based on the properties of the derivative around its first significant local maximum, generally picks out the first visually unambiguous activation (**Figure 2C**). As a result, it may be necessary to trim recordings where there are (large amplitude) artifacts prior to the activation of interest.

ii. TRs may be produced by the action of multiple active units (e.g., different receptor species) within the biological system under consideration, each having distinct properties and activation times. This leads to multiple delayed activations taking place over a broad range of time (**Figure 2D**). Due to the superposition of the drift in the baseline with these responses, it is entirely possible for recordings to be contaminated by strongly decreasing drift and for the expected maximum TR to not coincide with the actual maximum of the data. Thus, we have developed a method to search for the visually most likely point at which the TR reaches its maximum in the presence of a drift. For each one of the N_{local} significant local maxima of the first derivative along the activation phase, we find the location of the previous local minimum of the TV estimate of the data using

$$i^{min}_{local} = \left\{ \max_{n<r} \left\{ n : \hat{u} \leq 0 \right\} : r \in \mathrm{i}_{local} \right\}$$

as well as the location of the next local maximum at

$$i^{max}_{local} = \left\{ \min_{n>r} \left\{ n : \hat{u} \leq 0 \right\} : r \in \mathrm{i}_{local} \right\}.$$

From the positions of the local extrema of the data, we can estimate the value of the baseline drift from each local minimum to the next local maximum using the linear extrapolation

$$\nu_- = \left\{ \begin{array}{ll} \left(A\hat{u}\right)_q + \left(t_r - t_q\right) \frac{\hat{u}_{r-1} + \hat{u}_r}{2} & : q = \left(i^{min}_{local}\right)_s, \\ & r = \left(i^{max}_{local}\right)_s, s \in \{1, \ldots, N_{local}\} \end{array} \right\} \tag{9}$$

where A is the operator of antidifferentiation with $Au \approx F_{true}$ (see Equation 1). Based on this, we then estimate the average rate of activation for each significant local maximum of the derivative according to

$$\mu = \left\{ \begin{array}{ll} \frac{\left(A\hat{u}\right)_r - \left(\nu_-\right)_s}{t_r - t_q} & : q = \left(i^{min}_{local}\right)_s, r = \left(i^{max}_{local}\right)_s, \\ & s \in \{1, \ldots, N_{local}\} \end{array} \right\}$$

and select the first local maximum at the location

$$i^0_{max} = min \left\{ \begin{array}{l} r : \mu_r \geq mean\left(\mu\right) - 3 \cdot std(\mu) \text{ and } \nu_r \\ \quad > mean\left(\nu\right) - std\left(\nu\right) \end{array} \right\}.$$

which has an average rate and magnitude within a statistically acceptable range (that excludes small outliers). $t_{i^0_{max}}$ is the time point at which the derivative reaches a local maximum. It may differ from the one that corresponds to the local maximum immediately following the first activation time point $t_{i^0_{act}}$ (**Figure 2D**). This is because i^0_{act} depends solely on the derivative around its first local maximum, while i^0_{max} takes into account an approximation to the average rate of change around all local maxima; the local maximum after i^0_{act} should only differ from i^0_{max} when there is a succession of activations and the first does not have the largest rate of activation.

Starting from this local maximum of the derivative, we seek the location of the first point in time $t_{i^{end}_{max}}$ where the change in the signal drops below the estimated noise level σ_- (see Section Dynamic Determination of Total-Variational Parameters)

$$i^{end}_{max} = \min_{i \in i^0_{max},\ldots,N} \left\{ i : \hat{u}_\mathrm{i} \cdot \left(t_\mathrm{i} - t_{\mathrm{i}-1}\right) < \sigma_- \right\}, \tag{10}$$

and thus isolate the time interval in which the most significant activation occurs. Having isolated the most significant activation, we finally arrive at the location of the first estimate of the time where the response reaches its maximum

$$\mathrm{i}_{max} = min\left\{ \underset{\mathrm{i}\in \mathrm{i}^0_{\max},\ldots,\mathrm{i}^{\mathrm{end}}_{\max}}{\arg\max} \left(A\hat{u}\right)_i \right\}.$$

iii. Having identified the time at which activation is likely to begin $t_{\mathrm{i}^0_{\mathrm{act}}}$, we can now assume that the signal prior to this point is the drift. If the recording is of a sufficiently long duration, the response will return to baseline and the end of the recording should once again be dominated by the (secondary) drift (**Figure 2A**). To account for this drift, we need to estimate the time duration of deactivation. This is done in a manner nearly identical to how we determined the first time of activation $t_{\mathrm{i}^0_{\mathrm{act}}}$. However, due to the possibility of having MPR after the initiation of TR (**Figure 2E**), we cannot restrict ourselves to time intervals in which the first derivative is non-positive. To solve this issue, we define a set of time points after the presumed maximum of TR, located at $\mathrm{i}_{decay} = \{\mathrm{i}_{max}, \ldots, N\}$, and construct a measure

$$n_{end} = \max_{n\in\{3,2,1,0\}} \left\{ \begin{array}{r} n : \exists \left|\hat{u}_\mathrm{i}\right| - mean\left(\left|\hat{u}_{\mathrm{i}_{decay}}\right|\right) \\ > n \cdot std\left(\left|\hat{u}_{\mathrm{i}_{decay}}\right|\right), i \in \mathrm{i}_{decay} \end{array} \right\}$$

to quantify how far the derivative deviates from its mean during the decay. This measure allows us to robustly detect the location of the time point where TR is negligible

$$i^0_{end} = \min_{\mathrm{i}\in \mathrm{i}_{decay}} \left\{ i : \left|\hat{u}_\mathrm{i}\right| - mean\left(\left|\hat{u}_{\mathrm{i}_{decay}}\right|\right) > n_{end} \cdot std\left(\left|\hat{u}_{\mathrm{i}_{decay}}\right|\right) \right\}.$$

The time point $t_{\mathrm{i}^0_{\mathrm{end}}}$ is then used to determine the start of the new drift.

Drift Fitting

To isolate the TR from the drift, it is necessary to generate an accurate fit for the drift. This is achieved by employing a succession of least square fits that progressively incorporates more data and models that account for additional components of the signal (including activation and deactivation phases of TR). The first step in this successive least-square-fitting method is to obtain preliminary estimates of the parameters $\theta_{drift} = [a_1, \tau_1, m_1, a_2, \tau_2, m_2]$ of the drift model. The MATLAB implementation of the non-linear least squares method (Marquardt, 1963; Moré, 1978) is used. More specifically, we initially minimize the error function between the drift model and the data

$$S^0_{drift}\left(\theta_{drift}\right) = \sum_k \left(D\left(t_k; \theta_{drift}\right) - F_k\right)^2, \quad (11)$$

where the set of indices k is defined by $k = \{\{1,\ldots,i^0_{act}\} \bigcup \{i^0_{end},\ldots,N\}\} \bigcap j$ (i^0_{act} and i^0_{end} are as defined in Section Drift Delimitation while j is defined in Section Removal of Noise Spikes). We denote the first estimate of parameters obtained from the minimizaiton of Equation (11) by θ^0_{drift}. There is no guarantee, however, that the drift model described by Equation (8) can accurately represent the actual drift in the baseline. The emergence of drifting trends which are not related to the onset of TR (**Figure 2F**) may require the inclusion of more than two functions of the type described in Equation (7), yet the decision to include more drift terms, or alternatively to truncate the signal, would require manual intervention. To circumvent this limitation, we perform a second fit where the individual terms in Equation (11) are weighed according to a weight function w, and the first derivative of the data is taken into account. Although the set of parameters θ^0_{drift} will be able to produce the "general" trends in the data before and after TR, the presence of multiple drifting trends necessitates the use of the derivative estimate, $\hat{u}$, to identify the segment(s) of the signal that actually follow the trend described in Equation (8) and remove the effects of others. The inclusion of $\hat{u}$, however, in the sum of squared errors can lead to erroneous fittings when the drift is not well represented by Equation (8). The weight function w alleviates this problem by preventing the linear and exponential trends in Equation (7) from growing unjustifiably large.

To determine w, we require that it approaches zero when the match between $\hat{u}$ and $\dot{D}\left(t; \theta^0_{drift}\right)$ is minimal. To achieve this, we choose w to depend on $y = \left|\hat{u} - \dot{D}\left(t; \theta^0_{drift}\right)\right|$ as follows

$$w_k = exp\left(\frac{-y_k^2}{mean\left(y_k^2\right)}\right).$$

We then apply the non-linear least squares fitting procedure to minimize the error function

$$S_{drift}\left(\theta_{drift}\right) = \sum_k w_k \left[\begin{array}{r} \left(D\left(t_k; \theta_{drift}\right) - F_k\right)^2 \\ + \left(\hat{u}_k - \dot{D}\left(t_k; \theta_{drift}\right)\right)^2 \end{array} \right],$$

and use θ^0_{drift} as an initial condition for the fitting procedure.

Activation Fitting

For the activation phase of the response, we use the model

$$g_{act}\left(t; A_{act}, \beta, n_{act}, m_{act}\right) = A_{act}\frac{t^{n_{act}}}{t^{n_{act}} + \beta^{n_{act}}} + m_{act}\int_0^t \frac{x^{n_{act}}}{x^{n_{act}} + \beta^{n_{act}}}dx.$$

where A_{act} is the maximum of the Hill function, n_{act} is the Hill coefficient, β is the time at which the the Hill function reaches its half maximum, and m_{act} is the slope of the quasi-linear function that accounts for the trend which dominates at the end of activation. The Hill function allows for a rapid rise and switch in convexity due to many biological units being activated at once, whereas the linear trend allows for a delayed and slower rise induced by more units being progressively recruited into the generation of the signal. The values of A_{act} and m_{act} determine the magnitude of these two trends, whereas n_{act} and β affect primarily the timescales of switching between the two. The time at which the activation phase begins is denoted as t_{on}, and its estimate is confined to the time interval $\left[t_{i^0_{act}}, t_{i_{max}}\right]$. In order to obtain a preliminary estimate of the values of these parameters and those of the drift function, $\theta_{act} = [t_{on}, A_{act}, \beta, n_{act}, m_{act}, \theta_{drift}]$, we minimize the sum of square errors between the activation data and the model along with their derivatives, given by the function

$$S_{act}(\theta_{act}) = \sum_j \left[\begin{array}{l} \phi^2 \left(F_j - G_{act}\left(t_j; \theta_{act}\right)\right)^2 \\ + \phi\left(\hat{u}_j - \dot{G}_{act}\left(t_j; \theta_{act}\right)\right)^2 \end{array} \right] + S_{drift}\left(\theta_{drift}\right) \quad (12)$$

where ϕ is the noise-to-signal ratio as defined in Section Dynamic Determination of Total-Variational Parameters, and

$$G_{act}(t; \theta_{act}) = \begin{cases} D(t; \theta_{act}) + g_{act}(t - t_{on}; \theta_{act}) & \text{if } t_{on} \le t < t_{max} \\ D(t; \theta_{act}) & \text{otherwise} \end{cases}$$

is our activation data model that takes into account the effect of the drift D (defined in Equation 8) on the perceived activation. To abrogate the influence of noise spikes on measured parameters, the sum of squares in Equation (12) is evaluated at the set of indices j (defined in Section Dynamic Determination of Total-Variational Parameters) which are not dominated by noise. Moreover, we use $\hat{\theta}^{act}_{drift}$ to denote the set of drift parameters obtained from the minimization of Equation (12).

Signal Detection

In order for the algorithm to resolve whether there is a discernable TR present in $F_{true}(t)$, three conditions must be satisfied: (i) i_{local} must be a non-empty set, (ii) the initial drift estimate

$$D\left(t; \underset{\theta \in \left\{\hat{\theta}_{drift}, \hat{\theta}^{act}_{drift}\right\}}{\arg\min} S_{drift}(\theta)\right)$$

must be below the TV data estimate by a detection threshold of $4\sigma_-$ for at least six data points, and (iii) the difference between the TV data estimate and the initial drift estimate must not be a strictly increasing function of time after t^0_{max}. These three criteria allow for the detection of the TR and further analysis of its characteristics [Section Transient Response (TR) Model].

To evaluate the signal detection performance of the algorithm, 450 individual traces of ATP-induced calcium responses were used as a validation set. Manual results were then compared to automated detection of TRs to assess extent of agreement between the two methods. Classical signal detection nomenclature (i.e., true positive or negative and false positive or negative) was intentionally avoided due to lack of certainty in determining the true presence of TRs in more ambiguous cases. We found that the automated and manual methods agreed in detecting a TR in 88.1% of cases, and they disagreed in 11.9% of cases (**Figure 3A**). Further dissection of these results showed that 64.7% of disagreement arose from the algorithm reporting an absence of TR while visual evaluation suggested otherwise (**Figure 3A**), indicating that the algorithm has a tendency to be more conservative than user-mediated assessments.

To determine whether there were particular types of recordings that contributed to these disagreements, time-series traces were qualitatively divided into two groups: Clean signals with clearly defined responses were classified as "pronounced" (**Figure 3B**, top), and signals containing ambiguous signals with low signal to noise ratio or strong drift were classified as "obscure" (**Figure 3B**, bottom). The total frequency of disagreement was 3.6 times greater for obscure signals compared to those classified as pronounced (17.4 vs. 4.8%; **Figure 3C**). Regardless of the group, the algorithm signal detection remained more conservative compared to the manual method.

Transient Response (TR) Model

Transient cellular responses are generally complex with multiple time scales and amplitudes. They may, in fact, exhibit prolonged MPRs superimposed on a more acute response (see **Figure 2E**). For these reasons, a complete characterization of all possible TRs is unlikely to be attainable. In order to remain as general as possible, we propose modeling TR as a continuously differentiable piece-wise defined function that first increases during the activation phase and subsequently decreases during the deactivation phase. Due to the large number of parameters required for such a description and the automated nature of our fitting procedure, we decompose the fitting of the whole TR into a sequential fitting of the activation phase alone followed by a fit of both phases simultaeously. This yields significantly more reliable results with faster convergence rates over a wide gamut of input data, because it allows for information obtained during preliminary simple fits to be used in a progressively more complex manner. In order to capture the complex fluorescence response generated by the spatially separated units in a live cell, we use a combination of Hill functions and quasi-linear functions generated by the integral of the corresponding Hill functions.

Response Fitting

Following the least squares fitting of the activation data, we now seek to fit the entire recording (with the drift and TR) using a continuously differentiable function. A decreasing Hill Function is used to describe deactivation phase of the signal in a manner similar to Equation (11), as follows

$$g_{de}(t; A_{de}, \gamma, n_{de}) = A_{de} \frac{\gamma^{n_{de}}}{t^{n_{de}} + \gamma^{n_{de}}} + m_{de} \int_0^t \frac{\gamma^{n_{de}}}{x^{n_{de}} + \gamma^{n_{de}}} dx,$$

where A_{de} is the amplitude of the Hill function, n_{de} is the Hill coefficient, γ is the time at which the Hill function reaches its half maximum, and m_{de} is the slope of the quasi-linear function that accounts for the trend dominating at the beginning

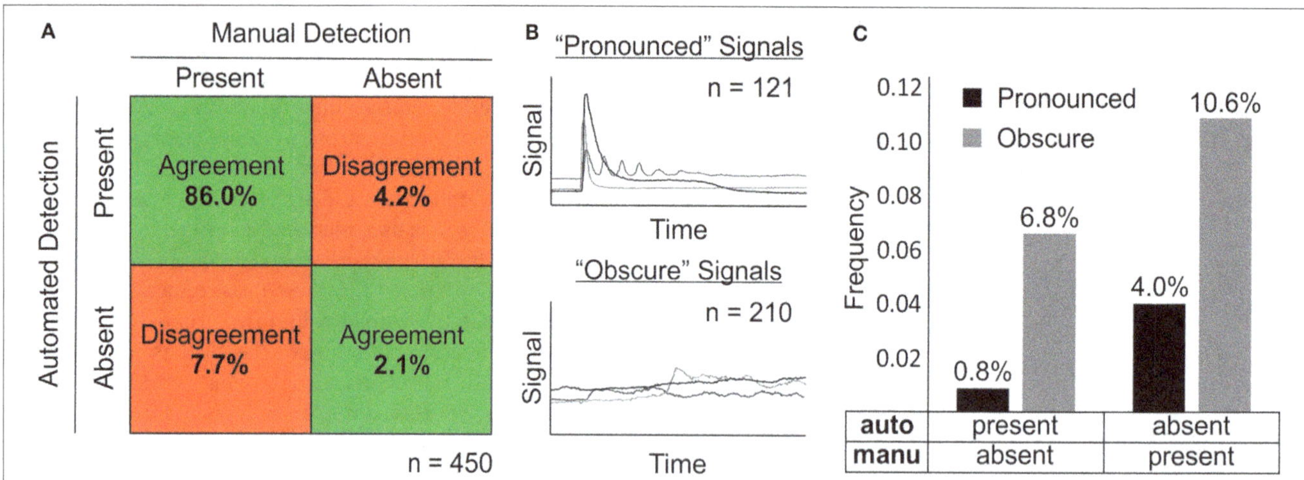

FIGURE 3 | Signal detection performance of algorithm. (A) Signal detection analysis demonstrates relative selectivity and sensitivity of the algorithm as compared to user-mediated manual detection. "Agreement" refers to instances where the detection of a response was consistent between algorithm and manual methods. "Disagreement" refers to the contrary. "Present" indicates cases where a response was detected using the indicated method, while "absent" refers to cases where a response was not detected. **(B)** Representative time-series for pronounced (top) and obscure (bottom) signals. **(C)** Analysis of disagreement cases for pronounced and obscure signals. Automated detection, auto; manual detection, manu; sample size, n.

of deactivation. The time when the response switches to the deactivation function is denoted by t_{de}. The parameter m_{de} is chosen such that the response function returns to zero by the end of the recording and is given by

$$m_{de} = \frac{g_{act}\left(\ell_{act}; A_{act}, \beta, n_{act}, m_{act}\right) - A_{de}\frac{\gamma^{n_{de}}}{(t_N - t_{de})^{n_{de}} + \gamma^{n_{de}}}}{\int_0^{t_N - t_{de}} \frac{\gamma^{n_{de}}}{x^{n_{de}} + \gamma^{n_{de}}} dx},$$

where $\ell_{act} = t_{de} - t_{on}$ is the time duration of the activation phase of the response. If differentiability is not enforced at the point t_{de}, where the two functions g_{act} and g_{de} meet, then the fitting may contain sharp edges indicative of unconverged solution. To solve this issue, the continuity of the first derivative of these two functions, particularly at t_{de}, is achieved through a third-order Hermite spline (Traub, 1964) on the time interval $[t_{de} - \varsigma_{act}, t_{de} + \varsigma_{de}]$, where ς_{act} and ς_{de} are two parameters that must satisfy $\varsigma_{act} < \ell_{act}$ and $\varsigma_{de} < 2\gamma$ (see Data Sheet 2 in Supplementary Materials). The overall response model is thus given by

$$g_{resp}\left(t; \theta_{resp}\right) = \begin{cases} g_{act}\left(t - t_{on}; \theta_{resp}\right) & if\ t_{on} \leq t < t_{de} - \varsigma_{act} \\ p_{Hermite}\left(t - (t_{de} - \rho_{act}); \theta_{resp}\right) & if\ t_{de} - \varsigma_{act} \leq t < t_{de} + \varsigma_{de} \\ g_{de}\left(t - t_{de}; \theta_{resp}\right) & if\ t_{de} + \varsigma_{de} \leq t \end{cases},$$

where $\theta_{resp} = [\theta_{act}, t_{de}, \varsigma_{act}, \varsigma_{de}, A_{de}, \gamma, n_{de}\theta_{drift}]$. Given the response model, we define the global data model as

$$G_{resp}\left(t; \theta_{resp}, \theta_{drift}\right) = \begin{cases} D\left(t; \theta_{resp}\right) + g_{resp}\left(t_i; \theta_{resp}\right) & if\ t_{on} \leq t \\ D\left(t; \theta_{resp}\right) & otherwise \end{cases}, \quad (13)$$

and minimize the error function

$$S_{resp}\left(\theta_{resp}\right) = \phi \sum_j \left[\phi\left(F_j - G_{resp}\left(t_j; \theta_{resp}\right)\right)^2 + \left(\hat{u}_j - \dot{G}_{resp}\left(t_j; \theta_{resp}\right)\right)^2 + \kappa \phi \left(D_2\left(t_k - t_2; a_2, \tau_2, m_2\right)\right)^2 \right] + \frac{\lambda}{\phi^2} S_{drift}\left(\theta_{drift}\right), \quad (14)$$

to obtain the fitting, where κ is a parameter quantifying the apparent coherence between the drift and reponse models (D and g_{resp}), given by

$$\kappa = \left| \frac{\left(A\hat{u}\right)_N - g_{act}\left(t_{max} - t_{on}; A_{act}, \beta, n_{act}, m_{act}\right)}{\left(A\hat{u}\right)_N - \min_{i \in i^0_{act}, \ldots, i_{max}} \left(A\hat{u}\right)_i} \right|,$$

λ is a parameter defined by

$$\lambda = \frac{\max_{i > i^0_{end}} \left|\hat{u}_i\right|}{\max_{i \in i^0_{act}, \ldots, i_{max}} \hat{u}_i},$$

and the values for the parameters A_{act}, n_{act}, β, m_{act}, and t_{on} are taken from the activation fitting. The first two terms in the sum of squares in Equation (14) are analgous to those in Equation (12), whereas the third term minimizes the AUC for the second drift function D_2. By including the coefficient κ in this third term, however, allows D_2 to become more significant when there is a large mismatch in the value of the baseline between $t_{i^0_{\text{act}}}$ and t_N (that cannot be explained by d_1).

Activation Parameter Validation: t_{onset}, $t_{10-90\%}$, and Amplitude

The three parameters, t_{onset}, $t_{10-90\%}$, and amplitude, are considered together because they describe what happens at the activation phase of the TR, with no regard for the deactivation phase or MPR. The time at which a TR is discernable from baseline is defined as t_{onset} and is estimated directly as the parameter t_{on} of the response function g_{resp}. There was a strong linear agreement between the manual and automated-estimates, with a linear slope of 1.06 and a correlation coefficient (r^2) of 0.94 (**Figure 4A**, left). On average, automated estimates of t_{onset} were 4.6% greater than manual estimates, with limits of agreement ranging between −10 and 19% difference (**Figure 4A**, right). The $t_{10-90\%}$ is also estimated numerically from the response function g_{resp}. Using the response function allows to overcome issues arising from the subsampling of rapid dynamics by numerically evaluating on a time grid 10 times finer than the input times. Manual estimation of this parameter is contingent upon accurate estimation of the baseline and peak occurrences, both of which present potential sources of error, particularly for noisy signals with drift. The relationship between manual and automated estimates had a linear slope of 0.77 and a correlation coefficient of 0.77 (**Figure 4B**, left). The higher degree of scatter away from the line of exact linear correlation is reflected by the wider Bland–Altman interval of agreement, ranging from −89 to 54% difference between manual and automated estimates. Overall, there was a −17% difference between all paired estimates of $t_{10-90\%}$, revealing that $t_{10-90\%}$ was manually overestimated compared to the automated estimates (**Figure 4B**, right). Amplitude estimates obtained by the manual and automated methods had a strong linear relationship with a slope of 1.02 and a correlation coefficient of 0.84 (**Figure 4C**, left). The limits of agreement, ranging from −26 to 25%, were narrow with a mean percent difference of −0.3% between all paired estimates of amplitude (**Figure 4C**, right).

TR Parameter Validation: AUC, FWHM, and τ_{decay}

Due to the inherent differences between TRs and MPRs and the approach taken with this algorithm, the AUC, FWHM, and t_{decay} are limited to describing TRs. Nevertheless, these parameters will be also reported in the presence of MPRs where they should be interpreted with the following considerations. (i) if a MPR is superimposed on a TR, the reported parameters describe the underlying TR, not the superimposed MPR. (ii) if TR presence is not detectable and MPR demonstrates a purely oscillatory response, the reported parameters characterize the first peak only. With these considerations in mind, manual evaluation of TR parameters was performed with a variety of signals, including MPRs. AUC estimates are manually determined using a geometric estimation of the area of a triangle whose vertices are at the start, peak, and end of the TR. Algorithmically, AUC was evaluated from the area under g_{resp}, using the trapezoidal rule (Rice, 1973) implemented in MATLAB using "trapz()." Comparing the manual and automated estimates demonstrated a linear relationship with a slope of 0.85 and a correlation coefficient of 0.79 (**Figure 5A**, left). On average, automated estimates were 4.6% larger than manual estimates with an interval of agreement ranging between −38 and 47% difference (**Figure 5A**, right). Considering the geometric-approach used for manual-estimation of AUC values, it is reasonable to assume that the error arose from manual limitations.

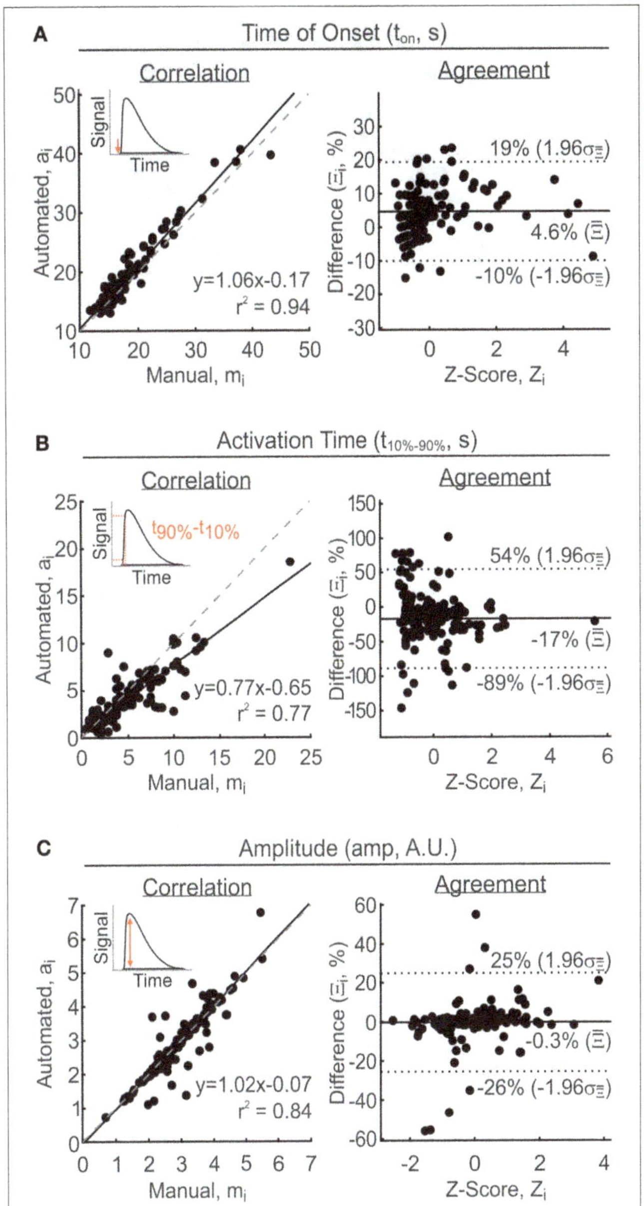

FIGURE 4 | Validation of activation parameter measurements. Parameters were estimated manually and then compared with automated-estimated values using correlation analysis (left: solid line, linear regression line; broken line, line of equality) and Bland-Altman analysis (right, solid line: mean percent difference, $\bar{\Xi}$; broken line: limits of agreement, $\bar{\Xi} \pm 1.96\,\sigma_{\bar{\Xi}}$). **(A)** Time of onset, t_{onset}. **(B)** Activation time, $t_{10-90\%}$. **(C)** Amplitude. Arbitrary units, A.U. Insets: Visual representations of parameters measured.

Due to the difficulty in determining the precise time at which the signal returns to its former baseline, it would be challenging to manually describe the duration and decay characteristics of the

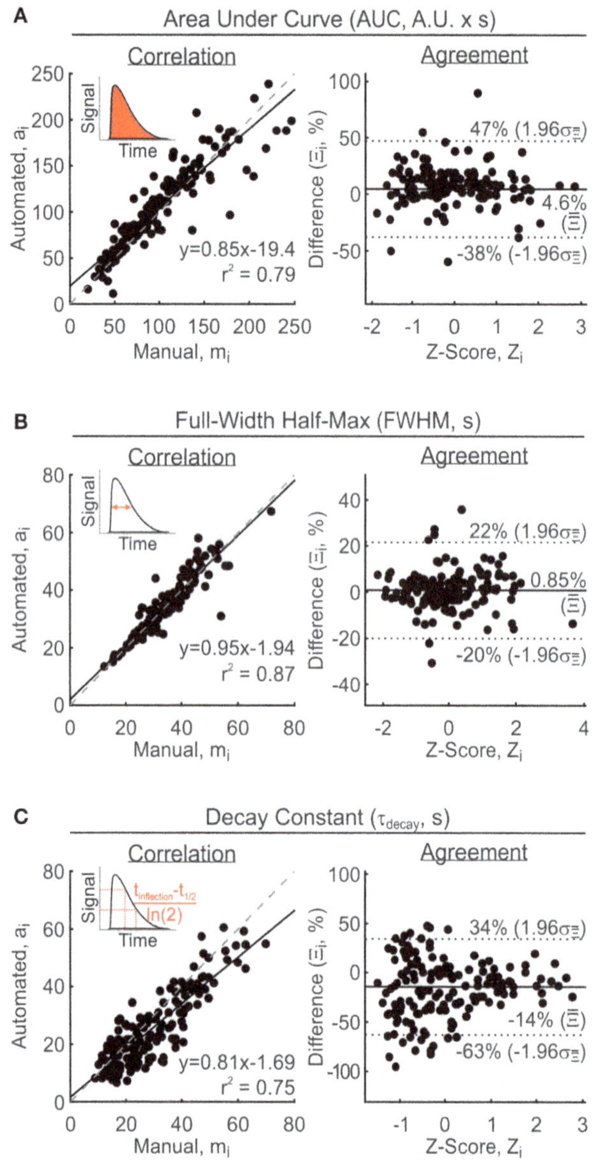

FIGURE 5 | Validation of TR parameter measurements. Parameters were estimated manually and then compared with automated-estimated values using correlation analysis (left: solid line, linear regression line; broken line, line of equality) and Bland-Altman analysis (right: solid line, mean percent difference, $\bar{\Xi}$; broken line, limits of agreement, $\bar{\Xi} \pm 1.96\ \sigma_{\bar{\Xi}}$). **(A)** Area under curve, AUC. **(B)** Full-width half-max, FWHM. **(C)** Decay constant, τ_{decay}. Arbitrary units, A.U. Insets: Visual representations of parameters measured.

response. Incomplete recordings and background drift are largely responsible for generating such behavior. The full-width half-max (FWHM) is defined as the time elapsed between the two half-max coordinates of a peak. Our analysis of FWHM revealed that the linear relationship between manual and automated-estimates was strong, with a correlation coefficient of 0.87 and slope of 0.95 (**Figure 5B**, left). The Bland Altman analysis demonstrated that the agreement interval ranged from −20 to 22% difference with a mean percent difference of 0.85% between all paired estimates (**Figure 5B**, right).

Finally, to manually estimate the decay constant, the time of the inflection point (ρ) is visually estimated and the general trend of the data following ρ is represented by a mono-exponential decay. The decay constant is then determined by the time it takes for the signal to reduce to approximately 37% of its initial value (1/e). Algorithmically, the time ρ is determined by solving for the inflection point in the deactivation function, given by

$$\rho = t_{de} + \gamma \left(\frac{n_{de} - 1}{n_{de} + 1} \right)^{1/n_{de}}.$$

The data following ρ is then fit to a mono-exponential decay function using least squares, to determine the time constant of decay. The slope of the linear agreement between manual and automated-estimates was 0.81 and the correlation coefficient was 0.75 (**Figure 5C**, left). The Bland Altman analysis revealed a strong systematic bias of −14% difference, with an agreement range of −63 to 34% difference, signifying that manual efforts to estimate the decay constant consistently overshot the values reported by the algorithm (**Figure 5C**, right).

Multi-Peaked Responses

To isolate the characteristic parameters of MPRs that are frequently superimposed on TRs (**Figure 2D**), the TR model g_{resp} must be refined to serve as a non-stationary baseline around which the MPR will oscillate. This refinement is necessary because it is often the case that the TR model g_{resp} will produce sub-optimal fittings where the data deviates significantly from the TR fitting. Therefore, to accurately characterize the TR and quantify the properties of truly oscillatory MPRs, it is first necessary to adapt the least squares fitting procedure of Section Response Fitting to remove the effects of data points not well represented by G_{resp}. This is done by first identifying large deviations representing MPRs from the G_{resp} -fitting (obtained by minimizing Equation 14), and then reweighing the sum of squares in Equation (14) to remove the influence of those deviations from the fits. We subsequently perform a secondary fitting of G_{resp} to determine more accurately the baseline, delineating the TR, where the MPR-associated deviations originate from. Finally, we analyze the MPRs by determining whether they represent oscillations and, if so, quantify their properties.

To identify these MPR-associated deviations from the newly defined baseline, we first employ the MATLAB "findpeaks()" function. This finds the peaks and troughs of the significant deviations in the TV data estimate, $A\hat{u}(t)$, from the TR estimate, $G_{resp}\left(t; \hat{\theta}_{resp}\right)$, truncated at its half-maximums, where $\hat{\theta}_{resp}$ is the optimal parameter set obtained from minimizing Equation (14). This truncation permits for the possibility that the onset of TR coincides with the first peak of the MPR-associated deviations. The implementation of peak-finding algorithm, on the other hand, allows for the specification of minimum heights and timing between peaks, set to be $6\sigma_{-}$ and 5 s, respectively. The algorithm yields the heights (E^{peak}, E^{trough}), the FWHM (ξ^{peak}, ξ^{trough}), and the times ($\tilde{t}^{peak}$, $\tilde{t}^{trough}$) of significant peaks and troughs of the deviations, respectively (see **Figure 2D**). In total, there are $\tilde{N} = N_{peak} + N_{trough}$ of these deviations, including N_{peak} peaks and N_{trough} troughs. If $\tilde{N} \leq 2$, then the only deviation in the signal is the TR and the algorithm can terminate. Without prior knowledge of the nature of the MPR-associated deviations, it is very difficult to determine whether they result from trends which are above, below, or symmetric to the TR. To resolve this issue, we assume that the estimated baseline from Section Response Fitting underlies the signal in the absence of deviations. To incorporate this assumption algorithmically, we define two bias parameters based on the relative heights of the first peak and trough, as follows

$$o_{peak} = \exp\left[-\left(\frac{E_1^{peak}}{E_1^{trough}}\right)^4\right] \quad \text{and}$$

$$o_{trough} = \exp\left[-\left(\frac{E_1^{trough}}{E_1^{peak}}\right)^4\right].$$

These quantities are then used to calculate weighting functions for the data based on the properties of peaks

$$\pi_i^{peak} = \sum_{r=1}^{N_{peak}} o_{peak} \exp\left(-\phi\left(\frac{o_{peak}\left(t_i - \tilde{t}_r^{peak}\right)}{\xi_r^{peak}}\right)^2\right) + \left(1 - o_{peak}\right)\left(1 - \exp\left(-\phi\left(\frac{o_{peak}\left(t_i - \tilde{t}_r^{peak}\right)}{\xi_r^{peak}}\right)^2\right)\right)$$

and troughs

$$\pi_i^{trough} = \sum_{r=1}^{N_{trough}} o_{trough} \exp\left(-\phi\left(\frac{o_{trough}\left(t_i - \tilde{t}_r^{trough}\right)}{\xi_r^{trough}}\right)^2\right) + \left(1 - o_{trough}\right)\left(1 - \exp\left(-\phi\left(\frac{o_{trough}\left(t_i - \tilde{t}_r^{trough}\right)}{\xi_r^{trough}}\right)^2\right)\right).$$

The weighting functions π_i^{peak}, π_i^{trough} quantify the relative reliability of the data around each deviation based on how close it is to the fitting function G_{resp} and on its duration. We can also assess the reliability of each time point of the recording (including the TR, the drift, and any MPR-associated deviations present) by how well its derivative matches $\dot{G}_{resp}\left(t; \hat{\theta}_{resp}\right)$. This is done using another weighting function, defined by

$$\Gamma_i = \exp\left(-\left(\frac{\hat{u}_i - \dot{G}_{resp}\left(t; \hat{\theta}_{resp}\right)}{2\, std\left(\hat{u}_i - \dot{G}_{resp}\left(t; \hat{\theta}_{resp}\right)\right)}\right)^2\right).$$

We combine these weighting functions using the criterion that for a data point to be reliable, it must have either a large value of π^{peak} or π^{trough}, and a large value of Γ. It is implemented in the weighting function Ω, as follows

$$\Omega_i = \Gamma_i\left(\pi_i^{peak} + \pi_i^{trough} - \min_i\left(\pi_i^{peak}\right) - \min_i\left(\pi_i^{trough}\right)\right).$$

With Ω_i, we can fit the TR reliably in the presence of significant deviations from the model of Equation (13). This is done by minimizing the error function

$$S_{resp}^{\Omega}\left(\theta_{resp}\right) = \phi \sum_j \left[\begin{array}{c} \phi\, \Omega_j^2\left(F_j - G_{resp}\left(t_j; \theta_{resp}\right)\right)^2 \\ +\Omega_j^2\left(\hat{u}_j - \dot{G}_{resp}\left(t_j; \theta_{resp}\right)\right)^2 \\ +\kappa\phi\left(D_2\left(t_k - t_2; a_2, \tau_2, m_2\right)\right)^2 \end{array}\right] + \frac{\lambda}{\phi^2} S_{drift}\left(\theta_{drift}\right). \tag{15}$$

Identifying Coherent Oscillations

Not all MPRs correspond to periodic oscillations (Thurley et al., 2014). To address this, the algorithm reports two sets of MPR parameters, the first to describe all the peaks detected within a MPR, and the second to describe the subset of coherent peaks present within the same MPR. This section focuses on how the subset of coherent oscillatory peaks is identified. We use a clustering algorithm which is an unsupervised learning technique that enables for the identification of natural groupings or patterns with a defined data set. By minimizing Equation (15), we obtain the most reliable estimate of the TR (specified by the model G_{resp} and its optimal parameter set $\hat{\theta}_{resp}^{\Omega}$), which we take to be the baseline of the MPR-associated deviations. Given the estimate $G_{resp}\left(t; \theta_{resp}^{\Omega}\right)$, we repeat the peak finding steps detailed in Section Multi-Peaked Responses. To determine whether or not the detected deviations represent oscillations, we use a Gaussian mixture model clustering technique. It groups together (in clusters) peaks and troughs with comparable periods, T (determined by the difference between two consecutive peak or trough times; i.e., $\tilde{t}^{peak}$ or $\tilde{t}^{trough}$) and FWHM, ξ. Two adjacent deviations are deemed to be coherent oscillations if they are grouped in the same cluster. In situations where the period or FWHM are modulated throughout time, Gaussian clustering technique may not be able to cluster all coherent oscillations adequately. We therefore process clusters by defining period- and FWHM-trends for all coherent oscillations. If this trend can accurately predict the period and FWHM of the first deviation of an adjacent cluster, then both clusters are deemed to form a set of coherent oscillations. This is repeated for all pairs of adjacent clusters, progressively updating the set of coherent oscillations with those previously deemed incoherent at prior steps.

This procedure use the Expectation Maximization (EM) clustering algorithm (McLachlan and Peel, 2005) to cluster the period and FWHM of a potential oscillatory MPRs, and to determine the optimal number of clusters using gap statistic (Tibshirani et al., 2001). It allows for a reliable separation of oscillatory data from recording artifacts or non-oscillatory MPRs with visually different properties. It also yields a set of $\hat{N}_{peak}$ peaks ($\hat{N}_{trough}$ troughs) occurring at times $\hat{t}^{peak}$ ($\hat{t}^{trough}$), which together represent coherent oscillations. Having determined the properties of the individual features making up the oscillations, they can be used to quantify the properties of the oscillations.

Characterizing Oscillatory Parameters

In order for the algorithm to report the oscillatory properties of a signal, the MPR-associated deviations must satisfy $\tilde{N} > 2$, for the set of MPR parameters describing all detected peaks. If these deviations form a coherent set of oscillations, a second set of MPR parameters characterizing this coherent oscillatory behavior is also reported. In both instances, the following parameters will be reported: the number of oscillations (N_{osc}), the average magnitude of the oscillations (defined as $E^{peak} + E^{trough}$), the average period of oscillation (T), the standard deviation of the periodicity (σ_T), the total time for which the oscillations persist (defined as $\ell_{osc} = \hat{t}^{peak}_{\hat{N}_{peak}} - \hat{t}^{peak}_1$), and the mean duty cycle parameter (given by ξ^{peak}/T) (Smedler and Uhlén, 2014).

MPR Parameter Validation: N_{osc}, E, T, l_{osc}, and ξ^{peak}/T

Manual estimates of N_{osc} is determined by counting the number of discernable peaks within the signal. The slope of correlation was 0.85 with an r^2-score of 0.78 (**Figure 6A**, left). The mean difference between manual and automated estimates was 15% with an interval of agreement ranging from −39 to 69% (**Figure 6A**, right). The peak magnitude of the oscillations, E, is manually estimated by the mean change between peak maxima and their subsequent trough minima, after correcting for a non-stationary baseline that is often a consequence of a concurrent TR. For most signals the non-stationary baseline can be manually estimated to be linear. However, there are a few cases where an estimate of an exponential baseline is required. The correlation between manual and automated estimates was relatively strong, with an r^2-value of 0.92 and a slope of 1.08 (**Figure 6B**, left). The agreement analysis on the other hand revealed relatively no bias, with a mean difference of −2% and limit of agreement ranging from −48 to 44% difference (**Figure 6B**, right). The periodicity is manually estimated by the average time between adjacent peaks. The linear relationship was slightly weaker with a r^2-value of 0.55 and slope of 0.65 (**Figure 6C**, left). The mean difference between manual and automated estimates was negligible, at only 0.3%, indicating an absence systematic bias, and the limits of agreement spanned from −54 to 54% difference (**Figure 6C**, right). The standard deviation of periodicity was obtained from the same set of periods used to estimate the mean period. The linear slope was 0.77 and the r^2-value was 0.77 (**Figure 6D**, left). Similar to periodicity, the mean difference for the standard deviation of periodicity was a negligible −0.4%, with limits of agreement ranging from −61 to 62% difference (**Figure 6D**, right). Oscillatory persistence is chosen to describe how long oscillations are sustained within a given recording, and is estimated as the elapsed time between the first and last discernable peaks in the MPR. The correlation between manual and automated estimates was supported by a r^2-value of 0.79 and slope of 0.85 (**Figure 6E**, left). The mean difference between paired estimates was only −5.3% with a limit of agreement between −47 and 37% difference (**Figure 6E**, right). Finally, the duty cycle is manually estimated by the ratio between ξ^{peak} and T. ξ^{peak} is manually determined by the mean FWHM of individual oscillatory peaks and the same T value obtained above is used to calculate x^{peak} /T. The linear relationship between manual and automated estimated of ξ^{peak} /T was decidedly weak with a slope of 0.51 and r^2 of 0.26 (**Figure 6F**, left). The Bland Altman analysis, however, suggests that there was a systematic bias that could explain the poorer correlation results. The mean difference between manual and automated estimated was −19% with a limit of agreement between −69 and 31% (**Figure 6F**, right).

The MPR parameter validation described above focuses on the set of parameter estimates describing all the peaks in the MPR, rather than the subset of coherent oscillations. This is because manual detection of each peak in the MPR is less subjective than detecting only the coherent peaks in the MPR. Since the algorithm sub-selects the coherent oscillatory peaks from the initial set of identified deviations, the performance reported for the characterization of all peaks extends to the subset of coherent oscillations. Furthermore, as expected, the standard deviation of the periodicity is consistently lower for coherent oscillations when compare to the σ_T reported for all peaks in the same MPR (i.e., more regular periodicities result in lower standard deviations).

To ensure confidence in the reported MPR parameters, users of this algorithm are urged to visually verify the quality of the signal fittings to determine whether the algorithm is characterizing their peaks of interest, as these may not always coincide with the most prevalent oscillatory component of the signal (Thurley et al., 2014). Furthermore, N_{osc} reported for all peaks and coherent peaks can be compared to be aware of how many peaks were omitted during the clustering step. Collectively, the information reported for MPRs is sufficient for the informed analysis of a diverse selection of MPRs, including those that exhibit coherent oscillations and those that do not.

Application to Pathophysiology

In the context of bone physiology, the deleterious consequences of disrupting extracellular nucleotide-mediated cross talk have been highlighted by the emergence of P2 receptor knockout mouse models (Lenertz et al., 2015). P2 receptors are particularly sensitive to changes in the extracellular milieu. Consequently, P2 receptor pathophysiology is often coupled to events that influence the extracellular composition, thereby compromising processes regulated by the P2 receptor network. In particular, changes in extracellular pH alter P2 receptor function (King et al., 1997; Gerevich et al., 2007; Wildman, 2009; Langfelder et al., 2015). Such conditions arise from pathological acidosis that is commonly caused by systemic acid-base disturbances, such as metabolic or respiratory acidosis (Krieger et al., 2004; Miller, 2012; Berend et al., 2014). More localized acidifications can also be associated with tumors (Martin and Jain, 1994; Kato et al., 2013). Since the skeleton is a common metastatic site for cancer, and participates in systemic buffering of protons, the effect of acidosis on the skeletal system is of particular interest. On the cellular level, acidosis promotes the activation of osteoclasts, resulting in elevated bone resorption which manifests itself in osteoporotic phenotype (Bushinsky and Frick, 2000; Krieger et al., 2003; Ahn et al., 2012; Gasser et al., 2014). However, it

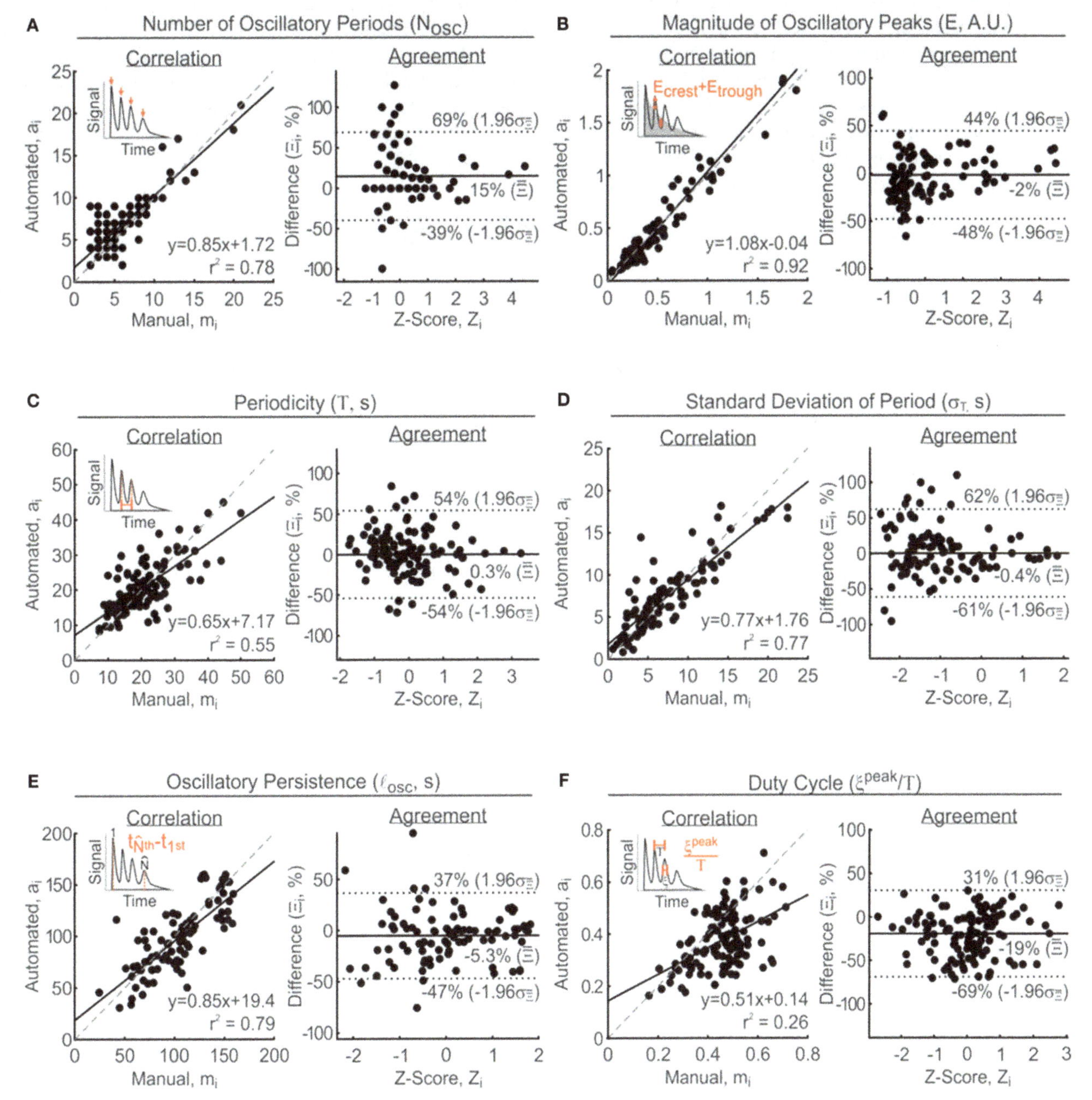

FIGURE 6 | Validation of MPR parameter measurements. Parameters were estimated manually and then compared with automated-estimated values using correlation analysis (left: solid line, linear regression line; broken line, line of equality) and Bland-Altman analysis (right: solid line, mean percent difference, $\bar{\Xi}$; broken line, limits of agreement, $\bar{\Xi} \pm 1.96\, \sigma_{\bar{\Xi}}$). **(A)** Number of oscillatory periods N_{OSC}. **(B)** Magnitude of oscillatory peaks, E. **(C)** Periodicity, T. **(D)** Standard Deviation of Period, σ_T. **(E)** Oscillatory persistence, l_{OSC}. **(F)** Duty cycle, ξ^{peak}/T. Arbitrary Units, A.U. Insets: Visual representations of parameters measured.

remains unclear whether the P2 receptor network plays a direct role in this cascade of events. The most immediate influence of acidosis on the P2-receptor network can be studied at the level of the $[Ca^{2+}]_i$ response evoked immediately upon application of a purinergic agonist.

We investigated the effect of acidosis on ATP-mediated $[Ca^{2+}]_i$ responses in bone-marrow derived osteoclast precursors to demonstrate the applicability of the developed algorithm. The application of ATP (100 nM to 10 mM) to the fura2-loaded osteoclast precursors evoked a $[Ca^{2+}]_i$ TR in a dose-dependent manner in control and acidosis conditions (**Figures 7A,B**). The response amplitudes under acidic conditions were virtually indistinguishable from the control for ATP concentrations up to 10 μM. However, above this threshold concentration, the amplitude of the control responses continued to increase with rising concentrations of ATP, while $[Ca^{2+}]_i$ responses under acidic conditions plateaued at 10 μM (**Figure 7C**). With respect to the AUC of the $[Ca^{2+}]_i$ responses, the observed differences between the two conditions were more gradual with a diverging trend beginning as low as 1 μM ATP and becoming more prominent at high ATP (**Figure 7D**). Finally, acidosis was found to have no significant effect on the periodicity of the oscillatory responses (**Figure 7E**).

These findings support that acidosis, while having no effect on ATP-mediated $[Ca^{2+}]_i$ responses at lower ATP concentration, significantly attenuates the magnitude of $[Ca^{2+}]_i$ transients responding to higher ATP concentrations (>10 μM ATP). Within the limited scope of this study that is focused on the development of a data analysis algorithm, we can only hypothesize on the mechanism by which these differences arise. One possibility is that the rise in extracellular $[H^+]$ has a significant influence on the electro-chemical gradient across the cellular membrane, which may consequently alter the extent of calcium flux across certain ionotropic P2X receptors. Since the oscillations are commonly driven by inositol triphosphate-mediated release of calcium from internal calcium stores (i.e., isolated from extracellular $[H^+]$), it may explain why the oscillatory behavior is not affected by acidosis. Alternatively, there may exist a subset of P2 receptors that are sensitive to fluctuations in extracellular $[H^+]$, while P2-receptors involved in oscillatory behavior and/or responses to lower ATP concentrations (≤10 μM) are resilient to such changes. Regardless of the underlying mechanism, these results highlight that the P2 receptor network can be differentially modulated by extracellular pH.

CONCLUSIONS

This paper presents an autonomous signal-processing algorithm capable of robustly removing signal-contaminating noise and delineating the various components seen in a calcium response, including non-stationary drift, TRs, and MPRs (possibly caused by flickers, puffs, and sparks) sampled with at least twice the Nyquist frequency. By fitting piece-wise defined model functions to data, the algorithm also extracts estimates for the parameters that are relevant to the characterization of cellular transient dynamics. Any time-series recordings can be used as an input for the algorithm, provided that they resemble a single or multi-peak transient response. As demonstrated in the validation process, manual estimation of certain parameters has an inherent degree of subjectivity and measurement error associated with

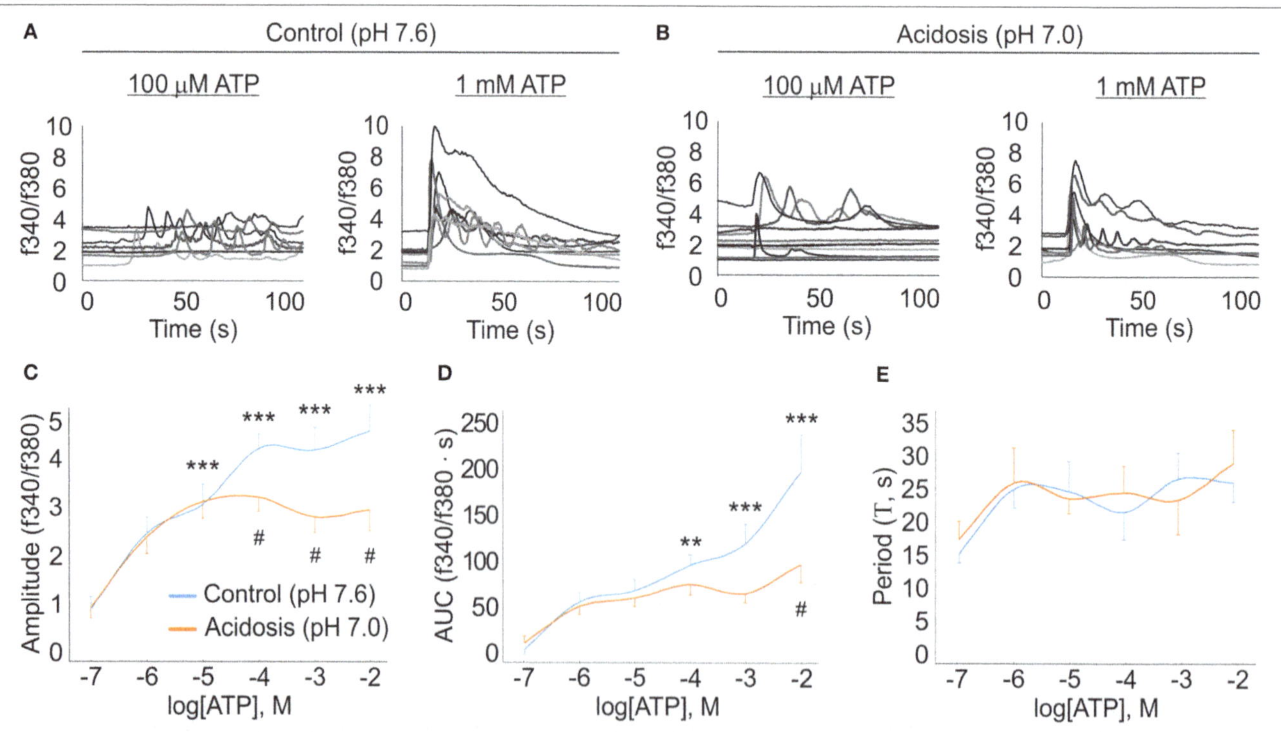

FIGURE 7 | Algorithm application in characterization of pathological states. ATP (100 nM–10 mM) was applied to Fura-2 loaded osteoclast precursors, under control (pH 7.6) and acidosis (pH 7.0) conditions, and $[Ca^{2+}]_i$ responses were recorded. Algorithm was used to obtain estimates for amplitude, AUC and periodicity. **(A)** Representative $[Ca^{2+}]_i$ response traces for 100 μM and 1 mM ATP under control conditions. **(B)** Representative $[Ca^{2+}]_i$ response traces for 100 μM and 1 mM ATP under acidosis conditions. **(C)** Amplitude dose-response curves. **(D)** AUC dose-response curves. **(E)** Period dose-response curves. For **(C–E)**, data are mean ± S.E.M. The effect of ATP under control conditions was examined using one-way ANOVA. The effect of acidosis on ATP-mediated responses was examined using two-way ANOVA. The Bonferroni test was used for post hoc multi-comparison analysis; $^{**}p < 0.01$; $^{***}p < 0.001$ indicate significant difference compared to the response to the lowest ATP concentration; $^{\#}p < 0.05$ indicates significant difference between responses to the same [ATP] in control and acidosis conditions.

it. In particular, the manual evaluations of AUC values, decay constant and the time of onset, as well as most of the MPR-parameter values, were found to rely on subjective estimates and thus lacked true accuracy and consistency. Because of such limitation in the validation method, manual-estimates are to be recognized as representative estimates, rather than accurate values for these parameters. Consequently, validation method applied here should be considered as a comparison against imperfect estimates.

Nevertheless, our analysis of the automated method has verified that the algorithm performs within acceptable margins of agreement when compared to manual analysis. Regarding the response detection capabilities, the algorithm behaves conservatively compared to manual assessments, especially when presented with low-magnitude TRs or ambiguous response signals. Most importantly, our algorithm has been validated against experimental $[Ca^{2+}]_i$ recording data, rather than simulated data, ensuring that the method is capable of handling variations in drift and noise that realistically reflect signal contaminations of experimental data acquisition. We have demonstrated that this automated methodology is effective in analyzing empirical data, providing quantitative insights about them and identifying differences between them.

A particularly unique feature of this algorithm is its capacity to characterize the magnitude and temporal characteristics of MPRs exhibiting stochastic and deterministic behavior. It is well established that a diverse amount of biochemical processes can be amplitude- and/or frequency-modulated (Adachi et al., 1999; Micali et al., 2015). To analyse such oscillatory data, the fast Fourier transform (FFT) is commonly used, which allows for the conversion of a signal from its time domain, into the frequency domain. Unfortunately, the variance in the frequency domain is proportional to the number of repetitive components in the time-domain. Therefore, if the oscillatory signals present few repetitive components then reliable resolution of the true periodicity of the signal is unachievable. To circumvent the limitations inherent to FFT, we apply the MATLAB "findpeaks()" function to identify peaks of interest. To isolate underlying coherent oscillations that are often present, we applied a clustering method. This is based on the principle of clustering deviations from baseline according to their temporal offset and respective FWHM. The advantage of this approach is that it allows for the reliable detection of periodic peaks, even in the presence of stochastic discharges, as is often the case in experimental recordings. Secondly, comparison of the set of MPR parameters for all peaks and subset of coherent peaks allows users to quantify the extent of stochastic activity within MPRs. Alternatively, the relationship between mean and standard deviation of periodicity in a MPR has been previously used to reveal the contribution of stochastic processes to the periodicity (Thurley et al., 2014). We anticipate this methodology will contribute to the comprehensive analysis of diverse MPRs.

Calcium signaling is by no means unique to the P2-receptor network, but rather represents the most ubiquitous and versatile messenger found in biological systems. All kinds of extracellular signals exploit calcium as a secondary messenger, including P2 agonists (i.e., ATP, ADP, UTP, and UDP), endothelin-1, oxotremorine-M, norephinephrine, thrombin, PDGF, bombensin (Balla et al., 1991; Palmer, 1994; Burnstock, 2004). The universal involvement of calcium ranges from basic physiological processes such as muscle contraction, neuronal discharge and pancreatic secretion, to early development events including mammalian egg fertilization and embryonic pattern formation (Berridge et al., 2000). Calcium signaling is also known to be impaired in various pathological states, as suggested for metabolic acidosis in this study, chronic renal failure (Massry et al., 1995), Alzheimer's (Brawek et al., 2014), Diabetes (Chen et al., 2015), and zinc deficiency (O'Dell and Browning, 2013). However, despite all that we know about calcium's role in biological processes, there remains ongoing debate on how calcium signals robustly encode information while still exhibiting a large degree of heterogeneity within and between various cellular populations. Many theories have been proposed to establish how information can be encoded. Some of these involve encoding information on the basis of calcium binding cooperativity (Larsen et al., 2004), amplitude and frequency modulation (De Pitta et al., 2009), changes in spike time variation (Thurley et al., 2014), and signal integration (Hannanta-anan and Chow, 2016). In order to reconcile these theories and establish a universal syntax for calcium-encoded information, tools such as this algorithm will aid in the large-scale analysis of experimental data sets required for the validation of mathematical models.

The consideration of signaling nuances that are specifically found in physiological signals, but may or may not be present in non-biological signals, was a critical step in the development of this algorithm. As demonstrated in this study, physiological signals were decomposed into their elementary components and mathematically generalized to enable for the computational reconstruction of a diverse range of signature forms. In doing so, we were able to provide a foundation for further modeling of the nonlinear multi-parametric physiological signals. This study demonstrates that the accurate description of complex physiological signals is non-trivial, but rather an extensive mathematical undertaking. Therefore, we believe that, beyond serving the purpose of a signal-processing tool, this algorithm will also contribute to future efforts to modeling physiological signals.

In summary, we have detailed an open-source MATLAB algorithm intended to facilitate the analysis of time-series recordings. With minimal user-input required, this tool dramatically decreases analysis time and ensures consistency in parameter characterization of complex physiological signals. This algorithm is capable of handling noise and drift and robustly characterizes the magnitude and kinetics of dynamic processes, outputting the amplitude, time of onset (t_{onset}), activation time ($t_{10-90\%}$), full-width half-max (FWHM), AUC, and decay constant (τ_{decay}). In the presence of MPR, six additional parameters are characterized which include number of oscillations (N_{osc}), magnitude of oscillatory peaks (E), periodicity (T), standard deviation of periodicity (σ_T), oscillatory persistence (l_{osc}), and the duty cycle (ξ^{peak}/T). This algorithm is not limited to any specific data-type, but $[Ca^{2+}]_i$

recordings represent an obvious application. In addition to calcium imaging, other imaging modalities such as adapted fluorescence resonance energy transfer (FRET) biosensors, real-time bioluminescence and voltage and current measurements can generate time-series data for which characterization of signal magnitude and kinetics can provide valuable information. As data acquisition becomes more efficient and data sets become increasingly complex, automated analysis will serve as an essential tool for conducting basic research and clinical screening.

AUTHOR CONTRIBUTIONS

Study conception and design: LM, NM, SK, AK. Algorithm development: LM, AK. Acquisition of data: NM. Analysis and interpretation of data: LM, NM. Drafting of Manuscript: LM, NM, SK, AK. All authors contributed to the critical revision of manuscript and approved the final version to be published.

ACKNOWLEDGMENTS

This research work was supported by the Natural Sciences and Engineering Research Council (NSERC) grants to AK and SK. The authors would like to acknowledge the partial financial support provided to LM by Dr. Michael Mackey through his NSERC fund.

REFERENCES

Abu Khamidakh, A. E., Juuti-Uusitalo, K., Larsson, K., Skottman, H., and Hyttinen, J. (2013). Intercellular Ca(2+) wave propagation in human retinal pigment epithelium cells induced by mechanical stimulation. *Exp. Eye Res.* 108, 129–139. doi: 10.1016/j.exer.2013.01.009

Adachi, Y., Kindzelskii, A. L., Ohno, N., Yadomae, T., and Petty, H. R. (1999). Amplitude and frequency modulation of metabolic signals in leukocytes: synergistic role of IFN-γ in IL-6- and IL-2-mediated cell activation. *J. Immunol.* 163, 4367–4374.

Ahn, H., Kim, J. M., Lee, K., Kim, H., and Jeong, D. (2012). Extracellular acidosis accelerates bone resorption by enhancing osteoclast survival, adhesion, and migration. *Biochem. Biophys. Res. Commun.* 418, 144–148. doi: 10.1016/j.bbrc.2011.12.149

Appleby, P. A., Shabir, S., Southgate, J., and Walker, D. (2015). Sources of variability in cytosolic calcium transients triggered by stimulation of homogeneous uro-epithelial cell monolayers. *J. R. Soc. Interface* 12:20141403. doi: 10.1098/rsif.2014.1403

Balla, T., Sim, S. S., Iida, T., Choi, K. Y., Catt, K. J., and Rhee, S. G. (1991). Agonist-induced calcium signaling is impaired in fibroblasts overproducing inositol 1,3,4,5-tetrakisphosphate. *J. Biol. Chem.* 266, 24719–24726.

Berend, K., de Vries, A. P., and Gans, R. O. (2014). Physiological approach to assessment of acid-base disturbances. *N. Engl. J. Med.* 371, 1434–1445. doi: 10.1056/NEJMra1003327

Berridge, M. J., Lipp, P., and Bootman, M. D. (2000). The versatility and universality of calcium signalling. *Nat. Rev. Mol. Cell Biol.* 1, 11–21. doi: 10.1038/35036035

Bland, J. M., and Altman, D. G. (1986). Statistical methods for assessing agreement between two methods of clinical measurement. *Lancet* 1, 307–310. doi: 10.1016/S0140-6736(86)90837-8

Boraschi-Diaz, I., and Komarova, S. V. (2016). The protocol for the isolation and cryopreservation of osteoclast precursors from mouse bone marrow and spleen. *Cytotechnology* 68, 105–114. doi: 10.1007/s10616-014-9759-3

Brawek, B., Schwendele, B., Riester, K., Kohsaka, S., Lerdkrai, C., Liang, Y., et al. (2014). Impairment of *in vivo* calcium signaling in amyloid plaque-associated microglia. *Acta Neuropathol.* 127, 495–505. doi: 10.1007/s00401-013-1242-2

Bray, M. A., Geisse, N. A., and Parker, K. K. (2007). Multidimensional detection and analysis of Ca2+ sparks in cardiac myocytes. *Biophys. J.* 92, 4433–4443. doi: 10.1529/biophysj.106.089359

Burnstock, G. (2004). Introduction: P2 receptors. *Curr. Top. Med. Chem.* 4, 793–803. doi: 10.2174/1568026043451014

Burnstock, G., and Verkhratsky, A. (2009). Evolutionary origins of the purinergic signalling system. *Acta Physiol. (Oxf.)* 195, 415–447. doi: 10.1111/j.1748-1716.2009.01957.x

Bushinsky, D. A., and Frick, K. K. (2000). The effects of acid on bone. *Curr. Opin. Nephrol. Hypertens.* 9, 369–379. doi: 10.1097/00041552-200007000-00008

Cao, D., Lin, G., Westphale, E. M., Beyer, E. C., and Steinberg, T. H. (1997). Mechanisms for the coordination of intercellular calcium signaling in insulin-secreting cells. *J. Cell Sci.* 110(Pt 4), 497–504.

Chambolle, A., Caselles, V., Novaga, M., Cremers, D., and Pock, T. (2009). *An Introduction to Total Variation for Image Analysis.* Available online at: http://cvgmt.sns.it/paper/2007/; https://hal.archives-ouvertes.fr/hal-00437581/en/; https://www.semanticscholar.org/paper/An-Introduction-to-Total-Variation-for-Image-Chambolle-Caselles/19d5136714aa8aa84129ca2b5f0fefca1ea6ce8d#citingPapers

Chartrand, R. (2007). "Nonconvex regularization for shape preservation," in *Paper Presented at the 2007 IEEE International Conference on Image Processing* (San Antonio, TX). doi: 10.1109/ICIP.2007.4378949

Chartrand, R. (2011). Numerical differentiation of noisy, nonsmooth data. *ISRN Appl. Math.* 2011, 1–11. doi: 10.5402/2011/164564

Chartrand, R., and Staneva, V. (2008). A quasi-newton method for total variation regulatrization of images corrupted by non-gaussian noise. *IET Image Process.* 2, 1–6. doi: 10.1049/iet-ipr:20080017

Chartrand, R., and Wohlberg, B. (2010). "Total-variation regularization with bound constraints," in *Proceedings of IEEE International Conference on Acoustics, Speech, and Signal Processing (ICASSP)* (Dallas, TX), 766–769.

Chen, H., Kold-Petersen, H., Laher, I., Simonsen, U., and Aalkjaer, C. (2015). Impaired endothelial calcium signaling is responsible for the defective dilation of mesenteric resistance arteries from db/db mice to acetylcholine. *Eur. J. Pharmacol.* 767, 17–23. doi: 10.1016/j.ejphar.2015.09.043

Churchill, G. C., Atkinson, M. M., and Louis, C. F. (1996). Mechanical stimulation initiates cell-to-cell calcium signaling in ovine lens epithelial cells. *J. Cell Sci.* 109, 355–365.

Clapham, D. E. (2007). Calcium signaling. *Cell* 131, 1047–1058. doi: 10.1016/j.cell.2007.11.028

De Pittà, M., Volman, V., Levine, H., and Ben-Jacob, E. (2009). Multimodal encoding in a simplified model of intracellular calcium signaling. *Cogn. Process.* 10(Suppl. 1), S55–S70. doi: 10.1007/s10339-008-0242-y

Dickinson, G. D., and Parker, I. (2013). Temperature dependence of IP3-mediated local and global Ca^{2+} signals. *Biophys. J.* 104, 386–395. doi: 10.1016/j.bpj.2012.12.024

Dupont, G., and Combettes, L. (2009). What can we learn from the irregularity of Ca^{2+} oscillations? *Chaos* 19, 037112. doi: 10.1063/1.3160569

Dupont, G., Combettes, L., Bird, G. S., and Putney, J. W. (2011). Calcium oscillations. *Cold Spring Harb. Perspect. Biol.* 3:a004226. doi: 10.1101/cshperspect.a004226

Ellefsen, K. L., Settle, B., Parker, I., and Smith, I. F. (2014). An algorithm for automated detection, localization and measurement of local calcium signals from camera-based imaging. *Cell Calcium* 56, 147–156. doi: 10.1016/j.ceca.2014.06.003

Frame, M. K., and de Feijter, A. W. (1997). Propagation of mechanically induced intercellular calcium waves via gap junctions and ATP receptors in rat liver epithelial cells. *Exp. Cell Res.* 230, 197–207. doi: 10.1006/excr.1996.3409

Francis, M., Waldrup, J. R., Qian, X., Solodushko, V., Meriwether, J., and Taylor, M. S. (2016). Functional tuning of intrinsic endothelial Ca2+ dynamics in swine coronary arteries. *Circ. Res.* 118, 1078–1090. doi: 10.1161/CIRCRESAHA.115.308141

Fritzsche, M., Fernandes, R. A., Colin-York, H., Santos, A. M., Lee, S. F., Lagerholm, B. C., et al. (2015). CalQuo: automated, simultaneous single-cell and population-level quantification of global intracellular Ca2+ responses. *Sci. Rep.* 5:16487. doi: 10.1038/srep16487

Gallagher, J. A., and Buckley, K. A. (2002). Expression and function of P2 receptors in bone. *J. Musculoskelet. Neuronal Interact.* 2, 432–439.

Gasser, J. A., Hulter, H. N., Imboden, P., and Krapf, R. (2014). Effect of chronic metabolic acidosis on bone density and bone architecture *in vivo* in rats. *Am. J. Physiol. Renal Physiol.* 306, F517–F524. doi: 10.1152/ajprenal.00494.2013

Gerevich, Z., Zadori, Z. S., Köles, L., Kopp, L., Milius, D., Wirkner, K., et al. (2007). Dual effect of acid pH on purinergic P2X3 receptors depends on the histidine 206 residue. *J. Biol. Chem.* 282, 33949–33957. doi: 10.1074/jbc.M705840200

Hannanta-anan, P., and Chow, B. Y. (2016). Optogenetic control of calcium oscillation waveform defines NFAT as an integrator of calcium load. *Cell Syst.* 2, 283–288. doi:10.1016/j.cels.2016.03.010

Hansen, M., Boitano, S., Dirksen, E. R., and Sanderson, M. J. (1993). Intercellular calcium signaling induced by extracellular adenosine 5′-triphosphate and mechanical stimulation in airway epithelial cells. *J. Cell Sci.* 106(Pt 4), 995–1004.

Isakson, B. E., Evans, W. H., and Boitano, S. (2001). Intercellular Ca2+ signaling in alveolar epithelial cells through gap junctions and by extracellular ATP. *Am. J. Physiol. Lung Cell. Mol. Physiol.* 280, L221–L228.

James, L. R., Andrews, S., Walker, S., de Sousa, P. R., Ray, A., Russell, N. A., et al. (2011). High-throughput analysis of calcium signalling kinetics in astrocytes stimulated with different neurotransmitters. *PLoS ONE* 6:e26889. doi: 10.1371/journal.pone.0026889

Jorgensen, N. R., Geist, S. T., Civitelli, R., and Steinberg, T. H. (1997). ATP- and gap junction-dependent intercellular calcium signaling in osteoblastic cells. *J. Cell Biol.* 139, 497–506. doi: 10.1083/jcb.139.2.497

Jørgensen, N. R., Teilmann, S. C., Henriksen, Z., Civitelli, R., Sørensen, O. H., and Steinberg, T. H. (2003). Activation of L-type calcium channels is required for gap junction-mediated intercellular calcium signaling in osteoblastic cells. *J. Biol. Chem.* 278, 4082–4086. doi: 10.1074/jbc.M205880200

Juhola, M., Penttinen, K., Joutsijoki, H., Varpa, K., Saarikoski, J., Rasku, J., et al. (2015). Signal analysis and classification methods for the calcium transient data of stem cell-derived cardiomyocytes. *Comput. Biol. Med.* 61, 1–7. doi: 10.1016/j.compbiomed.2015.03.016

Kaczmarek-Hájek, K., Lörinczi, E., Hausmann, R., and Nicke, A. (2012). Molecular and functional properties of P2X receptors–recent progress and persisting challenges. *Purinergic Signal.* 8, 375–417. doi: 10.1007/s11302-012-9314-7

Kato, Y., Ozawa, S., Miyamoto, C., Maehata, Y., Suzuki, A., Maeda, T., et al. (2013). Acidic extracellular microenvironment and cancer. *Cancer Cell Int.* 13:89. doi: 10.1186/1475-2867-13-89

King, B. F., Wildman, S. S., Ziganshina, L. E., Pintor, J., and Burnstock, G. (1997). Effects of extracellular pH on agonism and antagonism at a recombinant P2X2 receptor. *Br. J. Pharmacol.* 121, 1445–1453. doi: 10.1038/sj.bjp.0701286

Koshimizu, T., Koshimizu, M., and Stojilkovic, S. S. (1999). Contributions of the C-terminal domain to the control of P2X receptor desensitization. *J. Biol. Chem.* 274, 37651–37657. doi: 10.1074/jbc.274.53.37651

Krieger, N. S., Bushinsky, D. A., and Frick, K. K. (2003). Cellular mechanisms of bone resorption induced by metabolic acidosis. *Semin. Dial.* 16, 463–466. doi: 10.1046/j.1525-139X.2003.16100.x

Krieger, N. S., Frick, K. K., and Bushinsky, D. A. (2004). Mechanism of acid-induced bone resorption. *Curr. Opin. Nephrol. Hypertens.* 13, 423–436. doi: 10.1097/01.mnh.0000133975.32559.6b

Langfelder, A., Okonji, E., Deca, D., Wei, W. C., and Glitsch, M. D. (2015). Extracellular acidosis impairs P2Y receptor-mediated Ca(2+) signalling and migration of microglia. *Cell Calcium* 57, 247–256. doi: 10.1016/j.ceca.2015.01.004

Larsen, A. Z., Olsen, L. F., and Kummer, U. (2004). On the encoding and decoding of calcium signals in hepatocytes. *Biophys. Chem.* 107, 83–99. doi: 10.1016/j.bpc.2003.08.010

Lenertz, L. Y., Baughman, C. J., Waldschmidt, N. V., Thaler, R., and van Wijnen, A. J. (2015). Control of bone development by P2X and P2Y receptors expressed in mesenchymal and hematopoietic cells. *Gene* 570, 1–7. doi: 10.1016/j.gene.2015.06.031

Li, F., Shen, C., Fan, J., and Shen, C. (2007). Image restoration combining a total variational filter and a fourth-order filter. *J. Vis. Commun. Image Represent.* 18, 322–330. doi: 10.1016/j.jvcir.2007.04.005

Lock, J. T., Ellefsen, K. L., Settle, B., Parker, I., and Smith, I. F. (2015). Imaging local Ca^{2+} signals in cultured mammalian cells. *J. Vis. Exp.* e52516. doi: 10.3791/52516

Marquardt, D. W. (1963). An algorithm for least-squares estimation of nonlinear parameters. *J. Soc. Indust. Appl. Math.* 11, 431–441. doi: 10.1137/0111030

Martin, G. R., and Jain, R. K. (1994). Noninvasive measurement of interstitial pH profiles in normal and neoplastic tissue using fluorescence ratio imaging microscopy. *Cancer Res.* 54, 5670–5674.

Massry, S. G., Klin, M., Ni, Z., Tian, J., Kedes, L., and Smogorzewski, M. (1995). Impaired agonist-induced calcium signaling in hepatocytes from chronic renal failure rats. *Kidney Int.* 48, 1324–1331. doi: 10.1038/ki.1995.417

McLachlan, G., and Peel, D. (2005). "ML fitting of mixture models," in *Finite Mixture Models* (Hoboken, NJ: John Wiley and Sons, Inc.), 40–80. doi: 10.1002/0471721182.ch2

Micali, G., Aquino, G., Richards, D. M., and Endres, R. G. (2015). Accurate encoding and decoding by single cells: amplitude versus frequency modulation. *PLoS Comput. Biol.* 11:e1004222. doi: 10.1371/journal.pcbi.1004222

Miller, P. D. (2012). Unrecognized and unappreciated secondary causes of osteoporosis. *Endocrinol. Metab. Clin. North Am.* 41, 613–628. doi: 10.1016/j.ecl.2012.05.005

Nobile, M., Monaldi, I., Alloisio, S., Cugnoli, C., and Ferroni, S. (2003). ATP-induced, sustained calcium signalling in cultured rat cortical astrocytes: evidence for a non-capacitative, P2X7-like-mediated calcium entry. *FEBS Lett.* 538, 71–76. doi: 10.1016/S0014-5793(03)00129-7

O'Dell, B. L., and Browning, J. D. (2013). Impaired calcium entry into cells is associated with pathological signs of zinc deficiency. *Adv. Nutr.* 4, 287–293. doi: 10.3945/an.112.003624

Oh, S., Woo, H., Yun, S., and Kang, M. (2013). Non-convex hybrid total variation for image denoising. *J. Vis. Commun. Image Represent.* 24, 332–344. doi: 10.1016/j.jvcir.2013.01.010

Palmer, R. K., Yule, D. I., McEwen, E. L., Williams, J. A., Fisher, S. K. (1994). Agonist-specific calcium signaling and phosphoinositide hydrolysis in human SK-N-MCIXC neuroepithelioma cells. *J. Neurochem.* 63, 2099–2107. doi: 10.1046/j.1471-4159.1994.63062099.x

Patel, T. P., Man, K., Firestein, B. L., and Meaney, D. F. (2015). Automated quantification of neuronal networks and single-cell calcium dynamics using calcium imaging. *J. Neurosci. Methods* 243, 26–38. doi: 10.1016/j.jneumeth.2015.01.020

Picht, E., Zima, A. V., Blatter, L. A., and Bers, D. M. (2007). SparkMaster: automated calcium spark analysis with ImageJ. *Am. J. Physiol. Cell Physiol.* 293, C1073–C1081. doi: 10.1152/ajpcell.00586.2006

Rast, G., Weber, J., Disch, C., Schuck, E., Ittrich, C., and Guth, B. D. (2015). An integrated platform for simultaneous multi-well field potential recording and Fura-2-based calcium transient ratiometry in human induced pluripotent stem cell (hiPSC)-derived cardiomyocytes. *J. Pharmacol. Toxicol. Methods* 75, 91–100. doi: 10.1016/j.vascn.2015.04.005

Rice, S. O. (1973). Efficient evaluation of integrals of analytic functions by the trapezoidal rule. *Bell Syst. Tech. J.* 52, 707–722. doi: 10.1002/j.1538-7305.1973.tb01986.x

Romanello, M., and D'Andrea, P. (2001). Dual mechanism of intercellular communication in HOBIT osteoblastic cells: a role for gap-junctional hemichannels. *J. Bone Miner. Res.* 16, 1465–1476. doi: 10.1359/jbmr.2001.16.8.1465

Ruffinatti, F. A., Lovisolo, D., Distasi, C., Ariano, P., Erriquez, J., and Ferraro, M. (2011). Calcium signals: analysis in time and frequency domains. *J. Neurosci. Methods* 199, 310–320. doi: 10.1016/j.jneumeth.2011.05.009

Shabir, S., and Southgate, J. (2008). Calcium signalling in wound-responsive normal human urothelial cell monolayers. *Cell Calcium* 44, 453–464. doi: 10.1016/j.ceca.2008.02.008

Skupin, A., Kettenmann, H., Winkler, U., Wartenberg, M., Sauer, H., Tovey, S. C., et al. (2008). How does intracellular Ca(2+) oscillate: by chance or by the clock? *Biophys. J.* 94, 2404–2411. doi: 10.1529/biophysj.107.119495

Smedler, E., and Uhlen, P. (2014). Frequency decoding of calcium oscillations. *Biochim. Biophys. Acta* 1840, 964–969. doi: 10.1016/j.bbagen.2013.11.015

Smith, I. F., Wiltgen, S. M., and Parker, I. (2009). Localization of puff sites adjacent to the plasma membrane: functional and spatial characterization of Ca2+ signaling in SH-SY5Y cells utilizing membrane-permeant caged IP3. *Cell Calcium* 45, 65–76. doi: 10.1016/j.ceca.2008.06.001

Steele, E. M., and Steele, D. S. (2014). Automated detection and analysis of Ca(2+) sparks in x-y image stacks using a thresholding algorithm implemented within the open-source image analysis platform ImageJ. *Biophys. J.* 106, 566–576. doi: 10.1016/j.bpj.2013.12.040

Stoehr, A., Neuber, C., Baldauf, C., Vollert, I., Friedrich, F. W., Flenner, F., et al. (2014). Automated analysis of contractile force and Ca2+ transients in engineered heart tissue. *Am. J. Physiol. Heart Circ. Physiol.* 306, H1353–H1363. doi: 10.1152/ajpheart.00705.2013

Sun, F., Berry, D. J., Leong, D. A., and Veldhuis, J. D. (1997). Recruitment of individually (all-or-none) responding cells, rather than amplitude enhancement, is the single-cell mechanism subserving the dose-responsive activation of intracellular calcium second messenger signaling by the human luteinizing-hormone receptor. *Endocrine* 7, 219–226. doi: 10.1007/BF02778144

Moré, J. J. (1978). *The Levenberg-Marquardt Algorithm: Implementation and Theory.* Heidelberg: Springer.

Thurley, K., Tovey, S. C., Moenke, G., Prince, V. L., Meena, A., Thomas, A. P., et al. (2014). Reliable encoding of stimulus intensities within random sequences of intracellular Ca2+ spikes. *Sci. Signal.* 7, ra59. doi: 10.1126/scisignal.2005237

Tibshirani, R., Walther, G., and Hastie, T. (2001). Estimating the number of clusters in a data set via the gap statistic. *J. R. Stat. Soc. B* 63, 411–423. doi: 10.1111/1467-9868.00293

Traub, J. F. (1964). On lagrange-hermite interpolation. *J. Soc. Indust. Appl. Math.* 12, 886–891. doi: 10.1137/0112076

Vogel, C. (2002). *Computational Methods for Inverse Problem.* Philadelphia, PA: Society for Industrial and Applied Mathematics.

Volonté, C., Amadio, S., D'Ambrosi, N., Colpi, M., and Burnstock, G. (2006). P2 receptor web: complexity and fine-tuning. *Pharmacol. Ther.* 112, 264–280. doi: 10.1016/j.pharmthera.2005.04.012

von Kügelgen, I., and Hoffmann, K. (2016). Pharmacology and structure of P2Y receptors. *Neuropharmacology* 104, 50–61. doi: 10.1016/j.neuropharm.2015.10.030

Wildman, S. S. P. (2009). Changes in tubular fluid pH and nucleotide concentration alter P2 receptor-mediated regulation of renal ENaC. *FASEB J.* 23(1 Suppl.), 602–605.

Wong, L. C., Lu, B., Tan, K. W., and Fivaz, M. (2010). Fully-automated image processing software to analyze calcium traces in populations of single cells. *Cell Calcium* 48, 270–274. doi: 10.1016/j.ceca.2010.09.008

Xing, S., Grol, M. W., Grutter, P. H., Dixon, S. J., and Komarova, S. V. (2016). Modeling interactions among individual P2 receptors to explain complex response patterns over a wide range of ATP concentrations. *Front. Physiol.* 7:294. doi: 10.3389/fphys.2016.00294

Yan, Z., Khadra, A., Li, S., Tomic, M., Sherman, A., and Stojilkovic, S. S. (2010). Experimental characterization and mathematical modeling of P2X7 receptor channel gating. *J. Neurosci.* 30, 14213–14224. doi: 10.1523/JNEUROSCI.2390-10.2010

Zhao, Z., Walczysko, P., and Zhao, M. (2008). Intracellular Ca2+ stores are essential for injury induced Ca2+ signaling and re-endothelialization. *J. Cell. Physiol.* 214, 595–603. doi: 10.1002/jcp.21248

8

Mathematical Modeling of *Streptococcus pneumoniae* Colonization, Invasive Infection and Treatment

Elisa Domínguez-Hüttinger*[1,2]*, *Neville J. Boon*[1], *Thomas B. Clarke*[3]* and *Reiko J. Tanaka*[1]

[1] Department of Bioengineering, Imperial College London, London, UK, [2] Instituto de Ecología, Universidad Nacional Autónoma de México, Mexico City, Mexico, [3] Department of Medicine, Imperial College London, London, UK

Edited by:
Krasimira Tsaneva-Atanasova, University of Exeter, UK

Reviewed by:
Samantha Jane King, The Research Institute at Nationwide Children's Hospital, USA
Jason H. Yang, Massachusetts Institute of Technology, USA

***Correspondence:**
Elisa Domínguez-Hüttinger
elisa.dominguez@mail.ecologia.unam.mx
Thomas B. Clarke
thomas.clarke@imperial.ac.uk
Reiko J. Tanaka
r.tanaka@imperial.ac.uk

Streptococcus pneumoniae (*Sp*) is a commensal bacterium that normally resides on the upper airway epithelium without causing infection. However, factors such as co-infection with influenza virus can impair the complex *Sp*-host interactions and the subsequent development of many life-threatening infectious and inflammatory diseases, including pneumonia, meningitis or even sepsis. With the increased threat of *Sp* infection due to the emergence of new antibiotic resistant *Sp* strains, there is an urgent need for better treatment strategies that effectively prevent progression of disease triggered by *Sp* infection, minimizing the use of antibiotics. The complexity of the host-pathogen interactions has left the full understanding of underlying mechanisms of *Sp*-triggered pathogenesis as a challenge, despite its critical importance in the identification of effective treatments. To achieve a systems-level and quantitative understanding of the complex and dynamically-changing host-*Sp* interactions, here we developed a mechanistic mathematical model describing dynamic interplays between *Sp*, immune cells, and epithelial tissues, where the host-pathogen interactions initiate. The model serves as a mathematical framework that coherently explains various *in vitro* and *in vitro* studies, to which the model parameters were fitted. Our model simulations reproduced the robust homeostatic *Sp*-host interaction, as well as three qualitatively different pathogenic behaviors: immunological scarring, invasive infection and their combination. Parameter sensitivity and bifurcation analyses of the model identified the processes that are responsible for qualitative transitions from healthy to such pathological behaviors. Our model also predicted that the onset of invasive infection occurs within less than 2 days from transient *Sp* challenges. This prediction provides arguments in favor of the use of vaccinations, since adaptive immune responses cannot be developed *de novo* in such a short time. We further designed optimal treatment strategies, with minimal strengths and minimal durations of antibiotics, for each of the three pathogenic behaviors distinguished by our model. The proposed mathematical framework will help to design better disease management strategies and new diagnostic markers that can be used to inform the most appropriate patient-specific treatment options.

Keywords: *Streptococcus pneumoniae*, upper airway epithelium, data integration, hybrid systems, commensal bacteria, systems biology, antibiotics resistance

1. INTRODUCTION

Streptococcus pneumoniae (*Sp*) is a commensal bacterium that is part of the upper airway microbiota. While it normally resides on the upper airway epithelium without causing serious infection or tissue damaging inflammation (World Health Organization, 2012), factors such as co-infection with the influenza virus often result in the development of life-threatening infectious and inflammatory diseases, including pneumonia, meningitis or even sepsis (World Health Organization, 2012; McCullers, 2014), since these factors can cause a weakened immune response to *Sp* or tissue damage that may disrupt the normal interactions between *Sp* and host. The threat of *Sp* infection has been increasing despite interventions by widely available antibiotics, due to the increasing presence of multiple antibiotic-resistant *Sp* strains (Nuorti et al., 1998; McCullers et al., 2000). Reduced susceptibility to penicillin was detected in all WHO regions (World Health Organization, 2014) and the pneumococcus remains a major cause of morbidity and mortality, not solely from the lower lung infection Siegel and Weiser (2015). There is an urgent need to devise better intervention strategies that can effectively halt the onset or persistence of *Sp*-mediated pathology at its early stages using a minimal amount of antibiotics for a short duration, in order to avoid the emergence of further antibiotic-resistant *Sp* strains (Schrag et al., 2001; Prina et al., 2015b).

Identification and design of effective intervention strategies require systems-level and quantitative understanding of the complex and dynamically-changing host-pathogen interactions that can lead to either healthy *Sp* colonization or pathological conditions, such as infection or inflammation. This paper proposes a mathematical model of the host-pathogen interactions between *Sp* and the upper airway epithelium, the initial site of interaction between *Sp* and the host which is the first step in all disease tiggered by this bacterium (Siegel and Weiser, 2015). We analyse the model to systematically and quantitatively investigate the mechanisms by which the homeostatic interactions are disrupted, for example by a weakened barrier function (McCullers, 2014) or immune suppression (Didierlaurent et al., 2008), and cause the onset of infectious processes.

Previously proposed mathematical models (Smith et al., 2011; Shrestha et al., 2013; Smith et al., 2013; Mochan et al., 2014) considered *Sp* infections in the lung which is a normally sterile site of the airway epithelium. In this paper, we develop a mechanistic model of homeostatic interactions between the host's upper airway and *Sp* as a commensal bacterium, based on a variety of experimental data from *in vivo* and *in vitro* studies. Given that the tissue-damaging effects of neutrophil transmigration are responsible for part of the pathology of infection (Chin and Parkos, 2007; Zemans et al., 2009), we specifically model how impaired host-pathogen interactions lead to loss of epithelial homeostasis and serious infection. Our mechanistic model describes dynamic interplays between *Sp*, immune cells, and epithelial tissues by a hybrid system of ordinary differential equations (ODEs), and elucidates the mechanisms by which commensal bacteria cause infection.

Our model demonstrates a robust behavior of healthy clearance of asymptomatic pneumococcal colonization without overt disease. Perturbation of the model parameters, corresponding to virtual patient cohorts, demonstrates three clinically observed pathological behaviors (disease phenotypes) triggered by disrupted *Sp*-host interactions. Using this mathematical model of pneumococcal colonization, we further suggest optimal treatment regimens that minimize use of antibiotics to intervene the pathogenic processes for each patient cohort. As colonization is a prerequisite for all pneumococcal disease (Siegel and Weiser, 2015), studying and the modeling of colonization by *Sp* to understand its interaction with the host could be important for not only looking at invasive infections but also other types of interaction/infection of the pneumococcus and host.

2. RESULTS

2.1. Mathematical Model of *Sp* Colonization in the Upper Airway Epithelium

Our proposed mathematical model of *Sp* colonization (**Figures 1A,B**) is a system-level representation of the prominent interactions between *Sp*, the airway epithelium, and immune cells (a–j) in **Figures 1A,B**, that were identified based on the empirical evidence from numerous experimental *in vivo* and *in vitro* studies as detailed below.

Under homeostatic conditions, a population of commensal bacteria, *Sp*, resides in the lumen on the apical side of the airway epithelium, where they are contained by a competent epithelial barrier integrity (Beisswenger et al., 2007) (**Figures 1Aa,Ba**) and immune responses mediated by neutrophils and macrophages (Dick et al., 2008; Standish and Weiser, 2009) (**Figures 1Ab,Bb**). Through disrupted barrier, apically located *Sp* can translocate to reach the blood vessel (Beisswenger et al., 2007) (**Figures 1Ac,Bc**), where they are either killed by resident immune cells that circulate in the blood (Li et al., 2002; Li, 2004), or grow uncontrollably and result in invasive infection if the immune cells cannot contain the translocated *Sp* (Silverstein and Rabadan, 2012). The amount of the translocated *Sp* in the blood vessel is therefore a determinant of whether disrupted *Sp*-host interactions cause serious infection such as sepsis.

Translocation of *Sp* occurs through the airway epithelial barrier, whose integrity is regulated by the apically located bacteria load. The bacteria bind to Pattern-Recognition immune receptors, specifically Toll-like receptors (TLR2s) that are preferentially expressed on the apical side of the airway epithelial cells (Melkamu et al., 2009), and activate the TLR signaling cascade (**Figures 1Ad,Bd**). The activation of the TLR cascade in epithelial cells decrease the barrier integrity of the airway epithelium (**Figures 1Ae,Be**) by TLR-mediated activation of proteases that damage the epithelial cells (Oggioni et al., 2004; Schmeck et al., 2004; Attali et al., 2008; Tieu et al., 2009) and by reduction of the barrier recovery rate due to the increased expression of the transcriptional repressor SNAIL1, which inhibits the expression

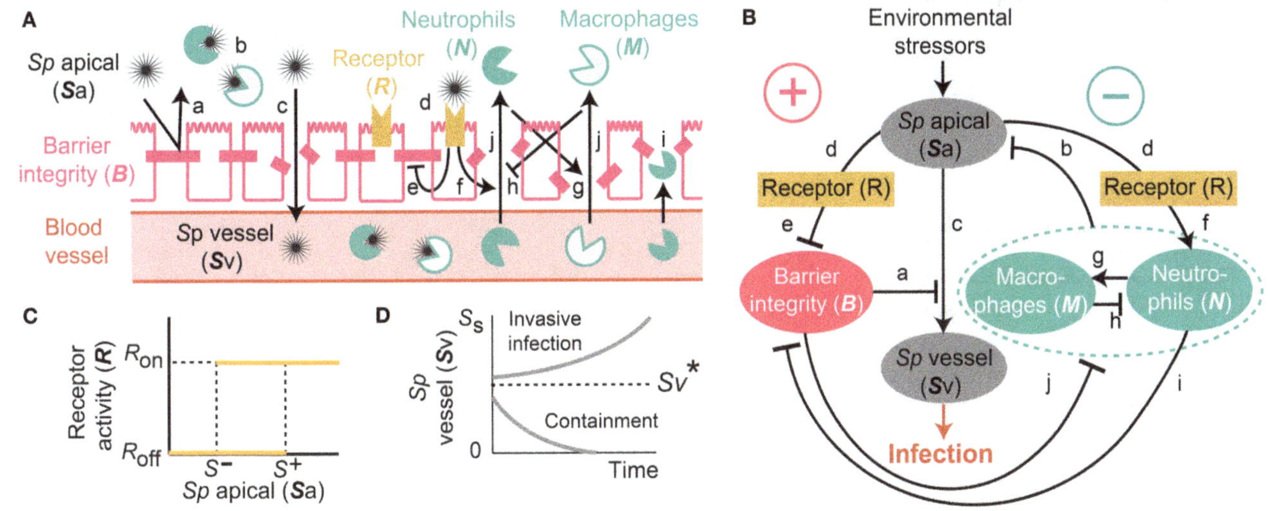

FIGURE 1 | A mechanistic model of *Sp.* colonization. (A) A schematic diagram of the processes included in the model. **(B)** The dynamic interplay between environmental stressors, barrier function and immune responses regulates infiltration of *Sp* to the blood vessel, which can result in infection. **(C)** *R*-switch for reversible activation of TLRs. **(D)** S_V-switch for the threshold behavior (invasive infection or containment) of the infiltrated *Sp* in the blood vessel.

of *claudin*, a component of the tight junctions (Clarke et al., 2011).

Active TLR signaling also induces recruitment of neutrophils from the neutrophil pool in the blood vessel, via the release of IL-17 (Zhang et al., 2009) that activates neutrophil-attracting interleukins IL-8 (Lindén, 2001) (**Figures 1Af,Bf**). The recruited neutrophils trigger transmigration of macrophages to the site of infection (Zhang et al., 2009), further potentiating the immune responses to the apically located pathogens (**Figures 1Ag,Bg**), whereas macrophages on the lumen restrict neutrophil transmigration (Zhang et al., 2009) by releasing neutrophil-repellent anti-inflammatory cytokines (Knapp et al., 2003) (**Figures 1Ah,Bh**). Transmigrating neutrophils release barrier degrading proteases (Chin et al., 2008) to reduce the barrier integrity (Nash et al., 1987; Nusrat et al., 1997; Zemans et al., 2009) (**Figures 1Ai,Bi**). The reduced barrier integrity in turn allows more transmigration of both neutrophils and macrophages from the blood vessel to the site of infection (Nash et al., 1987) (**Figures 1Aj,Bj**).

The model elucidates the main control structure of the system that maintains homeostatic interactions between commensal bacteria, *Sp*, and the host, via a dynamic interplay between environmental stressors, epithelial barrier integrity and immune responses (**Figure 1B**). At the apical side of the airway epithelium, *Sp* load is regulated via activation of TLRs, which induce immune responses that decrease the bacterial load but also reduce the epithelial barrier integrity. While the reduced barrier integrity enables transmigration of immune cells from the blood vessel for effective killing of *Sp* at the apical side of the epithelium barrier, it also allows transmigration of *Sp* from the apical side of the epithelium to the blood vessel, potentially causing systemic infection (sepsis). The dynamic interplay between the immune responses and the epithelial barrier integrity are further modulated by their mutual inhibition.

Our model further assumes two switches, an *R*-switch for TLR activation and an S_v-switch for the growth of the transmigrated bacteria in the blood vessel, based on the experimental evidence described below. The *R*-switch for *Sp*-mediated activation of TLRs reflects the observations that low concentrations of *Sp* do not cause activation of TLR signaling, while high concentrations lead to a sharp increase in TLR activity with hysteresis (He et al., 2009; Shalek et al., 2013; Sung et al., 2014). We model the *R*-switch by a perfect switch, which is a phenomenological representation of the bistable switch (Sung et al., 2014), and is described by the off- and on-states ($R = R_{\text{off}}$ and R_{on}) with the activation (S^+) and inactivation (S^-) thresholds for the critical concentrations of apically located *Sp* that abruptly and sharply turn on-or-off TLR activity (**Figure 1C** and Equation 2). The S_v-switch reflects the observations that transmigrated bacteria in the blood vessel (S_v) either overgrow (Benton et al., 1997) or are contained by resident immune cells depending on the bacterial concentration (Supplementary Figure 3B). We model the S_v-switch with a switching threshold of S_v^*, above which the infiltrated bacteria in the blood vessel grow exponentially (**Figure 1D**).

The resulting model is described by a hybrid system of five ODEs (Equation 1). The nominal values of the 24 model parameters (**Table 1**) were derived by fitting the model outcome to datasets from 11 independent studies, namely three *in vivo* studies (Benton et al., 1997; Zhang et al., 2009 and our own experiment) and eight *in vitro* studies (Nash et al., 1987; Coyne et al., 2002; Lagrou et al., 2003; Attali et al., 2008; Chin et al., 2008; Komori et al., 2011; Hathaway et al., 2012; Kwok et al., 2012), as detailed in the Supplementary Material. Our model therefore provides a

TABLE 1 | Nominal parameters of the model.

Parameter	Description	Value	References
N_V	Size of the neutrophil pool	10^8	Tanaka et al., 2015
δ_N	N degradation rate	6.1×10^{-2}/h	Tanaka et al., 2015
κ_B	Barrier recovery rate	4.6×10^{-2}/h	Coyne et al., 2002
κ_S	Bacteria growth rate	4.8×10^{-1}/h	Hathaway et al., 2012
$\bar{B}$	Nominal barrier integrity	1	
S^+	Activation threshold for R-switch	10^7 CFU/ml	Komori et al., 2011; Kwok et al., 2012
S^-	Deactivation threshold for R-switch	10^3 CFU/ml	Komori et al., 2011; Kwok et al., 2012
θ_S	Rate of bacterial transmigration through barrier	1.1×10^{-4}/h	Lagrou et al., 2003; Attali et al., 2008; Zhang et al., 2009
ϵ_{SB}	Inhibition rate of S_a transmigration by B	3.1	Lagrou et al., 2003; Attali et al., 2008; Zhang et al., 2009
ϵ_{BS}	Inhibition rate of B recovery by S_a	2.6×10 ml/CFU	Lagrou et al., 2003; Attali et al., 2008
ϕ_{SB}	Degradation rate of B by S_a	1.4×10^{-1}ml/CFU×h	Lagrou et al., 2003; Attali et al., 2008
ϵ_{NB}	Inhibition Rate of N recruitment by B	3.6×10	Nash et al., 1987; Chin et al., 2008
ϵ_{MB}	Inhibition rate of M recruitment by B	$= \epsilon_{NB}$	Nash et al., 1987; Chin et al., 2008
ϕ_{NB}	Degradation rate of B by N	4.0×10^{-8} ml/cells×h	Nash et al., 1987; Chin et al., 2008
μ_S	Saturation limit for S_a	3.7×10^4 CFU/ml	Zhang et al., 2009
ϕ_{NS}	Rate of S_a killing by N	6.1×10^{-4} ml/cells× h	Zhang et al., 2009
ϕ_{MS}	Rate of S_a killing by M	6.3×10^{-3} ml/cells× h	Zhang et al., 2009
K	Half-killing constant of S_v	1.3×10^4 CFU/ml	Benton et al., 1997 and Figure S3
δ_S	Rate of S_v killing by circulating immune cells	6.9×10^3 cells/ml×h	Benton et al., 1997 and Figure S3
α	Rate of N recruitment by S_a	$0.465 \times 150\text{x}10^{(-8)}$ ml/CFU×h	Zhang et al., 2009
ϵ_{NM}	Inhibition rate of N recruitment by M	1.6×10^{-1} ml/cells	Zhang et al., 2009
β	Rate of M recruitment by N	2.6×10^{-2} ml/cells× h	Zhang et al., 2009
Mv	Number of macrophage pool	3.0×10^{-1} cells/ml	Zhang et al., 2009
δ_M	M degradation rate	6.4×10^{-5}/h	Zhang et al., 2009

coherent mathematical framework to explain both *in vivo* and *in vitro* data.

2.2. Healthy Clearance of Asymptomatic *Sp* Colonization is Robustly Observed

One of the dataset used for the parameter estimation was obtained from *in vivo* studies in Zhang et al. (2009), where the mice were challenged with 10^7 CFU of *Sp* and recovered their healthy state, which is characterized by nonzero apical commensal bacterial load that does not trigger host responses. Our model was fitted to reproduce the experimental measurement in Zhang et al. (2009) for the apical bacterial load (S_a) and the concentrations of neutrophils (N) and macrophages (M) (**Figure 2A**). Both the experimental data and our model simulation demonstrate that the transient *Sp* challenge (increase of S_a) triggers a transient increase in N and a subsequent increase in M. These immune responses can bring S_a down to a homeostatic level, when the saturation limit for S_a is not high enough, which enabled the recovery of the mice from the bacterial challenge within 7 days without demonstrating invasive infection.

The simulation of our model with the data-calibrated nominal parameters further predicts the dynamics of three variables that were not measured in this experiment, TLR activity (R), barrier integrity (B) and infiltrated *Sp* in the blood vessel (S_v), thereby explaining the underlying mechanism of the healthy recovery from a bacterial challenge. Upon a *Sp* challenge, the apically located bacterial load becomes high enough ($S_a(0) > S^+$) to activate TLRs ($R = R_{on}$), which trigger recruitment of immune cells to the site of infection. These immune responses bring the initially high S_a down to below S^-, where the R-switch turns off ($R = R_{off}$) and stays off as S_a remains below S^+, as suggested by the focal point analysis (see Methods). The epithelial barrier integrity (B) continuously decreases while the R-switch is on ($R = R_{on}$), allowing bacteria to invade the blood vessel, as demonstrated by a rise in S_v. However, a healthy clearance of S_v is achieved without causing sepsis, since the peak of S_v remains below the threshold, S_v^*, of the S_v-switch. Note that both the experiments and our model simulation demonstrate that M stays high while B is kept high after N goes to zero, suggesting the importance of M as an immune mediator that does not compromise the barrier integrity.

The healthy recovery behavior described above is characterized in our model by convergence to the off-state of TLR activity ($R = R_{off}$) accompanied by the containment of S_v ($S_v < S_v^*$), and is robustly observed under perturbations to the parameter values. Among 10,000 simulations conducted by randomly sampling parameter values from an uniform distribution over two orders of magnitude around the nominal values, 83% of the simulations demonstrated a healthy recovery

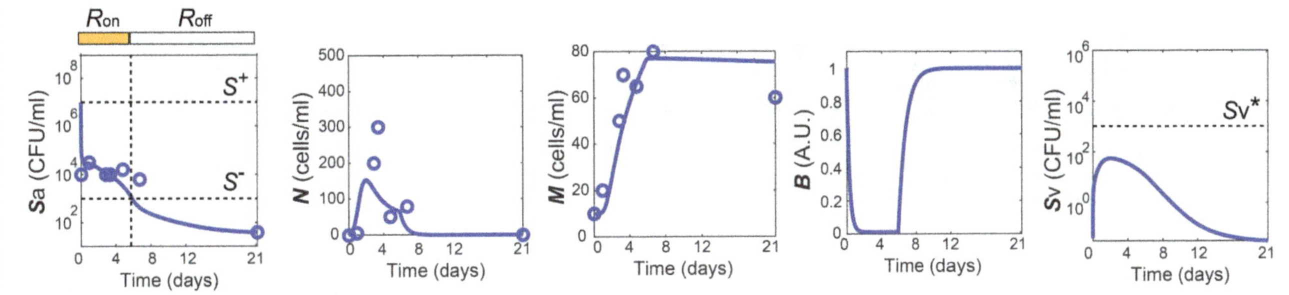

FIGURE 2 | Healthy recovery from a transient *Sp* challenge. Blue circles and solid lines represent the *in vivo* experimental data from Zhang et al. (2009) and the model prediction, respectively.

from a transient *Sp* challenge (**Figures 3A,B**). In 98% of the healthy recovery cases computationally observed, S_v reached its peak while $R = R_{on}$ (**Figure 3E**), suggesting that the appropriate host responses via TLR activation are responsible for containing S_v. The appropriate level of the barrier damage by active TLRs enables effective recruitment of immune cells that can reduce S_a, but prevents excessive transmigration of S_a to S_v, keeping the S_v lower than the threshold, S_v^*. The robust appearance of the healthy recovery in our model simulations confirms that that our model can coherently explain the mechanism behind the healthy recovery of the host from pneumococcal colonization, which can be effectively cleared by the natural host responses without any treatments.

2.3. Four Phenotypes Classified by the Double Switch

The remaining 17% of the simulations with parameters perturbed from their nominal values demonstrated systems dynamics that correspond to serious infection or inflammation. They are classified into three disease phenotypes, depending on the states of the S_v- and R-switches (**Figure 3A**). The state of the S_v-switch determines whether sepsis occurs due to invasive infection of $S_v(> S_v^*)$, and that of the R-switch determines whether *immunological scarring* occurs due to persistent host responses caused by $R = R_{on}$. Immunological scarring refers to the cumulative and long-term effects of immune response to pathogens, including tissue remodeling and altered immune responses to new pathogenic challenges, that persist after the pathogenic organism has been cleared (Fonseca et al., 2015). In our simulations, 13% demonstrated sepsis without immunological scarring ($S_v > S_v^*$ and $R = R_{off}$) and the other two disease phenotypes, immunological scarring ($R = R_{on}$) with and without sepsis, were observed 2% each (**Figure 3A**).

Immunological scarring is characterized by a persistent on-state of the R-switch due to S_a staying above S^- (**Figure 3C**). The R-switch triggers persistent host responses leading to sustained immune responses which are not strong enough to decrease S_a below S^- but cause persistent barrier damage. Note that the peak of N is much lower in the immunological scarring phenotype than that in the healthy recovery and sepsis phenotypes with $R = R_{off}$ (**Figures 3B,D**), resulting in weak immune responses that are not sufficient to decrease S_a. As a result, the host becomes vulnerable to a second bacterial attack due to the damaged barrier and the sub-threshold concentration of S_v ($< S_v^*$), which stay as silent remainders of the first pathogenic challenge.

Sepsis is characterized by outgrowth of S_v once it surpasses the threshold S_v^* (**Figure 3C**). In 99.7% of the sepsis phenotypes simulated by our model, the onset of sepsis occurs (when $S_v = S_v^*$ is achieved) while R is on (**Figure 3G**), suggesting that whether sepsis occurs or not is determined by the dynamics of S_v while R is on. It is similar with the healthy recovery case, where S_v reaches its peak below the threshold S_v^*, while R is on. Moreover, the duration of $R = R_{on}$ is much longer for the sepsis phenotype compared to the healthy recovery phenotype (**Figure 3H**), suggesting that persistent host response may allow excessive transmigration of *Sp* into the blood vessel above S_v^*.

When both the S_v- and R-switches are on, sepsis is accompanied by immunological scarring (**Figure 3E**), where the barrier is severely damaged and S_v continues increasing above S_v^*, while S_a remains above S^-.

The four phenotypes, including a healthy phenotype and three disease phenotypes, correspond to different patient cohorts observed in the clinic. *Healthy recovery from colonization* is the most common outcome of host-pneumococcal interactions (Austrian, 1986) and corresponds to patients who can clear their symptoms from transient infection without any antibiotics treatment. *Sepsis* corresponds to patients who would develop systemic infection as a consequence of dysregulated transepithelial crossing of bacteria if no treatment is applied (Clarke et al., 2011; Siegel and Weiser, 2015). *Immunological scarring* corresponds to tissue-damaging inflammation that prevails even after clearance of the pathogens (Periselneris et al., 2015). The long-term deleterious consequences of such sterile inflammation and the associated tissue restructuring/damage are considered to underlie many diseases, including pulmonary fibrosis associated to previous *Sp* infections (Knippenberg et al., 2015), chronic obstructive pulmonary disease (Garcha et al., 2012) and cancer (Elinav et al., 2013; Pradere et al., 2014). A sustained activation TLR is recognized to be an important molecular player responsible for this tissue damage (Pradere et al., 2014), as in our model. *Sepsis with immunological scarring* corresponds to patient cohorts who would develop a severe infection with long-term deleterious effects in absence of treatment (Leibovici, 2013).

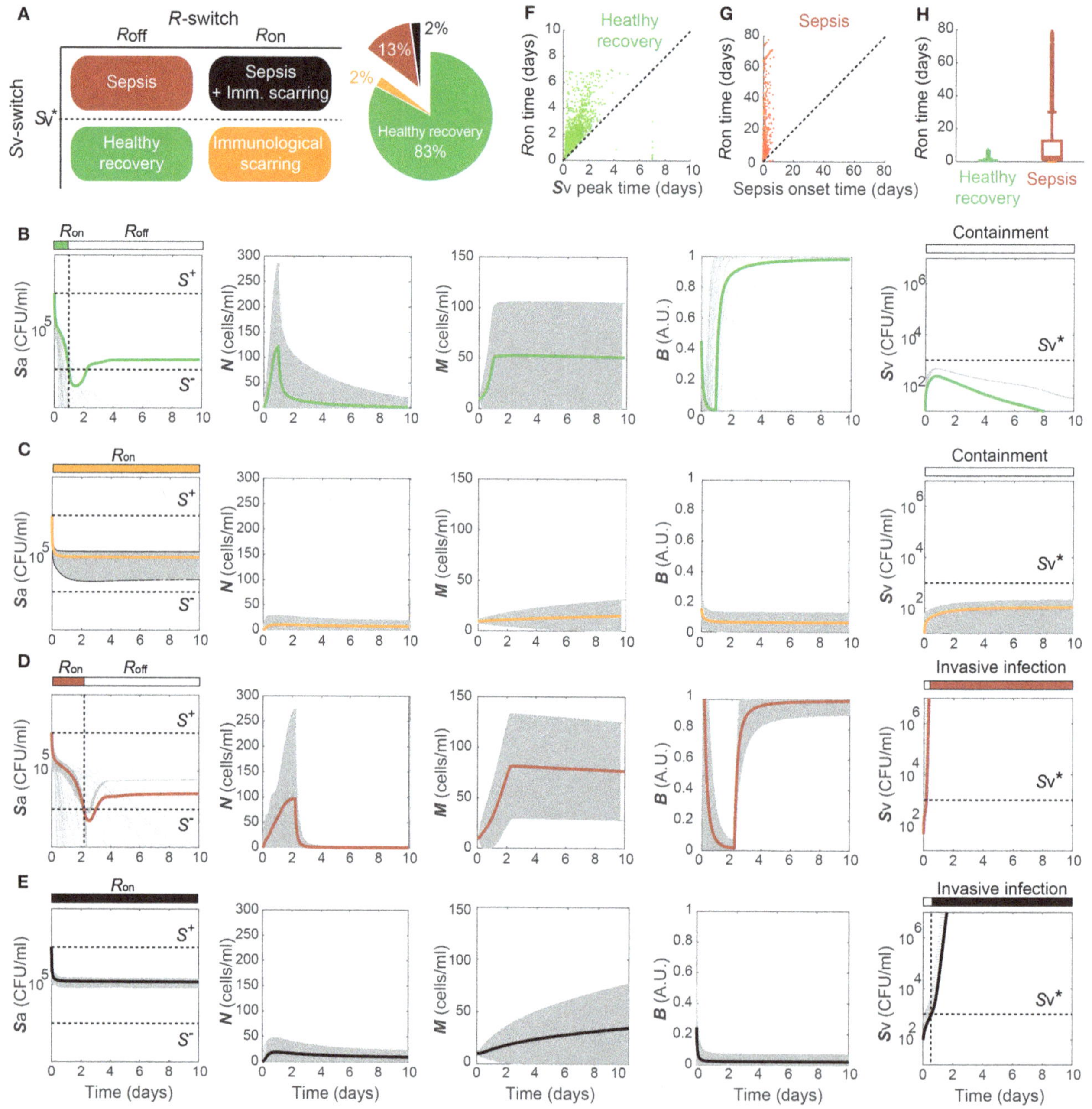

FIGURE 3 | Four phenotypes resulting from alterations in the *Sp*-host interactions. (A) Four (a healthy and three pathological) phenotypes determined by the states of *R*- and S_V-switches, and their respective frequency of observation in 10,000 simulations with varying parameters for our model. The states of the *R*- and S_V-switches determine whether host response persists causing immunological scarring, and whether sepsis occurs, respectively. **(B–E)** The dynamics of the four phenotypes: healthy recovery **(B)**, immunological scarring **(C)**, sepsis **(D)** and sepsis accompanied with scarring **(E)**. Solid lines and the gray shaded regions correspond to the mean dynamics and the ± standard deviation. **(F)** *R*-switch ON time vs. Time to reach the peak of S_V for the healthy recovery case. **(G)** *R*-switch ON time vs. Time for sepsis onset for the sepsis phenotype. **(H)** Boxplots representing the minimum, first quartile, median, third quartile, maximum and outlayers of the *R*-switch ON time for healthy recovery and sepsis cases.

2.4. Risk Factors for Disease Phenotypes

To identify the model parameters that affect the states of the *R*- and S_V-switches thereby determine the four phenotypes, we conducted the global parameter sensitivity analysis of our model with respect to *R* and S_V, respectively, using both Sobol and eFAST sensitivity indices (Marino et al., 2008; Cannavó, 2012).

The analysis identified the three most sensitive parameters for the propensity to turn on both the *R*-switch (to develop immunological scarring) and the S_V-switch (to develop sepsis)

(**Figure 4**): the rate of bacterial transmigration through the barrier (θ_S), the bacterial carrying capacity (μ_S), and the killing rate of bacteria by macrophages (ϕ_{MS}) further confirming the importance of macrophages. Simulations with systematic variations of these three parameters further suggest that they affect the occurrence of sepsis and of immunological scarring, as well as how quickly these occur after the *Sp* challenge (**Figure 5**).

These three sensitive parameters have a direct correspondence with risk factors for disease triggered by *Sp* that have been reported in the experimental literature. For example, increase in θ_S can be caused by co-infection, which damages the barrier directly or by having triggered previous immune responses (McCullers, 2014). Increase in μ_S is caused by previous infections, for example by influenza virus, that damage the tissue, increase nutrient contents (Siegel et al., 2014), or shift the microbiome composition affecting the dynamics of the different bacterial populations (McCullers, 2014). ϕ_{MS} can be affected for example by severe asthma (Liang et al., 2014).

Other parameters that are also affected by co-infection were not identified to be very sensitive for the propensity to develop sepsis (increase in S_v) or unresolved host responses (increase in *R*) (**Figure 4**). These parameters include S^+ which can increase as a consequence of TLR2 desensitization caused by a previous influenza virus infection (Didierlaurent et al., 2008), *M* that may increase as a consequence of previous infectious events (La Gruta et al., 2007; Yin et al., 2013), and the size of the neutrophil pool (N_v) which may decrease by chemotherapy or severe infections (Dick et al., 2008). The unsensitivity to the initial conditions can be partially explained by the existence of a unique stable steady state corresponding to the healthy recovery.

2.5. A Rapid Onset of Sepsis Triggered by a Transient *Sp* Challenge

In the septic behavior observed in 15% of the simulations (**Figures 3C,D**), the sepsis occurred (S_v increases above S_v^*) within 2 days post *Sp* challenge in 79% of the cases (**Figure 6**). When sepsis is accompanied with immunological scarring (*R* stays on and S_v increases above S_v^*, **Figure 3D**), the time to sepsis is longer than when it is not (**Figure 6**). The computationally predicted rapid onset of the sepsis is consistent with experimental observations in Andonegui et al. (2009) that the mice either survived or died within 36 h upon *Sp* challenge applied directly into the lumen of the lungs. The results suggest that rapid treatments within 36 h are crucial to prevent the onset of sepsis.

Increasing the immune activity, for example by activation of adaptive immune responses, could be an effective way to decrease the risk of sepsis onset, as it elevates the switching threshold, S_v^*, which depends on the strength of the resident immune cells. While the adaptive immunity could be activated naturally, the time for activation of the adaptive immune responses (which involves the *de novo* differentiation of naive T cells into mature T cells) by infiltrated pathogens was experimentally evaluated to

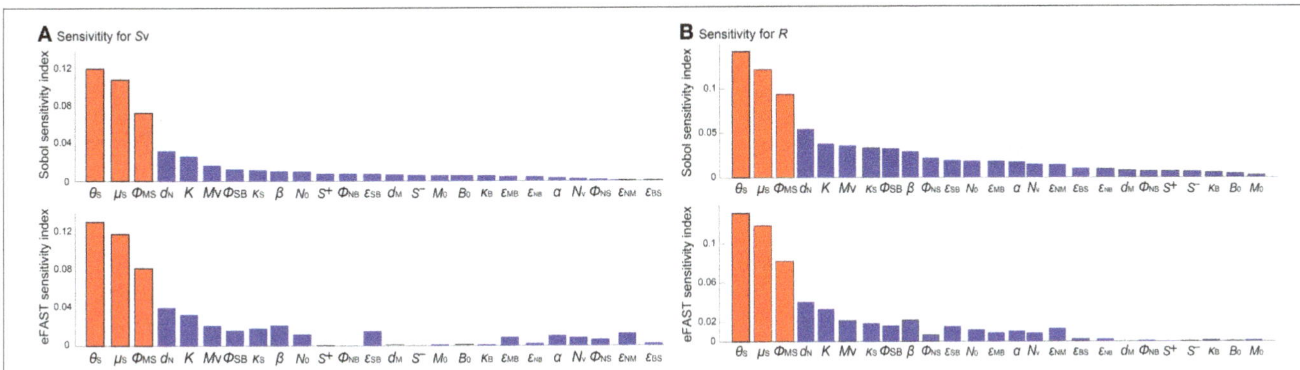

FIGURE 4 | Global sensitivity analysis of the model with respect to (A) S_V and **(B)** *R*, using the SOBOL and eFAST sensitivity indices.

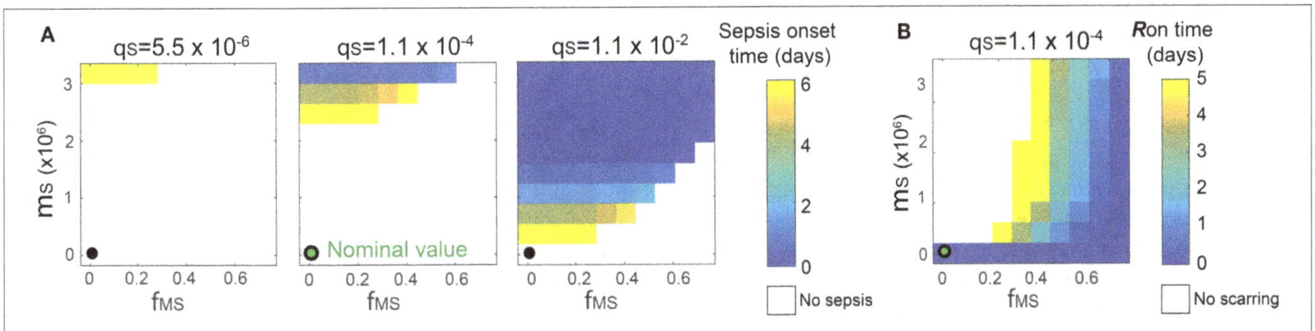

FIGURE 5 | Combinatorial effects of the three most sensitive parameters (θ_S, μ_S and ϕ_{MS}) on the sepsis onset time (A) and the R_{on} time **(B)**. The black circles correspond to the nominal values for (μ_S, ϕ_{MS}) and the nominal value for θ_S is 1.1×10^{-4}. Changes in θ_S do not affect the R_{on} time.

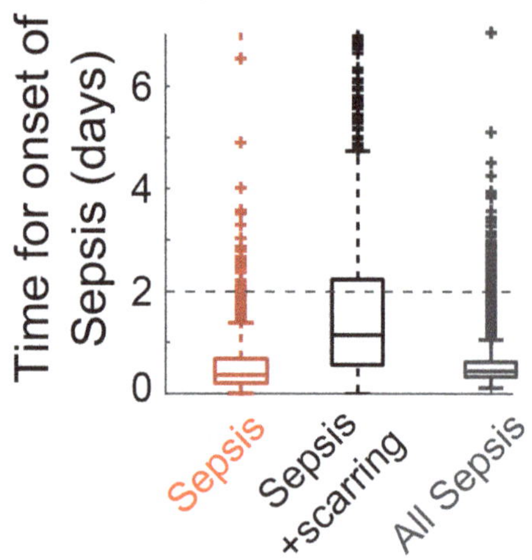

FIGURE 6 | Computationally predicted time for onset of sepsis.

be more than 2 days in mice (Zheng and Flavell, 1997). Such slow activation of the adaptive immunity therefore cannot prevent the onset of sepsis within 36 h.

These results suggest that prophylactic activation of the adaptive immune responses, for example by vaccinations, could be an effective strategy to prevent the incidence of sepsis, as demonstrated by the protective effects of *Sp* vaccination in mice (Cao et al., 2013). It is also consistent with the clinical suggestions to use vaccines as a preventive measurement against transient bacterial challenges in the high-risk patients (World Health Organization, 2012).

2.6. Optimal Antibiotics Treatment Regimens for Each of the Three Patient Cohorts

Using the proposed model, we investigate optimal treatment regimens and determine the minimal strength and duration of antibiotics treatment that are required to prevent or revert the pathological consequences of a transient *Sp* challenge. The minimal use of antibiotics is important for tackling the problem of antibiotics resistance (Nuorti et al., 1998), since the emergence of antibiotic-resistant *Sp* strains has been associated to the excessive use of antibiotics (Schrag et al., 2001; Prina et al., 2015b). We consider two different types of bactricidal antibiotics treatments in our modeling framework: apical application of antibiotics in the luminal side of the mucosa that decreases S_a and can thereby turn off the R-switch and stop the immunological scarring, and systemic application of antibiotics in the blood vessel that directly decreases S_v to prevent the onset of invasive infection (described in the Methods Section 4.3).

When the patients have immunological scarring without sepsis (**Figure 3C**), the treatment by apical application of antibiotics should aim to reduce S_a down below S^- to turn off the R-switch (**Figure 7A**). Once the R-switch is turned off, further use of antibiotics is no longer needed, as the healthy steady state with $R = R_{\text{off}}$ is locally attractive ($S_a < S^+$) for all the parameter combinations tested (over 10,000 simulations). The minimal treatment potency of apically applied antibiotics (minimal strength × minimal duration) to bring S_a down below S^- depends on the severity of the phenotype measured by the deviation of the high focal point from S^- ($R^2 = 0.46804$).

When the patients are susceptible for sepsis (**Figure 3D**), the treatment by systemic application of antibiotics should aim to reduce S_v to avoid reaching S_v^* and thereby causing invasive infection (**Figure 7B**). The minimal strength of systemically applied antibiotics allows the maximum of S_v to reach just below S_v^*, and the minimal duration of the treatment with the minimal strength corresponds to the time required for $R(t)$ to naturally turn off by S_a reaching S^-. The minimal treatment potency of systemic antibiotics to prevent sepsis depends on the severity of the phenotype measured by the time to reach S_v^* in the absence of treatments.

When the patients are susceptible to the combination of sepsis and immunological scarring (**Figure 3E**), they require antibiotics that are strong enough to be apically applied until S_a decreases below S^- to turn off the R-switch (**Figure 7C**). Our model simulations predicted that the apical treatment is enough to prevent invasive infection for 48% of these cases, since the reduction of S_a also reduces S_v, but the remaining 52% of the cases require additional application of comparatively small amounts of antibiotics directly in the blood vessel.

The distributions of the minimal treatment strengths and durations that we computationally predicted can be used as a guide to design safe and effective treatment options for the three patient cohorts. For example, our results suggest that antibiotics treatment for 20 days can prevent or revert most of immune scarring (**Figure 7A**) or invasive infection (**Figure 7B**), but that a much longer antibiotics treatment is needed for patients with a propensity for both sepsis and immune scarring (**Figure 7C**).

3. DISCUSSION

In this paper, we have proposed the first mathematical model of *Sp* colonization of the upper airway epithelium, and demonstrated that it robustly reproduces the healthy co-existence between this bacterium and the host. Our mathematical model is a hybrid system of ODEs, describing the interactions between the bacteria, immune cells and epithelial barrier function in a mechanistic, dynamical, quantitative and integrative way.

A key element of our model to determine the healthy and pathological phenotypes is a "double switch motif" (Domínguez-Hüttinger et al., in press). The first switch describes activation of the TLR2 signaling pathway by apically located bacteria (**Figure 1C**). It can reflect both the resting, homeostatic *Sp*-upper airway interactions (when $R = R_{\text{off}}$) characteristic of *Sp* as a commensal bacterium, and the transient host response to a *Sp* challenge (when $R = R_{\text{on}}$). Failure to inactivate this R-switch due to impaired host-pathogen interactions, for example by weakened immune responses (**Figures 4B, 5B**), can have long term consequences such as immunological scarring (**Figure 3C**) that require treatment to resolve it. The second switch distinguishes a transient growth of S_v that can be

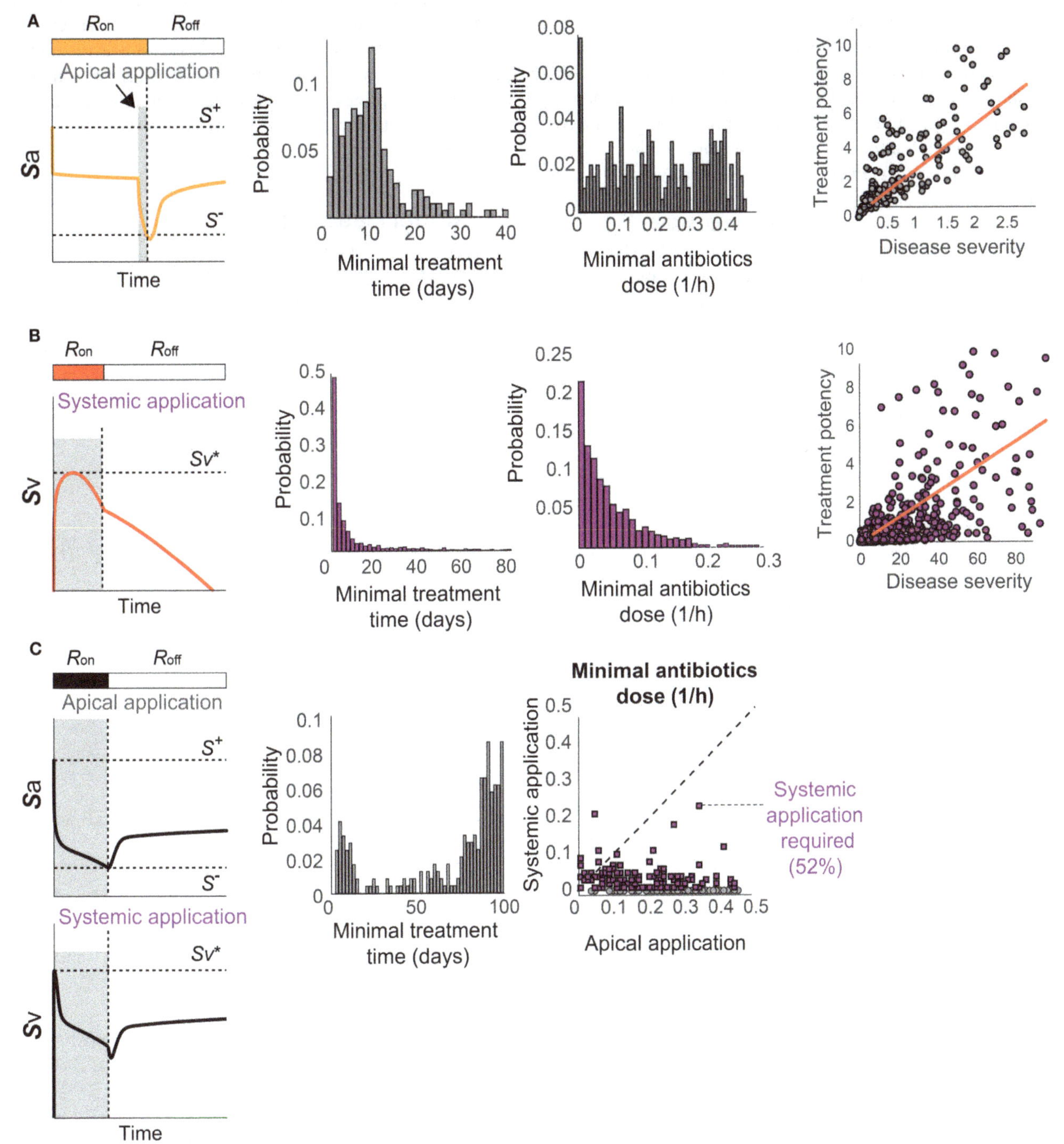

FIGURE 7 | Optimal antibiotics treatment strategies for patients with (A) Immunological scarring to turn off the *R*-switch, **(B)** sepsis, and **(C)** sepsis and immunological scarring. Gray shaded regions in the left column denote the minimal time of application of antibiotics in the apical side of the epithelium or in the blood vessel (systemic application).

contained without treatments (when $S_V \leq S_V^*$) and an invasive infection (when $S_V > S_V^*$) which would require a large dose of antibiotics treatments (**Figure 1D**). Our model simulations predict that invasive infection is developed within 36 h, in consistent with experimental observations (Andonegui et al., 2009) that were not used for development of our model. Our model analysis identified the most likely risk factors for an increased susceptibility to develop invasive infection, in response to transient *Sp* challenges (**Figures 4, 5**). Based on the state of this double-switch motif, we characterized four different phenotypes

(**Figure 3**), and identified those susceptible cohorts that require specific antibiotics treatment to prevent or revert the adverse effects of a *Sp* challenge. We further used our mathematical model to calculate the minimal strengths and durations of antibiotics application to effectively treat each of these disease phenotypes (**Figure 7**). These results suggest that the proposed quantitative and systems-level framework of *Sp* infection can be used to design optimal and personalized treatment strategies, as it can predict the minimal application times that are required to achieve prevention or remission for individual patient cohort.

While our mathematical model was constructed based on murine and *in vitro* experiments, future calibration of the model with human data could make the proposed mathematical modeling framework directly translatable to the clinic, to help stratification of patients and identification of patient-specific optimal treatment strategies. For example, our model analysis suggested that the efficacy of bacterial killing by immune cells (**Figures 4, 5**) could be used as a marker to distinguish vulnerable patient cohorts who would require preventive treatments *before* the onset of sepsis. This efficacy could be determined *ex vivo*, from serum or broncheoalveolar lavage fluid extracted from patients, to predict patient-specific responses characteristic of those vulnerable patient cohorts. A similar approach to stratify patients based on measurements of isolated components of a more complex physiological system has been shown to be effective for other complex diseases (Fey et al., 2015). The computational method demonstrated in this paper could then allows us to predict the minimal strength and duration of antibiotics application for individual patient cohort and for a specific antibiotics, given the experimentally determined information on the efficacy of the antibiotics (Mandell et al., 2007; Prina et al., 2015b) and the growth rate of the pneumococcal strain in the patients' serum. Our model will also enable us to investigate and design preventive strategies by early vaccines against invasive infection in patient cohorts who are identified to be high-risk. Extension of our modeling framework to human disease will also require systematic investigation of the dose-dependent outcome of *Sp*-airway interactions (Yershov et al., 2005). Another interesting future research direction includes the assessment of the long-term effects of immunological scarring on subsequent *Sp* challenges with different amplitudes and frequencies to identify the mechanisms behind the increased risk of developing serious infections after a first bacterial challenge (Habibzay et al., 2013). Finally, extending our model to incorporate the local spread of *Sp* from the upper airway epithelium to the lung and other sites on the respiratory epithelium that are normally sterile, for example by combining our model to the model of Smith et al. (2013), could allow us to investigate the association between a dysregulated colonization of the upper airway epithelium and the development of pneumococcal pneumionia.

The results of our mathematical model of commensal bacteria infection at the upper airway epithelium shed light on the mechanism behind a loss of homeostasis caused by dysregulation of the complex interactions between epithelial surfaces and microorganisms. A key element in this control structure is a "double-switch motif," which has been shown to govern other complex epithelial diseases, such as Atopic dermatitis (Domínguez-Hüttinger et al., in press) and cancer (Tian et al., 2013). Analysis of complex disease with a mechanistic, quantitative and systems-level framework as proposed here will help to reveal further general mechanisms underlying epithelium function in health and disease.

4. METHODS

4.1. Model Description

The proposed model for commensal bacterial infection describes the dynamics of bacterial load on the surface of the airway epithelium barrier, $S_a(t)$[CFU/ml], infiltrated bacterial load, $S_v(t)$[CFU/ml], concentrations of neutrophils and macrophages on the surface of the mucosal barrier, $N(t)$ and $M(t)$ [cells/ml], and the strength of barrier integrity, $B(t)$ relative to the maximum strength, by

$$\frac{dS_a(t)}{dt} = \frac{\kappa_S}{\mu_S}S_a(t)(1 - S_a(t)) - \frac{\theta_S}{1 + \epsilon_{SB}B(t)}S_a(t) - \phi_{NS}N(t)S_a(t) - \phi_{MS}M(t)S_a(t), \tag{1a}$$

$$\frac{dS_v(t)}{dt} = \kappa_S S_v(t) + \frac{\theta_S}{1 + \epsilon_{SB}B(t)}S_a(t) - \frac{\delta_S}{K + S_v(t)}S_v(t), \tag{1b}$$

$$\frac{dN(t)}{dt} = \alpha\frac{R(S_a(t))}{(1 + \epsilon_{NB}B(t))(1 + \epsilon_{NM}M(t))}N_v - \delta_N N(t), \tag{1c}$$

$$\frac{dM(t)}{dt} = \beta\frac{N(t)}{1 + \epsilon_{MB}B(t)}M_v - \delta_M M(t), \tag{1d}$$

$$\frac{dB(t)}{dt} = \frac{\kappa_B}{1 + \epsilon_{BS}R(S_a(t))}B(t)(\tilde{B} - B(t)) - \phi_{SB}R(S_a(t))B(t) - \phi_{NB}N(t)B(t). \tag{1e}$$

The variable $R(S_a(t))$ denotes the S_a-dependent TLR activation level described by a perfect switch,

$$R(S_a(t)) = \begin{cases} R_{\text{off}} & \text{if } S_a(t) < S^- \text{ or } \{S^- \leq S_a(t) < S^+ \text{ and } R(S_a(t^-)) = R_{\text{off}}\}, \\ R_{\text{on}} & \text{if } S_a(t) \geq S^+ \text{ or } \{S^- \leq S_a(t) < S^+ \text{ and } R(S_a(t^-)) = R_{\text{on}}\}, \end{cases} \tag{2}$$

where t^- is a time slightly before the time t. The dynamics of the TLR activity stabilizes within hours (Filewod et al., 2009; Witt et al., 2009; Hoffman et al., 2015).

The growth of the bacterial load on the apical side of the barrier, S_a, is modeled by a logistic equation (Smith et al., 2011), where the growth rate is limited by a carrying capacity (saturation term μ_S) that reflects the limited availability of nutrients in the epithelial lumen (Burnaugh et al., 2008). S_a is eradicated by immune cells, N and M, and transmigrates to the basal side of the epithelial barrier. The transmigrated bacteria, S_v, is assumed to grow exponentially in the blood vessel with abundant nutrients, but is killed by resident immune cells. The capacity to contain S_v is described by the saturated degradation of S_v, leading to the complete decay of S_v if it is below the threshold S_v^*, which corresponds to the unstable steady state of the ODE for S_v when $S_a = S^-$.

Recruitment of neutrophils (N) and macrophages (M) to the site of infection from their respective pool in the blood vessel (N_v) and the airway tissues (M_v) is inhibited by the epithelium barrier integrity (B), and is enhanced by the TLR activation (R) and the recruited neutrophils, respectively. Recruitment of N is further inhibited by M. N and M decay with a respective constant decay rate, as *de novo* production of the immune cells does not occur outside the bone marrow (Tak et al., 2013) and they do not divide in the epithelial tissue.

The self-recovery of the mucosal barrier to its homeostatic level (Nusrat et al., 1997; Coyne et al., 2002; Heijink et al., 2012) is modeled in a phenomenological manner, with the recovery rate being compromised by a decreased gene expression of epithelial cell differentiation markers (Clarke et al., 2011) induced by TLR activation. The barrier is directly damaged by transmigrating neutrophils and by proteases that are activated via TLR signaling (Chun and Prince, 2009). The switch-like activation of the TLR signaling is triggered by apically located bacteria, and is modeled by a phenomenological representation (Mochan et al., 2014; Domínguez-Hüttinger et al., in press). For simplicity, the inhibition by x is modeled phenomenologically by $\frac{1}{1+x}$.

4.2. Numerical Integration of the Hybrid Model

All the numerical model analysis was conducted using MATLAB version R2014a (The MathWorks, Inc., Natick, MA, USA). Numerical integration was conducted by `ode15s` from the initial conditions corresponding to a transient *Sp* challenges with $S_a(0) = 10^7$ CFU/ml, $S_v(0) = 0$ CFU/ml, $N(0) = 0$ cells/ml, $M(0) = 10$ cells/ml, and $B(0) = 1$, with $R(0) = 1$, as in the experiments in Zhang et al. (2009). The switch-dependent governing equation was chosen by the `event-location` function.

4.3. Modeling Antibiotics—Calculation of Minimal Strength and Minimal Duration of Antibiotics Treatment

We model the effects of bactericidal antibiotics, such as penicillin, ceftriaxone and amoxilin, which are commonly prescribed to treat pneumococcal infection (Mandell et al., 2007; Prina et al., 2015b) by $\frac{dS_v}{dt} = -VS_v(t)$ (systemic application of antibiotics) and $\frac{dS_a}{dt} = -AS_a(t)$ (apical application of antibiotics), where V and A represent a constant strength of antibiotics that kill $S_v(t)$ and $S_a(t)$, respectively. The strength of the antibiotics (with a unit of $1/h$) is described by $V = E_V D_V$ or $A = E_A D_A$, where E_V and E_A is the antibiotics killing efficacy and D_V and D_A is the amount applied, and can be chosen in the clinic by either selecting an antibiotic with a particular killing efficacy and/or adjusting the dose administrated. The killing efficacy of an antibiotic over a specific bacterial strain is commonly evaluated by the Minimal Inhibitory Concentration (MIC), the lowest concentration of antibiotics that will inhibit the visible growth of a bacterium after overnight incubation in a kinetic growth assay. From such experimental information, the antibiotics efficacy in our model, E_A (and E_V) can be calculated as $E_A = \frac{\kappa_S}{\mu_S}(1 - S_{MIC}(t)) \times \frac{1}{D_A^{MIC}}$, where $S_{MIC}(t)$ is the concentration of *Sp* exposed to an antibiotics dose of $D_A^{MIC} = \text{MIC}$, and hence does not increase further. This expression is obtained from the steady state equation $\frac{dS_{MIC}(t)}{dt} = \kappa_S \mu_S(1 - S_{MIC}(t)) - E_A D_A^{MIC} S_{MIC}(t) = 0$, which holds for the value of $S_{MIC}(1\text{day}) = S_a$ in these experiments.

Minimal strength of antibiotics treatment to achieve remission was determined by checking whether $S_v(t) < S_v^*$ or $S_a(t) < S^-$ is achieved while gradually increasing V or A, respectively, by an increment of 0.01. The minimal duration of treatments corresponds to the time required to achieve $S_a = S^-$. The minimal antibiotics treatment regimens to revert immune scarring without invasive infection (**Figure 7A**) were calculated under the assumptions that the treatment starts when S_a reached its steady state, after a transient S_a challenge that was modeled with initial conditions of $S_a(0) = 10^7$ CFU/ml, $S_v(0) = 0$ CFU/ml, $N(0) = 0$ cells/ml, $M(0) = 10$ cells/ml, and $B(0) = 1$, with $R(0) = 1$. For the regimens to prevent invasive infection (**Figure 7B**), as well as to reverse immunological scarring and to prevent sepsis (**Figure 7C**), we assumed that the treatment starts at the time of the S_a challenge.

4.4. Robustness Analysis and Parameter Sensitivity Analysis

We varied all the model parameters over one order of magnitude (0.1 - 10 times) around the nominal values, and the initial conditions $N(0)$, $M(0)$ and $B(0)$ within the intervals [0 1,000], [0 100] and [0 1], respectively. Robustness of the healthy behavior was tested by simultaneously varying all the parameters which were sampled from uniform distributions for 10,000 iterations. The global parameter sensitivity was evaluated by 10,000 iterations using the Global Sensitivity Analysis Toolbox for Matlab (Cannavó, 2012), with respect to the final concentrations of S_v and R at 7 days post *Sp* challenge with $S_a(0) = 10^7$ CFU/ml.

4.5. Focal Point Analysis

We conducted a focal point analysis to determine the long-term behavior of the model in absence of a *Sp* challenge. Following the methodology in Oyarzún et al. (2012), we considered two subsystems for Equation (1), defined by fixing R to either $R = R_{off}$ or to $R = R_{on}$, and evaluated the local stability of their steady states.

ETHICS STATEMENT

Blood used in pneumococcal growth assays was obtained from healthy volunteers who had given written consent. Ethical approval for this work was obtained from the Tissue Management Committee of the ICHTB (Project: R14053, ICHTB HTA license: 12275, REC Wales approval: 12/WA/0196). Tissue samples were provided by the Imperial College Healthcare NHS Trust Tissue Bank. Other investigators may have received samples from these same tissues. The research was supported by the National Institute for Health research (NIHR) Biomedical Research Centre based at Imperial College Healthcare NHS Trust and Imperial College London. The views expressed are those of the author(s) and not necessarily those of the NHS, NIHR, or the Department of Health.

AUTHOR CONTRIBUTIONS

ED, TC, NB, and RT designed the research, developed the mathematical model and analyzed data. ED performed the computational experiments. TC performed the *Sp* growth experiments. ED, TC, and RT wrote the paper.

FUNDING

ED acknowledges funding from the Mexican Council for Science and Technology (CONACyT, Ph.D. scholarship 212800) and from the National Autonomous University of Mexico (UNAM, postdoctoral scholarship). TC is a Sir Henry Dale Fellow jointly funded by the Wellcome Trust and Royal Society (grant No. 107660/Z/15/Z). RT acknowledges EPSRC Career Acceleration Fellowship (EP/G007446/1).

ACKNOWLEDGMENTS

We acknowledge George Buckle for his inputs on the parameter derivation of the model. Work with human blood was conducted in collaboration with Dr. Andrew Edwards (Imperial College London).

REFERENCES

Andonegui, G., Goring, K., Liu, D., McCafferty, D. M., and Winston, B. W. (2009). Characterization of *S. pneumoniae* pneumonia-induced multiple organ dysfunction syndrome: an experimental mouse model of gram-positive sepsis. *Shock* 31, 423–428. doi: 10.1097/SHK.0b013e318188c273

Attali, C., Durmort, C., Vernet, T., and Di Guilmi, A. M. (2008). The interaction of *Streptococcus pneumoniae* with plasmin mediates transmigration across endothelial and epithelial monolayers by intercellular junction cleavage. *Infect. Immun.* 76, 5350–5356. doi: 10.1128/IAI.00184-08

Austrian, R. (1986). Some aspects of the pneumococcal carrier state. *J. Antimicrob. Chemother.* 18(Suppl. A), 35–45. doi: 10.1093/jac/18.Supplement_A.35

Beisswenger, C., Coyne, C. B., Shchepetov, M., and Weiser, J. N. (2007). Role of p38 MAP kinase and transforming growth factor-beta signaling in transepithelial migration of invasive bacterial pathogens. *J. Biol. Chem.* 282, 28700–28708. doi: 10.1074/jbc.M703576200

Benton, K., Paton, J. C., and Briles, D. E. (1997). Differences in virulence for mice among Streptococcus pneumoniae strains of capsular types 2,3,4,5, and 6 are not attributable to Differences in pneumolysin production. *Infect. Immun.* 65, 1237–1244.

Burnaugh, A. M., Frantz, L. J., and King, S. J. (2008). Growth of *Streptococcus pneumoniae* on human glycoconjugates is dependent upon the sequential activity of bacterial exoglycosidases. *J. Bacteriol.* 190, 221–230. doi: 10.1128/JB.01251-07

Cannavó, F. (2012). Sensitivity analysis for volcanic source modeling quality assessment and model selection. *Comput. Geosci.* 44, 52–59. doi: 10.1016/j.cageo.2012.03.008

Cao, J., Zhang, X., Gong, Y., Zhang, Y., Cui, Y., Lai, X., et al. (2013). Protection against pneumococcal infection elicited by immunization with multiple pneumococcal heat shock proteins. *Vaccine* 31, 3564–3571. doi: 10.1016/j.vaccine.2013.05.061

Chin, A. C., Lee, W. Y., Nusrat, A., Vergnolle, N., and Parkos, C. A. (2008). Neutrophil-mediated activation of epithelial protease-activated receptors-1 and -2 regulates barrier function and transepithelial migration. *J. Immunol.* 181, 5702–5710. doi: 10.4049/jimmunol.181.8.5702

Chin, A. C., and Parkos, C. A. (2007). Pathobiology of neutrophil transepithelial migration: implications in mediating epithelial injury. *Annu. Rev. Pathol.* 2, 111–143. doi: 10.1146/annurev.pathol.2.010506.091944

Chun, J., and Prince, A. S. (2009). TLR2-induced calpain cleavage of epithelial junctional proteins facilitates leukocyte transmigration. *Cell Host Microbe* 5, 47–58. doi: 10.1016/j.chom.2008.11.009

Clarke, T. B., Francella, N., Huegel, A., and Weiser, J. N. (2011). Invasive bacterial pathogens exploit TLR-mediated downregulation of tight junction components to facilitate translocation across the epithelium. *Cell Host Microbe* 9, 404–414. doi: 10.1016/j.chom.2011.04.012

Coyne, C. B., Vanhook, M. K., Gambling, T. M., Johnny, L., Boucher, R. C., Johnson, L. G., et al. (2002). Regulation of airway tight junctions by proinflammatory cytokines. *Mol. Biol. Cell* 13, 3218–3234. doi: 10.1091/mbc.E02-03-0134

Dick, E. P., Prince, L. R., and Sabroe, I. (2008). *Ex vivo*-expanded bone marrow CD34+ derived neutrophils have limited bactericidal ability. *Stem Cells* 26, 2552–2563. doi: 10.1634/stemcells.2008-0328

Didierlaurent, A., Goulding, J., Patel, S., Snelgrove, R., Low, L., Bebien, M., et al. (2008). Sustained desensitization to bacterial Toll-like receptor ligands after resolution of respiratory influenza infection. *J. Exp. Med.* 205, 323–329. doi: 10.1084/jem.20070891

Domínguez-Hüttinger, E., Christodoulides, P., Miyauchi, K., Irvine, A. D., Okada-Hatakeyama, M., Kubo, M., et al. (in press). Mathematical modeling of atopic dermatitis reveals "double switch" mechanisms underlying four common disease phenotypes. *J. Allergy Clin. Immunol.* doi: 10.1016/j.jaci.2016.10.026

Elinav, E., Nowarski, R., Thaiss, C. A., Hu, B., Jin, C., and Flavell, R. A. (2013). Inflammation-induced cancer: crosstalk between tumours, immune cells and microorganisms. *Nat. Rev. Cancer* 13, 759–771. doi: 10.1038/nrc3611

Fey, D., Halasz, M., Dreidax, D., Kennedy, S. P., Hastings, J. F., Rauch, N., et al. (2015). Signaling pathway models as biomarkers: patient-specific simulations of JNK activity predict the survival of neuroblastoma patients. *Sci. Signal.* 8, 1–16. doi: 10.1126/scisignal.aab0990

Filewod, N. C., Pistolic, J., and Hancock, R. E. (2009). Low concentrations of LL-37 alter IL-8 production by keratinocytes and bronchial epithelial cells in response to proinflammatory stimuli. *FEMS Immunol. Med. Microbiol.* 56, 233–240. doi: 10.1111/j.1574-695X.2009.00571.x

Fonseca, D. M., Hand, T. W., Han, S. J., Gerner, M. Y., Zaretsky, A. G., Byrd, A. L., et al. (2015). Microbiota-dependent sequelae of acute infection compromise tissue-specific immunity. *Cell* 163, 354–366. doi: 10.1016/j.cell.2015.08.030

Garcha, D. S., Thurston, S. J., Patel, A. R., Mackay, A. J., Goldring, J. J., Donaldson, G. C., et al. (2012). Changes in prevalence and load of airway bacteria using quantitative PCR in stable and exacerbated COPD. *Thorax* 67, 1075–1080. doi: 10.1136/thoraxjnl-2012-201924

Habibzay, M., Weiss, G., and Hussell, T. (2013). Bacterial superinfection following lung inflammatory disorders. *Future Microbiol.* 8, 247–256. doi: 10.2217/fmb.12.143

Hathaway, L. J., Brugger, S. D., Morand, B., Bangert, M., Rotzetter, J. U., Hauser, C., et al. (2012). Capsule type of *Streptococcus pneumoniae* determines growth phenotype. *PLoS Pathog.* 8:e1002574. doi: 10.1371/journal.ppat.1002574

He, D., Su, Y., Usatyuk, P. V., Spannhake, E. W., Kogut, P., Solway, J., et al. (2009). Lysophosphatidic acid enhances pulmonary epithelial barrier integrity and protects endotoxin-induced epithelial barrier disruption and lung injury. *J. Biol. Chem.* 284, 24123–24132. doi: 10.1074/jbc.M109.007393

Heijink, I. H., Brandenburg, S. M., Postma, D. S., and van Oosterhout, A. J. (2012). Cigarette smoke impairs airway epithelial barrier function and cell-cell contact recovery. *Eur. Respir. J.* 39, 419–428. doi: 10.1183/09031936.00193810

Hoffman, D. R., Kroll, L. M., Basehoar, A., Reece, B., Cunningham, C. T., and Koenig, D. W. (2015). Immediate and extended effects of abrasion on stratum corneum natural moisturizing factor. *Skin Res. Technol.* 21, 366–372. doi: 10.1111/srt.12201

Knapp, S., Leemans, J. C., Florquin, S., Branger, J., Maris, N. A., Pater, J., et al. (2003). Alveolar macrophages have a protective antiinflammatory role during murine pneumococcal pneumonia. *Am. J. Respir. Crit. Care Med.* 167, 171–179. doi: 10.1164/rccm.200207-698OC

Knippenberg, S., Ueberberg, B., Maus, R., Bohling, J., Ding, N., Tort Tarres, M., et al. (2015). *Streptococcus pneumoniae* triggers progression of pulmonary fibrosis through pneumolysin. *Thorax* 70, 636–646. doi: 10.1136/thoraxjnl-2014-206420

Komori, M., Nakamura, Y., Ping, J., Feng, L., Toyama, K., Kim, Y., et al. (2011). Receptor 2 in the mouse middle ear epithelial cells. *Pediatr. Res.* 69, 101–105. doi: 10.1203/PDR.0b013e3182055237

Kwok, S. K., Cho, M. L., Her, Y. M., Oh, H. J., Park, M. K., Lee, S. Y., et al. (2012). TLR2 ligation induces the production of IL-23/IL-17 via IL-6, STAT3 and NF-kB pathway in patients with primary Sjogren's syndrome. *Arthritis Res. Ther.* 14:R64. doi: 10.1186/ar3780

Lagrou, K., Peetermans, W. E., Verhaegen, J., Jorissen, M., and Van Eldere, J. (2003). Disruption of nasopharyngeal epithelium by pneumococci is density-linked. *Eur. J. Clin. Invest.* 33, 340–345. doi: 10.1046/j.1365-2362.2003.01144.x

La Gruta, N. L., Kedzierska, K., Stambas, J., and Doherty, P. C. (2007). A question of self-preservation: immunopathology in influenza virus infection. *Immunol. Cell Biol.* 85, 85–92. doi: 10.1038/sj.icb.7100026

Leibovici, L. (2013). Long-term consequences of severe infections. *Clin. Microbiol. Infect.* 19, 510–512. doi: 10.1111/1469-0691.12160

Li, Y. (2004). Determination of the critical concentration of neutrophils required to block bacterial growth in tissues. *J. Exp. Med.* 200, 613–622. doi: 10.1084/jem.20040725

Li, Y., Karlin, A., Loike, J. D., and Silverstein, S. C. (2002). A critical concentration of neutrophils is required for effective bacterial killing in suspension. *Proc. Natl. Acad. Sci. U.S.A.* 99, 8289–8294. doi: 10.1073/pnas.122244799

Liang, Z., Zhang, Q., Thomas, C. M., Chana, K. K., Gibeon, D., Barnes, P. J., et al. (2014). Impaired macrophage phagocytosis of bacteria in severe asthma. *Respir. Res.* 15:72. doi: 10.1186/1465-9921-15-72

Lindén, A. (2001). Role of interleukin-17 and the neutrophil in asthma. *Int. Arch. Allergy Immunol.* 126, 179–184. doi: 10.1159/000049511

Mandell, L., Wunderink, R. G., Anzueto, A., Bartlett, J. G., Campbell, G. D., Dean, N. C., et al. (2007). Infectious diseases society of America/American thoracic society consensus guidelines on the management of community-acquired pneumonia in adults. *Clin. Infect. Dis.* 44(Suppl. 2), S27–S72. doi: 10.1086/511159

Marino, S., Hogue, I. B., Ray, C. J., and Kirschner, D. E. (2008). A methodology for performing global uncertainty and sensitivity analysis in systems biology. *J. Theor. Biol.* 254, 178–196. doi: 10.1016/j.jtbi.2008.04.011

McCullers, J. A. (2014). The co-pathogenesis of influenza viruses with bacteria in the lung. *Nat. Rev. Microbiol.* 12, 252–262. doi: 10.1038/nrmicro3231

McCullers, J. A., English, B. K., and Novak, R. (2000). Isolation and characterization of vancomycin-tolerant *Streptococcus pneumoniae* from the cerebrospinal fluid of a patient who developed recrudescent meningitis. *J. Infect. Dis.* 181, 369–373. doi: 10.1086/315216

Melkamu, T., Squillace, D., Kita, H., and O'Grady, S. M. (2009). Regulation of TLR2 expression and function in human airway epithelial cells. *J. Membr. Biol.* 229, 101–113. doi: 10.1007/s00232-009-9175-3

Mochan, E., Swigon, D., Ermentrout, B., Lukens, S., and Clermont, G. (2014). A mathematical model of intrahost pneumococcal pneumonia infection dynamics in murine strains. *J. Theor. Biol.* 353, 44–54. doi: 10.1016/j.jtbi.2014.02.021

Nash, S., Stafford, J., and Madara, J. L. (1987). Effects of polymorphonuclear leukocyte transmigration of cultured intestinal epithelial monolayers. *J. Clin. Invest.* 80, 1104–1113. doi: 10.1172/JCI113167

Nuorti, P., Butler, J. C., Crutcher, J., Guevara, R., Welch, D., Holder, P., et al. (1998). An outbreak of multidrug- resistantant pneumococcal pneumonia and bacteremia among unvaccinated nursing rome residents. *N. Engl. J. Med.* 338, 1861–1868. doi: 10.1056/NEJM199806253382601

Nusrat, A., Parkos, C. A., Liang, T. W., Carnes, D. K., and Madara, J. L. (1997). Neutrophil migration across model intestinal epithelia: monolayer disruption and subsequent events in epithelial repair. *Gastroenterology* 113, 1489–1500. doi: 10.1053/gast.1997.v113.pm9352851

Oggioni, M. R., Memmi, G., Maggi, T., Chiavolini, D., Iannelli, F., and Pozzi, G. (2004). Pneumococcal zinc metalloproteinase ZmpC cleaves human matrix metalloproteinase 9 and is a virulence factor in experimental pneumonia. *Mol. Microbiol.* 49, 795–805. doi: 10.1046/j.1365-2958.2003.03596.x

Oyarzún, D., Chaves, M., and Hoff-Hoffmeyer-Zlotnik, M. (2012). Multistability and oscillations in genetic control of metabolism. *J. Theor. Biol.* 295, 139–153. doi: 10.1016/j.jtbi.2011.11.017

Periselneris, J., José, R. J., and Brown, J. (2015). Targeting inflammatory responses to *Streptococcus pneumoniae*. *New Horizons Transl. Med.* 2, 167–174. doi: 10.1016/j.nhtm.2015.09.002

Pradere, J.-P., Dapito, D. H., and Schwabe, R. F. (2014). The Yin and Yang of Toll-like receptors in cancer. *Oncogene* 33, 3485–3495. doi: 10.1038/onc.2013.302

Prina, E., Ranzani, O. T., and Torres, A. (2015b). Community-acquired pneumonia. *Lancet* 386, 1097–1108. doi: 10.1016/S0140-6736(15)60733-4

Schmeck, B., Gross, R., Guessan, P. D. N., Hocke, A. C., Hammerschmidt, S., Mitchell, T. J., et al. (2004). Streptococcus pneumoniae- induced caspase 6-dependent apoptosis in lung epithelium *Streptococcus pneumoniae*- induced caspase 6-dependent apoptosis in lung epithelium. *Infect. Immun.* 72, 4940–4947. doi: 10.1128/IAI.72.9.4940-4947.2004

Schrag, S. J., Peña, C., Fernández, J., Sánchez, J., Gómez, V., Pérez, E., et al. (2001). Effect of short-course, high-dose amoxicillin therapy on resistant pneumococcal carriage: a randomized trial. *JAMA* 286, 49–56. doi: 10.1001/jama.286.1.49. Available online at: http://jamanetwork.com/journals/jama/fullarticle/193977

Shalek, A. K., Satija, R., Adiconis, X., Gertner, R. S., Gaublomme, J. T., Raychowdhury, R., et al. (2013). Single-cell transcriptomics reveals bimodality in expression and splicing in immune cells. *Nature* 498, 236–240. doi: 10.1038/nature12172

Shrestha, S., Foxman, B., Dawid, S., Aiello, A. E., Davis, B. M., Berus, J., et al. (2013). Time and dose-dependent risk of pneumococcal pneumonia following influenza: a model for within-host interaction between influenza and Streptococcus pneumoniae. *Interf. Focus* 10:20130233. doi: 10.1098/rsif.2013.0233

Siegel, S. J., Roche, A. M., and Weiser, J. N. (2014). Influenza promotes pneumococcal growth during coinfection by providing host sialylated substrates as a nutrient source. *Cell Host Microbe* 16, 55–67. doi: 10.1016/j.chom.2014.06.005

Siegel, S. J., and Weiser, J. N. (2015). Mechanisms of bacterial colonization of the respiratory tract. *Annu. Rev. Microbiol.* 69, 425–444. doi: 10.1146/annurev-micro-091014-104209

Silverstein, S. C., and Rabadan, R. (2012). How many neutrophils are enough (redux, redux)? *J. Clin. Invest.* 122, 2776–2779. doi: 10.1172/JCI63939

Smith, A. M., Adler, F. R., Ribeiro, R. M., Gutenkunst, R. N., McAuley, J. L., McCullers, J. A., et al. (2013). Kinetics of coinfection with influenza A virus and Streptococcus pneumoniae. *PLoS Pathog.* 9:e1003238. doi: 10.1371/journal.ppat.1003238

Smith, A. M., McCullers, J. A., and Adler, F. R. (2011). Mathematical model of a three-stage innate immune response to a pneumococcal lung infection. *J. Theor. Biol.* 276, 106–116. doi: 10.1016/j.jtbi.2011.01.052

Standish, A. J., and Weiser, J. N. (2009). Human neutrophils kill Streptococcus pneumoniae via serine proteases. *J. Immunol.* 183, 2602–2609. doi: 10.4049/jimmunol.0900688

Sung, M. H., Li, N., Lao, Q., Gottschalk, R. A., Hager, G. L., and Fraser, I. D. C. (2014). Switching of the relative dominance between feedback mechanisms in lipopolysaccharide-induced NF-κB signaling. *Sci. Signal.* 7:ra6. doi: 10.1126/scisignal.2004764

Tak, T., Tesselaar, K., Pillay, J., Borghans, J. A., and Koenderman, L. (2013). What's your age again? Determination of human neutrophil half-lives revisited. *J. Leukoc. Biol.* 94, 595–601. doi: 10.1189/jlb.1112571

Tanaka, R. J., Boon, N. J., Vrcelj, K., Nguyen, A., Vinci, C., Armstrong-James, D., et al. (2015). *In silico* modeling of spore inhalation reveals fungal persistence following low dose exposure. *Sci. Rep.* 5:13958. doi: 10.1038/srep13958

Tian, X. J., Zhang, H., and Xing, J. (2013). Coupled reversible and irreversible bistable switches underlying TGFβ-induced epithelial to mesenchymal transition. *Biophys. J.* 105, 1079–1089. doi: 10.1016/j.bpj.2013.07.011

Tieu, D. D., Kern, R. C., and Schleimer, R. P. (2009). Alterations in epithelial barrier function and host defense responses in chronic rhinosinusitis. *J. Allergy Clin. Immunol.* 124, 37–42. doi: 10.1016/j.jaci.2009.04.045

Witt, J., Barisic, S., Schumann, E., Allgöwer, F., Sawodny, O., Sauter, T., et al. (2009). Mechanism of PP2A-mediated IKK beta dephosphorylation: a systems biological approach. *BMC Syst. Biol.* 3:71. doi: 10.1186/1752-0509-3-71

World Health Organization (2012). *Measuring Impact of Streptococcus pneumoniae and Haemophilus Influenzae Type b Conjugate Vaccination*. Tech. rep.

World Health Organization (2014). *Antimicrobial Resistance: Fact Sheet Number 194*. Available online at: http://www.who.int/mediacentre/factsheets/fs194/en/

Yershov, A. L., Jordan, B. S., Guymon, C. H., and Dubick, M. A. (2005). Relationship between the inoculum dose of *Streptococcus pneumoniae* and pneumonia onset in a rabbit model. *Eur. Respir. J.* 25, 693–700. doi: 10.1183/09031936.05.00091904

Yin, L., Xu, S., Cheng, J., Zheng, D., Limmon, G. V., Leung, N. H. N., et al. (2013). Spatiotemporal quantification of cell dynamics in the lung following influenza virus infection. *J. Biomed. Opt.* 18:046001. doi: 10.1117/1.JBO.18.4.046001

Zemans, R. L., Colgan, S. P., and Downey, G. P. (2009). Transepithelial migration of neutrophils: mechanisms and implications for acute lung injury. *Am. J. Respir. Cell Mol. Biol.* 40, 519–535. doi: 10.1165/rcmb.2008-0348TR

Zhang, Z., Clarke, T. B., and Weiser, J. N. (2009). Cellular effectors mediating Th17-dependent clearance of pneumococcal colonization in mice. *J. Clin. Invest.* 119, 1899–1909. doi: 10.1172/jci36731

Zheng, W., and Flavell, R. (1997). The transcription factor GATA-3 is necessary and sufficient for Th2 cytokine gene expression in CD4 T cells. *Cell* 89, 587–596. doi: 10.1016/S0092-8674(00)80240-8

9

Toward Precision Healthcare: Context and Mathematical Challenges

Caroline Colijn[1,2*], *Nick Jones*[1,2], *Iain G. Johnston*[2,3], *Sophia Yaliraki*[2,4] and *Mauricio Barahona*[1,2*]

[1] *Department of Mathematics, Imperial College London, London, UK,* [2] *EPSRC Centre for Mathematics of Precision Healthcare, Imperial College London, London, UK,* [3] *School of Biosciences, University of Birmingham, Birmingham, UK,* [4] *Department of Chemistry, Imperial College London, London, UK*

Precision medicine refers to the idea of delivering the right treatment to the right patient at the right time, usually with a focus on a data-centered approach to this task. In this perspective piece, we use the term "precision healthcare" to describe the development of precision approaches that bridge from the individual to the population, taking advantage of individual-level data, but also taking the social context into account. These problems give rise to a broad spectrum of technical, scientific, policy, ethical and social challenges, and new mathematical techniques will be required to meet them. To ensure that the science underpinning "precision" is robust, interpretable and well-suited to meet the policy, ethical and social questions that such approaches raise, the mathematical methods for data analysis should be transparent, robust, and able to adapt to errors and uncertainties. In particular, precision methodologies should capture the complexity of data, yet produce tractable descriptions at the relevant resolution while preserving intelligibility and traceability, so that they can be used by practitioners to aid decision-making. Through several case studies in this domain of precision healthcare, we argue that this vision requires the development of new mathematical frameworks, both in modeling and in data analysis and interpretation.

Edited by:
Krasimira Tsaneva-Atanasova, University of Exeter, UK

Reviewed by:
Biswapriya Biswavas Misra, Texas Biomedical Research Institute, USA
Marcus Kaiser, Newcastle University, UK

***Correspondence:**
Caroline Colijn
c.colijn@imperial.ac.uk
Mauricio Barahona
m.barahona@imperial.ac.uk

Keywords: precision medicine, precision healthcare, data science, precision public health, mathematical modeling

INTRODUCTION: PRECISION MEDICINE AND ITS CHALLENGES

The phrase "precision medicine" describes the idea of delivering the right treatment to the right person at the right time. Precision approaches aim to achieve a medical revolution: individualized therapies based on quantitative, patient-specific datasets, integrated via algorithmic analyses that can aid patient stratification, monitoring, and treatment design. These approaches have been broadly supported in the US under President Obama's Precision Medicine Initiative (Collins and Varmus, 2015; White House Precision Medicine Initiative, 2016), by the Gates Foundation (Cisneros, 2016), and by the Chan Zuckerberg Initiative (Chan Zuckerberg Initiative — Advancing human potential and promoting equal opportunity, 2017). Although some of the aspects of this vision date back to the inception of the Human Genome Project, precision medicine now expands beyond the restrictions of genomics to encompass a wide range of data sources increasingly available to clinicians. The idea of embedding diagnostics and treatment with omics and other medical and physiological datasets at the heart of medicine has been variously described as systems medicine, personalized medicine, computational systems biomedicine, P4 (Predictive,

Preventative, Personalized, Participatory) medicine, and precision medicine, to name just a few (Duffy, 2016).

The development of the field has been underpinned by some striking successes, particularly in cancer (Derks et al., 2014; Hiley et al., 2014; Arnedos et al., 2015; Friedman et al., 2015; Navin, 2015; Rubin, 2015; Stover and Wagle, 2015; Wills and Mead, 2015; Cloney, 2017), where molecular profiling is increasingly routine in lung, breast, and colorectal cancers, as well as in leukemias and melanomas (Larry Jameson and Longo, 2015). In asthma, the heterogeneity in clinical response has been shown to overlap with differences in a number of predictive biomarkers, allowing patient stratification for tailored therapies (Muraro et al., 2016). Precision approaches can have immediate benefits for drug repurposing and treatment: the link between type 2 diabetes and early stage Alzheimer's, in which there is often impaired glucose metabolism in the brain, is giving rise to a body of research for new therapeutics that includes repurposing existing drugs (Yarchoan and Arnold, 2014). Cardiovascular disease is another natural domain for precision medicine, as chronic, pervasive problems like diabetes, obesity, and hypertension (with a significant socio-economic and life style component) are directly linked to severe disease including heart failure (Antman and Loscalzo, 2016). These highly prevalent conditions are themselves diverse, multifactorial, and co-occurrent in many individuals, yet mechanism-based markers that predict the development of hypertension can already be identified based on functional genetic and epi-genetic markers (El Shamieh and Visvikis-Siest, 2012; Zhang et al., 2015). In the domain of infectious diseases, precision technologies can also be used to identify pathogens and to determine susceptibility to antimicrobial agents, guiding prescription, e.g., CD4+ cell counts and viral loads can guide HIV therapies (Barnett et al., 2008). Beyond single infections, the function of the microbiome is being probed for disease associations (Gilbert et al., 2016) and metabolomics and integrated omics' tools are revealing disease phenotypes (Chen et al., 2012; Dorrestein et al., 2014).

While there are plenty of potential "low hanging fruits" yet to be plucked, for precision medicine to maximize its impact as envisioned, a number of significant challenges need to be met across multiple domains. Some of these challenges are technical and relate to data collection, processing, storing, and sharing (Garber and Tunis, 2009; Servant et al., 2014; Palmisano et al., 2016; Sboner and Elemento, 2016), and have broad scientific, clinical, social, and ethical ramifications (Juengst et al., 2012; Khoury et al., 2012; Castaneda et al., 2015; Schork, 2015; Cohn et al., 2016). Indeed, advances in sequencing, metabolomics, biomarker discovery, genetics and single-cell technologies, alongside computing, and data science, have brought a strong impetus to the development of the scientific toolkit, data management systems, and regulatory framework for precision medicine. Data collection is currently taking place across the traditional channels of hospitals, community health care settings, and public health bodies, but also increasingly in a decentralized manner via social media analytics and wearable devices. The adoption of systematic formats for Electronic Health Records has improved data collection and consistency, but a considerable effort in data processing and integration still needs to take place (Garber and Tunis, 2009; Servant et al., 2014; Palmisano et al., 2016; Sboner and Elemento, 2016). Storing and accessing extremely high volumes of data is difficult and a concerted effort must be developed to enable clinicians, policy-makers, and academics to access these datasets, thus reducing the need for custom bioinformatics expertise. The question of whether data management is done by public or private organizations, and whether researchers and other users will need to pay to use data is an additional area of concern. Further, efforts to harness large datasets will require the development of sophisticated graphical user interfaces and visualization, data quality management, and data storage (Duffy, 2016). Programmes like the UK Biobank (2016), an open resource collecting de-identified data on health and well-being from 500,000 volunteers, and making it available for research, will be instrumental in meeting these challenges. There is also the issue of obtaining informed consent about the storage and use of data, when the uses are dynamic and expanding (Khoury et al., 2016). Finally, as yet there are no centralized resources collecting datasets, modeling and software analysis tools, and pipelines for precision medicine, which would facilitate method-sharing and allow interested researchers to join the effort.

Precision approaches, as they develop, must also accommodate the ethical and transparent use of data. Recently, O'Neil has coined the phrase "Weapons of Math Destruction" (WMD) (O'Neil, 2016) to describe how black-box algorithms can create pernicious and damaging feedback loops, with unfair consequences to individuals, often without much effort placed on identifying and correcting errors (O'Neil, 2016). Hood and Friend (2011) present the vision that "in the not-too-distant future, each patient will be surrounded by a 'virtual cloud' of billions of data points that will uniquely define their past medical history and current health status. Furthermore, it will be possible to mine the billions of data points from hundreds of millions of individuals to generate algorithms to help predict the future clinical needs for each patient." Hence, although precision medicine under this broad vision could have the beneficial potential to identify diseases earlier, to reduce burdens of treatment, and to improve screening by reducing false positives, and ultimately improve health, a sceptic might imagine a scenario in which these same predictions are used to produce quality-adjusted life year estimates, affecting which treatments are covered for whom, and guiding hiring, lending or health insurance decisions (O'Neil, 2016). With the amount and breadth of data available, there is the danger that such decisions could yield negative discrimination according to e.g., postcode lotteries, socio-economic factors, social network data, past healthcare interactions, judicial and law enforcement history.

How can such scenarios be avoided? Part of the answer must come from policy and regulation to ensure openness and fair use of data (Noveck, 2015). Yet, in addition, we need to develop the kind of mathematics and statistics for data science that will keep the "human in the loop" so that decision-making can be transparent and based on interpretable features and evidence. In doing so, we will need to develop methods that can track back and be updated in response to errors, taking full account of uncertainties, thus avoiding the over-reliance on complex computational decision black boxes. With this aim of model intelligibility, an important goal is to extract improved data-driven descriptions at the appropriate intermediate scales between the fully individualized level, which carries the risk of non-transparent and damaging over-use of data, and descriptions that are too coarse, which lead to insufficient precision in the face of individual variation. There is virtue in reaching a nuanced, data-informed middle-ground between these extremes: one that considers the individual in a population context and includes the role of human judgment.

It is thus essential to build theoretical understanding at the appropriate scale. One of the advantages of an integrated precision approach in medicine is to refine disease classification, increasing and finessing the number of groups of patients to reflect the true diversity of major diseases like cancers, so as to target treatment appropriately. While biomarker information can substantially improve clinical trial design as well as treatment (Trusheim et al., 2011), stratification also greatly increases the number of categories for which clinical trials may need to be carried out, reducing reproducibility and posing challenges to evidential policy (Khoury et al., 2012). Conversely, data-centric approaches may identify common mechanisms and treatments across disparate diseases, reducing stratification. Such approaches could potentially deliver dramatic cost efficiency. For these reasons, precision approaches must act at the right scale, which will often be intermediate–between "one size fits all" medicine and fully individualized therapies.

However, we do not yet have the ideal tools at hand to identify relevant features and integrate them to obtain interpretable predictions, optimized therapies, and new policies–even if merged datasets describing multivariate aspects of individuals' health across time (including, e.g., genomic, proteomic, metabolomic, brain images, social, and behavioral data) collected with informed consent were available. In addition, such combined genetic, genomic, proteomic, metabolomic, or single-cell data will only provide highly enriched and noisy snapshots taken at a few times—at best, we will have sparse noisy samples of the underlying process of disease, and sparse samples of the context of each patient.

To reveal the potential of such datasets in medicine, we must thus develop mathematical frameworks that are able to describe high-dimensional, dynamic, noisy, sparsely-sampled processes. Ideally, we must then be able to extract concise descriptions (coarse-grained at the right resolution) which are intelligible and actionable, and which link co-occurrences of events, co-morbidities, and time patterns in disease and in health-related processes. This area poses a set of core mathematical challenges: creating transparent, replicable descriptions in healthcare, which make use of large diverse datasets, placing individuals in context, and which use dynamical information across time at the correct scales. These mathematical challenges must be researched in parallel with precision medicine, ideally spanning the individual- and population-level perspectives.

In our view, these constitute deep additional challenges to mathematical modeling and data analysis that will need to be met in order for precision approaches to meet their promise. In the remainder of this perspective, we lay out a vision for what we term *precision healthcare*, its aims and its mathematical challenges. We do not aim to write a review of precision medicine; many reviews of tools and methods in different medical domains are available (see for example Chen and Snyder, 2013; Rosell and Karachaliou, 2013; Hiley et al., 2014; Ignatiadis and Dawson, 2014; Arnedos et al., 2015 among many others), as well as perspectives from a variety of viewpoints (Mirnezami et al., 2012; Roychowdhury and Chinnaiyan, 2013; Ciardiello et al., 2014; Ignatiadis and Dawson, 2014; Servant et al., 2014; Arnett and Claas, 2016; Rost et al., 2016; Vargas and Harris, 2016).

Why Precision Healthcare?

For most of these challenges, population-level thinking coupled with mathematical data science analytics can help translate the benefits of precision medicine to address broader effects at the group level, including concerns regarding health equity and ethics. We use the phrase *"precision healthcare"* to encompass this vision that integrates the population and individual perspectives. Precision healthcare thus aims to build tools that make use of the increasing array of data sources, allowing for their continuous refinement in the face of new data, and whose predictions are aimed at and respond to the requirements of healthcare practitioners (clinicians, the public, policy thinkers, and other stakeholders).

This vision will require the use of an array of mathematical tools to unify individual-level precision medicine with public health, placing high-dimensional individual data and refined interventions in their social network context. Indeed, in many instances, individual health cannot be separated from its behavioral and social context. For example, highly targeted interventions against a cancer can be undermined by metabolic diseases caused by dietary behaviors which, in turn, co-vary with social network structure and other societal constructs. An adjuvant therapy for cancer might thus be to influence the diet and behavior of the patient taking into account their close social contacts.

The scenario by Hood and Friend (2011) mentioned above can thus be thought of as the analysis of a virtual cloud of a large number of high-dimensional feature vectors corresponding to the different individuals. Dynamical datasets in this scenario would correspond to a large collection of paths in such a space. If the technical and policy challenges to collect and integrate such data into a single accessible point of access were surmounted, methods for dimensionality reduction could be applied to reduce the relevant features to a few "components" which could then be used to "cluster" (or classify) the data into groups of similar individuals according to their paths. This is an area of current active research, ranging from the direct application of classic methods such as principal components analysis (PCA), support vector machines (SVMs), and independent component analysis (ICA) with all their myriad of variants, through manifold learning to the revivified use of neural networks for such classification tasks (Mallat, 2016). Developing ways to cope with noisy data and noisy labels is an ongoing challenge in machine learning (Xiao et al., 2015) and across precision medicine, as omics datasets can be extremely noisy.

However, specific requirements in the precision healthcare setting make such tasks especially difficult. The datasets are dynamic and usually sparsely sampled. The processes involved are high-dimensional, highly nonlinear, noisy, and uncertain. The dimensionality reduction framework for such datasets should ideally achieve competing objectives: preserve, to some extent, the meaning of the original descriptive variables (without mixing all features into conglomerates) while extracting concise (i.e., sparse) representations in terms of few relevant extracted features. Ideally, it should be possible to adjust the level of detail (i.e., the resolution scale) of such models depending on the quality of the data and the needs of the practitioner. Finally, the mathematical framework should deliver robust outcomes, and include the possibility of restricting and conditioning the extracted models to incorporate additional and complementary data without the need for refitting.

Indeed, in the process of harnessing these large-scale data, a great degree of caution is required. Most biomedical research is plagued by a flood of false positive results due to experiments of insufficient discriminatory power (Ioannidis, 2005). The translational impact of this trend is starkly illustrated by recent failures to reproduce landmark cancer studies and low success rates in clinical trials (Prinz et al., 2011; Begley and Ellis, 2012). In particular, the quest for (publishable) p-values over (meaningful) effect sizes (Goodman, 1999; Ziliak and McCloskey, 2008) has led to the likely incorrect linking of many genetic features with diseases (Johnston, 2016). Selecting appropriate mathematical models can help increase the statistical power of large-scale experimental data, allowing rigorous statistical treatments to discriminate likely from spurious effects, and quantifying the sizes of effects so that the scientific, as well as the nominally statistical, significance of observations can be better understood.

The interface of individual-level personalized medicine and public health will thus need to develop new mathematical tools to formulate and analyse mathematical questions for data-rich characterization of disease progression and transmission, controlled intervention, and healthcare provision. Key areas that we see in the remit of precision healthcare include: statistics for noisy, incomplete, heterogeneous data; stochastic modeling; inference and control of network dynamics; mathematical approaches to exploit complex structure in large datasets, and methods to couple imaging and omics. More broadly, a central distinction between precision medicine and precision healthcare is that the former treats individuals, whereas the latter treats individuals explicitly embedded in a society or broader context. Precision healthcare thus aims to link "big data" tools to explore individual agents with an understanding of how those individuals behave collectively and respond to society-wide initiatives.

Some Proposed Case Studies in Precision Healthcare

We now describe a number of demonstrative examples, illustrating some of the tools that come under the umbrella of precision healthcare. These range from systems precision medicine approaches focusing on the representation of complex dynamic data, to precision healthcare approaches including both retrospective analysis and real-time interventions that are rooted in complex individual and population data.

Gene Therapies for Mitochondrial Diseases

A combination of new maths, statistics, and large-scale experimental data has led to recognition of the importance of personalized therapeutic approaches in cutting-edge gene therapies addressing the inheritance of mitochondrial diseases. These diseases (e.g., mitochondrial encephalomyopathy, lactic acidosis, and stroke-like episodes—MELAS, myoclonic epilepsy with ragged red fibers—MERRF, Leber's hereditary optic neuropathy—LHON) result from mutations in mitochondrial DNA (mtDNA) which are passed from mother to child (DiMauro and Davidzon, 2005). Mitochondrial replacement therapies aim to prevent this inheritance by replacing mutated mother mtDNA with mtDNA from a third party woman, but technological limitations in the procedure can lead to small amounts of mother mtDNA being amplified leading to disease (Burgstaller et al., 2015). Classically the risk of differential proliferation has been considered minimal, but evidence harnessed with statistical modeling and large-scale data from mouse models has shown it is common (Burgstaller et al., 2014). Further, quantitative modeling on large-scale human mtDNA datasets has confirmed that this risk is present in heterogeneous human populations (Røyrvik et al., 2016), as supported by experimental observations (Hyslop et al., 2016; Yamada et al., 2016). The personalized aspect stems from the fact that the risk of differential proliferation depends on the genetic details of the mother's and third-party's mtDNA, which vary throughout global human populations according to geography and ancestry. Appropriate modeling can elucidate the biological details of why these proliferative differences arise, make probabilistic statements about the probability and timescales of therapeutic outcomes, and describe the mtDNA differences likely to arise in human populations. In the future, precision healthcare strategies could allow us to propose suitable third-party donors to optimize successful fertility strategies.

Pathways of Disease Progression in High-Dimensional Spaces

Recent mathematical and statistical developments in the study of evolution have shed light on the emergence of efficient photosynthesis (Williams et al., 2013) and the reduction of organelle genomes (Johnston and Williams, 2016) by modeling evolution as the acquisition (or loss) of a set of L-discrete traits. Evolution in this picture takes place on an L-dimensional hypercube, with each vertex corresponding to a given pattern of trait presence/absence and each edge corresponding to an evolutionary innovation. Observations of evolutionary intermediates can then be used, as in a hidden Markov model, to infer likely trajectories through this space. This paradigm can be developed to infer likely pathways of disease progression (generalizing statistical studies on disease progression; Hjelm et al., 2006; Pagel and Meade, 2006; Loohuis et al., 2014; Beerenwinkel et al., 2015), picturing the "space of symptoms" as an analogous hypercube, and disease progression as paths over its edges. Large-scale and longitudinal patient datasets can be used to infer likely sets of "evolutionary" trajectories through this space, so that probabilistic statements can be made about the likely next step for any given individual patient–and thus a personalized optimal therapeutic strategy. Interestingly, this approach can be linked with descriptions based on *continuous* variables, where similarity graphs are obtained from distance matrices by using graph-theoretical sparsifications that preserve the topological and geometrical structure of the data (Beguerisse-Diaz et al., 2013). The structure of the similarity graphs from the data can then be analyzed using multiscale community detection algorithms leading to highly nonlinear clustering of symptoms and individuals describing the observed pathways of disease progression (Schaub et al., 2012).

Social Networks in Health Policy

Twitter provides a platform to interact directly with a large audience, and to sample and address public opinion and responses around specific issues and questions. However, it is critical to understand the different communities and conversations on Twitter, so as to target them appropriately. For example, a recent example following conversations on diabetes (Beguerisse-Díaz et al., 2017) used a unified mathematical framework (Delvenne et al., 2010; Beguerisse-Diaz et al., 2013; Beguerisse-Díaz et al., 2014; Lambiotte et al., 2015) that brings notions from stochastic processes on graphs and optimization to the analysis of Twitter networks. In this particular study, 2.5 million diabetes-related tweets were analyzed and found to fall within five broad thematic groups: health information, news, social interaction, commercial, and humor. Indeed, humorous messages and references to popular culture appear consistently, more than any other type of tweet, revealing the specific characteristics of social media interactions. The analysis of the temporal "hub" and "authority" scores of Twitter users revealed that the hub landscape is diffuse whereas the landscape of authorities is highly persistent. The Twitter authorities comprise not only bloggers, advocacy groups and NGOs related to diabetes, but also for-profit entities without specific diabetes expertise which influence the online exchanges. The top authorities fall into seven interest groups, as derived from their Twitter follower network revealing the flow of information with specific audiences. A similar analysis was carried out on the network of retweets generated by the debate surrounding the proposed adoption of the "care.data" (https://www.england.nhs.uk/ourwork/tsd/care-data/) scheme of personalized health care records by NHS England (Amor et al., 2016). In that case, a series of interest groups and conversations were identified revealing the different roles of users within and across communities, including the limited reach of some of the public policy accounts in the debate. Such findings could be used by public health professionals and policy makers to use social media as an engagement tool and to inform policy design. A similar analysis have been carried out in Beguerisse-Díaz et al. (2017) following other social movements.

Spreading of Vaccine Sentiment and Spreading of Vaccine Preventable Disease

Vaccine hesitancy and a vaccine preventable disease can be thought of as two distinct types of processes and they propagate through distinct media. We suppose that sentiment is spread socially (and is influenced by media outlets) but involves the slow evolution of beliefs rather than something as simple as the infectious propagation of a meme. In contrast, disease spread need not respect social network structure. An integrated intervention would not only target vaccination where the disease has been reported and vaccine coverage rates are low (de Figueiredo et al., 2015) but also where it is predicted to appear. Predictions would be based on integrated multi-variate "precision" data. Similarly, negative vaccine sentiment (Larson et al., 2016) could be targeted not only where it has been reported but also where it is predicted to appear given the social network structure. The coupling between belief dynamics and epidemiology now has an established theoretical presence (Wang et al., 2016) and importantly it has been observed that anti-vaccination behavior is socially clustered (Onnela et al., 2016) thereby undermining herd immunity (Salathé and Bonhoeffer, 2008); it is important to coordinate a public health response that can incorporate belief and behavior dynamics as well as the spread of infection.

Influencing Beliefs and Influencing Networks

Health outcomes for chronic conditions are modulated by health behavior, which in turn might be expected to show covariation sensitive to underlying social network structure (Centola, 2010; Shalizi and Thomas, 2011; Christakis and Fowler, 2013). It has further been suggested, independent of unhealthy behavior, that social position can modulate health outcomes (Snyder-Mackler et al., 2016). There are thus a number of possible types of social interventions to improve health: (1) influencing modes of thinking to encourage critical appraisal of apparently acceptable but unhealthy behavior (changing the models that individuals use) (2) influencing health beliefs about particular topics (changing the data individuals access) (3) influencing network structure (but not social co-ordinates) to build bridges between communities for the exchange of health behavior (4) influencing the social co-ordinates of individuals (or sectors of society) and thereby altering their network neighborhood (or the gross social network structure). Changes to (3) and (4) might also affect possible physiological consequences of status comparisons (Pickett and Wilkinson, 2015). While we can cite examples of each class of intervention, these can be remarkably challenging to effect: for example showing some extreme vaccine sceptics information about the consequences of vaccine preventable disease can *increase* their vaccine scepticism (Nyhan et al., 2014); overwhelming evidence has been presented of health inequities (Marmot and Commission on Social Determinants of Health, 2007; Adler et al., 2016) but the problem persists. Challenges (1–4) constitute challenges in contemporary network science and its interface with optimal control: ideal interventions will optimally control processes on networks and optimally influence the network structure itself (Liu and Barabási, 2016).

Genomic Epidemiology for Outbreak Reconstruction

Recent advances in sequencing technologies have driven changes in many biological domains, including epidemiology (Jombart et al., 2014; Kao et al., 2014; Colijn and Cohen, 2016). It is now feasible to obtain DNA or RNA sequences from viruses, bacteria and other pathogens, and to use these data to detect drug resistance, optimize treatments for individual patients (Vanderkooi et al., 2005; Perez et al., 2016), and to understand how pathogens are spreading and evolving by tracking small variations in the pathogen as it moves between individuals. To understand transmission, isolates are collected from patients alongside clinical data such as times of symptom onset. The isolates are sequenced and processed with bioinformatics tools, capturing even small levels of variation between patients (e.g., in a multiple sequence alignment). These can be integrated with evolutionary models to infer phylogenetic trees, describing patterns of shared ancestry among the isolates. An epidemiological model is used to define how likely a set of infection events are. This incorporates clinical information–for example, it is very unlikely that an individual would transmit an infection years before showing any symptoms, or while living in another area. Finally, mathematical models that link the phylogenetic and epidemiological information are used to compute the joint likelihood of the genetic data and the set of transmission events. This is embedded in a Bayesian approach, so the result is a posterior collection of transmission trees (who infected whom, and when), consistent with the data. There is a rapidly-growing body of work on these inference problems (Hall et al., 2015; Worby et al., 2015; De Maio et al., 2016; Klinkenberg et al., 2016; Worby et al., 2016; Didelot et al., 2017); **Figure 1** is based on the approach in Didelot et al. (2017). There are natural precision healthcare applications of these tools: if more transmission is inferred to have occurred in particular locations, interventions such as improved ventilation and cleaning, early screening and active case fining can be directed there. If risk factors such as community membership, age, or co-morbidities are identified, these can be managed similarly. But perhaps the most exciting applications of these tools will happen when sequencing can be done in a matter of hours or even days. Identifying where there are likely missing cases could allow us to identify cases early, treat them, and prevent onward transmission. Real-time sequencing and infection-tracing has already had impact in the recent Ebola epidemic (Quick et al., 2016), setting the stage for this direction in public health (Gardy et al., 2015).

DISCUSSION

We have framed *precision healthcare* to describe the development of precision approaches which, while capturing the complexity of individual data and its societal context, extract reduced dimensionality descriptions at the relevant resolution while preserving a measure of intelligibility of the models. This can enable practitioners in the loop to use these precision approaches effectively. Such methods should be transparent, robust and able to adapt to errors and uncertainties. In bridging from the individual to the population, the methodologies should take advantage of the multivariate data sources at the heart of precision medicine, yet take the social context and population levels into account. Through several case studies in this domain of precision healthcare, we argue that this vision requires the development of new mathematical frameworks, both in modeling and in data analysis and interpretation.

Recently, "precision public health" has been characterized as delivering the right intervention to the right population at the right time (Desmond-Hellmann, 2016; Khoury et al., 2016), mirroring the oft-cited characterization of precision medicine. With support from the Gates Foundation (Cisneros, 2016), precision public health aims to apply precision (data-centered) approaches to improve the health of populations and to reduce health disparities. Public health thinkers are concerned about precision medicine's current emphasis on individual approaches, its focus on extending the use of costly genetics and other omics', and the development of tailored drug treatments (Khoury et al., 2016). Bayer and Galea report that the number of NIH projects with "public" or "population" in the title has dramatically declined, and that in 2014, research areas described with the words "genetic," "genome," or "gene" received 50% more funding than those with "prevention". They are concerned that the focus on precision medicine is misguided (Bayer and Galea, 2015), and argue that improving health requires addressing persistent social realities that are not covered by access to clinical medicine (Adler et al., 2016). Persistent social inequalities can also be expected to be a major barrier in bringing advances from omics-based precision medicine to low-income countries, although recent use of rapid genomics-based tools in the Ebola outbreak (Quick et al., 2016) points to the potential to develop precision-based approaches for low-income settings.

Precision public health places emphasis on addressing such disparities, and (as with public health more generally) on prevention. In many ways, data-centered approaches have already been adopted by epidemiologists and public health practitioners and, as precision public health incorporates more individual-level data, it will require the envisaged scientific tools of *precision healthcare*. These methodologies will allow public health methods to integrate data on vaccine belief and social context with individual health records, genetic data, other biomarkers, and individual risk factors. Importantly, it is realistic to envision that the use of mobile and social network technologies will enable public health interventions typically considered at the level of populations to instead be tailored to individuals. We believe that an important aspect of the success of precision public health will depend on meeting the mathematical challenges we have outlined as precision healthcare. Identifying the right population for the right intervention will require data analysis, stratification, and modeling at the right scale: too fine, and there would be impractically many populations; too coarse, and the precision advantage is lost. It will require intelligible, transparent methods that can be communicated to public health practitioners, easily updated in the face of new data and human judgment. It will require using the right data to answer the right question, and avoiding mis-use of data to treat some populations unfairly.

It is no longer the case that the timescales of individual disease progression and the timescale of changes in health policy or social behavior are distinct. Chronic conditions from cancer to diabetes are managed over years and decades. Years and decades

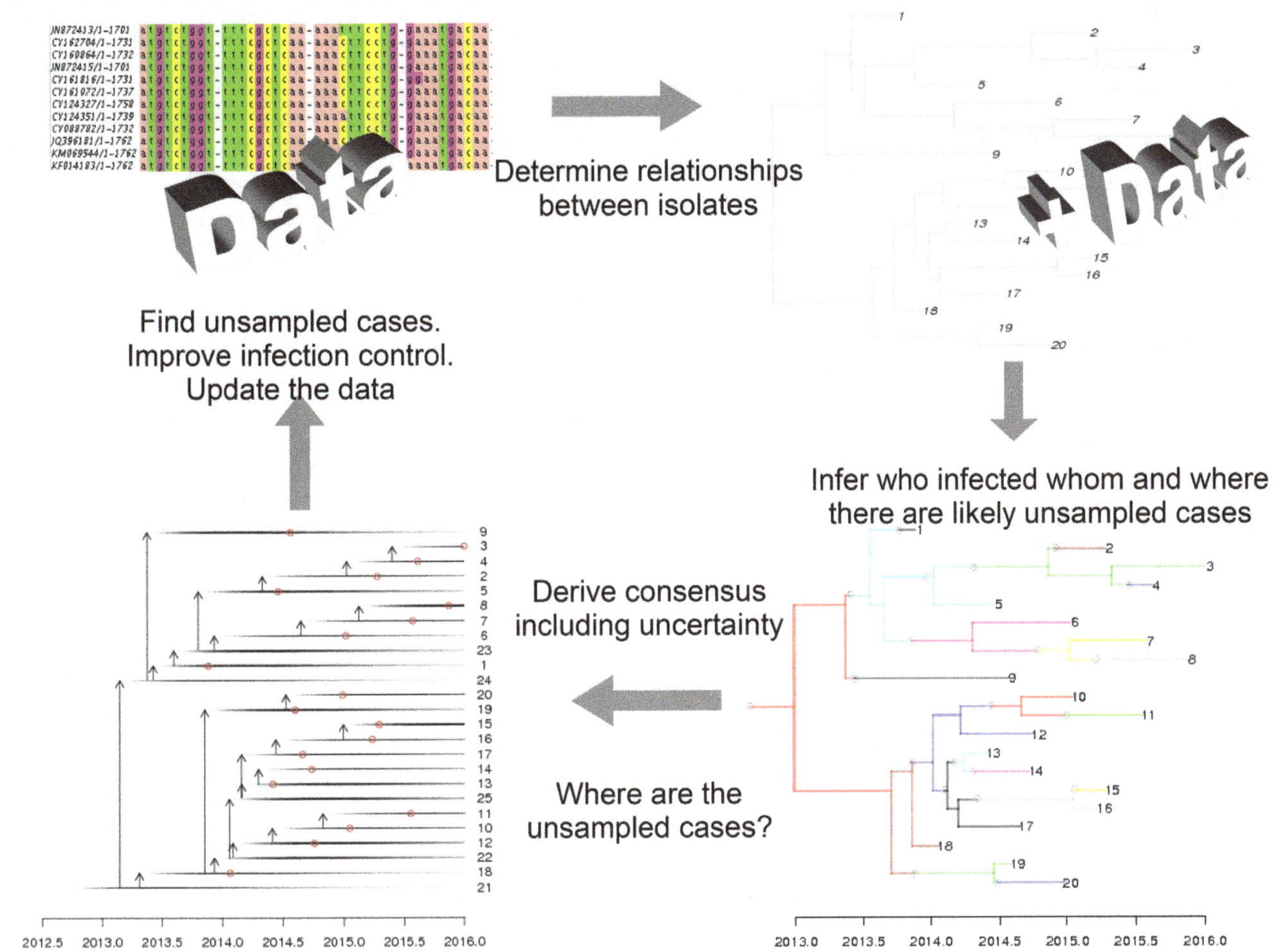

FIGURE 1 | Mathematical modeling is central to genomic epidemiology. In precision healthcare, models that link clinical, epidemiological, and sequencing data and produce interpretable results—such as predictions for where cases have been missed—can be used to direct public health interventions.

are equally the timescales on which other chronic problems are resolved: detrimental individual beliefs about healthy behavior, or disadvantageous social policies. The comparability of timescales of chronic diseases and chronic social problems, combined with the increase of chronic disease in the population, presents both policy, and mathematical challenges: parsimonious and predictive model choice for these slow coupled processes is an open challenge with important implications for the design of public health protocols and policies. Such problems are specific to precision healthcare: While precision medicine might integrate multiple individual-level datasets to improve treatment for a diabetic patient, it does not aim to consider the changing relevant environment and behavior (including beliefs about diet and obesity, food quality and availability, urban environments, and access to exercise). The research outlined above on social networks and health policy also exemplifies precision healthcare: it has a core set of mathematical challenges that are directly linked to healthcare (vs. medicine) and integrates opinion, engagement, delivery, and policy. While precision medicine and healthcare naturally have some overlap, the coupling of scales from individual information to societal behavior and intervention will be characteristic of precision healthcare. However, even with the best intentions, a version of precision healthcare that is highly dependent on advanced tools might be used to reduce, rather than enhance, health equity. A key challenge for precision healthcare is thus to create technologies and practices to drive us toward health equity.

AUTHOR CONTRIBUTIONS

CC, wrote first draft. NJ, MB, IJ, SY contributed text, edited drafts, contributed to the planning stages and co-authored the paper.

ACKNOWLEDGMENTS

We thank Paul Matthews, Beth Noveck, Gianluca Fontana, Sabine Vuik, Michael Schneider, Charles Coombes, and Ara Darzi for their insightful comments and ideas. All authors acknowledge funding from EPSRC grant EP/N014529/1.

REFERENCES

Adler, N. E., Glymour, M. M., and Fielding, J. (2016). Addressing social determinants of health and health inequalities. *JAMA* 316, 1641–1642. doi: 10.1001/jama.2016.14058

Amor, B., Vuik, S., Callahan, R., Darzi, A., Yaliraki, S. N., and Barahona, M. (2016). "Community detection and role identification in directed networks: understanding the twitter network of the care.data debate," in *Dynamic Networks and Cyber-Security*, eds N. Adams and N. Heard (World Scientific), 111–136. Available online at: http://www.worldscientific.com/worldscibooks/10.1142/q0022

Antman, E. M., and Loscalzo, J. (2016). Precision medicine in cardiology. *Nat. Rev. Cardiol.* 13, 591–602. doi: 10.1038/nrcardio.2016.101

Arnedos, M., Vicier, C., Loi, S., Lefebvre, C., Michiels, S., Bonnefoi, H., and Andre, F. (2015). Precision medicine for metastatic breast cancer—limitations and solutions. *Nat. Rev. Clin. Oncol.* 12, 693–704. doi: 10.1038/nrclinonc.2015.123

Arnett, D. K., and Claas, S. A. (2016). Precision medicine, genomics, and public health. *Diabetes Care* 39, 1870–1873. doi: 10.2337/dc16-1763

Barnett, D., Walker, B., Landay, A., and Denny, T. N. (2008). CD4 immunophenotyping in HIV infection. *Nat. Rev. Microbiol.*, 6(11 Suppl.), S7–S15. doi: 10.1038/nrmicro1998

Bayer, R., and Galea, S. (2015). Public health in the Precision-Medicine era. *N. Engl. J. Med.* 373, 499–501. doi: 10.1056/NEJMp1506241

Beerenwinkel, N., Schwarz, R. F., Gerstung, M., and Markowetz, F. (2015). Cancer evolution: mathematical models and computational inference. *Syst. Biol.* 64, e1–e25. doi: 10.1093/sysbio/syu081

Begley, C. G., and Ellis, L. M. (2012). Drug development: raise standards for preclinical cancer research. *Nature* 483, 531–533. doi: 10.1038/483531a

Beguerisse-Díaz, M., Garduno-Hernández, G., Vangelov, B., Yaliraki, S. N., and Barahona, M. (2014). Interest communities and flow roles in directed networks: the twitter network of the uk riots. *J. R. Soc. Interface* 11, 20140940. doi: 10.1098/rsif.2014.0940

Beguerisse-Díaz, M., McLennan, A. K., Garduño-Hernández, G., Barahona, M., and Ulijaszek, S. J. (2017). The 'who' and 'what' of #diabetes on twitter. *Digit. Health* 3:2055207616688841. doi: 10.1177/2055207616688841

Beguerisse-Diaz, M., Vangelov, B., and Barahona, M. (2013). "Finding role communities in directed networks using role-based similarity, markov stability and the relaxed minimum spanning tree," in *IEEE Global Conference on Signal and Information Processing (GlobalSIP), 2013* (London: IEEE), 937–940.

Burgstaller, J. P., Johnston, I. G., Jones, N. S., Albrechtová, J., Kolbe, T., Vogl, C., et al. (2014). MtDNA segregation in heteroplasmic tissues is common *in vivo* and modulated by haplotype differences and developmental stage. *Cell Rep.* 7, 2031–2041. doi: 10.1016/j.celrep.2014.05.020

Burgstaller, J. P., Johnston, I. G., and Poulton, J. (2015). Mitochondrial dna disease and developmental implications for reproductive strategies. *Mol. Hum. Reprod.* 21, 11–22. doi: 10.1093/molehr/gau090

Castaneda, C., Nalley, K., Mannion, C., Bhattacharyya, P., Blake, P., Pecora, A., et al. (2015). Clinical decision support systems for improving diagnostic accuracy and achieving precision medicine. *J. Clin. Bioinform.* 5:4. doi: 10.1186/s13336-015-0019-3

Centola, D. (2010). The spread of behavior in an online social network experiment. *Science* 329, 1194–1197. doi: 10.1126/science.1185231

Chan Zuckerberg Initiative — Advancing human potential and promoting equal opportunity (2017). *Chan Zuckerberg Initiative — Advancing Human Potential and Promoting Equal Opportunity.* Available online at: https://chanzuckerberg.com/ (Accessed January 24, 2017).

Chen, R., Mias, G. I., Li-Pook-Than, J., Jiang, L., Lam, H. Y. K., Chen, R., et al. (2012). Personal omics profiling reveals dynamic molecular and medical phenotypes. *Cell* 148, 1293–1307. doi: 10.1016/j.cell.2012.02.009

Chen, R., and Snyder, M. (2013). Promise of personalized omics to precision medicine. *Wiley Interdiscip. Rev. Syst. Biol. Med.* 5, 73–82. doi: 10.1002/wsbm.1198

Christakis, N. A., and Fowler, J. H. (2013). Social contagion theory: examining dynamic social networks and human behavior. *Stat. Med.* 32, 556–577. doi: 10.1002/sim.5408

Ciardiello, F., Arnold, D., Casali, P. G., Cervantes, A., Douillard, J.-Y., Eggermont, A., et al. (2014). Delivering precision medicine in oncology today and in future—the promise and challenges of personalised cancer medicine: a position paper by the european society for medical oncology (ESMO). *Ann. Oncol.* 25, 1673–1678. doi: 10.1093/annonc/mdu217

Cisneros, L. (2016). *White House and Gates Foundation to Convene Precision Public Health Summit at UCSF.*
Available online at: https://www.ucsf.edu/news/2016/05/403091/white-house-and-gates-foundation-convene-precision-public-health-summit-ucsf (Accessed November 30, 2016).

Cloney, R. (2017). Cancer genomics: single-cell RNA-seq to decipher tumour architecture. *Nat. Rev. Genet.* 18, 2–3. doi: 10.1038/nrg.2016.151

Cohn, E. G., Henderson, G. E., and Appelbaum, P. S. (2016). Distributive justice, diversity, and inclusion in precision medicine: what will success look like? *Genet. Med.* 19, 157–159. doi: 10.1038/gim.2016.92

Colijn, C., and Cohen, T. (2016). Whole-genome sequencing of mycobacterium tuberculosis for rapid diagnostics and beyond. *Lancet Respir. Med.* 4, 6–8. doi: 10.1016/S2213-2600(15)00510-X

Collins, F. S., and Varmus, H. (2015). A new initiative on precision medicine. *N. Engl. J. Med.* 372, 793–795. doi: 10.1056/NEJMp1500523

de Figueiredo, A., Johnston, I., Smith, D., and others (2015). Changing socioeconomic determinants of childhood vaccines: a global analysis over three decades. *Lancet Global Health.* doi: 10.1016/S2214-109X(15)70139-7. Available online at: http://www.thelancet.com/journals/langlo/article/PIIS2214-109X(15)70139-7/abstract?cc=y=

De Maio, N., Wu, C.-H., and Wilson, D. J. (2016). SCOTTI: efficient reconstruction of transmission within outbreaks with the structured coalescent. *PLOS Comput. Biol.* 12:e1005130. doi: 10.1371/journal.pcbi.1005130

Delvenne, J.-C., Yaliraki, S. N., and Barahona, M. (2010). Stability of graph communities across time scales. *Proc. Natl. Acad. Sci.* 107, 12755–12760. doi: 10.1073/pnas.0903215107

Derks, S., Cleven, A. H. G., Melotte, V., Smits, K. M., Brandes, J. C., Azad, N., et al. (2014). Emerging evidence for CHFR as a cancer biomarker: from tumor biology to precision medicine. *Cancer Metastasis Rev.* 33, 161–171. doi: 10.1007/s10555-013-9462-4

Desmond-Hellmann, S. (2016). Progress lies in precision. *Science* 353, 731–731. doi: 10.1126/science.aai7598

Didelot, X., Fraser, C., Gardy, J., and Colijn, C. (2017). Genomic infectious disease epidemiology in partially sampled and ongoing outbreaks. *Mol. Biol. Evol.* doi: 10.1093/molbev/msw275. [Epub ahead of print].

DiMauro, S., and Davidzon, G. (2005). Mitochondrial dna and disease. *Ann. Med.* 37, 222–232. doi: 10.1080/07853890510007368

Dorrestein, P. C., Mazmanian, S. K., and Knight, R. (2014). Finding the missing links among metabolites, microbes, and the host. *Immunity* 40, 824–832. doi: 10.1016/j.immuni.2014.05.015

Duffy, D. J. (2016). Problems, challenges and promises: perspectives on precision medicine. *Brief. Bioinform.* 17, 494–504. doi: 10.1093/bib/bbv060

El Shamieh, S., and Visvikis-Siest, S. (2012). Genetic biomarkers of hypertension and future challenges integrating epigenomics. *Clin. Chim. Acta* 414, 259–265. doi: 10.1016/j.cca.2012.09.018

Friedman, A. A., Letai, A., Fisher, D. E., and Flaherty, K. T. (2015). Precision medicine for cancer with next-generation functional diagnostics. *Nat. Rev. Cancer* 15, 747–756. doi: 10.1038/nrc4015

Garber, A. M., and Tunis, S. R. (2009). Does comparative-effectiveness research threaten personalized medicine? *N. Engl. J. Med.* 360, 1925–1927. doi: 10.1056/NEJMp0901355

Gardy, J., Loman, N. J., and Rambaut, A. (2015). Real-time digital pathogen surveillance - the time is now. *Genome Biol.* 16:155. doi: 10.1186/s13059-015-0726-x

Gilbert, J. A., Quinn, R. A., Debelius, J., Xu, Z. Z., Morton, J., Garg, N., et al. (2016). Microbiome-wide association studies link dynamic microbial consortia to disease. *Nature* 535, 94–103. doi: 10.1038/nature18850

Goodman, S. N. (1999). Toward evidence-based medical statistics. 1: The p value fallacy. *Ann. Intern. Med.* 130, 995–1004.

Hall, M., Woolhouse, M., and Rambaut, A. (2015). Epidemic reconstruction in a phylogenetics framework: transmission trees as partitions of the node set. *PLoS Comput. Biol.* 11:e1004613. doi: 10.1371/journal.pcbi.1004613

Hiley, C., de Bruin, E. C., McGranahan, N., and Swanton, C. (2014). Deciphering intratumor heterogeneity and temporal acquisition of driver events to refine precision medicine. *Genome Biol.* 15:453. doi: 10.1186/s13059-014-0453-8

Hjelm, M., Höglund, M., and Lagergren, J. (2006). New probabilistic network models and algorithms for oncogenesis. *J. Comput. Biol.* 13, 853–865. doi: 10.1089/cmb.2006.13.853

Hood, L., and Friend, S. H. (2011). Predictive, personalized, preventive, participatory (p4) cancer medicine. *Nat. Rev. Clin. Oncol.* 8, 184–187. doi: 10.1038/nrclinonc.2010.227

Hyslop, L. A., Blakeley, P., Craven, L., Richardson, J., Fogarty, N. M., Fragouli, E., et al. (2016). Towards clinical application of pronuclear transfer to prevent mitochondrial DNA disease. *Nature.* 534, 383–386. doi: 10.1038/nature18303

Ignatiadis, M., and Dawson, S.-J. (2014). Circulating tumor cells and circulating tumor DNA for precision medicine: dream or reality? *Ann. Oncol.* 25, 2304–2313. doi: 10.1093/annonc/mdu480

Ioannidis, J. P. (2005). Why most published research findings are false. *PLoS Med.* 2:e124. doi: 10.1371/journal.pmed.0020124

Johnston, I. G. (2016). Multiple hypothesis correction is vital and undermines reported mtDNA links to diseases including aids, cancer, and huntingdon's. *Mitochondrial DNA A DNA Mapp. Seq. Anal.* 27, 3423–3427. doi: 10.3109/19401736.2015.1022732

Johnston, I. G., and Williams, B. P. (2016). Evolutionary inference across eukaryotes identifies specific pressures favoring mitochondrial gene retention. *Cell Syst.* 2, 101–111. doi: 10.1016/j.cels.2016.01.013

Jombart, T., Aanensen, D. M., Baguelin, M., Birrell, P., Cauchemez, S., Camacho, A., et al. (2014). OutbreakTools: a new platform for disease outbreak analysis using the R software. *Epidemics* 7, 28–34. doi: 10.1016/j.epidem.2014.04.003

Juengst, E. T., Flatt, M. A., and Settersten, R. A. Jr. (2012). Personalized genomic medicine and the rhetoric of empowerment. *Hastings Cent. Rep.* 42, 34–40. doi: 10.1002/hast.65

Kao, R. R., Haydon, D. T., Lycett, S. J., and Murcia, P. R. (2014). Supersize me: how whole-genome sequencing and big data are transforming epidemiology. *Trends Microbiol.* 22, 282–291. doi: 10.1016/j.tim.2014.02.011

Khoury, M. J., Gwinn, M. L., Glasgow, R. E., and Kramer, B. S. (2012). A population approach to precision medicine. *Am. J. Prev. Med.* 42, 639–645. doi: 10.1016/j.amepre.2012.02.012

Khoury, M. J., Iademarco, M. F., and Riley, W. T. (2016). Precision public health for the era of precision medicine. *Am. J. Prev. Med.* 50, 398–401. doi: 10.1016/j.amepre.2015.08.031

Klinkenberg, D., Backer, J., Didelot, X., Colijn, C., and Wallinga, J. (2016). New method to reconstruct phylogenetic and transmission trees with sequence data from infectious disease outbreaks. *bioRxiv.* doi: 10.1101/069195

Lambiotte, R., Delvenne, J.-C., and Barahona, M. (2015). Random walks, markov processes and the multiscale modular organization of complex networks. *IEEE Trans. Netw. Sci. Eng.* 1, 76–90. doi: 10.1109/TNSE.2015.2391998

Larry Jameson, J., and Longo, D. L. (2015). Precision Medicine—Personalized, problematic, and promising. *Obstet. Gynecol. Surv.* 70, 612. doi: 10.1097/01.ogx.0000472121.21647.38

Larson, H. J., de Figueiredo, A., Xiahong, Z., Schulz, W. S., Verger, P., Johnston, I. G., et al. (2016). The state of vaccine confidence 2016: Global insights through a 67-country survey. *EBioMedicine* 12, 295–301. doi: 10.1016/j.ebiom.2016.08.042

Liu, Y.-Y., and Barabási, A.-L. (2016). Control principles of complex systems. *Rev. Mod. Phys.* 88:035006. doi: 10.1103/revmodphys.88.035006

Loohuis, L. O., Caravagna, G., Graudenzi, A., Ramazzotti, D., Mauri, G., Antoniotti, M., et al. (2014). Inferring tree causal models of cancer progression with probability raising. *PLoS ONE* 9:e108358. doi: 10.1371/journal.pone.0108358

Mallat, S. (2016). Understanding deep convolutional networks. *Philos. Trans. R. Soc. Lond. A Math. Phys. Eng. Sci.* 374:20150203. doi: 10.1098/rsta.2015.0203

Marmot, M., and Commission on Social Determinants of Health (2007). Achieving health equity: from root causes to fair outcomes. *Lancet* 370, 1153–1163. doi: 10.1016/S0140-6736(07)61385-3

Mirnezami, R., Nicholson, J., and Darzi, A. (2012). Preparing for precision medicine. *N. Engl. J. Med.* 366, 489–491. doi: 10.1056/nejmp1114866

Muraro, A., Lemanske, R. F. Jr., Hellings, P. W., Akdis, C. A., Bieber, T., Casale, T. B., et al. (2016). Precision medicine in patients with allergic diseases: airway diseases and atopic dermatitis—PRACTALL document of the european academy of allergy and clinical immunology and the american academy of allergy, asthma & immunology. *J. Allergy Clin. Immunol.* 137, 1347–1358. doi: 10.1016/j.jaci.2016.03.010

Navin, N. E. (2015). The first five years of single-cell cancer genomics and beyond. *Genome Res.* 25, 1499–1507. doi: 10.1101/gr.191098.115

Noveck, B. S. (2015). *Smart Citizens, Smarter State: The Technologies of Expertise and the Future of Governing.* Boston, MA: Harvard University Press.

Nyhan, B., Reifler, J., Richey, S., and Freed, G. L. (2014). Effective messages in vaccine promotion: a randomized trial. *Pediatrics* 133, e835–e842. doi: 10.1542/peds.2013-2365

O'Neil, C. (2016). *Weapons of Math Destruction: How Big Data Increases Inequality and Threatens Democracy.* New York, NY: Crown Publishing Group.

Onnela, J.-P., Landon, B. E., Kahn, A.-L., Ahmed, D., Verma, H., O'Malley, A. J., et al. (2016). Polio vaccine hesitancy in the networks and neighborhoods of malegaon, India. *Soc. Sci. Med.* 153, 99–106. doi: 10.1016/j.socscimed.2016.01.024

Pagel, M., and Meade, A. (2006). Bayesian analysis of correlated evolution of discrete characters by reversible-jump Markov chain Monte Carlo. *Am. Nat.* 167, 808–825. doi: 10.1086/503444.

Palmisano, A., Zhao, Y., Li, M.-C., Polley, E. C., and Simon, R. M. (2016). OpenGeneMed: a portable, flexible and customizable informatics hub for the coordination of next-generation sequencing studies in support of precision medicine trials. *Brief. Bioinform.* doi: 10.1093/bib/bbw059. [Epub ahead of print].

Perez, F., Chakhtoura, N. G. E., Papp-Wallace, K. M., Wilson, B. M., and Bonomo, R. A. (2016). Treatment options for infections caused by carbapenem-resistant enterobacteriaceae: can we apply "precision medicine" to antimicrobial chemotherapy? *Expert Opin. Pharmacother.* 17, 761–781. doi: 10.1517/14656566.2016.1145658

Pickett, K. E., and Wilkinson, R. G. (2015). Income inequality and health: a causal review. *Soc. Sci. Med.* 128, 316–326. doi: 10.1016/j.socscimed.2014.12.031

Prinz, F., Schlange, T., and Asadullah, K. (2011). Believe it or not: how much can we rely on published data on potential drug targets? *Nat. Rev. Drug Dis.* 10:712. doi:10.1038/nrd3439-c1

Quick, J., Loman, N. J., Duraffour, S., Simpson, J. T., Severi, E., Cowley, L., et al. (2016). Real-time, portable genome sequencing for ebola surveillance. *Nature* 530, 228–232. doi: 10.1038/nature16996

Rosell, R., and Karachaliou, N. (2013). Lung cancer: maintenance therapy and precision medicine in NSCLC. 10, 549–550. doi: 10.1038/nrclinonc.2013.152

Rost, B., Radivojac, P., and Bromberg, Y. (2016). Protein function in precision medicine: deep understanding with machine learning. *FEBS Lett.* 590, 2327–2341. doi: 10.1002/1873-3468.12307

Roychowdhury, S., and Chinnaiyan, A. M. (2013). Advancing precision medicine for prostate cancer through genomics. *J. Clin. Oncol.* 31, 1866–1873. doi: 10.1200/jco.2012.45.3662

Røyrvik, E., Burgstaller, J. P., and Johnston, I. G. (2016). mtdna diversity in human populations highlights the merit of haplotype matching in gene therapies. *Mol. Hum. Reprod.* 22, 809–817. doi: 10.1093/molehr/gaw062

Rubin, M. A. (2015). Health: make precision medicine work for cancer care. *Nature* 520, 290–291. doi: 10.1038/520290a

Salathé, M., and Bonhoeffer, S. (2008). The effect of opinion clustering on disease outbreaks. *J. R. Soc. Interface* 5, 1505–1508. doi: 10.1098/rsif.2008.0271

Sboner, A., and Elemento, O. (2016). A primer on precision medicine informatics. *Brief. Bioinform.* 17, 145–153. doi: 10.1093/bib/bbv032

Schaub, M. T., Delvenne, J.-C., Yaliraki, S. N., and Barahona, M. (2012). Markov dynamics as a zooming lens for multiscale community detection: non clique-like communities and the field-of-view limit. *PLoS ONE* 7:e32210. doi: 10.1371/journal.pone.0032210

Schork, N. J. (2015). Personalized medicine: time for one-person trials. *Nature* 520, 609–611. doi: 10.1038/520609a

Servant, N., Roméjon, J., Gestraud, P., La Rosa, P., Lucotte, G., Lair, S., et al. (2014). Bioinformatics for precision medicine in oncology: principles and application to the SHIVA clinical trial. *Front. Genet.* 5:152. doi: 10.3389/fgene.2014.00152

Shalizi, C. R., and Thomas, A. C. (2011). Homophily and contagion are generically confounded in observational social network studies. *Sociol. Methods Res.* 40, 211–239. doi: 10.1177/0049124111404820

Snyder-Mackler, N., Sanz, J., Kohn, J. N., Brinkworth, J. F., Morrow, S., Shaver, A. O., et al. (2016). Social status alters immune regulation and response to infection in macaques. *Science* 354, 1041–1045. doi: 10.1126/science.aah3580

Stover, D. G., and Wagle, N. (2015). Precision medicine in breast cancer: genes, genomes, and the future of genomically driven treatments. *Curr. Oncol. Rep.* 17:15. doi: 10.1007/s11912-015-0438-0

Trusheim, M. R., Burgess, B., Hu, S. X., Long, T., Averbuch, S. D., Flynn, A. A., et al. (2011). Quantifying factors for the success of stratified medicine. *Nat. Rev. Drug Discov.* 10, 817–833. doi: 10.1038/nrd3557

UK Biobank (2016). *UK Biobank*. Available online at: https://www.ukbiobank.ac.uk/ (Accessed November 30, 2016).

Vanderkooi, O. G., Low, D. E., Green, K., Powis, J. E., McGeer, A., and Toronto Invasive Bacterial Disease Network (2005). Predicting antimicrobial resistance in invasive pneumococcal infections. *Clin. Infect. Dis.* 40, 1288–1297. doi: 10.1086/429242

Vargas, A. J., and Harris, C. C. (2016). Biomarker development in the precision medicine era: lung cancer as a case study. *Nat. Rev. Cancer.* 16, 525–537. doi: 10.1038/nrc.2016.56

Wang, Z., Bauch, C. T., Bhattacharyya, S., d'Onofrio, A., Manfredi, P., Perc, M., et al. (2016). Statistical physics of vaccination. *Phys. Rep.* 664, 1–113. doi: 10.1016/j.physrep.2016.10.006

White House Precision Medicine Initiative (2016). *White House Precision Medicine Initiative.* Available online at: https://www.whitehouse.gov/precision-medicine (Accessed November 29, 2016).

Williams, B. P., Johnston, I. G., Covshoff, S., and Hibberd, J. M. (2013). Phenotypic landscape inference reveals multiple evolutionary paths to c4 photosynthesis. *Elife* 2:e00961. doi: 10.7554/eLife.00961

Wills, Q. F., and Mead, A. J. (2015). Application of single-cell genomics in cancer: promise and challenges. *Hum. Mol. Genet.* 24, R74–R84. doi: 10.1093/hmg/ddv235

Worby, C. J., Lipsitch, M., and Hanage, W. P. (2015). Shared genomic variants: identification of transmission routes using pathogen deep sequence data [Internet]. *bioRxiv.* doi: 10.1101/032458

Worby, C. J., O'Neill, P. D., Kypraios, T., Robotham, J. V., De Angelis, D., Cartwright, E. J. P., et al. (2016). Reconstructing transmission trees for communicable diseases using densely sampled genetic data. *Ann. Appl. Stat.* 10, 395–417. doi: 10.1214/15-aoas898

Xiao, T., Xia, T., Yang, Y., Huang, C., and Wang, X. (2015). "Learning from massive noisy labeled data for image classification," in *Proceedings of the IEEE Conference on Computer Vision and Pattern Recognition*, 2691–2699. Available online at: cv-foundation.org.

Yamada, M., Emmanuele, V., Sanchez-Quintero, M. J., Sun, B., Lallos, G., Paull, D., et al. (2016). Genetic drift can compromise mitochondrial replacement by nuclear transfer in human oocytes. *Cell Stem Cell* 18, 749–754. doi: 10.1016/j.stem.2016.04.001

Yarchoan, M., and Arnold, S. E. (2014). Repurposing diabetes drugs for brain insulin resistance in Alzheimer disease. *Diabetes* 63, 2253–2261. doi: 10.2337/db14-0287

Zhang, W., Wang, L., Chen, Y., Tang, F., Xue, F., and Zhang, C. (2015). Identification of hypertension predictors and application to hypertension prediction in an urban han chinese population: a longitudinal study, 2005-2010. *Prev. Chronic Dis.* 12:E184. doi: 10.5888/pcd12.150192

Ziliak, S. T., and McCloskey, D. N. (2008). *The Cult of Statistical Significance: How the Standard Error Costs us Jobs, Justice, and Lives.* Ann Arbor, MI: University of Michigan Press.

10

Patient Similarity: Emerging Concepts in Systems and Precision Medicine

Sherry-Ann Brown *

Department of Cardiovascular Diseases, Mayo Clinic, Rochester, MN, USA

Keywords: patient similarity, patient similarity analytics, computational medicine, big data analytics, clinical decision support

Edited by:
Krasimira Tsaneva-Atanasova, University of Exeter, UK

Reviewed by:
Pietro Lio, University of Cambridge, UK
Gustavo Glusman, Institute for Systems Biology, USA

***Correspondence:**
Sherry-Ann Brown
brown.sherryann@mayo.edu

INTRODUCTION

Healthcare data generates a huge *volume* of information in *various* formats at high *velocity* with sometimes questionable *veracity* (Barkhordari and Niamanesh, 2015) (4V). As a result, big data tools such as patient similarity are necessary to facilitate analytics, which reduces costs (Srinivasan and Arunasalam, 2013) and improves healthcare systems (Jee and Kim, 2013). Patient similarity investigates distances between a variety of components of patient data, and determines methods of clustering patients, based on short distances between some of their characteristics. Although patient similarity is in its early stages, ultimately information about diseases, risk factors, lifestyle habits, medication use, co-morbidities, molecular and histopathological information, hospitalizations, or death are compared with laboratory investigations, imaging, and other clinical data assessing medical evidence of human behavior (**Figure 1**). Such analytics consist of efficient computational analyses with patient stratification by multiple co-occurrence statistics, based on clinical characteristics. Algorithms create subgroups of patients based on similarities among their electronic avatars. Among electronic avatars found to be similar, subgroups of patients can be evaluated by further stratification guided by individual diagnoses, risk factors, medications, and so on. Because of the multiple networks of subgroups of patients, patient similarity can be considered an application of network medicine, with the output termed "patient similarity networks." Thus, data mining extracts clinically relevant information hidden in clinical notes and embedded in other areas of the electronic health record (EHR) coupled with International Classification of Disease codes. The result is a systematic individualized analysis of a subset of patients that can improve outcome prediction and help guide management for a particular patient currently being cared for by a clinician (Lee et al., 2015). The communication or output from the algorithms can be used to identify and predict disease correlations and occurrence, and potentially for clinical decision support at the point of care. Patient similarity analytics are not restricted to global findings from large clinical trials consisting of somewhat heterogeneous patient populations (Roque et al., 2011). In this way, patient similarity represents a paradigm shift that introduces disruptive innovation to optimize personalization of patient care. Some promising examples are regarding mental and behavioral disorders (Roque et al., 2011), infectious diseases (Li et al., 2015), cancers (Wu et al., 2005; Teng et al., 2007; Chan et al., 2010, 2015; Klenk et al., 2010; Cho and Przytycka, 2013; Li et al., 2015; Wang, 2015; Bolouri et al., 2016; Wang et al., 2016), endocrine (Li et al., 2015; Wang, 2015), and metabolic diseases (Zhang et al., 2014; Ng et al., 2015). Others involve diseases of the nervous system (Lieberman et al., 2005; Carreiro et al., 2013; Cho and Przytycka, 2013; Qian et al., 2014; Buske et al., 2015a; Li et al., 2015; Bolouri et al., 2016; Wang et al., 2016), eyes (Buske et al., 2015a; Li et al., 2015), skin (Buske et al., 2015a; Li et al., 2015), heart (Wu et al., 2005; Tsymbal et al., 2007; Syed and Guttag, 2011; Buske et al., 2015a; Li et al., 2015; Panahiazar et al., 2015a,b; Wang, 2015; Björnson et al., 2016), liver (Chan et al., 2015), intestines (Buske et al., 2015a), musculoskeletal system

(Buske et al., 2015a), congenital malformations (Buske et al., 2015a), and various other conditions or factors influencing health status (Gotz et al., 2012; Subirats et al., 2012; Ng et al., 2015).

PATIENT SIMILARITY IN SYSTEMS MEDICINE

Patient similarity is just starting to spread its wings and has the potential to transform Systems Medicine, which is Systems Biology applied to health care. Systems Biology studies the characteristics of cells, tissues, organisms, or other comprehensive biological units as whole systems. Systems Biology seeks to determine how changes in one part of the system can affect the behavior of the whole system, and often focuses on predictive modeling of the system in a perturbed state. Patient similarity analytics could be developed to bring together characteristics of the patient as a whole human system, and compare these to a multitude of similar patients. Accordingly, patient similarity analytics should in the near future incorporate genomics, transcriptomics, proteomics, microbiomics, and other "omics" and diverse components of systems medicine. In addition, simulation of physiology at the level of the molecule, cell, tissue, organ, and organism should be consolidated as a comprehensive similarity feature to give a broader view of interactions among organ systems. Patient similarity analytics could provide predictive models of a patient's outcome in the setting of disease perturbations or diagnoses relevant to the index patient. Making adjustments in the query data that serve as input for the predictive models would allow for assessment of how new diagnoses or therapies could impact the overall behavior and phenotype of the whole patient.

Beyond the reasoning above, integrating the majority of these systems medicine tools into patient similarity analytics is potentially the next frontier in Systems Medicine, for at least a few reasons. First, patient similarity analytics embrace a systems view by assessing a myriad of characteristics for hundreds or thousands of patients to produce a meaningful and useful result. Second, patient similarity analytics are analogous to various "omics" that in part compose Systems Biology. Just as transcriptomics refers to generation of messenger RNA expression profiles (Briefing, 1999), one could consider a term similaromics referring to generation or identification of patients similar to an index patient. Similaromics is also akin to phenomics, proteomics, and genomics, among others. Phenomics refers to cataloging the observable characteristics conferred by a gene and proteomics describes the generation of proteins expressed by a cell (Briefing, 1999). One might argue that patient similarity is not quite analogous to genomics, since an individual's genome is thought to be constant throughout their lifetime. However, this is no longer necessarily the case, due to the current progress of genome editing tools. Indeed, patient similarity is analogous to these various omics, all with the potential to change over the lifetime of the individual. Thus, just as a genome is the complement of all DNA within a cell, a similarome is the complement of patients found to be similar to an index patient. Within the similarome, one can further distinguish subgroups of patients that are most similar to an index patient, based on preferentially assigning preeminence to comorbidities or medications of most interest or relevance to the index patient, e.g., during a focused shared decision-making session with a clinician. Similar to genotyping then, which determines the presence or absence of a particular gene feature, simotyping would allocate the presence or absence of a particular similarity feature, for example, a diagnosis of diabetes. In this context then, a similarity-wide association study (SiWAS) has the

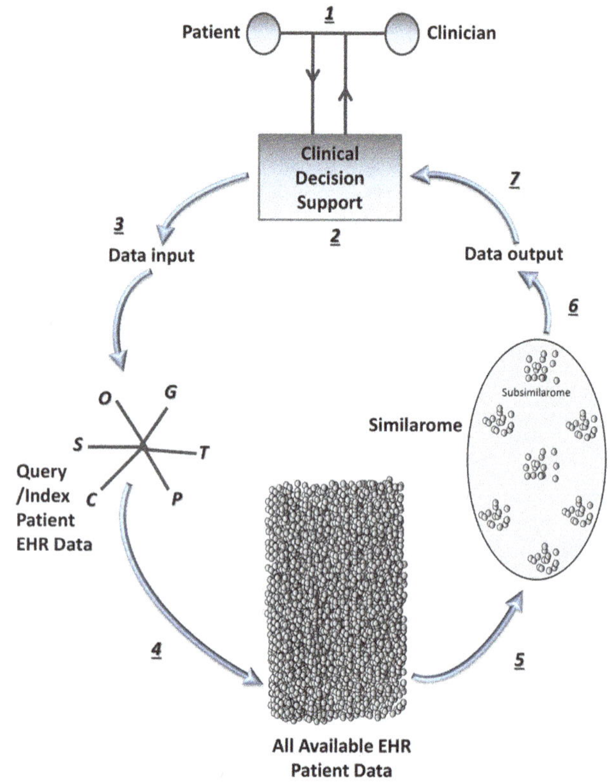

FIGURE 1 | The patient similarity analytics loop in systems medicine. Once a query patient is selected, the patient and clinician (e.g., physician or other health professional) in partnership can enter the "patient similarity analytics loop" (step 1), which is iterative as patient characteristics evolve over time and new patients become available for inclusion in the similarome. In step 2, query information is entered via a clinical decision support tool interface. In step 3, this information combines with data from the query or index patient's EHR to form the data input for the patient similarity algorithms. Each "omic" or systems medicine data type or tool (Brown, 2015b) functions as a predictor variable vector, all of which are incorporated into the multidimensional feature space for the patient. In step 4, the entire available EHR patient populous is interrogated with a patient similarity network analysis tool; efficient data mining is completed using patient similarity algorithms. In step 5, similarity data is arranged, yielding a similarome (cohort of patients most similar to the query/index patient), with subsimilaromes (subgroups of patients most similar to the query/index patient based on prioritizing various comorbidities/medications, etc.). Step 6 involves data collating and information retrieval. In step 7, the similarome (which includes subsimilaromes) is presented to the patient-clinician partnership via the clinical decision support tool interface for clinical decision-making at the point-of-care. C, Clinical information; G, Genomics; O, Other systems medicine data types or tools; P, Proteomics; S, Social network data; T, Transcriptomics.

goal of discovering clusters of patients similar to an index patient and identifying similar features that associate with specific outcomes, such as complications, procedures, hospitalizations, or death. For example, investigating whether in patients most similar to an index patient diabetes is more likely to associate with non-healing leg ulcers, critical limb ischemia, or gangrene leading to limb amputation.

Third, patient similarity analytics have the potential to bring together a variety of omics and other systems medicine tools, if we can do so in a way that is effective, accurate, consistent, and computationally efficient (Brown, 2015a). Indeed, several groups have proposed methods of aggregating omics and monitoring these over time for individual patients, and perhaps even using comprehensive patient avatars. Integrating these methods with patient similarity has the potential to launch systems medicine further into a future where medicine is even more precisely individualized. Patient similarity will likely become and persist as a useful tool in systems medicine.

MATHEMATICS IN PATIENT SIMILARITY ANALYTICS

For illustration of the utility of patient similarity in medicine, only briefly presented here are a few selected examples of patient similarity analytics used for diabetes and cancer, which are common chronic or terminal diseases, respectively, currently addressed in public health. In some studies, a patient similarity metric is determined as follows (Lee et al., 2015; Li et al., 2015). A patient can be represented by a Euclidean vector. Predictor variables such as laboratory test results or vital signs can define a multi-dimensional feature space. The cosine of the angle between two patients' vectors can define the associated patient similarity metric. A dot product can facilitate the calculation. This can be termed the "cosine similarity," defining the patient similarity metric as follows:

$$PSM\,(P_1, P_2) = \frac{P_1 \cdot P_2}{||P_1||\;||P_2||}, = \frac{\sum_{i=1}^{n} P_{1i} \times P_{2i}}{\sqrt{\sum_{i=1}^{n} P_{1i}{}^2} \times \sqrt{\sum_{i=1}^{n} P_{2i}{}^2}}$$

where P_{1i} and P_{2i} represent a single predictor variable vector for two separate patients, $\cdot$ represents the dot product, and $||\;||$ represents the Euclidean vector magnitude, as shown. Since the patient similarity metric is an angle cosine, it normalizes between -1 (considered minimum possible similarity) and 1 (considered maximum possible similarity). As expected, two predictor variable vectors pointing in the exact opposite direction to each other would have a 180° angle between them, and would therefore calculate to a patient similarity metric of -1. Conversely, two perfectly overlapping vectors would have an angle of 0° between them, and would therefore calculate to a patient similarity metric of 1. Accordingly, before calculating the total patient similarity metric, the product for each predictor variable vector would be normalized to the range of -1 to 1 in the multidimensional feature space, if continuous (Lee et al., 2015). The product for categorical/binary predictor variable vectors would be assigned a value of -1 or 1. The patient similarity metric would be calculated for each patient in a given data set, relative to an index patient P_1. The N most similar patients to the index patient would be utilized as a training data set for testing in a validation data set, with prediction of prognosis, morbidity, or mortality. After successful validation, the predictive model could be used for epidemiologic or clinical studies. For example, am algorithm using cosine similarity successfully identified three subgroups of patients with diabetes (Li et al., 2015). The first subgroup included patients with diabetic nephropathy (diabetes-related kidney disease) and diabetic retinopathy (diabetes-related eye disease). The second subgroup included several patients with cancer and cardiovascular diseases. The third subgroup included many patients who also had cardiovascular diseases, along with neurological diseases, allergies, and HIV infection. Various single nucleotide polymorphisms mapped to these three subgroups that were confirmed in the EHR, suggesting clinical relevance for patient similarity in precision medicine. Jaccard similarity, another metric that can be leveraged after assigning binary attributes to each patient's multifeature vector space, was useful to analyze features underlying deviant responses to therapeutics in patients with diabetes (Zhang et al., 2014).

Alternatively, unsupervised clustering of patients based on their clinical predictor variables could be used to produce a patient-patient network. The network could be organized using L-infinity centrality, which is the maximum distance from each point from any other point in a given data set. L-infinity centrality produces a detailed and succinct description of any data set yielding more information than scatter plots (Lum et al., 2013). Large values for L-infinity centrality correspond to data points at large distances from the center of the data set (Li et al., 2015). Other pattern analysis and cluster algorithms (Daemen and De Moor, 2009; Chan et al., 2010; Liu et al., 2013a; Mabotuwana et al., 2013; Sundar et al., 2014), or algorithms incorporating distance metric learning (Wang et al., 2011; Bian and Tao, 2012), locally supervised metric learning (Sun et al., 2012; Ng et al., 2015), local spline regression (Wang et al., 2012), or visual analytics (Tsymbal et al., 2009; Ebadollahi et al., 2010; Gotz et al., 2011; Perer, 2012; Heer and Perer, 2014; Bolouri et al., 2016; Ozery-Flato et al., 2016), can also be used for patient similarity to predict diabetes onset, develop treatment recommendations tailored to each patient, or predict survival after chemotherapy (Chan et al., 2010; Liu et al., 2013a; Ng et al., 2015; Ozery-Flato et al., 2016), among other applications. SNOMED CT and other medical terminology frameworks can be used to facilitate communication across platforms in various studies (Melton et al., 2006). There are also algorithms to incorporate a time series into patient similarity analysis, to predict trends over time among patients (Wu et al., 2005; Hartge et al., 2006; Ebadollahi et al., 2010; Carreiro et al., 2013; Alaa et al., 2016). For example, a patient similarity time series algorithm has been used to fine-tune radiation treatment planning for patients with head and neck cancers (Wu et al., 2005).

CHALLENGES IN PATIENT SIMILARITY

There are certain challenges in patient similarity, such as network bottlenecks, low hardware performance (processing power and memory), and data locality (Osman et al., 2013; Karapiperis and Verykios, 2014; Barkhordari and Niamanesh, 2015). Given the observational or retrospective nature of patient similarity, interpretation of data analysis will be imperfect. Confounder control and treatment selection bias are inherent limitations in such studies. However, groups have developed strategies to manage the potential for confounders, such as restriction, stratification, matching, inverse probability weighting, and covariate adjustment (Gallego et al., 2015). Several groups have also proposed solutions for other challenges that enable large scale patient indexing and accurate and efficient clinical data retrieval (Wang, 2015). Some have devised algorithms to address the complexity of clinical data and limited transparency of many existing clinical case retrieval decision support systems (Tsymbal et al., 2009), as well as integration of data from various heterogeneous omics studies (Wang et al., 2014, 2016; Gligorijević et al., 2016) and physician input and feedback (Wang et al., 2011; Sun et al., 2012; Fei and Sun, 2015). Others have produced algorithms that address scalability and uncertainty, by requiring parallel or distributed algorithm implementations built to scale, and enhancing interpretability by conveying the certainty of results presented (Feldman et al., 2015). One such algorithm or platform is scalable and distributable patient similarity (ScaDiPaSi), a dynamic method for investigating patient similarity that spreads the algorithm over several self-sufficient hardware nodes to process query data from various sources of different formats simultaneously (Barkhordari and Niamanesh, 2015). Another tool, MapReduce, employs several optimization techniques, such as job scheduling and cascading work flows over multiple interdependent hardware nodes (Dean and Ghemawat, 2008). Use of all of these technological solutions for patient similarity in precision medicine will be facilitated by bridging gaps among different scientific, technological, and medical cultures, through interdisciplinary collaborations among experts in medicine, biology, informatics, engineering, public health, economics, and the social sciences (Kuhn et al., 2008).

CONCLUSION

Various patient similarity algorithms have been deployed and have been found beneficial by improving clinical efficiency (Wang et al., 2015), enabling secure identification of similar patients and records sharing by clinicians and rare disease scientists (Buske et al., 2015a,b), predicting patients' prognosis or trajectory over time (Ebadollahi et al., 2010; Subirats et al., 2012; Wang et al., 2012; Gallego et al., 2015), providing clinical decision support (Daemen et al., 2009; Wang et al., 2011; Subirats et al., 2012; Sun et al., 2012; Gottlieb et al., 2013; Liu et al., 2013b; Gallego et al., 2015), tailoring individual treatments (Zhang et al., 2014), preventing unexpected adverse drug reactions (Hartge et al., 2006; Yang et al., 2014), flagging patients deserving more attention due to poor response to therapies (Zhang et al., 2014; Ozery-Flato et al., 2016), and pursuing comparative effectiveness studies (Wang et al., 2011), among other applications. In general, clinical guidelines often do not supply evidence on risks, secondary therapy effects, and long-term outcomes (Gallego et al., 2015). In this setting, patient similarity analytics can provide a cheaper, portable alternative or in fact adjunct to evidence-based clinical guidelines and randomized controlled trials, particularly if trial data are unavailable for conditions or patient characteristics specific to a query individual (Longhurst et al., 2014; Gallego et al., 2015). Synthesizing current patient similarity algorithms with systems medicine tools could provide actionable insights in precision medicine.

AUTHOR CONTRIBUTIONS

SB conceived, analyzed, designed, drafted, critically revised, approved, and agreed to be accountable for this submitted work.

ACKNOWLEDGMENTS

The author is grateful to Dr. Joerg Herrmann of Mayo Clinic in Rochester, Minnesota for reading the manuscript.

REFERENCES

Alaa, A., Yoon, J., Hu, S., and van der Schaar, M. (2016). "Personalized risk scoring for critical care patients using mixtures of Gaussian Process Experts," in *Proceedings of the 33 rd International Conference on Machine Learning*, vol 48 (New York, NY).

Barkhordari, M., and Niamanesh, M. (2015). ScaDiPaSi: an effective scalable and distributable MapReduce-Based method to find patient similarity on huge healthcare networks. *Big Data Res.* 2, 19–27. doi: 10.1016/j.bdr.2015.02.004

Briefing (1999). Proteomics, transcriptomics: what's in a name? *Nature* 402:715.

Bian, W., and Tao, D. (2012). Constrained empirical risk minimization framework for distance metric learning. *IEEE Trans. Neural Netw. Learn. Syst.* 23, 1194–1205. doi: 10.1109/TNNLS.2012.2198075

Björnson, E., Borén, J., and Mardinoglu, A. (2016). Personalized cardiovascular disease prediction and Treatment-A review of existing strategies and novel systems medicine tools. *Front Physiol.* 7:2. doi: 10.3389/fphys.2016.00002

Bolouri, H., Zhao, L. P., and Holland, E. C. (2016). Big data visualization identifies the multidimensional molecular landscape of human gliomas. *Proc. Natl. Acad. Sci. U.S.A.* 113, 5394–5399. doi: 10.1073/pnas.1601591113

Brown, S. A. (2015a). Principles for developing patient avatars in precision and systems medicine. *Front Genet.* 6:365 doi: 10.3389/fgene.2015.00365

Brown, S. A. (2015b). Building SuperModels: emerging patient avatars for use in precision and systems medicine. *Front. Physiol.* 6:318. doi: 10.3389/fphys.2015.00318

Buske, O. J., Girdea, M., Dumitriu, S., Gallinger, B., Hartley, T., Trang, H., et al. (2015b). PhenomeCentral: a portal for phenotypic and genotypic matchmaking of patients with rare genetic diseases. *Hum Mutat.* 36, 931–940. doi: 10.1002/humu.22851

Buske, O. J., Schiettecatte, F., Hutton, B., Dumitriu, S., Misyura, A., Huang, L., et al. (2015a). The Matchmaker Exchange API: automating patient matching through the exchange of structured phenotypic and genotypic profiles. *Hum Mutat.* 36, 922–927. doi: 10.1002/humu.22850

Carreiro, A., Madeira, S., and Francisco, A. (2013). "Unravelling communities of ALS patients using network mining," in *KDD-DMH'13* (Chicago, IL).

Chan, L., Chan, T., Cheng, L., and Mak, W. (2010). "Machine learning of patient similarity: a case study on predicting survival in cancer patient after locoregional chemotherapy," in *2010 IEEE International Conference on Bioinformatics and Biomedicine Workshops* (Hong Kong), 467–470.

Chan, L. W., Liu, Y., Chan, T., Law, H. K., Wong, S. C., Yeung, A. P., et al. (2015). PubMed-supported clinical term weighting approach for improving inter-patient similarity measure in diagnosis prediction. *BMC Med Inform Decis Mak.* 15:43. doi: 10.1186/s12911-015-0166-2

Cho, D. Y., and Przytycka, T. M. (2013). Dissecting cancer heterogeneity with a probabilistic genotype-phenotype model. *Nucleic Acids Res.* 41, 8011–8020. doi: 10.1093/nar/gkt577

Daemen, A., and De Moor, B. (2009). Development of a kernel function for clinical data. *Conf. Proc. IEEE Eng. Med. Biol. Soc.* 2009, 5913–5917. doi: 10.1109/iembs.2009.5334847

Daemen, A., Gevaert, O., Ojeda, F., Debucquoy, A., Suykens, J. A., Sempoux, C., et al. (2009). A kernel-based integration of genome-wide data for clinical decision support. *Genome Med.* 1:39. doi: 10.1186/gm39

Dean, J., and Ghemawat, S. (2008). MapReduce: simplified data processing on large clusters. *Commun, A.C.M.* 51, 107–113. doi: 10.1145/1327452.1327492

Ebadollahi, S., Sun, J., Gotz, D., Hu, J., Sow, D., and Neti, C. (2010). Predicting patient's trajectory of physiological data using temporal trends in similar patients: a system for near-term prognostics. *AMIA Annu. Symp. Proc.* 2010, 192–196.

Fei, W., and Sun, J. (2015). PSF: a unified patient similarity evaluation framework through metric learning with weak supervision. *IEEE J. Biomed. Health Inform.* 19, 1053–1060. doi: 10.1109/JBHI.2015.2425365

Feldman, K., Davis, D., and Chawla, N. V. (2015). Scaling and contextualizing personalized healthcare: a case study of disease prediction algorithm integration. *J. Biomed. Inform.* 57, 377–385. doi: 10.1016/j.jbi.2015.07.017

Gallego, B., Walter, S. R., Day, R. O., Dunn, A. G., Sivaraman, V., Shah, N., et al. (2015). Bringing cohort studies to the bedside: framework for a 'green button' to support clinical decision-making. *J. Comp. Eff. Res.* 11, 1–7. doi: 10.2217/cer.15.12

Gligorijević, V., Malod-Dognin, N., and Pržulj, N. (2016). Integrative methods for analyzing big data in precision medicine. *Proteomics* 16, 741–758. doi: 10.1002/pmic.201500396

Gottlieb, A., Stein, G. Y., Ruppin, E., Altman, R. B., and Sharan, R. (2013). A method for inferring medical diagnoses from patient similarities. *BMC Med.* 11:194. doi: 10.1186/1741-7015-11-194

Gotz, D., Stavropoulos, H., Sun, J., and Wang, F. (2012). ICDA: a platform for Intelligent Care Delivery Analytics. *AMIA Annu. Symp. Proc.* 2012, 264–273.

Gotz, D., Sun, J., Cao, N., and Ebadollahi, S. (2011). Visual cluster analysis in support of clinical decision intelligence. *AMIA Annu. Symp. Proc.* 2011, 481–490.

Hartge, F., Wetter, T., and Haefeli, W. E. (2006). A similarity measure for case based reasoning modeling with temporal abstraction based on cross-correlation. *Comput. Methods Progr. Biomed.* 81, 41–48. doi: 10.1016/j.cmpb.2005.10.005

Heer, J., and Perer, A. (2014). Orion: a system for modeling, transformation and visualization of multidimensional heterogeneous networks. *Informat. Visualizat.* 13, 111–133. doi: 10.1177/1473871612462152

Jee, K., and Kim, G. H. (2013). Potentiality of big data in the medical sector: focus on how to reshape the healthcare system. *Healthc Inform. Res.* 19, 79–85. doi: 10.4258/hir.2013.19.2.79

Karapiperis, D., and Verykios, V. (2014). A distributed near-optimal LSH-based framework for privacy-preserving record linkage. *Sci. Inf. Syst.* 11, 745–763. doi: 10.2298/CSIS140215040K

Klenk, S., Dippon, J., Fritz, P., and Heidemann, G. (2010). "Determining patient similarity in medical social networks," in *MEDEX 2010 Proceedings*, Vol. 572 (Raleigh, NC), 6–13.

Kuhn, K., Knoll, A., Mewes, H.-W., Schwaiger, M., Bode, A., Broy, M., et al. (2008). Informatics and medicine: from molecules to populations. *Methods Inf. Med.* 47, 283–295. doi: 10.3414/ME9117

Lee, J., Maslove, D. M., and Dubin, J. A. (2015). Personalized mortality prediction driven by electronic medical data and a patient similarity metric. *PLoS ONE* 10:e0127428. doi: 10.1371/journal.pone.0127428

Li, L., Cheng, W. Y., Glicksberg, B. S., Gottesman, O., Tamler, R., Chen, R., et al. (2015). Identification of type 2 diabetes subgroups through topological analysis of patient similarity. *Sci. Transl. Med.* 7:311ra174. doi: 10.1126/scitranslmed.aaa9364

Lieberman, M. A., Winzelberg, A., Golant, M., Wakahiro, M., DiMinno, M., Aminoff, M., et al. (2005). Online support groups for Parkinson's patients: a pilot study of effectiveness. *Soc. Work Health Care.* 42, 23–38. doi: 10.1300/J010v42n02_02

Liu, H., Xie, G., Mei, J., Shen, W., Sun, W., and Li, X. (2013a). An efficacy driven approach for medication recommendation in type 2 diabetes treatment using data mining techniques. *Stud. Health Technol. Inform.* 192, 1071.

Liu, L., Tang, J., Agrawal, A., Liao, W., and Choudhary, A. (2013b). *Mining Diabetes Complication and Treatment Patterns for Clinical Decision Support.* San Francisco, CA.

Longhurst, C. A., Harrington, R. A., and Shah, N. H. (2014). A 'green button' for using aggregate patient data at the point of care. *Health Aff. (Millwood).* 33, 1229–1235. doi: 10.1377/hlthaff.2014.0099

Lum, P. Y., Singh, G., Lehman, A., Ishkanov, T., Vejdemo-Johansson, M., Alagappan, M., et al. (2013). Extracting insights from the shape of complex data using topology. *Sci. Rep.* 3:1236. doi: 10.1038/srep01236

Mabotuwana, T., Lee, M. C., and Cohen-Solal, E. V. (2013). An ontology-based similarity measure for biomedical data-application to radiology reports. *J. Biomed. Inform.* 46, 857–868. doi: 10.1016/j.jbi.2013.06.013

Melton, G. B., Parsons, S., Morrison, F. P., Rothschild, A. S., Markatou, M., and Hripcsack, G. (2006). Inter-patient distance metrics using SNOMED CT defining relationships. *J. Biomed. Informat.* 39, 697–705. doi: 10.1016/j.jbi.2006.01.004

Ng, K., Sun, J., Hu, J., and Wang, F. (2015). Personalized predictive modeling and risk factor identification using patient similarity. *AMIA Jt. Summits Transl. Sci. Proc.* 2015, 132–136.

Osman, A., El-Refaey, M., and Elnaggar, A. (2013). "Towards real-time analytics in the cloud," in *2013 IEEE Ninth World Congress on Services (SERVICES)* (Santa Clara, CA: IEEE).

Ozery-Flato, M., Ein-Dor, L., Parush-Shear-Yashuv, N., Aharonov, R., Neuvirth, H., Kohn, M. S., et al. (2016). Identifying and investigating unexpected response to treatment: a diabetes case study. *Big Data* 4, 148–159. doi: 10.1089/big.2016.0017

Panahiazar, M., Taslimitehrani, V., Pereira, N. L., and Pathak, J. (2015b). Using EHRs for heart failure therapy recommendation using multidimensional patient similarity analytics. *Stud. Health Technol. Inform.* 210, 369–373.

Panahiazar, M., Taslimitehrani, V., Pereira, N., and Pathak, J. (2015a). Using EHRs and machine learning for heart failure survival analysis. *Stud Health Technol Inform.* 216, 40–44.

Perer, A. (2012). *Healthcare Analytics for Clinical and Non-Clinical Settings. CHI'12.* Austin, TX.

Qian, B., Wang, X., Cao, N., Li, H., and Jiang, Y. (2014). A relative similarity based method for interactive patient risk prediction. *Data Min. Knowl. Disc.* 29, 1070–1093.

Roque, F. S., Jensen, P. B., Schmock, H., Dalgaard, M., Andreatta, M., Hansen, T., et al. (2011). Using electronic patient records to discover disease correlations and stratify patient cohorts. *PLoS Comput. Biol.* 7:e1002141. doi: 10.1371/journal.pcbi.1002141

Srinivasan, U., and Arunasalam, B. (2013). Leveraging big data analytics to reduce healthcare costs. *IT Professional.* 15, 21–28. doi: 10.1109/MITP.2013.55

Subirats, L., Ceccaroni, L., and Miralles, F. (2012). Knowledge representation for prognosis of health status in rehabilitation. *Future Internet* 4, 762–775. doi: 10.3390/fi4030762

Sun, J., Wang, F., Hu, J., and Edabollahi, S. (2012). Supervised patient similarity measure of heterogeneous patient records. *SIGKDD Explorations.* 14, 16–24. doi: 10.1145/2408736.2408740

Sundar, I. K., Yao, H., Huang, Y., Lyda, E., Sime, P. J., Sellix, M. T., et al. (2014). Serotonin and corticosterone rhythms in mice exposed to cigarette smoke and in patients with COPD: implication for COPD-associated neuropathogenesis. *PLoS ONE* 9:e87999. doi: 10.1371/journal.pone.0087999

Syed, Z., and Guttag, J. (2011). Unsupervised similarity-based risk stratification for cardiovascular events using long-term time-series data. *J. Mach. Learn. Res.* 12, 999–1024.

Teng, C., Shapiro, L., Kalet, I., Rutter, C., and Nurani, R. (2007). "Head and neck cancer patient similarity based on anatomical structural geometry," in *2007 IEEE International Conference on Bioinformatics and Biomedicine Workshops* (San Jose, CA).

Tsymbal, A., Huber, M., Zillner, S., Hauer, T., and Zhuo, K. (2007). "Visualizing patient similarity in clinical decision support," in *LWA 2007: Lernen - Wissen - Adaption, Workshop Proceedings*, Vol. 6 (Halle: Martin-Luther-University Halle-Wittenberg), 304–311.

Tsymbal, A., Zhou, S. K., and Huber, M. (2009). Neighborhood graph and learning discriminative distance functions for clinical decision support. *Conf. Proc. IEEE Eng. Med. Biol. Soc.* 2009, 5617–5620. doi: 10.1109/iembs.2009.5333784

Wang, B., Mezlini, A. M., Demir, F., Fiume, M., Tu, Z., Brudno, M., et al. (2014). Similarity network fusion for aggregating data types on a genomic scale. *Nat. Methods* 11, 333–337. doi: 10.1038/nmeth.2810

Wang, F. (2015). Adaptive semi-supervised recursive tree partitioning: the ART towards large scale patient indexing in personalized healthcare. *J. Biomed Inform.* 55, 41–54. doi: 10.1016/j.jbi.2015.01.009

Wang, F., Hu, J., and Sun, J. (2012). "Medical prognosis based on patient similarity and expert feedback," in *21st International Conference on Pattern Recognition* (ICPR 2012) (Tsukuba), 1799–1802.

Wang, F., Sun, J., and Ebadollahi, S. (2011). "Integrating distance metrics learned from multiple experts and its application in patient similarity assessment," in *Proceedings of the 2011 SIAM International Conference on Data Mining* (Philadelphia, PA: Society for Industrial and Applied Mathematics), 59–70. doi: 10.1137/1.9781611972818.6

Wang, H., Zheng, H., Wang, J., Wang, C., and Wu, F. (2016). Integrating Omics Data With a Multiplex Network-Based Approach for the Identification of Cancer Subtypes. *IEEE Transact. Nanobiosci.* 15, 335–342. doi: 10.1109/TNB.2016.2556640

Wang, Y., Tian, Y., Tian, L. L., Qian, Y. M., and Li, J. S. (2015). An electronic medical record system with treatment recommendations based on patient similarity. *J. Med. Syst.* 39:55. doi: 10.1007/s10916-015-0237-z

Wu, H., Salzberg, B., Sharp, G., Jiang, S., Shirato, H., and Kaeli, D. (2005). "Subsequence matching on structured time series data," in *Proceedings of the ACM SIGMOD International Conference on Management of Data* (Baltimore, MD), 682–693.

Yang, F., Yu, X., and Karypis, G. (2014). "Signaling adverse drug reactions with novel feature-based similarity model," in *IEEE Conference on Bioinformatics and Biomedicine* (Belfast).

Zhang, P., Wang, F., Hu, J., and Sorrentino, R. (2014). Towards personalized medicine: leveraging patient similarity and drug similarity analytics. *AMIA Jt. Summits Transl. Sci. Proc.* 2014, 132–136.

11

Parameter Identifiability of Fundamental Pharmacodynamic Models

David L. I. Janzén[1,2,3], Linnéa Bergenholm[1,2*], Mats Jirstrand[3], Joanna Parkinson[4], James Yates[5], Neil D. Evans[1] and Michael J. Chappell[1]*

[1] *Biomedical and Biological Systems Laboratory, School of Engineering, University of Warwick, Coventry, UK,* [2] *Drug Metabolism and Pharmacokinetics, Cardiovascular and Metabolic Diseases, iMED, AstraZeneca, Gothenburg, Sweden,* [3] *Fraunhofer-Chalmers Centre, Chalmers Science Park, Gothenburg, Sweden,* [4] *Early Clinical Development, Quantitative Clinical Pharmacology, iMED, AstraZeneca, Gothenburg, Sweden,* [5] *Oncology, iMED, AstraZeneca, Cambridge, UK*

Issues of parameter identifiability of routinely used pharmacodynamics models are considered in this paper. The structural identifiability of 16 commonly applied pharmacodynamic model structures was analyzed analytically, using the input-output approach. Both fixed-effects versions (non-population, no between-subject variability) and mixed-effects versions (population, including between-subject variability) of each model structure were analyzed. All models were found to be structurally globally identifiable under conditions of fixing either one of two particular parameters. Furthermore, an example was constructed to illustrate the importance of sufficient data quality and show that structural identifiability is a prerequisite, but not a guarantee, for successful parameter estimation and practical parameter identifiability. This analysis was performed by generating artificial data of varying quality to a structurally identifiable model with known true parameter values, followed by re-estimation of the parameter values. In addition, to show the benefit of including structural identifiability as part of model development, a case study was performed applying an unidentifiable model to real experimental data. This case study shows how performing such an analysis prior to parameter estimation can improve the parameter estimation process and model performance. Finally, an unidentifiable model was fitted to simulated data using multiple initial parameter values, resulting in highly different estimated uncertainties. This example shows that although the standard errors of the parameter estimates often indicate a structural identifiability issue, reasonably "good" standard errors may sometimes mask unidentifiability issues.

Edited by:
Krasimira Tsaneva-Atanasova, University of Exeter, UK

Reviewed by:
Yasunori Aoki, National Institute of Informatics, Japan
Xin Lai, Universitätsklinikum Erlangen, Germany

***Correspondence:**
David L. I. Janzén
D.L.I.Janzen@warwick.ac.uk
Linnéa Bergenholm
E.L.Bergenholm@warwick.ac.uk

Keywords: structural identifiability, practical parameter identifiability, mixed effects models, pharmacodynamic models, fixed effects models

INTRODUCTION

Pharmacodynamic (PD) models quantify processes involved in drug action such as distribution to the effect site, receptor binding and signal transduction. PD models are valuable in making predictions of drug effects in un-tested scenarios such as outcomes across different populations or with new dosing schedules. Such predictions may not always be valid: In particular, there may be issues related to parameter identifiability. Within the concept of parameter identifiability, there are two distinct types: structural identifiability (Bellman and Åström, 1970) and practical identifiability (Raue et al., 2009).

As suggested by the name, structural identifiability concerns the inherent identifiability of the parameters in a model given its structure and observed outputs (Bellman and Åström, 1970). If a model is structurally unidentifiable, this means that at least one parameter can have any value without changing the model output (albeit with possible readjustment of remaining parameters). A well-known structurally unidentifiable problem is the linear model commonly used for estimating bioavailability F and volume of distribution V from plasma concentrations measured after oral drug administration, the most simple case being the one compartment PK model with first order absorption, where the plasma drug concentration C following a single dose is defined according to

$$C(t) = \frac{F \cdot DOSE \cdot k_a}{V\left(k_a - k_e\right)}\left(e^{-k_e \cdot t} - e^{-k_a \cdot t}\right) \quad (1)$$

where F is the bioavailability of the drug, $DOSE$ is the orally administered dose, V is the volume of distribution, k_a is the rate of absorption and k_e is the rate of elimination. It has been shown that only the fraction $\frac{F}{V}$ can be identified, and any estimate of F will therefore inversely correlate to V and both values will be biologically meaningless (Cheung et al., 2013). Importantly, predictions of $C(t)$ are still valid as these depend on the identifiable fraction $\frac{F}{V}$. While structural identifiability is a property of the postulated model structure given a set of outputs, practical identifiability is related to the experimental data. In particular, it is a measure of the amount of information contained in the experimental data and how this information is translated to parameter uncertainty and subsequent prediction uncertainty.

Parameter identifiability is unfortunately often only investigated and considered at the level of practical identifiability using more simple measurements such as standard errors or correlation matrices rather than more sophisticated approaches such as the profile likelihood approach (Raue et al., 2009). This is problematic for several reasons. The primary reason is that it cannot be guaranteed that the estimated parameter values are uniquely determined by just looking at the estimation results. In addition, if the structural identifiability of a model is unknown, it means that the source of uncertainty in the parameter estimates may be either due to the experimental data, the model structure, or both (**Figure 1**). Thus, increasing the quality of the data may or may not improve the precision of the parameter estimates. However, if structural identifiability analysis has concluded that the model is identifiable, the uncertainty in the model parameters is directly linked to the quality of the data and how well the model can describe them. In this scenario, the uncertainty of the model parameters can be improved by increasing the quality of the data. However, there will always be uncertainties in the parameter estimates even if the model is structurally identifiable and the quality and quantity of the experimental data are relatively high. An approach to further strengthen the plausibility of the model predictions under such conditions is to divide the experimental data into two parts: data used for parameter estimation and data used for model validation, i.e., by estimating the unknown parameters using a subset of the experimental data and using the resulting estimates to predict the validation data.

To further exemplify the importance of structural identifiability, consider the two following biological examples. In Evans et al. (2004), a model which aims to describe the activity of an anti-cancer agent named topotecan and its delivery to nuclear target DNA is presented. Prior to parameter estimation it was found that a subset of the model parameters was unidentifiable but if additional experimental measurements were made, in this case determining volume ratios, then the model would become structurally identifiable. In Evans et al. (2001), a parent-metabolite model for ivabradine is considered. The model was shown to be structurally unidentifiable with either intravenous, oral or combined intravenous and oral administration. It was also shown that by either fixing the volume parameter for the central compartment, or with a particular simplification of the model structure, then the model becomes identifiable for the given observations. If a formal structural identifiability analysis had not been performed then these two research projects would have most likely continued without these insights with the potential risk of missleading outcomes.

While PD models are highly diverse, many basic processes involved in drug action are similar across drugs and systems, such as distribution from the plasma to the target tissues, interaction with a target such as receptor binding or altered rates of production or loss of a target. These general processes have been described using semi-mechanistic models. For example, the effect compartment model (Sheiner et al., 1979) has been used to describe short delays in drug action due to distributional delays using a hypothetical "effect compartment." Similarly, receptor binding models (Danhof et al., 2007; Gabrielsson et al., 2011), turnover models (Gabrielsson et al., 2011) and the operational model (Black and Leff, 1983; Danhof et al., 2007) have been used to describe the processes of drug binding and signaling. However, despite frequent use, relatively few PD models have been analyzed from a structural identifiability perspective. An example of a published structural identifiability analysis is for an approximation of the receptor binding model. Receptor binding often occurs over very fast timescales relative to the PK, and sometimes also with respect to the effects elicited by the receptor once bound. In such cases, the receptor binding model may be approximated by a quasi- or pseudo-steady state approximation. When using such an approximation, it has been shown that the individual on and off rates of drug binding to the receptor cannot be uniquely identified (Chappell, 1996). Another example is the target-mediated drug disposition model (Mager and Jusko, 2001) applicable to the modeling of biologics, which has been shown to be structurally identifiable (Eudy et al., 2015). However, the identifiability of the effect compartment model and the operational model have, to our knowledge, not previously been analyzed. Furthermore, mixed effects ("population") models are often used to account for and quantify known sources of variability in data sets, such as between-subject variability (BSV). Such models are combined structural and statistical models, with additional statistical parameters describing the variance of a postulated distribution of the model parameter values across e.g., subjects. The structural identifiability of

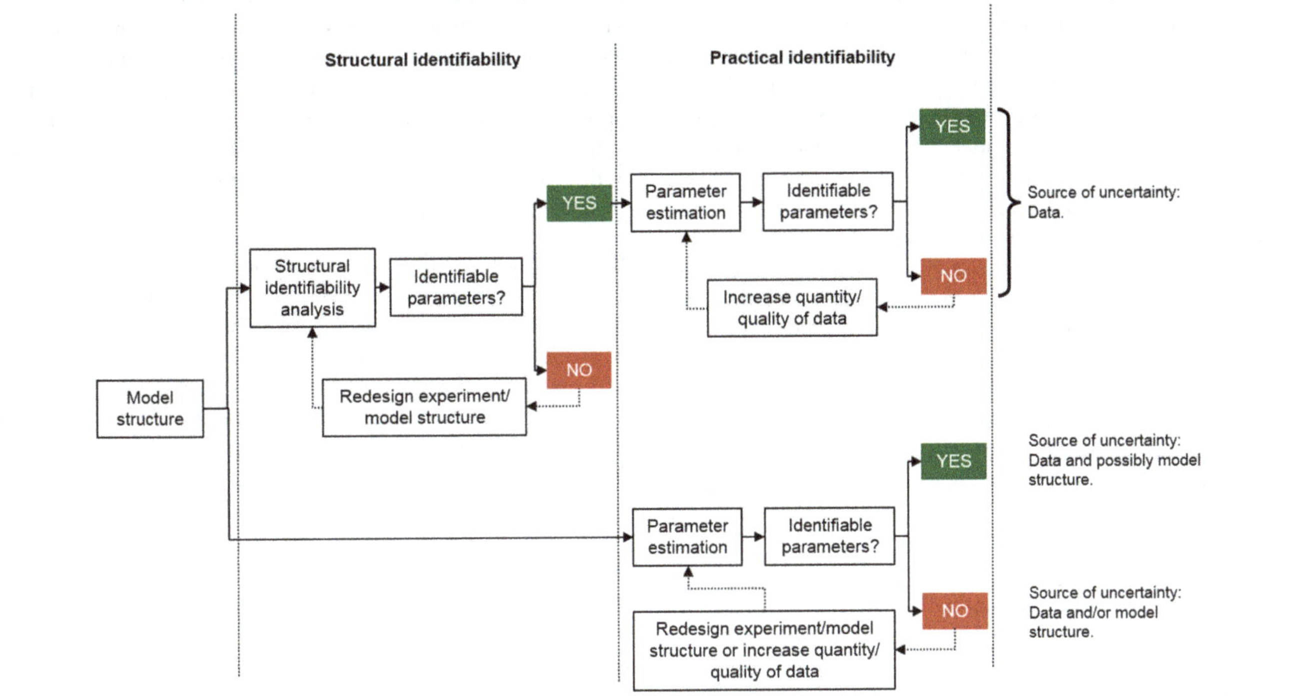

FIGURE 1 | Schematic comparing the model development process including or excluding a structural identifiability analysis. If the structural identifiability of a model is known, the standard errors in the parameter estimates reflect the uncertainty in the data and and how well the model can describe them. However, if the structural identifiability is unknown, the standard errors in the parameter estimates may reflect both issues with the model structure and the data.

mixed effects models describing BSV has not previously been analyzed.

The primary goal of this paper is to illustrate the concept and importance of parameter identifiability, both from a structural and practical perspective. Structural identifiability analysis is performed on a family of 16 commonly used PD models to serve as a database for modelers in the pharmaceutical domain. Both fixed-effects models and the corresponding mixed-effects (population) models are analyzed. Pharmacodynamic models describing combinations of none to three different mechanisms of delays in drug action are analyzed: (i) delays in drug distribution to the site of action applying the effect compartment model, (ii) delays in signal transduction, build-up or loss of effect applying turnover models and (iii) delays due to slow dissociation to the target applying receptor binding models. These and similar models are extensively used within mechanism-based PD models in pharmaceutical research (Ploeger et al., 2009; Peletier and Gabrielsson, 2012). In addition, the problem of structural identifiability and its relation to practical identifiability will be illustrated through a set of examples using both simulated data and real experimental data.

METHODS

Structural identifiability analysis has been performed on all models written in state-space form. A fixed-effects state-space model is written on the following form

$$\dot{\boldsymbol{x}}(t) = \boldsymbol{f}(\boldsymbol{x}(t), \boldsymbol{u}(t), \boldsymbol{\theta}), \quad \boldsymbol{x}(t_0) = \boldsymbol{x}_0 \tag{2}$$

$$y(t) = \boldsymbol{h}(\boldsymbol{x}(t), \boldsymbol{u}(t), \boldsymbol{\theta}) \tag{3}$$

where $\boldsymbol{x}(t) \in \mathbb{R}^n$ is the state (e.g., plasma concentration of the drug, bound and unbound receptors etc.) $\boldsymbol{u}(t) \in \mathbb{R}^q$ is the input (IV bolus, IV infusion etc.), $\boldsymbol{\theta} \in \mathbb{R}^p$ is the vector of model parameters (e.g., clearance rate, maximum saturation, etc.), $y(t) \in \mathbb{R}^m$ is the output (measurement of plasma concentration, drug effects) and $\boldsymbol{f}$ and $\boldsymbol{h}$ are smooth functions as C^∞ with respect to the functional arguments.

A mixed-effects model is written on one of the forms

$$\dot{\boldsymbol{x}}_i(t) = \boldsymbol{f}(\boldsymbol{x}_i(t), \boldsymbol{u}_i(t), \boldsymbol{\phi}_i) \quad \boldsymbol{x}_i(t_0) = \boldsymbol{x}_0(\boldsymbol{\phi}_i) \tag{4}$$

$$y_i(t) = \boldsymbol{h}(\boldsymbol{x}_i(t), \boldsymbol{u}_i(t), \boldsymbol{\phi}_i) \tag{5}$$

where $\boldsymbol{\phi}_i = g(\boldsymbol{\theta}, \eta_i, C_i)$ are the parameters for the i:th subject, $\boldsymbol{\eta}_i \sim N(\boldsymbol{0}, \boldsymbol{\Omega})$ are the random effects variables where $\boldsymbol{\Omega}$ is the variance-covariance matrix of the random effects $\boldsymbol{\eta}_i$, $\boldsymbol{\theta}$ are the population parameters and $\boldsymbol{C}_i$ are the covariates for the different subjects in the population.

Structural Identifiability: Definition

As mentioned in the introduction, structural identifiability is a theoretical concept with direct practical relevance. This is because if a model is structurally unidentifiable, some of the model parameters may take on arbitrary numerical values while the model may still describe the experimental data equally well. In a numerical structural identifiability analysis different numerical values are sought that will result in identical model responses. In an analytical structural identifiability analysis, more general conclusions can often be drawn since in such as analysis symbolic relationsships between the model parameters can be derived allowing for suggestions of reparametrization and/or additional measurements required to render an unidentifable model to become identifiable. Since different values of the unidentifiable parameters result in identical responses or predictions any subsequent biological interpretations of the estimates of those unidentifiable parameters (e.g., clearance, IC_{50}) are effectively meaningless in a biological context. It is because of this that structural identifiability is often referred to as a prerequisite to successful parameter estimation. In other words, if a structural identifiability analysis (in which perfect experimental conditions e.g., noise-free and continuous measurements, are assumed) has shown that some of the model parameters can not be determined, it follows directly that these parameters can never be determined in the less ideal case, i.e., under real experimental conditions for discrete measurements with noise present.

To define exactly what is meant by structural identifiability there now follows a more rigorous mathematical definition of the concept in the context of fixed-effects models.

Let the generic parameter vector $\boldsymbol{\theta}$ belong to a feasible parameter space $\boldsymbol{\Theta}$, i.e., $\boldsymbol{\theta} \in \Theta$. Let $y(t, \boldsymbol{\theta})$ be the output function from the state-space model. Further, consider a parameter vector $\bar{\boldsymbol{\theta}}$ where $y(t, \boldsymbol{\theta}) = y(t, \bar{\boldsymbol{\theta}})$ for all t. If this equality, in a neighborhood $\boldsymbol{N} \subset \Theta$ of $\boldsymbol{\theta}$, implies that $\boldsymbol{\theta} = \bar{\boldsymbol{\theta}}$ then the model is *structurally locally identifiable.* If $\boldsymbol{N} = \Theta$ then the model is *structurally globally identifiable.* If a model is structurally unidentifiable, then every neighborhood of $\boldsymbol{\theta}$ contains a $\bar{\boldsymbol{\theta}} \neq \boldsymbol{\theta}$ such that $y(t, \boldsymbol{\theta}) = y(t, \bar{\boldsymbol{\theta}})$ for all t.

Since the mixed-effects models to be considered in this paper are also analyzed from a structural identifiability perspective it must first be defined what is meant by the identifiability of such models. Since mixed-effects models yield individual predictions, in contrast to single predictions in the fixed-effects case, the previous definition is not immediately applicable to mixed-effects models. Instead, a generalized version of the definition of structural identifiability is used. In this new definition, first presented in Janzén et al. (2016), a model is defined to be structurally identifiable if the distribution of the output from the model determines both the structural and statistical parameters, i.e., the parameters in the vector $\boldsymbol{\theta}$ and the variance parameters in $\boldsymbol{\Omega}$ denoting the variance of the random effects η respectively. Now follows a more rigorous definition of structural identifiable for mixed-effects models.

Let $p(y_{\{\boldsymbol{\theta},\boldsymbol{\Omega}\}}, t)$ denote the distribution of the output signals y at time t. Let the generic parameter vector and matrix $\{\boldsymbol{\theta}, \boldsymbol{\Omega}\}$ belong to a feasible parameter space $\{\boldsymbol{\theta}, \boldsymbol{\Omega}\} \in \boldsymbol{\Theta}$, and consider the following two sets of parameters $\{\boldsymbol{\theta}, \boldsymbol{\Omega}\}$ and $\{\bar{\boldsymbol{\theta}}, \bar{\boldsymbol{\Omega}}\}$. If $p(y_{\{\boldsymbol{\theta},\boldsymbol{\Omega}\}}, t) = p(y_{\{\bar{\boldsymbol{\theta}},\bar{\boldsymbol{\Omega}}\}}, t)$ for all t implies that $\{\boldsymbol{\theta}, \boldsymbol{\Omega}\} = \{\bar{\boldsymbol{\theta}}, \bar{\boldsymbol{\Omega}}\}$ in a neighborhood $N \subset \Theta$ then the model is *structurally locally identifiable*, and if $N = \Theta$ the model is *structurally globally identifiable.* For a *structurally unidentifiable* parameter, θ_i, or $\omega_i \in \boldsymbol{\Omega}$, every neighborhood N around θ_i, or ω_i, has a parameter vector/matrix $\bar{\boldsymbol{\theta}}$, or $\bar{\boldsymbol{\Omega}}$, where $\theta_i \neq \bar{\theta}_i$, or $\omega_i \neq \bar{\omega}_i$, give rise to the same distribution of identical input-output relations.

Investigated Model Structures

The model structures investigated to determine structural identifiability were all combinations of sub-models representing receptor binding, a hypothetical effect compartment and direct or indirect transduction (see **Figure 2**). In total, 16 different model structures were investigated (**Table 1**). Both fixed-effects and mixed-effects versions of each model were analyzed from a structural identifiability perspective.

Structural Identifiability: Example

To exemplify the structural identifiability analysis, a summary of the analysis of the structural identifiability of Model 13 (**Table 1**) is provided. This model is a dynamic receptor binding model with an effect compartment and linear transduction. The details of the structural identifiability analysis for this model is available in the Supplementary Materials. The mathematical model has the following structure

$$\begin{aligned} \dot{C}_e &= k_{e0}(C_p - C_e) \\ \dot{RC} &= k_{on}(R_{tot} - RC)C_e - k_{off}RC \\ E &= k_e RC \end{aligned} \tag{6}$$

with the unknown parameter vector $\boldsymbol{\theta} = (k_{e0}, k_e, k_{on}, k_{off})$ and where C_p is the concentration in the blood plasma and is in this case a known input signal, C_e is a state representing the concentration in the hypothetical effect compartment, RC is the receptor complex, E is the observed effect and R_{tot} representing the percentage of total number of receptors, which is fixed at 100%.

The approach chosen to study structural identifiability here is the input-output approach, for which details can be found in Bearup et al. (2013). A general outline of the method is given here followed by an example of how a structural identifiability analysis is performed.

The input-output approach used in this paper was chosen for three reasons. The first reason was because the input-output approach can be used to show whether a model is globally or locally identifiable, or unidentifiable. Some of the other methods that are available for performing a structural identifiability analysis can only be used to show whether a model is at least locally identifiable or unidentifiable. The second reason was that there is a direct extension from non-population (fixed-effects) models to population (mixed-effects) models when it comes to structural identifiability analysis using the input-output approach as will be explained further below. The third reason is because the method is applicable to both linear and nonlinear models.

The main idea behind the input-output form approach is to transform the model to a form from which the identifiability problem can more easily be studied. This is performed by iteratively computing higher order time derivatives of the output function and using subsequent substitution to eliminate all state variables in order to express the system as a monomial solely in terms of the output functions(s) and its (their) derivatives. As the assumption of perfect experimental conditions is made, it follows that the output function and its higher order derivatives are assumed to be known. In other words, a model rewritten on an input-output form is a single equation with the output

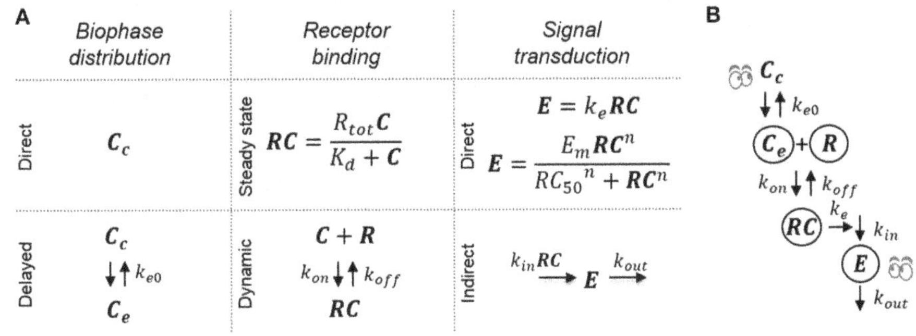

FIGURE 2 | Schematic of the investigated pharmacodynamic models. (A) The 16 investigated models are constructed by combining the following submodels: Direct or delayed biophase concentration through distribution to a hypothetical effect compartment, dynamic or direct receptor binding using the steady-state approximation and direct proportional or sigmoid signal transduction or delayed signal transduction applying a turnover model. **(B)** Example of a full model where all three processes are assumed to be dynamic and cause delay between plasma concentration and drug effect.

function and its higher order derivatives being known and the model parameters (that enters as the monomial coefficients) being unknown. Determining whether a model is structurally identifiable or otherwise is then a case of showing whether the resultant input-output equation has a single, finite, or an infinite number of solutions for the parameters in the coefficient expressions.

By iteratively differentiating the output signal and eliminating the state variables the model can be rewritten in the following input-output form

$$- R_{tot}{}^2 C_p k_e^2 k_{e0} k_{on} - 2\, R_{tot}\, C_p k_e k_{e0} k_{on} E + R_{tot}\, k_e k_{e0} k_{off} E - C_p k_{e0} k_{on} E^2 + R_{tot}\, k_e k_{e0} \dot{E} + R_{tot}\, k_e k_{off} \dot{E} + k_{e0} k_{off} E^2 + R_{tot}\, k_e \ddot{E} + k_{e0} E \dot{E} + E \ddot{E} - \dot{E}^2 = 0. \quad (7)$$

The structural identifiability of a model can then be studied by considering the coefficients in the input-output form of the model. Introducing an alternative parameter vector $\bar{\boldsymbol{\theta}}$ and collecting the coefficients in the input-output form as

$$\sum_{k=1}^{l} c_k(\boldsymbol{\theta}, \bar{\boldsymbol{\theta}}) \phi_k(E(t,\boldsymbol{\theta}), \dot{E}(t,\boldsymbol{\theta}), \ddot{E}(t,\boldsymbol{\theta}), \ldots) = 0 \quad (8)$$

permits determination of whether the model is structurally identifiable or otherwise, given that the $\phi_k(\cdot)$ are linearly independent. The analysis shows that $\boldsymbol{\theta} = \bar{\boldsymbol{\theta}}$, meaning that model 13 is structurally globally identifiable (details are given in the Supplementary Materials).

Similarly, a mixed-effects version of the model can be studied by using the coefficients in the input-output relation. As outlined and discussed in detail in Janzén et al. (Under review), since individual estimates are obtained in a mixed-effects model a distribution, assuming an infinite number of subjects (i.e., ideal experimental conditions in a mixed-effects context), of $c_k(\boldsymbol{\theta})$ is in turn obtained. This distribution is directly linked to the distribution of the output functions. By introducing the random effects on the coefficients from the input-output form, according to the statistical sub-model, functions of random variables are derived. By studying whether the distributions of the generated functions of random variables determine both the fixed effects and the random effects related parameters, conclusions regarding whether the mixed-effects model is structurally identifiable or otherwise can be made. The mixed-effects version of model 13 with lognormally distributed random effects on all model parameters with a diagonal covariance matrix is also structurally globally identifiable. This follows from the fact that the structural model has been shown to be structurally globally identifiable (detail in the Supplementary Material) and the statistical parameters are uniquely determined by the lognormal distribution.

It is worth mentioning that structural identifiability analysis using analytical techniques such as the input-output approach may encounter certain limitations in terms of model size and complexity. In general, the more complex a model is in terms of the state-space dimensions and number of unknown parameters, then the more computationally demanding the subsequent analysis may become. If an analytical approach is not possible due to symbolic computational intractability then a hybrid symbolic/numerical analysis approach is an alternative, see the profile likelihood approach (Raue et al., 2009) or the Exact Arithmetic Rank approach (Karlsson et al., 2012). For an extensive comparison between the profile likelihood approach, the Exact arithmetic Rank approach and a differential algebra approach implemented in a software called DAISY, see Raue et al. (2014).

Practical Identifiability

Once the structural identifiability of the postulated model has been determined, parameter estimation can be performed. As with the structural identifiability example, Model 13 (**Table 1**) was selected for the simulation study to investigate the influence of varying data quality on the practical identifiability of the parameters. This model includes two different sources of delay, one from distribution to the effect site, where the rate is controlled by the parameter k_{e0}, and also through slow receptor dynamics, where the off-rate is controlled by the parameter k_{off}. The possibility to distinguish the two different delays in practice

TABLE 1 | Summary of the 16 PD fixed effects and mixed effects models for which the structural identifiability was investigated.

N	Model equations	I/O	ICs	Fixed effects models	Mixed effects models	
				Fixed effect parameters	Fixed effect parameters[a]	Random effect parameters[b]
1	$E = ke\frac{R_{tot}C_p}{K_d + C}$	C_p/E		R_{tot}, ke, K_d	ke, K_d	η_{ke}, η_{Kd}
2	$E = \frac{E_m(R_{tot}C_p)^n}{(K_d + C_p)^n RC_{50}^n + (R_{tot}C_p)^n}$	C_p/E		$R_{tot}, E_m, RC_{50}, n, K_d$	E_m, RC_{50}, n, K_d	$\eta_{Em}, \eta_{RC50}, \eta_n, \eta_{Kd}$
3	$\dot{E} = k_{in}(1 + ke\frac{R_{tot}C_p}{K_d + C_p}) - k_{out}E$	C_p/E	$E(0) = k_{out}/k_{in}$	$R_{tot}, k_{in}, k_{out}, ke, K_d$	k_{in}, k_{out}, ke, K_d	$\eta_{kin}, \eta_{kout}, \eta_{ke}, \eta_{Kd}$
4	$\dot{E} = k_{in} - k_{out}(1 + ke\frac{R_{tot}C_p}{K_d + C_p})E$	C_p/E	$E(0) = k_{out}/k_{in}$	$R_{tot}, k_{in}, k_{out}, ke, K_d$	k_{in}, k_{out}, ke, K_d	$\eta_{kin}, \eta_{kout}, \eta_{ke}, \eta_{Kd}$
5	$\dot{RC} = k_{on}(R_{tot} - RC)C_p - k_{off}RC$ $E = k_eRC$	C_p/E	$RC(0) = 0$	$R_{tot}, k_{on}, k_{off}, k_e$	k_{on}, k_{off}, k_e	$\eta_{kon}, \eta_{koff}, \eta_{ke}$
6	$\dot{RC} = k_{on}(R_{tot} - RC)C_p - k_{off}RC$ $E = \frac{E_mRC^n}{RC_{50}^n + RC^n}$	C_p/E	$RC(0) = 0$	$R_{tot}, k_{on}, k_{off}, E_m, RC_{50}, n$	$k_{on}, k_{off}, E_m, RC_{50}, n$	$\eta_{RC50}, \eta_{kon}, \eta_{koff}, \eta_{Em}, \eta_n$
7	$\dot{RC} = k_{on}(R_{tot} - RC)C_p - k_{off}RC$ $\dot{E} = k_{in}(1 + k_eRC) - k_{out}E$	C_p/E	$RC(0) = 0$ $E(0) = k_{out}/k_{in}$	$R_{tot}, k_{on}, k_{off}, k_{in}, k_{out}, k_e$	$k_{on}, k_{off}, k_{in}, k_{out}, k_e$	$\eta_{kon}, \eta_{koff}, \eta_{kin}, \eta_{kout}, \eta_{ke}$
8	$\dot{RC} = k_{on}(R_{tot} - RC)C_p - k_{off}RC$ $\dot{E} = k_{in} - k_{out}(1 + k_eRC)E$	C_p/E	$RC(0) = 0$ $E(0) = k_{out}/k_{in}$	$R_{tot}, k_{on}, k_{off}, k_{in}, k_{out}, k_e$	$k_{on}, k_{off}, k_{in}, k_{out}, k_e$	$\eta_{kon}, \eta_{koff}, \eta_{kin}, \eta_{kout}, \eta_{ke}$
9	$\dot{C_e} = k_{e0} * (C_p - C_e)$ $E = ke\frac{R_{tot}C_e}{K_d + C_e}$	C_p/E	$C_e(0) = 0$	k_{e0}, R_{tot}, ke, K_d	k_{e0}, ke, K_d	$\eta_{ke0}, \eta_{ke}, \eta_{Kd}$
10	$\dot{C_e} = k_{e0} * (C_p - C_e)$ $E = \frac{E_m(R_{tot}C_e)^n}{(K_d + C_e)^n RC_{50}^n + (R_{tot}C_e)^n}$	C_p/E	$C_e(0) = 0$	$k_{e0}, R_{tot}, E_m, RC_{50}, n, K_d$	$k_{e0}, E_m, RC_{50}, n, K_d$	$\eta_{RC50}, \eta_{ke0}, \eta_{Em}, \eta_n, \eta_{Kd}$
11	$\dot{C_e} = k_{e0} * (C_p - C_e)$ $\dot{E} = k_{in}(1 + ke\frac{R_{tot}C_e}{K_d + C_e}) - k_{out}E$	C_p/E	$C_e(0) = 0$ $E(0) = k_{out}/k_{in}$	$k_{e0}, R_{tot}, k_{in}, k_{out}, ke, K_d$	$k_{e0}, k_{in}, k_{out}, ke, K_d$	$\eta_{ke0}, \eta_{kin}, \eta_{kout}, \eta_{ke}, \eta_{Kd}$
12	$\dot{C_e} = k_{e0} * (C_p - C_e)$ $\dot{E} = k_{in} - k_{out}(1 + ke\frac{R_{tot}C_e}{K_d + C})E$	C_p/E	$C_e(0) = 0$ $E(0) = k_{out}/k_{in}$	$k_{e0}, R_{tot}, k_{in}, k_{out}, ke, K_d$	$k_{e0}, k_{in}, k_{out}, ke, K_d$	$\eta_{ke0}, \eta_{kin}, \eta_{kout}, \eta_{ke}, \eta_{Kd}$
13	$\dot{C_e} = k_{e0} * (C_p - C_e)$ $\dot{RC} = k_{on}(R_{tot} - RC)C_e - k_{off}RC$ $E = k_eRC$	C_p/E	$C_e(0) = 0$ $RC(0) = 0$	$k_{e0}, R_{tot}, k_{on}, k_{off}, k_e$	$k_{e0}, k_{on}, k_{off}, k_e$	$\eta_{ke0}, \eta_{kon}, \eta_{koff}, \eta_{ke}$
14	$\dot{C_e} = k_{e0} * (C_p - C_e)$ $\dot{RC} = k_{on}(R_{tot} - RC)C_e - k_{off}RC$ $E = \frac{E_mRC^n}{RC_{50}^n + RC^n}$	C_p/E	$C_e(0) = 0$ $RC(0) = 0$	$k_{e0}, R_{tot}, RC_{50}, k_{on}, k_{off}, E_m, n$	$k_{e0}, RC_{50}, k_{on}, k_{off}, E_m, n$	$\eta_{ke0}, \eta_{RC50}, \eta_{kon}, \eta_{koff}, \eta_{Em}, \eta_n$
15	$\dot{C_e} = k_{e0} * (C_p - C_e)$ $\dot{RC} = k_{on}(R_{tot} - RC)C_e - k_{off}RC$ $\dot{E} = k_{in}(1 + k_eRC) - k_{out}E$	C_p/E	$C_e(0) = 0$ $RC(0) = 0$ $E(0) = k_{out}/k_{in}$	$k_{e0}, R_{tot}, k_{on}, k_{off}, k_{in}, k_{out}, k_e$	$k_{e0}, k_{on}, k_{off}, k_{in}, k_{out}, k_e$	$\eta_{ke0}, \eta_{kon}, \eta_{koff}, \eta_{kin}, \eta_{kout}, \eta_{ke}$
16	$\dot{C_e} = k_{e0} * (C_p - C_e)$ $\dot{RC} = k_{on}(R_{tot} - RC)C_e - k_{off}RC$ $\dot{E} = k_{in} - k_{out}(1 + k_eRC)E$	C_p/E	$C_e(0) = 0$ $RC(0) = 0$ $E(0) = k_{out}/k_{in}$	$k_{e0}, R_{tot}, k_{on}, k_{off}, k_{in}, k_{out}, k_e$	$k_{e0}, k_{on}, k_{off}, k_{in}, k_{out}, k_e$	$\eta_{ke0}, \eta_{kon}, \eta_{koff}, \eta_{kin}, \eta_{kout}, \eta_{ke}$

N, Model number; I/O, Model inputs/outputs; ICs, Initial conditions. [a] R_{tot} was fixed at 100 when analysing the mixed effects models. [b] Each mixed-effects model was assumed to have a diagonal covariance matrix $\mathbf{\Omega}$ with lognormally distributed random effects.

under varying data quality was investigated in a simulation study. R_{tot} was fixed to 1 following the results of the structural identifiability analysis to ensure the structural identifiability of the model. True parameter values were assigned to each model parameter: $k_e = 1$, $k_{e0} = 0.2$, $k_{off} = 0.02$ and $k_{on} = 0.05$ amounts per minute. All parameters were assumed to vary between subjects following a log-normal distribution as this ensures positive rates for all subjects, with standard deviation $\sigma = 0.3$ amounts per minute to represent differences in a population. The model is summarized in **Figure 3A**.

The simulation study was performed in MATLAB 2013b (The MathWorks, Inc., 2016) and Monolix 4.3.2 (Lixoft, 2012) as outlined in **Figure 3B**. (1) PK data were simulated without variability or noise, applying an intravenous bolus dose of 20 mg/kg to a hypothetical typical individual with volume of distribution 1 and rate of elimination 0.2 mg/kg. (2) Model 13 with the selected "true" parameter values was used to simulate data sets of varying size and quality. Three factors were changed that influence the information available in the data: (i) different sampling densities $\Delta t = 1, 2, 5, 10, 15, 20, 25, 30$ min. (ii) different additive noise levels $\sigma = 0.05, 0.15, 0.5$ response units and (iii) different numbers of subjects $n = 100, 40, 12$. (3) Parameters were estimated using each simulated data set, with the following initial guess selected for the optimization algorithm: $k_e = 1$, $k_{e0} = 0.1$, $k_{off} = 0.01$ and $k_{on} = 0.01$ units per minute for the structural parameters and 0.3 units per minute for the

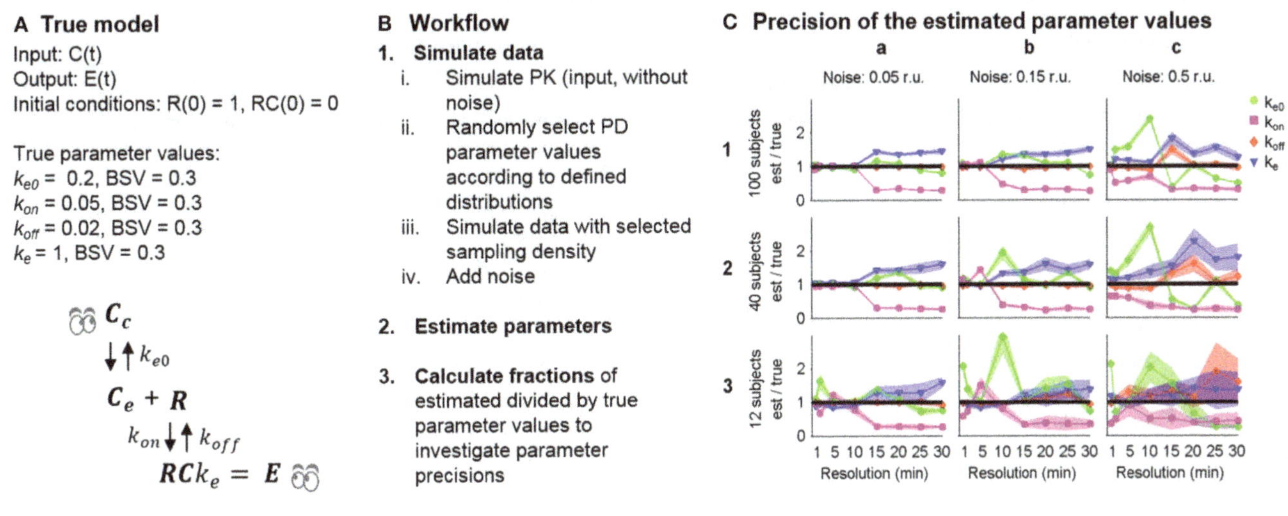

FIGURE 3 | (A) Model 13 and the selected true parameter values. BSV is between-subject variability. **(B)** Workflow for the simulation study. **(C)** Accuracy of the typical parameter estimates for the combined effect compartment/dynamic receptor model fitted to simulated data. The accuracy of each parameter (y-axis: estimated/true parameter value, line of unity marked by black line) and its uncertainty (normalized standard error, filled lighter area) is given at each data resolution level (x-axis: time between samples increasing from 1, 2, 5, 10, 15, 20, 25 to 30 min) for simulated data for 100, 40 and 12 subjects (row 1, 2, 3) adding additive noise with standard deviations 0.05, 0.15, and 0.5 response units (r.u, column a, b, c) respectively.

standard deviations. (4) The ratio between the final parameter estimates and the true parameter values were calculated and compared for the typical parameters to investigate the effects of varying sampling frequency, noise levels and number of subjects on parameter accuracy.

RESULTS

Structural Identifiability Analysis

The results of the structural identifiability analysis applying the input-output approach are summarized in **Table 2**, including the structural identifiability results and the reparameterization solutions to achieve structurally identifiable models.

All fixed effects versions of the models were in their original parameterization shown to be structurally unidentifiable. For all of the models, the source of the unidentifiability problem was the parameters R_{tot} and either RC_{50} (Models 2, 6, 10, 14) or k_e (remaining models) (see **Table 2**). The analysis showed that these parameters are unidentifiable and therefore any numerical estimates of them are effectively meaningless from a biological perspective. Furthermore, it was shown that even though R_{tot} and k_e or RC_{50} are unidentifiable, the product $R_{tot}k_e$ and fraction R_{tot}/RC_{50} are globally identifiable. The remaining parameters in the analyzed models were all shown to be globally identifiable. Therefore, three methods may be applied to ensure structurally globally identifiable models: (1) A new parameter may be defined as $R_{tot}k_e$, representing the effect when all targets are bound, and R_{tot}/RC_{50}, representing the transducer ratio, to replace the unidentifiable parameters. (2) R_{tot} or (3) k_e and RC_{50} may be fixed to known or assumed numerical values. However, this affects the units and interpretation of the non-fixed parameter. For example, R_{tot} may be fixed at 100%, resulting in changed units for k_e to units per percent bound receptor.

As discussed in Janzén et al. (Under review), if the structural model is structurally globally identifiable, and if the statistical sub-model is structurally globally identifiable, it follows that the mixed-effects model is also structurally globally identifiable. The statistical sub-model for the random effects considered in this paper takes the form of the structurally globally identifiable lognormal distribution. Therefore, the mixed-effects versions of the models in **Table 1** are structurally globally identifiable following the reparameterization or fixing of R_{tot} or k_e.

Simulation Study of Practical Identifiability

In the simulation study, increasing noise, reducing sampling frequency and reducing the number of subjects all led to worse parameter estimation results (**Figure 3**). At the lowest noise level (column a), the model parameters were well estimated up to a sampling density of $\Delta t = 10$, while increasing the sampling interval above this level led to over- and underestimation of k_e and k_{on} respectively. At the intermediate noise level (column b), similar results were obtained, although problems occurred at smaller sampling intervals. At the highest noise level (column c), the parameter estimation was unsuccessful for all estimation runs except for 100 subjects and 1 min sampling interval. The simulation study shows a trend of decreasing accuracy to estimate the true parameters when the amount and quality of the data decreases. Some of the model parameters vary more than others when the data become worse in terms of noise levels, the number of measurements and the number of subjects. For instance, k_{off} was estimated reasonably well, except for the very worst case 3c, while the estimates for k_e and k_{on} are poor in 1a. It can also be seen that the uncertainty in the parameter estimates (standard errors) generally widens with either increased noise, reduced sampling density or reduced number of number subjects. Interestingly, high precision (small

TABLE 2 | Results of the structural identifiability analysis of the mixed-effects models 1–16 in Table 1.

Model description		Structural identifiability results		
		Fixed effects		Random effects[a]
N	Distr. | Binding | Transd.	SU parameters	SI parameters & combinations	SI parameters
1	Direct | SS | Linear	R_{tot}, k_e	$R_{tot}k_e, K_d$	η_{ke}, η_{Kd}
2	Direct | SS | Sigmoid	R_{tot}, RC_{50}	$R_{tot}/RC_{50}, k_e, K_d, n$	$\eta_{RC50}, \eta_{ke}, \eta_{Kd}, \eta_n$
3	Direct | SS | Indirect	R_{tot}, k_e	$R_{tot}k_e, k_{in}, k_{out}, K_d$	$\eta_{kin}, \eta_{kout}, \eta_{ke}, \eta_{Kd}$
4	Direct | SS | Indirect	R_{tot}, k_e	$R_{tot}k_e, k_{in}, k_{out}, K_d$	$\eta_{kin}, \eta_{kout}, \eta_{ke}, \eta_{Kd}$
5	Direct | Dynamic | Linear	R_{tot}, k_e	$R_{tot}k_e, k_{on}, k_{off}$	$\eta_{Rtot/RC50}, \eta_{kon}, \eta_{koff}, \eta_{Em}$
6	Direct | Dynamic | Sigmoid	R_{tot}, RC_{50}	$R_{tot}/RC_{50}, k_{on}, k_{off}, E_m, n$	$\eta_{RC50}, \eta_{kon}, \eta_{koff}, \eta_{Em}, \eta_n$
7	Direct | Dynamic | Indirect	R_{tot}, k_e	$R_{tot}k_e, k_{on}, k_{off}, k_{in}, k_{out}$	$\eta_{kon}, \eta_{koff}, \eta_{kin}, \eta_{kout}, \eta_{ke}$
8	Direct | Dynamic | Indirect	R_{tot}, k_e	$R_{tot}k_e, k_{on}, k_{off}, k_{in}, k_{out}$	$\eta_{kon}, \eta_{koff}, \eta_{kin}, \eta_{kout}, \eta_{ke}$
9	Delay | SS | Linear	R_{tot}, k_e	$R_{tot}k_e, k_{e0}, K_d$	$\eta_{ke0}, \eta_{ke}, \eta_{Kd}$
10	Delay | SS | Sigmoid	R_{tot}, RC_{50}	$R_{tot}/RC_{50}, k_{e0}, k_{on}, k_{off}, E_m, n$	$\eta_{ke0}, \eta_{RC50}, \eta_{Em}, \eta_n, \eta_{Kd}$
11	Delay | SS | Indirect	R_{tot}, k_e	$R_{tot}k_e, k_{e0}, k_{in}, k_{out}, K_d$	$\eta_{ke0}, \eta_{kin}, \eta_{kout}, \eta_{ke}, \eta_{Kd}$
12	Delay | SS | Indirect	R_{tot}, k_e	$R_{tot}k_e, k_{e0}, k_{in}, k_{out}, K_d$	$\eta_{ke0}, \eta_{kin}, \eta_{kout}, \eta_{ke}, \eta_{Kd}$
13	Delay | Dynamic | Linear	R_{tot}, k_e	$R_{tot}k_e, k_{e0}, k_{on}, k_{off}$	$\eta_{ke0}, \eta_{kon}, \eta_{koff}, \eta_{ke}$
14	Delay | Dynamic | Sigmoid	R_{tot}, RC_{50}	$R_{tot}/RC_{50}, k_{e0}, k_{on}, k_{off}, k_{in}, E_m, n$	$\eta_{ke0}, \eta_{RC50}, \eta_{kon}, \eta_{koff}, \eta_{Em}, \eta_n$
15	Delay | Dynamic | Indirect	R_{tot}, k_e	$R_{tot}k_e, k_{e0}, k_{on}, k_{off}, k_{in}, k_{out}$	$\eta_{ke0}, \eta_{kon}, \eta_{koff}, \eta_{kin}, \eta_{kout}, \eta_{ke}$
16	Delay | Dynamic | Indirect	R_{tot}, k_e	$R_{tot}k_e, k_{e0}, k_{on}, k_{off}, k_{in}, k_{out}$	$\eta_{ke0}, \eta_{kon}, \eta_{koff}, \eta_{kin}, \eta_{kout}, \eta_{ke}$

SU, Structurally unidentifiable; SI, Structurally identifiable. [a] R_{tot} was fixed at 100 when analysing the mixed effects models.
Structurally identifiable and unidentifiable parameters and a suggested reparameterization are provided for the corresponding fixed effects models. Random effects were evaluated for the reparameterized models.

standard errors) is in many optimizations acquired despite low accuracy in the parameter estimates.

Case Study: Analysing Cardiac (Side) Effects

A case study was conducted in order to exemplify the process of model development, including structural identifiability analysis. Side effects of potential new drugs on the heart must be evaluated by monitoring changes in the duration of specific intervals monitored in the electrocardiogram (ECG), such as the QT interval (defined by the Q and T peaks in the ECG) which corresponds to the duration of the ventricular action potential. The main part of the QT interval constitutes the ventricular repolarization phase, corresponding to the JT interval (defined by the J point and T peak in the ECG), and prolongations are strongly linked to inhibition of the cardiac ion channel hERG (Pollard et al., 2010). In this example, model 10 (**Table 1**) was applied to link inhibition of the hERG ion channel *in vitro* to prolongation of the JT interval following treatment with the anti-arrhythmic compound and mixed ion channel blocker AZD1305, a proprietary AstraZeneca compound. Model 10 was selected since an identifiable version of this model has been used previously to fit this type of data (Jonker et al., 2005) and following evaluation of additional structures, for example model 2 (without the effect compartment).

Methods

Clinical study and PK and QT interval data are described in Parkinson et al. (2013). This phase I study was performed in accordance with the ethical principles of the Declaration of Helsinki and is consistent with the International Conference on Harmonisation (ICH)/Good Clinical Practice. JT intervals were calculated by subtracting QRS from QT. *In vitro* data were acquired from the original data collected by Carlsson et al. (2009). Methods for PKPD model development are detailed in Bergenholm et al. (2016). Baseline variability of JT intervals was minimized applying a circadian rhythm and RR correction models (Chain et al., 2011; Bergenholm et al., 2016). The PK and PD were modeled sequentially, and Model 10 (**Table 1**) was selected to describe the drug effect. K_d was estimated prior to the PKPD modeling using the I_{max} model, where the inhibition in % is calculated according to

$$I(C) = 100 * C/(IC_{50} + C) \tag{9}$$

where IC_{50} corresponds to the drug concentration resulting in 50% inhibition, substituting K_d in Model 10. Parameter estimations were performed using the stochastic approximation expectation maximization (SAEM) algorithm as implemented in Monolix 4.3.2 (Lixoft, 2012).

Results

The estimated IC_{50} of hERG was 0.37 ± 0.04 μM with between cell variability of 0.19 ± 0.09 μM. Fitting all parameters of the operational model led to high uncertainty and correlation between R_{tot} and RC_{50} (**Table 3**). Structural identifiability analysis of this model showed that only the fraction R_{tot}/RC_{50} is identifiable (see **Table 2**) and the model was therefore reparameterized with $\tau = R_{tot}/RC_{50}$, resulting in a structurally identifiable model. Estimation of the reduced model resulted in

TABLE 3 | Estimated parameter values for the original and re-parameterized Model 10 fitted to AZD1305 PK-hERG-JT interval data.

Parameter	Unit	Unidentifiable model		Identifiable model	
		Estimate (SE)	BSV % (SE)	Estimate (SE)	BSV % (SE)
E_m	ms	172 (23.9)	18.7 (9.09)	162 (18.9)	20.6 (7.67)
RC_{50}	μM	0.753 (173)	13.3 (15300)	–	–
n		2.02 (0.24)	35.1 (7.5)	2.1 (0.219)	36.4 (7.69)
R_{tot}	μM	1.1 (252)	13.2 (15400)	–	–
τ		–	–	1.55 (0.163)	15.2 (8.17)
IC_{50}	μM	0.37 (fixed)	0.19 (fixed)	0.37 (fixed)	0.19 (fixed)
k_{e0}	h^{-1}	9.37 (2.96)	125 (24)	9.42 (2.91)	123 (23.4)
Residuals	ms	6.64 (0.155)	–	6.64 (0.155)	–
−2LL		7662		7670	

SE, Standard error; BSV, Between-subject variability; −2LL, −2 LogLikelihood.

similar parameter values for all of the identifiable parameters, similar goodness of fit values and residuals and good precision in the population estimate of τ (**Table 3**). The fits to the generated data can be seen in **Figure 4**.

Discussion

Both the full and reparametererized versions of model 10 described the data well. However, standard errors and correlations of R_{tot} and RC_{50} correctly indicated identifiability issues with the former. The estimated parameters were converted to the traditional E_{max} and EC_{50} parameters, which describe the maximal effect and the drug concentration at half-maximum effect respectively. E_{max} and EC_{50} were calculated according to

$$E_{max} = \frac{E_m \tau^n}{1+\tau^n} \tag{10}$$

$$EC_{50} = \frac{IC_{50}}{(2+\tau^n)^{1/n}-1} \tag{11}$$

and resulted in an estimated E_{max} of 117 ms and 116 ms and EC_{50} of 0.36 and 0.35 μM respectively for the full and reparameterized models. This highlights that identifiable parts of a structurally unidentifiable model are still informative. The estimated E_{max} is similar to that in previous hERG-QT modeling of dofetilide (Jonker et al., 2005), while the estimated hERG block at 10 ms JT prolongation was slightly higher (18 vs. 9%). This may be caused by AZD1305-induced calcium block (Carlsson et al., 2009), as the calcium current depolarizes the cardiac cells (Amin et al., 2010), counter-acting the repolarization by hERG. The structural identifiability analysis showed that two model parameters could not be estimated. This led to model reduction. Performing this analysis prior to parameter estimation ensures the theoretical possibility of estimating all parameters in the model. Estimating the parameters of the unidentifiable model could have been avoided, reducing the number of iterations in the optimization. Also, ensuring structural identifiability improves confidence in the biological interpretation of the estimated parameter values.

DISCUSSION

Unidentifiability issues can cause many different types of problems if not mitigated when models are used to quantify, predict and understand the effects of potential drugs. Most importantly, the biological/physiological interpretations of structurally or practically unidentifiable parameters are not valid. This may lead to wrong conclusions, for example when unknowingly comparing unidentifiable parameters to rate candidate drugs or for comparison with competitors. Also, any predictions based on the profiles of unmeasured states of the system may be meaningless if the parameters directly or indirectly related to those states are unidentifiable. For example, if the effect of interest in a toxicity or efficacy study depends on the concentration in a compartment for which the profile is linked to structurally unidentifiable parameters, it may be impossible to separate the distribution to this compartment and the drug effect. Unidentifiability issues may also cause technical problems, as the parameter estimation step may take a very long time, or fail (crash), if a structurally unidentifiable model is used (depending on what form of optimization routine is used).

We have investigated the structural identifiability of 16 fundamental pharmacodynamic models and identified parameterizations that are structurally identifiable both for fixed effects- and mixed-effects- versions of the models, as summarized in **Table 2**. For all of the investigated models, the total amount of receptor in the system was fixed (to e.g., 1 or 100%) in order to achieve structural identifiability. This implies that some parameters for the "signal transduction" are relative. For example, the units of a proportional signal transduction are effect units per fraction bound/inhibited receptor if R_{tot} is fixed to 1. This analysis shows that given sufficient data quality, it is, in theory, possible to distinguish between different sources of delay from the data. Thus, it is possible to differentiate delays that are compound-specific (e.g., distribution, drug-receptor binding kinetics) from delays that are system-specific (e.g., turnover of receptors) to compare compounds and simulate untested systems. The investigated models have been used successfully and repeatedly in practice (Ploeger et al., 2009; Peletier and Gabrielsson, 2012), and our results confirm the general assumption of structural identifiability. This provides confidence in the theoretical soundness of using these models.

Next, we estimated parameters of the unidentifiable and reparameterized versions of Model 5 (**Tables 1, 2**) in to investigate the possible consequences of estimating unidentifiable models (**Figure 5**). Three separate runs of parameter estimation were performed. Parameters in the unidentifiable version of the model were estimated in two different runs using different initial estimates. For the third parameter estimation run, the model was reparameterized following insights from the structural identifiability analysis.

Investigating the estimated parameters shows that standard errors of unidentifiable parameters differ significantly between the two estimation runs, and are larger than the standard error of the product of the parameters. For one of the estimation runs the magnitude of the standard errors (37.6 and 33.7%) did not clearly indicate a structural identifiability problem. In

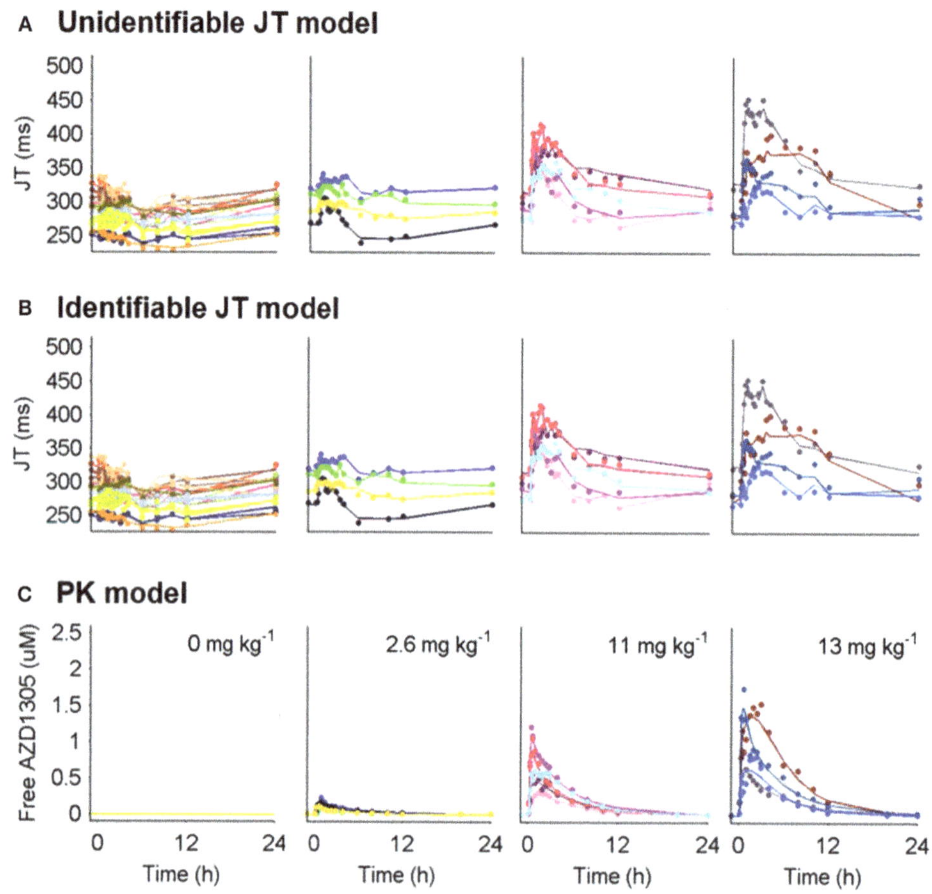

FIGURE 4 | PK and JT interval data (markers) and model predictions (lines) for humans treated with placebo and 3 selected doses of AZD1305. (A) Model predictions by the unidentifiable JT model. **(B)** Model predictions by the identifiable JT model. **(C)** Individual PK model parameters predicting the PK in each subject were used to drive the PD response. Individual subjects are separated by color.

A Structural identifiability analysis

Input: C
Output: E
Initial conditions: $R(0) = R_{tot}$, $RC(0) = 0$
Parameters: k_{on}, k_{off}, k_e

$$C + R \underset{k_{off}}{\overset{k_{on}}{\rightleftharpoons}} RC, \quad RCk_e = E$$

Identifiable parameters: k_{on}, k_{off}
Unidentifiable parameters: k_e, R_{tot}
Identifiable combinations: $k_e{*}R_{tot}$

B Estimation of structurally unidentifiable (SU) models

1. SU model

	Start	Estimate	RSE	Residuals
R_{tot}	10	11.9	37.6	7.66
k_e	10	16.4	33.7	
k_{on}	1	3.71	22.3	
k_{off}	1	4.85	22.5	

2. SU model (different start guess)

	Start	Estimate	RSE	Residuals
R_{tot}	100	87	163	7.66
k_e	1	2.17	154	
k_{on}	1	3.45	22.9	
k_{off}	1	4.76	22.8	

3. Reparameterised model (R_{tot} = 100)

	Start	Estimate	RSE	Residuals
k_e	1	2.03	15.8	7.67
k_{on}	1	3.61	22.6	
k_{off}	1	4.86	22.9	

FIGURE 5 | (A) Results of the structural identifiability analysis of Model 5. **(B)** Optimization results following estimation of unidentifiable and identifiable versions of Model 5 using example data.

the second estimation run the standard errors (163 and 154%) did indicate a structural identifiability problem. However, for both estimation runs the estimated correlations between R_{tot} and k_e were −0.9 and −0.99 respectively, indicating a potential structural identifiability problem in both cases. Alternatively, analysing the models using the profile likelihood approach (Raue et al., 2009) would also potentially indicate a problem with structural and practical identifiability. Although estimation of an unidentifiable model in theory should lead to infinitely large uncertainty for the structurally unidentifiable parameters due to a flat likelihood function in the directions representing those parameters, this did not happen in practice. The reason why this did not happen can be explained by measurement and numerical noise. In real-world problems, the likelihood function is never completely flat which introduces false local minima where the optimization routine may become "stuck" depending on the initial guesses used for the model parameters and the optimization algorithm itself. This example shows the potential danger of using practical identifiability analysis as a tool to deduce structural identifiability. For the first set of initial guesses for the parameters, the reported RSE-values are unreasonably high indicating a structural identifiability issue. However, the RSE-values reported using a different set of initial guesses for the model parameters do not indicate that there is any structural identifiability problem. The results of these estimations were used to draw some general conclusions. These are as follows:

- Different initial guesses of the model parameters may lead to different estimates of structurally unidentifiable parameters.
- Large standard errors may indicate that a parameter is structurally (or practically) unidentifiable but **unidentifiable parameters may also appear well-determined**.
- Reparameterizing the structurally unidentifiable model to become identifiable leads to similar residuals (and likelihood) and improved parameter precision of the new parameter(s).
- Identifiable parameters can still be well-determined when other parameters are unidentifiable.

Similar findings regarding masking of structural unidentifiability, i.e., estimation of seemingly reasonable RSE-values of structurally unidentifiable parameters, has been reported in the conference contribution (Aoki et al., 2015) and in the follow-up paper (Aoki et al., 2014). These findings were reported using NONMEM, rather than Monolix, which indicates that estimation of misleading RSE-values under structural unidentifiability conditions is not a software specific issue but instead a general numerical computational instability issue. In these two publications, a numerical approach called preconditioning is suggested. In short, this approach involves reparametrization of the model in such a way so that the subsequent numerical computations of the RSE-values reportedly becomes more stable and thus more reliable under structurally unidentifiable conditions.

It is important to remember that having a structurally identifiable model is only a prerequisite for successful parameter estimation. In other words, that parameters are identifiable with ideal data (continuous, noise-free data from an infinite number of subjects in the mixed effects model case) does not guarantee that they will be practically identifiable with a finite number of noisy data points from a finite number of subjects.

The effects of practical identifiability were investigated in a simulation study, where the quality of the data was varied from good to worse, but the structural model was known to be identifiable (Model 13). Conclusions from this example are that:

- A structurally identifiable model does not guarantee reliable parameter estimates.
- Data must contain information over relevant time scales for the investigated system.
- Noise levels, sampling density and the number of subjects (mixed-effects models) are all important in order to be able to estimate parameters with reasonably high precision.

When the data do not contain information on the time scale of the rate parameters in the system, the model should be reduced to only account for effects over the relevant time scales. This applies even when all parameters are structurally identifiable.

CONCLUSIONS

Parameter identifiability should be investigated to ensure both structural and practical identifiability. Our work confirms the structural identifiability of a set of fundamental pharmacodynamic models, and provides examples of estimation results with unidentifiable models. The investigated models have been proven to have a sound theoretical basis in terms of structural identifiability and thus are reliable in this respect. This in turn increases the reliability of using such models in clinical pharmacology and therapeutics.

AUTHOR CONTRIBUTIONS

All authors participated in the design of the research study. DJ performed the structural identifiability analyses. LB and DJ performed the practical identifiability analyses. LB conducted the analyses for the case study. All authors participated in analysing the results. All authors participated in writing the manuscript.

ACKNOWLEDGMENTS

The case study was conducted specifically for this publication, using previously published clinical data (Parkinson et al., 2013). This work is funded through the Marie Curie FP7 People ITN European Industrial Doctorate (EID) project No.316736, IMPACT (Innovative Modeling for Pharmacological Advances through Collaborative Training).

REFERENCES

Amin, A. S., Tan, H. L., and Wilde, A. A. (2010). Cardiac ion channels in health and disease. *Heart Rhythm* 7, 117–126. doi: 10.1016/j.hrthm.2009.08.005

Aoki, Y., Nordgren, R., and Hooker, A. C. (2014). Preconditioning of nonlinear mixed effects models for stabilisation of variance-covariance matrix computations. *AAPS J.* 18, 505–518. doi: 10.1208/s12248-016-9866-5

Aoki, Y., Nordgren, R., and Hooker, A. C. (2015). *Preconditioning of Nonlinear Mixed Effect Models for Stabilization of the Covariance Matrix Computation.* Available online at: http://www.page-meeting.org/?abstract=3586

Bearup, D. J., Evans, N. D., and Chappell, M. J. (2013). The input-output relationship approach to structural identifiability analysis. *Comput. Methods Prog. Biomed.* 109, 171–181. doi: 10.1016/j.cmpb.2012.10.012

Bellman, R., and Åström, K. J. (1970). On structural identifiability. *Math. Biosci.* 7, 329–339. doi: 10.1016/0025-5564(70)90132-X

Bergenholm, L., Collins, T., Evans, N. D., Chappell, M. J., and Parkinson, J. (2016). PKPD modelling of PR and QRS intervals in conscious dogs using standard safety pharmacology data. *J. Pharm. Toxicol. Methods* 79, 34–44. doi: 10.1016/j.vascn.2016.01.002

Black, J. W., and Leff, P. (1983). Operational models of pharmacological agonism. *Proc. R. Soc. B Biol. Sci.* 220, 141–162. doi: 10.1098/rspb.1983.0093

Carlsson, L., Andersson, B., Linhardt, G., and Löfberg, L. (2009). Assessment of the ion channel-blocking profile of the novel combined ion channel blocker AZD1305 and its proarrhythmic potential versus dofetilide in the methoxamine-sensitized rabbit *in vivo*. *J. Cardiovasc. Pharmacol.* 54, 82–89. doi: 10.1097/FJC.0b013e3181ac62c9

Chain, A. S., Krudys, K. M., Danhof, M., and Della Pasqua, O. (2011). Assessing the probability of drug-induced QTc-interval prolongation during clinical drug development. *Clin. Pharmacol. Ther.* 90, 867–875. doi: 10.1038/clpt.2011.202

Chappell, M. J. (1996). Structural identifiability of models characterizing saturable binding: comparison of pseudo-steady-state and non-pseudo-steady-state model formulations. *Math. Biosci.* 133, 1–20. doi: 10.1016/0025-5564(95)00064-X

Cheung, S. Y., Yates, J. W., and Aarons, L. (2013). The design and analysis of parallel experiments to produce structurally identifiable models. *J. Pharmacokinet. Pharmacodyn.* 40, 93–100. doi: 10.1007/s10928-012-9291-z

Danhof, M., de Jongh, J., De Lange, E. C., Della Pasqua, O., Ploeger, B. A., and Voskuyl, R. A. (2007). Mechanism-based pharmacokinetic-pharmacodynamic modeling: biophase distribution, receptor theory, and dynamical systems analysis. *Ann. Rev. Pharmacol. Toxicol.* 47, 357–400. doi: 10.1146/annurev.pharmtox.47.120505.105154

Eudy, R. J., Riggs, M. M., and Gastonguay, M. R. (2015). A priori identifiability of target-mediated drug disposition models and approximations. *AAPS J.* 17, 1280–1284. doi: 10.1208/s12248-015-9795-8

Evans, N., Moyse, H., Lowe, D., Briggs, D., Higgins, R., Mitchell, D., et al. (2013). Structural identifiability of surface binding reactions involving heterogeneous analyte : application to surface plasmon resonance experiments. *Automatica* 49, 48–57. doi: 10.1016/j.automatica.2012.09.015

Evans, N. D., Errington, R. J., Shelley, M., Feeney, G. P., Chapman, M. J., Godfrey, K. R., et al. (2004). A mathematical model for the *in vitro* kinetics of the anti-cancer agent topotecan. *Math. Biosci.* 189, 185–217. doi: 10.1016/j.mbs.2004.01.007

Evans, N. D., Godfrey, K. R., Chapman, M. J., Chappell, M. J., Aarons, L., and Dufull, S. D. (2001). An identifiability analysis of a parent-metabolite pharmacokinetic model for ivabradine. *J. Pharmacokinet. Pharmacodyn.* 28, 93–105. doi: 10.1023/A:1011521819898

Gabrielsson, J., Fjellström, O., Ulander, J., Rowley, M., and Van Der Graaf, P. H. (2011). Pharmacodynamic-pharmacokinetic integration as a guide to medicinal chemistry. *Curr. Top. Med. Chem.* 11, 404–418. doi: 10.2174/156802611794480864

Janzén, D. L. I., Jirstrand, M., Chappell, M. J., and Evans, N. D. (2016). Three novel approaches to structural identifiability analysis in mixed-effects models. *Comput. Methods Prog. Biomed.* doi: 10.1016/j.cmpb.2016.04.024. [Epub ahead of print].

Jonker, D. M., Kenna, L. A., Leishman, D., Wallis, R., Milligan, P. A., and Jonsson, E. N. (2005). A pharmacokinetic-pharmacodynamic model for the quantitative prediction of dofetilide clinical QT prolongation from human ether-a-go-go-related gene current inhibition data. *Clin. Pharmacol. Therapeut.* 77, 572–582. doi: 10.1016/j.clpt.2005.02.004

Karlsson, J., Anguelova, M., and Jirstrand, M. (2012). "An efficient method for structural identifiability analysis of large dynamic systems," in *16th IFAC Symposium on system identification* (Brussels), 941–946.

Lixoft (2012). *Monolix 4.3.2.*

Mager, D. E., and Jusko, W. J. (2001). General pharmacokinetic model for drugs exhibiting target-mediated drug disposition. *J. Pharmacokinet. Pharmacodyn.* 28, 507–532. doi: 10.1023/A:1014414520282

Parkinson, J., Visser, S. A., Jarvis, P., Pollard, C., Valentin, J. P., Yates, J. W., et al. (2013). Translational pharmacokinetic-pharmacodynamic modeling of QTc effects in dog and human. *J. Pharmacol. Toxicol. Methods* 68, 357–366. doi: 10.1016/j.vascn.2013.03.007

Peletier, L. A., and Gabrielsson, J. (2012). Dynamics of target-mediated drug disposition: characteristic profiles and parameter identification. *J. Pharmacokinet. Pharmacodyn.* 39, 429–451. doi: 10.1007/s10928-012-9260-6

Ploeger, B. A., van der Graaf, P. H., and Danhof., M. (2009). Incorporating receptor theory in mechanism-based pharmacokinetic-pharmacodynamic (PK-PD) modeling. *Drug Metab. Pharmacokinet.* 24, 3–15. doi: 10.2133/dmpk.24.3

Pollard, C. E., Abi Gerges, N., Bridgland-Taylor, M. H., Easter, A., Hammond, T. G., and Valentin, J. P. (2010). An introduction to QT interval prolongation and non-clinical approaches to assessing and reducing risk. *Br. J. Pharmacol.* 159, 12–21. doi: 10.1111/j.1476-5381.2009.00207.x

Raue, A., Karlsson, J., Saccomani, M. P., Jirstrand, M., and Timmes, J. (2014). Comparison of approaches for parameter identifiability analysis of biological systems. *Bioinformatics* 30, 1440–1448. doi: 10.1093/bioinformatics/btu006

Raue, A., Kreutz, C., Maiwald, T., Bachmann, J., Schilling, M., Klingmuller, U., et al. (2009). Structural and practical identifiability analysis of partially observed dynamical models by exploiting the profile likelihood. *Bioinformatics* 25, 1923–1929. doi: 10.1093/bioinformatics/btp358

Sheiner, L. B., Stanski, D. R., Vozeh, S., Miller, R. D., and Ham, J. (1979). Simultaneous modeling of pharmacokinetics and pharmacodynamics: application to d-tubocurarine. *Clin. Pharmacol. Therapeut.* 25, 358–371. doi: 10.1002/cpt1979253358

The MathWorks, Inc. (2016). *The MathWorks, Inc.* Natick, MA: Matlab 2013b.

12

Personalized Medication Response Prediction for Attention-Deficit Hyperactivity Disorder: Learning in the Model Space vs. Learning in the Data Space

Hin K. Wong[1], *Paul A. Tiffin*[2*], *Michael J. Chappell*[3], *Thomas E. Nichols*[1], *Patrick R. Welsh*[4], *Orla M. Doyle*[5], *Boryana C. Lopez-Kolkovska*[1], *Sarah K. Inglis*[6], *David Coghill*[7], *Yuan Shen*[8] *and Peter Tiño*[8]

[1] Warwick Manufacturing Group, Institute of Digital Healthcare, University of Warwick, Coventry, UK, [2] Mental Health and Addiction Research Group, Department of Health Sciences, University of York, York, UK, [3] School of Engineering, University of Warwick, Coventry, UK, [4] School of Psychology, Newcastle University, Newcastle upon Tyne, UK, [5] Centre for Neuroimaging Sciences, King's College London, London, UK, [6] Division of Maternal and Child Health Sciences, Ninewells Hospital and Medical School, University of Dundee, Dundee, UK, [7] Departments of Paediatrics and Psychiatry, University of Melbourne, Melbourne, VIC, Australia, [8] School of Computer Science, University of Birmingham, Birmingham, UK

Edited by:
Zbigniew R. Struzik,
University of Tokyo, Japan

Reviewed by:
Joaquim Radua,
Fidmag Sisters Hospitallers, Spain
Qibin Zhao,
RIKEN Brain Science Institute, Japan

***Correspondence:**
Paul A. Tiffin
paul.tiffin@york.ac.uk

Attention-Deficit Hyperactive Disorder (ADHD) is one of the most common mental health disorders amongst school-aged children with an estimated prevalence of 5% in the global population (American Psychiatric Association, 2013). Stimulants, particularly methylphenidate (MPH), are the first-line option in the treatment of ADHD (Reeves and Schweitzer, 2004; Dopheide and Pliszka, 2009) and are prescribed to an increasing number of children and adolescents in the US and the UK every year (Safer et al., 1996; McCarthy et al., 2009), though recent studies suggest that this is tailing off, e.g., Holden et al. (2013). Around 70% of children demonstrate a clinically significant treatment response to stimulant medication (Spencer et al., 1996; Schachter et al., 2001; Swanson et al., 2001; Barbaresi et al., 2006). However, it is unclear which patient characteristics may moderate treatment effectiveness. As such, most existing research has focused on investigating univariate or multivariate correlations between a set of patient characteristics and the treatment outcome, with respect to dosage of one or several types of medication. The results of such studies are often contradictory and inconclusive due to a combination of small sample sizes, low-quality data, or a lack of available information on covariates. In this paper, feature extraction techniques such as latent trait analysis were applied to reduce the dimension of on a large dataset of patient characteristics, including the responses to symptom-based questionnaires, developmental health factors, demographic variables such as age and gender, and socioeconomic factors such as parental income. We introduce a Bayesian modeling approach in a "learning in the model space" framework that combines existing knowledge in the literature on factors that may potentially affect treatment response, with constraints imposed by a treatment response model. The model is personalized such that the variability among subjects is accounted for by a set of subject-specific parameters. For remission classification, this approach compares favorably with conventional methods such as support vector machines and mixed effect models on a range of performance measures. For instance, the proposed approach achieved an area under receiver operator characteristic curve of 82–84%, compared to 75–77% obtained from conventional regression or machine learning ("learning in the data space") methods.

Keywords: attention-deficit hyperactivity disorder, Bayesian inference, machine learning, methylphenidate, mixed effects model, personalized medicine, prognosis, treatment response

1. INTRODUCTION

The ability to predict treatment response (or non-response) in patients with mental health issues is potentially beneficial to both clinicians and patients in a number of ways. First, any treatment is accompanied by the risk of adverse effects—where non-response is a probable outcome then the risks of treatment may outweigh the benefits. Second, prediction of treatment response may guide both the dose and choice of medication. For example, where adverse events are dose-dependent then a clinician may chose to abandon a treatment course if a patient was a probable non-responder. Third, response prediction helps to calibrate both clinician and patient expectations of treatment outcomes. Finally, identifying non-responders may prompt a re-appraisal of the diagnosis and formulation of a patient's problem—misdiagnosis being one potential cause of non-response. These benefits certainly apply to Attention-Deficit Hyperactive Disorder (ADHD), which is one of the most common developmental disorders among school-aged children with an estimated prevalence of 5% in the general population worldwide (American Psychiatric Association, 2013). Stimulants, particularly methylphenidate (MPH), are the first-line option in the treatment of ADHD (Reeves and Schweitzer, 2004; Dopheide and Pliszka, 2009). Stimulants are prescribed to an increasing number of children and adolescents in the US and the UK every year (Safer et al., 1996; McCarthy et al., 2009), though recent studies suggest that this trend is tailing off e.g., Holden et al. (2013). The beneficial effects of stimulant medication on the core symptoms of ADHD have been demonstrated by numerous clinical trials, reviews and meta-analyses (Banaschewski et al., 2006; Greenhill et al., 2006; van der Oord et al., 2008; Storebø et al., 2015). Nevertheless, adverse effects of the medications are also common (Storebø et al., 2015). The findings from previous research suggest that around 70% of children demonstrate a clinically significant treatment response to stimulant medication (Spencer et al., 1996; Schachter et al., 2001; Swanson et al., 2001; Barbaresi et al., 2006). However, it is unclear which patient characteristics may moderate treatment effectiveness and whether non-response can be predicted.

To date, achieving accurate predictions of the clinical outcomes for patients with ADHD has proven elusive—most of the literature has focused on investigating the potential correlations between a set of patient characteristics and the outcome following treatment with one or more types of medication. Information relating to patient characteristics has mostly been in the form of subjective questionnaire ratings, clinical notes and qualitative psychometric data; for example, the ratings from symptom-based questionnaires such as the Swanson, Nolan, and Pelham (SNAP) questionnaire (Swanson et al., 1983; Atkins et al., 1985; Swanson, 1992; Bussing et al., 2008), along with demographic variables such as age, sex and social economic background. The results from such studies are often contradictory and inconclusive due to small sample sizes and/or limited availability and quality of data, especially in the temporal (longitudinal) domain.

Along with more conventional statistical approaches, machine learning has also shown promise in predicting treatment response or prognosis in healthcare applications. Indeed, recently a random forest regression analysis was used to predict outcome in a group of patients affected by Obsessive Compulsive Disorder (OCD) from a relatively small pool of questionnaire items, with a reported error rate of 24.6% (Askland et al., 2015). Likewise, there has been a previous attempt to use machine learning techniques to predict treatment response in ADHD (Kim et al., 2015); support vector machine classification from this study was reported as 84.6% accurate (not to be confused with the balanced accuracy measure used in this paper). However, in addition to demographic and clinical questionnaire-derived data, the study used genetic as well as neuroimaging and neuropsychological information as inputs. Such data are unlikely to be readily available to clinicians in routine practice.

In this paper we investigate whether the inclusion of prior knowledge relating to the potential mechanism behind the presentation of a mental health condition and characteristics of individual patients can add value in predicting treatment. Thus, we hypothesized that a pragmatic machine learning approach based on a mechanistic or parametric model (a "learning in the model space" framework) for treatment response prediction may offer an advantage over more conventional methods (Brodersen et al., 2011; Doyle et al., 2013; Chen et al., 2014; Shen et al., 2016). This method represents each newly observed patient through a model; the models are personalized such that individual differences are accounted for by a set of subject-specific parameters. In the case of ADHD, developing a plausible mechanistic model is not straightforward—despite decades of research, the underlying mechanism for the disorder is not well understood. In addition, any mechanistic model would have to be based on data that are likely to be available in good, but routine, clinical practice.

This paper documents, within the "learning in the model space" framework, a Bayesian linear regression model for the prediction of treatment response in a cohort of children diagnosed and treated for ADHD in the UK. The performance of this new approach is then compared with conventional regression and machine learning methods ("learning in the data space") to assess whether or not the new approach offers benefits, and if so under what circumstances.

2. MATERIALS AND METHODS

2.1. Participants

The children enrolled in the study were drawn from the ADHD Drug Use and Chronic Effects (ADDUCE) cohort study (The ADDUCE Consortium, 2016), covered by a data sharing agreement with patient consent. The participants were from the UK NHS Tayside region who had attended the ADHD treatment clinics held at Dundee and Perth, UK. 262 families of eligible children were contacted, of which 181 (70%) were recruited and data on 173 of them were obtained for the purpose of this study. In addition, data were available on 94 healthy controls. Out of the 173 patients (whose baseline data are available), 157 of them started dose optimization studies and therefore longitudinal (temporal) data are available (See Section 3.1). To be eligible for the ADHD group, children had to be 6–17 years of age, have a clinical diagnosis of ADHD (see below), have had no previous medical history of methylphenidate use (medication-naïve) and have parental and child consent/assent to commence. The criteria for the healthy control group were similar apart from them having no current or previous psychiatric diagnoses. The recruitment was carried out over a 30-month period from January 2012–August 2014.

All patients in the ADHD group had already been clinically diagnosed with ADHD; this diagnosis was based on the clinical judgment of the assessing physician, informed by structured interviews with parents/carers, information provided by the child's school, direct observation of the child at the clinic, and at times, in their educational setting. Thus, the physician had to be satisfied that the child fulfilled the diagnostic criteria for a hyperkinetic disorder according to the International Classification of Diseases 10th edition (ICD-10) (World Health Organization, 2010), or ADHD as defined by the Diagnostic and Statistical Manual 4th edition (DSM-IV) (American Psychiatric Association, 2000). This means that the child had to demonstrate disabling and pervasive inattentiveness, hyperactivity, and impulsivity across a range of settings. The clinic was designed to implement a "dose optimization titration" scheme of medication in children diagnosed with ADHD. This involved giving increasing doses of methylphenidate (as the first line medication) at roughly weekly intervals until remission from symptoms was achieved or problematic adverse effects were encountered. If remission was not achieved with a first line medication within recommended dosage limits, or if problematic side-effects were encountered then a second line drug was initiated, and again, increased in dosage, as before.

2.2. Assessment

A range of baseline social and demographic factors was recorded, including parental marital status, family composition, and socioeconomic status as indicated by the Scottish Index of Multiple Deprivation (SIMD) 2012 (APS Group Scotland, 2012) derived from the family home postcode. A history of any previous psychiatric or non-psychiatric medication exposure was recorded, as were any physical health issues. Verbal and non-verbal intellectual functioning was estimated from parental reports and any educational issues noted. Problems with anxiety and low mood were rated using the short form of the Mood and Feelings Questionnaire (MFQ) with the parents, and where appropriate, the child as informants (Angold et al., 1996). Dystonia and abnormal movements were recorded using the Abnormal Involuntary Movement Scale (AIMS) (Guy, 1974, pp. 534–537). Oppositional and ADHD symptoms and behaviors were rated, according to parental report, using the Swanson, Nolan, and Pelham (SNAP-IV) questionnaire (Swanson et al., 1983). Any substance used by the participants was recorded using the Substance Use Questionnaire (SUQ). Fine motor issues were recorded using the Developmental Coordination Disorder Questionnaire 2007 (DCDQ'07). Several sections of the Development and Well-Being Assessment (DAWBA) were used (Goodman et al., 2000); these were (1) Rapidly Changing Mood (child and parent versions), (2) Tic disorders, including the Tourette syndrome, (3) Awkward and troublesome behavior. Tic severity (where present) was also rated using the Yale Global Tic Severity Scale (YGTSS) (Leckman et al., 1989). Possible behaviors associated with an underlying Autism Spectrum Disorder (ASD) were evaluated using the Social Communication Questionnaire (SCQ) (Rutter et al., 2003). The Strengths and Difficulties Questionnaire (SDQ) (Goodman, 1997-07) was used to rate parental perceived levels of pro-social behavior, hyperactivity/impulsivity, conduct problems, emotional symptoms and peer relationship problems. The overall clinical impression was recorded using the Clinical Global Impression—Severity scale (CGI-S) (Guy, 1974, pp. 218–222) and Children's Global Assessment Scale (CGAS) (Shaffer et al., 1983).

Responses to medication, in terms of levels of ADHD symptoms, were reported by parents and recorded using the SNAP-IV questionnaire at each visit. Likewise, any potential adverse effects and co-morbidity problems were reported using the standard clinic proforma, along with weight, height and blood pressure of the child at each visit.

2.3. Feature Extraction/Factor Analysis

The aforementioned questionnaires included a large number of items with categorical (binary or ordinal) response formats. Thus, in order to facilitate model development by reducing the dimensionality of the data whilst minimizing the loss of information, a series of factor (latent trait) analyses were conducted.

The key questionnaires used in the modeling process were the SCQ, the SDQ, and the SNAP-IV (see the previous section). In particular, the SNAP-IV scores served as the outcome variables, which indicated whether symptomatic remission had been achieved, following the dose-optimized titration of medication. The factor analyses sought to identify the dimensionality underlying the responses to the questionnaires and, consequently, the standardized factor scores represented the level of trait for each patient in that underlying dimension or construct.

In order to estimate the dimensionality, the sample of 173 patients and 94 healthy controls was randomly divided into two roughly equal exploratory and confirmatory datasets. A parallel analysis (Horn, 1965), adapted for categorical data, was then implemented in the freeware FACTOR (Lorenzo-Seva and Ferrando, 2006) using unweighted least squares (ULS) estimation method. A weighted "promax" rotation was deployed to achieve factor simplicity (Abdi, 2003). The maximum number of plausible factors (latent variables) was assumed to be indicated at the point where the eigenvalues of the factors in randomly generated data exceeded those observed in the real data. A series of exploratory factor analyses (EFAs–adapted for categorical dependent variables) were then conducted to aid interpretation of the factors. Oblique "geomin" rotation was used (Asparouhov and Muthén, 2009), assuming that, as in almost all psychological measures, underlying latent traits would be correlated with each other to some extent (Thurstone, 1931). A series of confirmatory factor analyses (CFAs) were then conducted using the held-back, confirmatory data (see Section 3.1 on cross-validation), in order to ensure that the factor structures derived fitted the data adequately. All EFAs and CFAs were conducted in the Mplus software version 7.1, using robust weighted least squares with mean and variance adjustment (WLSMV) as the estimation method (Muthén et al., 1997). Remission was defined by a child having a reported factor score in the hyperactive and inattentive dimensions (both elicited from factor analysis) equivalent to a mean item score in the SNAP-IV of one or less, which is conventionally taken to indicate symptomatic remission (Hechtman, 2005; Chou et al., 2012). The resulting symptom score thresholds are only slightly different for inattentiveness and hyperactivity (−0.97 vs. −0.92).

3. MODELING APPROACH

The causal factor model, shown in **Figure 1**, was derived using a rapid review approach to appraise and synthesize the existing evidence (Khangura et al., 2012). This model also took into account the nature of the data available in the cohort and was modified accordingly. The goal is not for the causal model to be comprehensive or definitive, but to identify from the literature

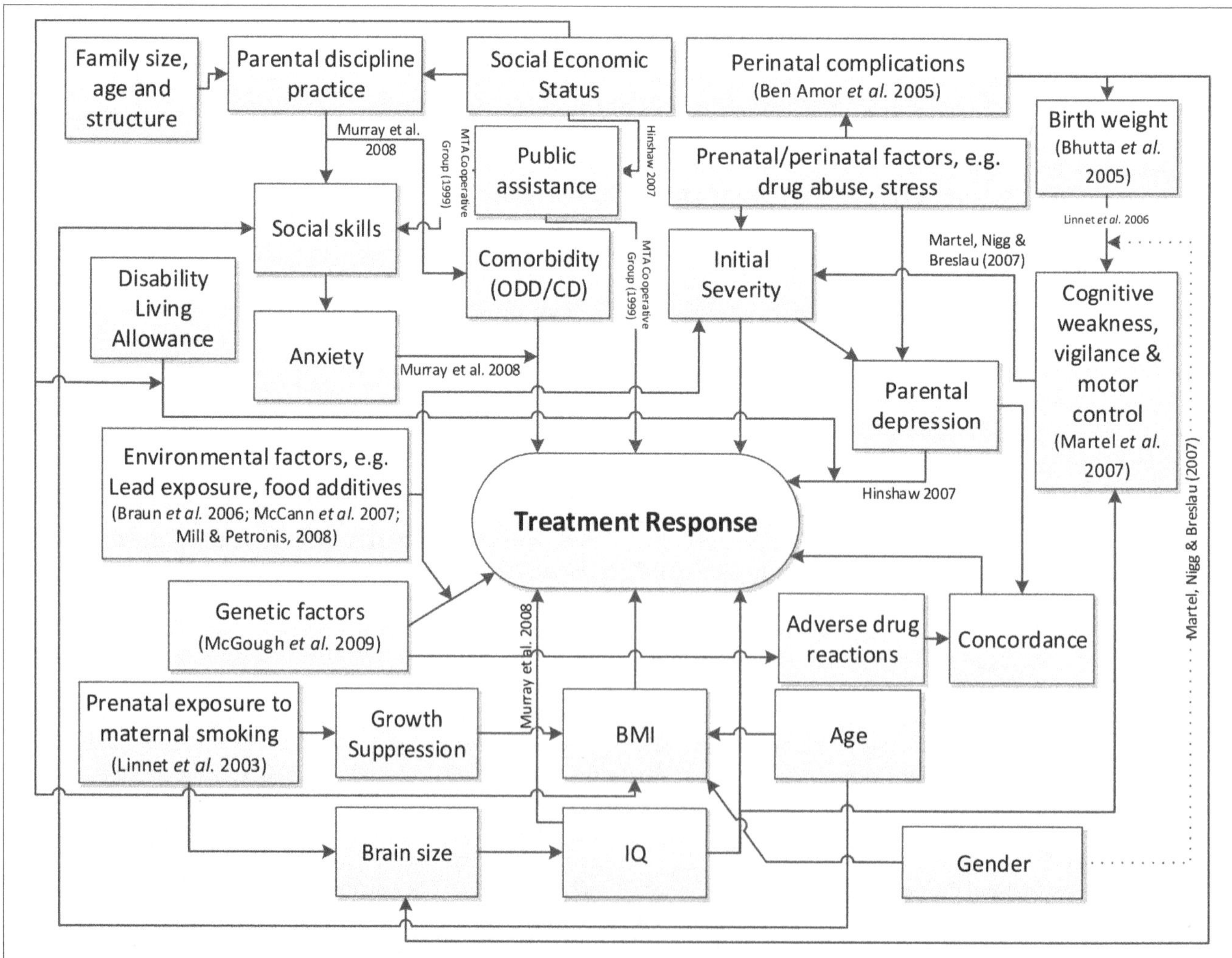

FIGURE 1 | High level causal factor model of treatment response prediction in ADHD.

as many potential factors relating to treatment response as there are available from the dataset, as well as helping to elicit the Bayesian prior distributions (Section 3.2.1). Model development was based on a literature review. This involved running searches in the EMBASE, MEDLINE and PsycINFO databases using the synonyms for ADHD (e.g., hyperkinesis) combined with terms relating to treatment outcome or response, and the names of the medications (both scientific and trademarks, full and abbreviated) prescribed in the cohort. The medications include 1) immediate release methylphenidate (IR-MPH, e.g., Ritalin®), 2) long-acting methylphenidate (XR-MPH, e.g., Concerta XL®, Equasym XL®, Medikinet XL®), 3) dextroamphetamine (DEX, e.g., Dexedrine®) including its prodrug lisdexamfetamine dimesylate (e.g., Elvanse®), and 4) atomoxetine (ATOM, e.g., Strattera®). Secondary sources were followed up. The quality of trial-based studies could be appraised using the CONSORT checklist (Schulz et al., 2010) and observational studies via the STROBE guidance (von Elm et al., 2007). Two of the authors (HKW and PAT) then made a judgment, based on the findings reported in the literature and the perceived likelihood of bias or uncertainty as to what extent variables in the model might be related to treatment response in ADHD. The model derived was consequently used to populate prior distributions for the patient-specific model parameters (i.e., the hyperpriors). Where the evidence was uncertain or inconsistent, the variances (i.e., imprecision) of the hyperpriors were increased.

Not every piece of information mentioned in Section 2.2 was used for the purpose of modeling, because of insufficient data or multicollinearity between the variables. The causal factor model was then simplified based on the breadth of available data from the cohort, leading to a much reduced model as shown in **Figure 2**. Some factors were combined through another layer of feature extraction; for example, the *motor* and *control* latent factors, themselves also obtained from applying feature extraction to the DCDQ'07 questionnaire data (see Section 2.2), were combined with the *non-verbal communication* factor from the SCQ questionnaire to obtain a developmental adversity factor. Some factors were not obtained from standard questionnaires; for example, the *perinatal adversity* factor (see **Figure 2**) was constructed from birth weight and gestation age; the *family size and socioeconomic status* factor combined the number of siblings, parental house ownership (owned, mortgaged or rented) and the SIMD 2012 index (APS Group Scotland, 2012).

3.1. Data

There were 267 subjects whose baseline characteristics were measured (173 clinically diagnosed with ADHD and 94 healthy controls) at the first clinical appointment. Of the 173 non-controls, 157 were enrolled in dose optimization titration studies with parental consent, for whom longitudinal (temporal) data are available. The 157 patients with longitudinal data were randomized and 10-fold cross-validation partitions were constructed. Subjects were partitioned into 10 subgroups of roughly equal size in a patient-coherent fashion, i.e., data from a single patient only appeared in a single fold.

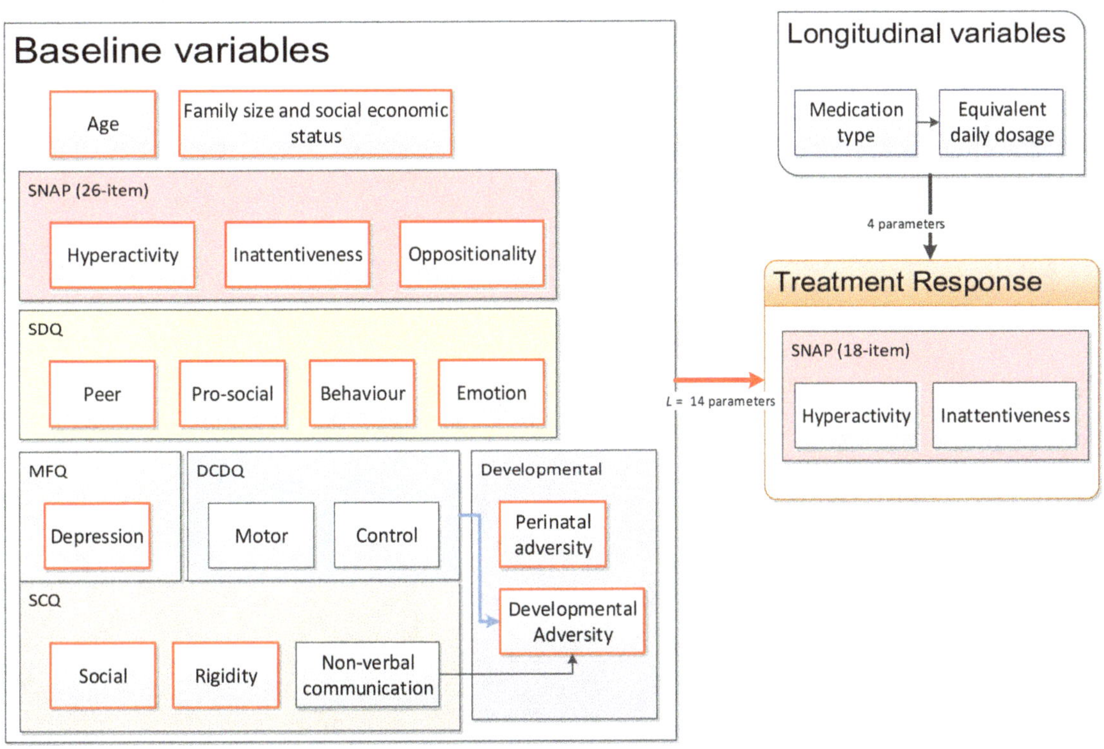

FIGURE 2 | Reduced causal factor model of treatment response prediction in ADHD.

For all models investigated in this paper, a single fold was used as the validation dataset and the remaining nine folds were combined to serve as the training dataset. This process was iterated until each fold had served as validation data exactly once.

3.1.1. Baseline Characteristics

We labeled the patient subjects by the indexing variable $s = 1, 2, \ldots, N$. A set of L patient-specific baseline continuous latent factors, encoded in a row vector $\mathbf{b}_s \in \mathbb{R}_{1\times L}$ was obtained by performing feature extraction as described in Section 2.3 over the questionnaires detailed in Section 2.2. Referring to **Figure 2**, $L = 14$ factors were used for the baseline. Data from the controls in addition to the training dataset were utilized during feature extraction to ensure that the resulting latent factor models can sufficiently encompass the entire range of characteristics from ADHD patients to normal children. The resulting continuous latent factors would, in theory, be sufficiently representative of the information conveyed by the categorical questionnaire response variables.

To ensure that validation data were strictly not used for the model building, feature extraction was first performed using only training data from each of the folds (plus all the controls). This resulted in 10 sets of factor scores corresponding to each fold. The factor model structures (e.g., the number of factors per questionnaire) over the folds did not change across the folds, as statistical fit indices and Chi-square difference tests did not suggest that any changes were necessary. The factor models were then used to estimate the baseline factor scores for the validation sets in each of the folds.

Each of the 10 cross-validation runs resulted in a set of corresponding continuous latent factors, which were used as inputs to subsequent models. The models were trained and validated using the same training-validation partitioning used in the feature extraction process.

3.1.2. Longitudinal Data

Each of the 157 subjects with longitudinal data visited the clinic a varying number of times—from titration, stabilization to continuing care; the number of doctor's appointments, A_s, varies from 1 to 22. At each appointment, the parent or guardian of the patient was asked to fill in an 18-item SNAP-IV questionnaire, which measures the degree of inattentiveness and hyperactivity. The responses were entered into a factor model (identified through feature extraction) to extract a continuous symptom score for inattentiveness and hyperactivity. We denote the appointment number by the indexing variable a so that $a = 1, 2, \ldots, A_s$. Let the independent "input" variables $m_{a,1}$, $m_{a,2}$, $m_{a,3}$, $m_{a,4}$ be the four types of medications, respectively, IR-MPH, XR-MPH, DEX, and ATOM for subject s at appointment a. Using datasheets for the medicines used, the dosages of DEX and ATOM were normalized to an equivalent daily dosage (EDD) of IR-MPH. For all a and s, this results in input and output matrices of the form:

$$\text{Input: } \mathbf{M}_s = \begin{bmatrix} m_{1,1} & m_{1,2} & m_{1,3} & m_{1,4} \\ m_{2,1} & m_{2,2} & m_{2,3} & m_{2,4} \\ \vdots & \vdots & \vdots & \vdots \\ m_{a,1} & m_{a,2} & m_{a,3} & m_{a,4} \\ \vdots & \vdots & \vdots & \vdots \\ m_{A_s,1} & m_{A_s,2} & m_{A_s,3} & m_{A_s,4} \end{bmatrix} \tag{1}$$

$$\text{Output: } \mathbf{y}_s = \begin{bmatrix} r_1 & r_2 & r_3 & \ldots & r_{A_s} \end{bmatrix}^{\mathsf{T}}, \tag{2}$$

where r is the symptom severity measure and can be either the inattentiveness factor score or the hyperactivity factor score.

The combined EDDs of medications (for the 4 types) prescribed over the appointment number for all patients are plotted as a boxplot in **Figure 3**. One can observe that as forced titration progressed over the appointments, the overall dosage level increased.

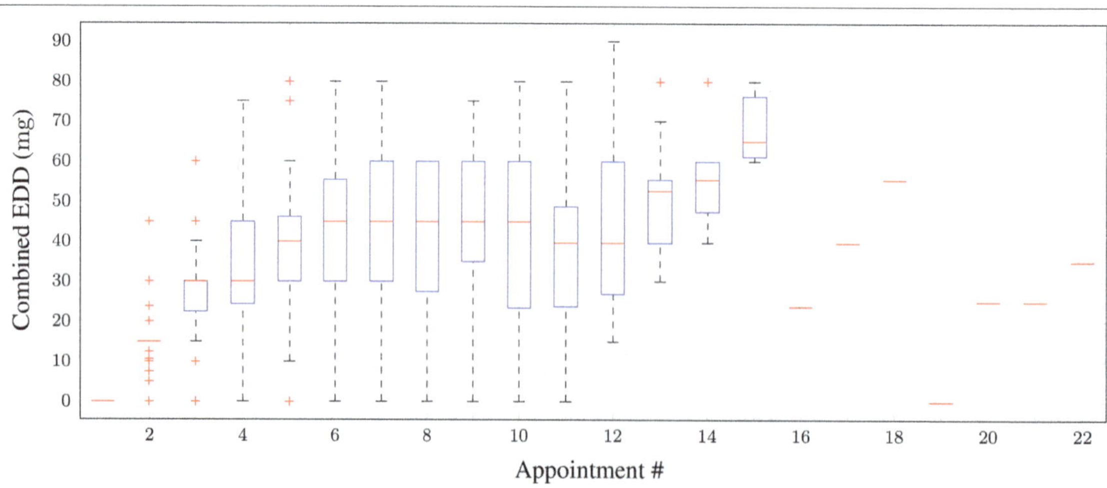

FIGURE 3 | Boxplot of combined equivalent (in IR-MPHunits) daily dosages of medications taken for all patients vs. appointment number. Red horizontal lines: median; boxes: interquartile range; whiskers: 95% confidence intervals; red crosses: outliers.

Figure 4 shows the distribution of inattentiveness and hyperactivity symptom factor scores for the patients for each appointment. The lower the factor scores, the less severe the symptoms are. In terms of a general trend, one can clearly see an effective and quick reduction in symptom levels over the first 5 appointments, as stimulant medication prescription ramps up during forced titration. The symptom scores cease to improve for appointments 6–8, after which a slight increase can be observed. This hints at adherence or persistence issues, but the available data do not allow further investigation—as such issues are not consistently reported by the parent/guardian or recorded in the clinical notes. While the model has no mechanism for modeling such effects, the adaptive learning nature of the Bayesian algorithm is able to self-correct and compensate for small deviations, for example, by "learning" to weight down the dose-response parameter for a given medication when the patient has a low adherence.

3.2. Treatment Response Model Formulation

The treatment outcome is modeled as a linear combination of the baseline variables and the medication dosage,

$$\mathbf{y}_s = \mathbf{X}_s \omega_s + \epsilon \tag{3}$$

where

$$\mathbf{X}_s = \begin{bmatrix} \mathbf{b}_s & & 1 \\ \vdots & \mathbf{M}_s & \vdots \\ \mathbf{b}_s & & 1 \end{bmatrix}, \tag{4}$$

$\epsilon \sim \mathcal{N}(0, \sigma_\epsilon, s^2) \in \mathbb{R}_{A_s \times 1}$ is an error term and $\omega_s \in \mathbb{R}_{P \times 1}$ is the *subject-specific* parameter vector moderating the effect of the baseline variables $\mathbf{b}_s$ on the treatment response, i.e., it accounts for how large an effect each of the various baseline variables or medication types has on treatment outcome. "Subject-specific" means that the parameter vector was allowed to be different for each subject so that patients with similar baseline characteristics can still have a different prediction outcome. The number of free parameters required is $P = L + 4 + 1 = 19$ for $L = 14$ (see **Figure 2**).

The baseline variables remain unchanged over different appointments while the medication dosage may vary according to the titration regime specified by the clinician. Hence, every row of the matrix $\mathbf{X}_s$ contains the same baseline characteristic vector for an individual patient, combined with the medication dosage vector. The row number in $\mathbf{X}_s$ corresponds to the appointment number.

Because the number of visits A_s of the subjects was usually fewer than the dimension of the parameter space P, the problem is mathematically underdetermined. Hence, classical least squares regression methods would fail without an appropriate regularization (Goodfellow et al., 2016). To this end, we employ a Bayesian formulation. In particular, the Bayesian linear regression was used to model the temporal evolution of the dose-response relationship for each patient.

In essence, a Bayesian approach allows prior or expert knowledge to be encoded into the problem formulation and this enables a probabilistic solution to be found despite the limited data available.

3.2.1. Prior Distributions and Knowledge

The joint prior probability density function $\Pr(\omega, \sigma^2)$ is given in Equation (A3, Supplementary Material). In this exercise, the causal treatment response model based on the literature was used to constrain the prior of ω and its covariance matrix Λ_0^{-1}.

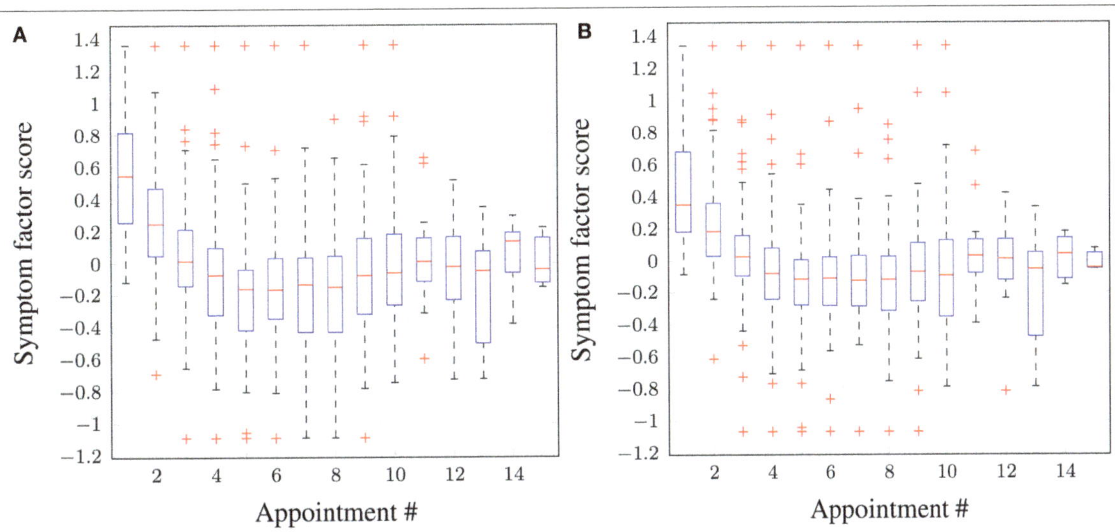

FIGURE 4 | Boxplots of symptom scores across all patients vs. appointment number. (A) Inattentiveness, **(B)** hyperactivity. Red horizontal lines: median; boxes: interquartile range; whiskers: 95% confidence intervals; red crosses: outliers.

When eliciting the prior, quantitative information from the literature was not used, e.g., setting the mean of the prior distribution of the parameter vector omega to a specific numerical value. This is because the demography, sample sizes and effect sizes across literature vary and there is no correct way to normalize them. Instead, only the sign (direction) of the effect was encoded. For example, there is evidence that methylphenidate improves treatment response in the literature (a positive dose results in lower symptom score), therefore a negative value of -1 was specified for the columns of $\boldsymbol{\tau}_0$ corresponding to $m_{a,1} \dots m_{a,4}$ in Equation (1). For positive associations with symptom scores, $+1$ was used instead. The same magnitude is used in other factors (i.e., either $+1$ or -1).

Each cohort study or clinical trial from the rapid review was appraised, respectively, using the STROBE and the CONSORT checklists by counting the number of pass and fail items out of the total. Evidence from the literature was marked as good quality when both the checklist score was similar to other studies on the same topic (within 20% from the best) and effect sizes were statistically significant as reported by the authors for the sample size used. The diagonals of the covariance matrix $\boldsymbol{\Lambda}_0^{-1}$ were assigned an initial value of one; for every contradicting evidence (the effect sizes are opposite in direction) satisfying these criteria, 0.5 was added to the corresponding variance in the covariance matrix. A higher value may be specified if necessary, to ensure that the prior distribution of parameter omega spans both positive and negative sides sufficiently—within one standard deviation of $\boldsymbol{\tau}_0$. On the other hand, if the effect sizes are positive the variance was reduced by 0.1 for each supporting studies, at the same time ensuring the variance does not go below 0.5. The (lack of) proposed existence of causal links between the variables in the causal model (see **Figure 1**) ensures the sparsity of the covariance matrix $\boldsymbol{\Lambda}_0^{-1}$.

While these numbers may not be completely objective, the amount of data available means that the sensitivity of the results to the prior is low—sensitivity analysis shows that the effect of scaling the prior covariance between 50 and 150% of its original values changes the errors by about 5% of the training root mean squared (rms) error, and 3% for the validation rms error.

3.2.2. Posterior Distributions

The posterior distribution is given in Equation (A4, Supplementary Material), where the parameters of the distributions are obtained through Equation (A5) in Supplementary Material. In the training set, Bayesian learning uses data from all of the appointments that a subject had, in which case $n = A_s$ where, as before, A_s is the total number of visits or appointments a subject has and data are available for. We introduce the simplified notations after Bayesian update has been applied to the training set using Equation (A5) in Supplementary Material, so that

$$\boldsymbol{\Lambda}_{A_s}^{-1} \to \hat{\boldsymbol{\Lambda}}_s^{-1}, \quad \boldsymbol{\tau}_{A_s} \to \hat{\boldsymbol{\tau}}_s, \quad \alpha_{A_s} \to \hat{\alpha}_s \text{ and } \beta_{A_s} \to \hat{\beta}_s. \quad (5)$$

This notation will be used in later sections. The reader should be reminded that the parameters are derived from each subject and hence are different across subjects. Notice that in a prediction exercise (instead of retrospective regression formulated here), the learning can be applied incrementally for each future observation with each update using just the new observation.

3.3. Virtual Patient Profile

When a new patient (denoted by $s = *$) is received, one can measure their baseline variables $\mathbf{b}_*$, but not their model parameter space $\boldsymbol{\omega}_*$. The goal is to estimate a *virtual patient profile* that is believed to best describe the new patient using only the available baseline measurements. To do this one derives the mathematical mapping functions from the baseline characteristics of a patient to their posterior parameters $\mathbf{b}_* \mapsto \Pr(\boldsymbol{\omega}_*)$ and $\Pr(\boldsymbol{\sigma}_*^2)$, such that a prediction can be made from the baseline variables. These functions are forged using machine learning on the existing pool of training data. Since $\Pr(\boldsymbol{\omega}_s)$ is parameterized by $(\boldsymbol{\tau}_s, \boldsymbol{\Lambda}_s^{-1})$ and likewise $\Pr(\boldsymbol{\sigma}_s^2)$ by $(\boldsymbol{\alpha}_s, \boldsymbol{\beta}_s)$, one has to learn the mappings from the baseline variables to the parameters. The learnt mathematical mapping functions can then be used to obtain estimates of $(\hat{\boldsymbol{\tau}}_*, \hat{\boldsymbol{\Lambda}}_*^{-1})$ and $(\hat{\alpha}_*, \hat{\beta}_*)$, which represent the virtual patent profile for the new patient in the model space.

Due to the conjugate nature of the priors, one does not need to derive the hyperparameters $\hat{\alpha}_*$ and $\hat{\beta}_*$ from Equation (5) for the purpose of having point estimates for the treatment response prediction. However, these hyperparameters are necessary in order to derive the posterior distribution of the predicted value—commonly referred to as the posterior predictive distribution. Knowing the distribution allows us to approximate the uncertainties of the estimates, e.g., 95% confidence intervals. Two methods were proposed for learning the mappings from the baseline variables to the virtual patient profile and they are introduced in the following subsections.

3.3.1. Method 1: Generalized Linear Regression

To determine the mappings, one finds the functions: a) $f_{\boldsymbol{\tau}}(\mathbf{b}_*) \approx \hat{\boldsymbol{\tau}}_*$ and b) $f_{\mathbf{u}}(\mathbf{b}_*) \approx \hat{\boldsymbol{\Lambda}}_*^{-1}$ where $\mathbf{u}_s$ is a row vector containing non-zero elements of the upper (or lower) triangular part of $\hat{\boldsymbol{\Lambda}}_s^{-1}$. Since $\hat{\boldsymbol{\Lambda}}^{-1}$ is a covariance matrix (hence symmetric), knowledge of the lower/upper half of the off-diagonal elements plus the diagonal elements is sufficient to fully recreate the matrix.

The mappings are learnt from the training data, in which the posterior distributions of $\hat{\boldsymbol{\tau}}_s$ and $\hat{\boldsymbol{\Lambda}}_s^{-1}$ are already available through Equation (A5) in Supplementary Material. Linear regression models of the form $\mathbf{Y} = \hat{\mathbf{P}}\mathbf{B}$ were used to model the two mappings, where the matrix $\mathbf{B}$ has rows of $\mathbf{b}_s$ vectors—one for each subject in the training set—and similarly $\mathbf{Y}$ is composed of rows of a) $\boldsymbol{\tau}_s$ for determining $f_{\boldsymbol{\tau}}$ or b) $\mathbf{u}_s$ for determining $f_{\mathbf{u}}$. The least squares solutions for the models are given by the Moore-Penrose pseudo-inverse,

$$\hat{\mathbf{Q}} = \left(\mathbf{B}^{\mathsf{T}}\mathbf{B}\right)^{-1}\mathbf{B}^{\mathsf{T}}\mathbf{Y}. \quad (6)$$

For prediction, $f_{\boldsymbol{\tau}}$ and $f_{\mathbf{u}}$ can both be formulated as $f(\mathbf{b}_*) = \mathbf{b}_*\hat{\mathbf{Q}}$.

The posterior estimates for the hyperparameters for a new patient are taken as the averaged values of $\hat{\alpha}_s$ and $\hat{\beta}_s$ across all subjects in the training set, resulting in $\hat{\alpha}_* = 5.5$ and $\hat{\beta}_* = 1.7$.

3.3.2. Method 2: Gaussian Kernel Weighted Averaging

An alternative method is to find $\hat{\boldsymbol{\tau}}_*$ and $\hat{\boldsymbol{\Lambda}}_*^{-1}$ using a weighted average of $\hat{\boldsymbol{\tau}}_{s'}$ and $\hat{\boldsymbol{\Lambda}}_{s'}^{-1}$, with $s' \in \mathbb{S}_*$ being a subset of subjects in the existing training pool whose baseline variables ($\mathbf{b}_{s'}$) were "similar" to those of the new patient ($\mathbf{b}_*$). Highly similar subjects will have a higher influence on the value of $\hat{\boldsymbol{\tau}}_*$ and $\hat{\boldsymbol{\Lambda}}_*^{-1}$. The "(dis)similarity" d_s is measured using the pairwise euclidean distance between $\mathbf{b}_s$ and $\mathbf{b}_*$, such that

$$d_s = (\mathbf{b}_* - \mathbf{b}_s)(\mathbf{b}_* - \mathbf{b}_s)^{\mathsf{T}}. \tag{7}$$

This is then sorted and the 17.5% of subjects in $\mathbb{S}_*$ with the smallest "dissimilarity" values are kept; this percentage value was chosen as it resulted in the lowest validation error. The weighting w_s was taken as the normalized Gaussian kernel

$$w_s = \frac{\exp(-\lambda \cdot d_s)}{\sum_{\forall s' \in \mathbb{S}_*} \exp(-\lambda \cdot d_{s'})} \tag{8}$$

where the parameter value $\lambda = 1.15$ was chosen as it again resulted in the lowest validation error. Using Equations (7) and (8), one can estimate $\hat{\boldsymbol{\tau}}_*$ and $\hat{\boldsymbol{\Lambda}}_*^{-1}$ as

$$\hat{\boldsymbol{\tau}}_* = \sum_{\forall s \in \mathbb{S}_*} w_s \boldsymbol{\tau}_s, \tag{9a}$$

$$\hat{\boldsymbol{\Lambda}}_*^{-1} = \sum_{\forall s \in \mathbb{S}_*} w_s \boldsymbol{\Lambda}_s^{-1}, \tag{9b}$$

and similarly the estimates of the hyperparameters are calculated using

$$\hat{\alpha}_* = \sum_{\forall s \in \mathbb{S}_*} w_s \alpha_s, \tag{9c}$$

$$\hat{\beta}_* = \sum_{\forall s \in \mathbb{S}_*} w_s \boldsymbol{\beta}_s. \tag{9d}$$

3.4. Prediction Using the Posterior Predictive Distribution

When a new subject visits the clinician, their $\mathbf{b}_*$ vector may be measured and used to approximate $\hat{\boldsymbol{\tau}}_*$, $\hat{\boldsymbol{\Lambda}}_*^{-1}$, $\hat{\alpha}_*$ and $\hat{\beta}_*$ using either of the methods in the previous subsections. Given a hypothetical medication input $\mathbf{x}_*$, the treatment response for the new subject can then be predicted through the posterior predictive distribution

$$\Pr(y_*) = \mathrm{t}_\nu\left(\mathbf{x}_*\hat{\boldsymbol{\tau}}_*, \frac{\hat{\beta}_*}{\hat{\alpha}_*}\left(\mathbf{I} + \mathbf{x}_*\hat{\boldsymbol{\Lambda}}_*\mathbf{x}_*^{\mathsf{T}}\right)\right) \tag{10}$$

where the number of degrees of freedom for the Student's t-distribution is given by $\nu = 2\hat{\alpha}_*$.

Equation (10) may be used to predict the treatment response for this new patient over their course of the treatment directly without learning; that is to treat each appointment as independent and the parameters are not updated. On the other hand, it is possible to perform incremental Bayesian learning over the course of treatment, by using Equation (A5) in Supplementary Material to update the parameters given the treatment outcome measured for each new visit and the associated inputs. Through incremental learning, the model corrects for discrepancies between the true profile and the virtual patient profile of the new patient. As such, one would expect the prediction to improve as data from more visits to the clinic become available. The implementation of these methods is discussed in more detail in Section 4.4.

At some point, the profile of the new patient in terms of the treatment outcome, inputs, and baseline characteristics can be added to the pool of existing patient profiles (training set) to improve the model's generalizability for future patients.

3.5. Training and Validation

As discussed in Section 3.1, 157 patients with longitudinal data were randomized and 10-fold cross-validation partitions were constructed resulting in 10-folds of training-validation data partitions.

First, for each fold, the framework detailed in Section 3.2 was followed and Bayesian linear regression was performed to fit patient-specific parameters to each patient in the training dataset. Second, either of the methods specified in Section 3.3 was used in order to construct virtual patient profiles for each patient in the validation set, using the patient-specific parameters. Finally, the procedure outlined in Section 3.4 was followed in order to obtain a prediction for patients in the validation set; effectively treating each patient as new.

3.6. Dichotomous Remission Prediction

Although the model was initially formulated to predict a continuous scale of symptom scores, one can explore dichotomizing the outcome into patients who have shown reduced symptoms and those who have not. The justification is that clinicians and doctors are less likely to be interested in a predicted SNAP-IV score or symptom severity scale as opposed to a simple "yes/no" answer as to whether the patient will be in remission for a given medication. A simple way to adapt the current model to do this is to apply a threshold to the continuous symptom score prediction, below which the patient is predicted to be in remission.

Some of the literature loosely defines remission in ADHD as having a large majority of SNAP-IV responses rated in category 0 (not at all) or 1 (a little) (Hechtman, 2005; Chou et al., 2012). Therefore in this paper, the thresholds were chosen such that the approximate continuous symptom score corresponds to the raw responses from the 18-item SNAP-IV questionnaire all lying in category 1. The resulting symptom score thresholds are only slightly different for inattentiveness and hyperactivity (−0.97 vs. −0.92, see Section 2.3).

Using these thresholds, it was found that the proportion of visits when measurements were taken indicates that remission was relatively rare, 160 out of a total of 1,147 (13.95%) for the inattentiveness score and 139 (12.1%) for the hyperactivity score. This is expected, as forced titration initially starts with a low medication dosage and one would not expect an effective reduction in symptom ratings to remission levels before the dosage was ramped up in later appointments; in addition, because of medical persistence issues patients can drop out before the clinicians are able to find an effective dose.

Note that all the methods in this paper were first used to predict the continuous symptom scores by regressing the baseline variables, medication prescribed to the treatment response at following appointments. Dichotomized remission prediction only occurs at a later stage. Right censoring (where patients prematurely drop out of dose optimization stage without achieving remission) is therefore not an issue; repression methods can utilize the remaining appointment information to model treatment response regardless of whether remission was achieved or not.

4. PERFORMANCE METRICS

To facilitate a comparison between the performance of the different approaches, several performance metrics were used. For the regression tasks, one is interested in the deviation in the predicted symptom scores against the true symptom scores; whereas for the remission classification tasks, one is interested in the performance of the classifiers with regard to the probabilities or ratios of true positive, false positive, true negative and false negative cases.

4.1. Regression Task

The root mean squared (rms) error measure is defined as the square of the averaged squared error across the 10-folds, across subjects and across all appointments for each individual, i.e.,

$$\sqrt{\text{rms} = \frac{1}{10|\mathbb{S}|}\sum_{k=1}^{10}\sum_{\forall s\in\mathbb{S}}\sum_{a=1}^{A_s} A_s^{-1}\left|y_s - \hat{y}_s\right|^2} \tag{11}$$

where $\mathbb{S}$ is the set of all subjects considered (e.g., those in the validation set), $|\mathbb{S}|$ denotes the number of subjects in $\mathbb{S}$; y_s, $\hat{y}_s$, and A_s are, respectively, the true outcome symptom score, the fitted or predicted outcome symptom score, and the total number of appointments for the individual subject s.

4.2. Classification Task

4.2.1. Sensitivity and Specificity

The sensitivity (SEN, also known as the true positive rate or recall) is defined as $N_{\text{TP}}/N_{\text{P}}$ where N_{TP} is the number of *true positives*—appointments where measurements indicated remission and were correctly predicted as such; and N_{P} is the actual number of positive cases, i.e., the number of appointments where the corresponding subjects were indeed in remission. This is reported in addition to the specificity (SPC, also known as the true negative rate or fall-out), defined as $N_{\text{TN}}/N_{\text{N}}$, where N_{TN} is the number of *true negatives*—those *not* in remission and correctly predicted as such; and N_N is the actual number of negative cases (Fletcher and Fletcher, 2005). Note that if one lets N_{FP} and N_{FN} be the number of false positives and false negatives respectively, then $N_{\text{P}} = N_{\text{TP}} + N_{\text{FN}}$ and $N_{\text{N}} = N_{\text{TN}} + N_{\text{FP}}$ (Fletcher and Fletcher, 2005).

Sensitivity characterizes the ability of a classifier to rule out false negative predictions (type-II errors) given that a condition is true. On the other hand, specificity measures the ability of a classifier to rule out false positive predictions (type-I errors) given that a condition is false. In this exercise, the sensitivity measure is more important; due to the rarity of remission, and the goal is to try to predict what level of medication is required to achieve remission, the ability of a classifier to recall remission cases (ruling out type-II errors) is more important than ruling out type-I errors.

4.2.2. PPV and NPV

The positive predictive value (PPV, also known as the precision) is the proportion of true positives in the *predicted* positive cases and is the probability of remission given a positive prediction by the algorithm. As such, the PPV is a measure of the "quality" of a given positive prediction. PPV is given by $N_{\text{TP}}/(N_{\text{TP}} + N_{\text{FP}})$. Conversely, the negative predictive value (NPV) is the proportion of true negatives in the *predicted* negative cases, and is the probability of non-remission given a negative prediction. NPV is given by $N_{\text{TN}}/(N_{\text{TN}} + N_{\text{FN}})$ (Fletcher and Fletcher, 2005).

By the argument outlined above, the PPV is more important for this exercise than the NPV.

4.2.3. Balanced Accuracy

The overall accuracy of a dichotomous predictor is defined by

$$\text{Accuracy} = (N_{\text{TP}} + N_{\text{TN}})/N,$$

where $N = 1,147$ is the total number of appointments across all subjects.

However, the overall accuracy measure is known to be problematic when the prevalence of success/failure is low (Alberg et al., 2004), i.e., the data are imbalanced (see also the end of Section 3.6). Due to this, some of the literature uses the balance accuracy (BAC) measure, defined as the average of sensitivity and specificity (Brodersen et al., 2010). This is the accuracy measure used throughout this paper. Note that, numerically, the BAC is closely related to the Youden's J-statistic (Youden, 1950), also known as "informedness" or "DeltaP′" (Powers, 2011), since it is equal to sensitivity plus specificity minus one.

4.2.4. ROC and AUC

The receiver operating characteristic (ROC) curve is commonly used in the medical and the machine learning community to evaluate the performance of binary classifiers Fawcett (2006). It plots the true positive rate (sensitivity) against the false positive rate (one minus specificity) for a given classifier. A curve is obtained when its classification performance can be tuned through setting a threshold or changing a parameter, trading off the true positive rate against the false positive rate. Binary classifiers that can achieve good compromise between sensitivity and specificity have a large area-under-the-curve (AUC), and this single metric may be used to compare the performance between the different classifiers Bradley (1997).

4.3. Trading Off Sensitivity and Specificity

From Section 3.6, the proportion of appointments without remission is (100 − 13.95)% = 86.05%. Therefore, given this statistic, one would expect that a null model guessing the result randomly would have a sensitivity of 13.95% and a specificity of 86.05%. Simply using point estimates of the continuous symptom score from the learning in the model space approach and thresholding them to give dichotomous predictions of remission results in classifiers with low sensitivity values between 22 and 28% and high specificity values of 94–97%. Due to the low number of remission cases compared to the non-remission cases, the classification is biased against predicting the remission cases, leading to low sensitivity (but high specificity). A classifier can be tuned to improve its sensitivity performance by trading off specificity to a certain degree. A good compromise would be maximizing both sensitivity and specificity equally, which is in essence maximizing the BAC or the Youden's J-statistic in Section 4.2.3.

The thresholds for remission are defined by the SNAP-IV symptom factor scores as in Section 2.3, and this defines the ground truth of whether a patient is in remission or not. However, one can take advantage of the fact that the predicted continuous symptom scores from the learning in the model space approach form full posterior distributions with uncertainties associated, and the levels of uncertainty are known (e.g., see the error bars in **Figures 5, 6**). One may define a critical value as the lower bound of the prediction, above which the probability of the prediction being correct is x%. Instead of the remission thresholds comparing against the point estimates, they may be compared against the point estimates minus a critical value. The larger the critical value, the higher the prediction score has to be in order to be classified as not in remission. This in effect is equivalent to raising the threshold, classifying more and more cases into remission, which increases the sensitivity and lowers the specificity. The range of "thresholds" or classifier parameter settings that makes this trade-off can be used to generate a ROC plot (Section 4.2.4). A similar trade-off can be made with classical machine learning algorithms and will be discussed in Section5.

The training data are used to find an optimal classifier setting in order to achieve the highest BAC, and the same classifier setting is then used to classify the validation data. This ensures that the validation data are not used to minimize the validation error.

4.4. Benchmarking and Implementation

The Bayesian learning in the model space approach relies on prior knowledge (Section 3.2.1) and virtual patient profiles in the model space (Section 3.3), as well as iterative learning (Bayesian update) in order to function. To assess whether these components contribute to the prediction capability of the model, several implementation strategies are investigated, namely:

1. **Appointment-independent prediction (`AI`):** Treating each appointment as independent (as the first appointment) and giving a prediction only using the virtual patient profile;

2. **Incremental Bayesian linear regression (`BR`):** The first prediction is performed in exactly the same manner as the appointment-independent case. Then, Bayesian linear regression using elicited priors (see Section 3.2.1) is applied progressively. That is, the effect/outcome of medication prescribed in appointment 1, then observed at appointment 2, is used in the regression model. Then, at appointment 3, outcomes from appointments 1 and 2 are used. Similarly, at appointment 4, outcomes from appointments 1, 2 and 3 are used, and so on. This means that except for the first prediction, the incremental Bayesian linear regression learns from scratch the patient-specific parameters (at each appointment) using the elicited priors. This essentially disregards any information already learnt from the current training set (the virtual patient profiles), treats the validation set as a new "training" set, and performs basic Bayesian linear regression fitting. However, instead of all appointment outcomes being available for each new patient, as is the case during the training phase, one simulates the fact that information is progressively collected during the course of treatment for new patients. Since the virtual patient profiles are not utilized, this serves as a benchmark reference to evaluate the effectiveness of the constructed virtual patient profiles when compared with the next case; this represents a method that can be implemented even when no training data exist.

3. **Incremental Bayesian learning/update (`BU`):** The first appointment is predicted as for the previous two cases, but then when the true value is observed (in appointment 2), it is fed-back into the Bayesian learning model, i.e., Equation (A5) in Supplementary Material. This updated model is then used to generate a prediction. This progressive updating continues up to the most recent appointment. The crucial difference between this method and the `BR` method is that, here, the priors used were derived from the virtual patient profiles, as opposed to the elicited priors used in the `BR` method. When compared to the `BR` case, this highlights whether the model space offers any utility in aiding the prediction of treatment response.

The Bayesian approach was implemented *ad hoc* in MATLAB software with custom routines. The performance of the prediction was compared across the three different implementations above, using the validation data.

5. COMPARISON WITH CONVENTIONAL METHODS

To provide context to the results achieved using the learning in the model space approach, the performance of conventional linear regression methods and machine learning methods was also investigated.

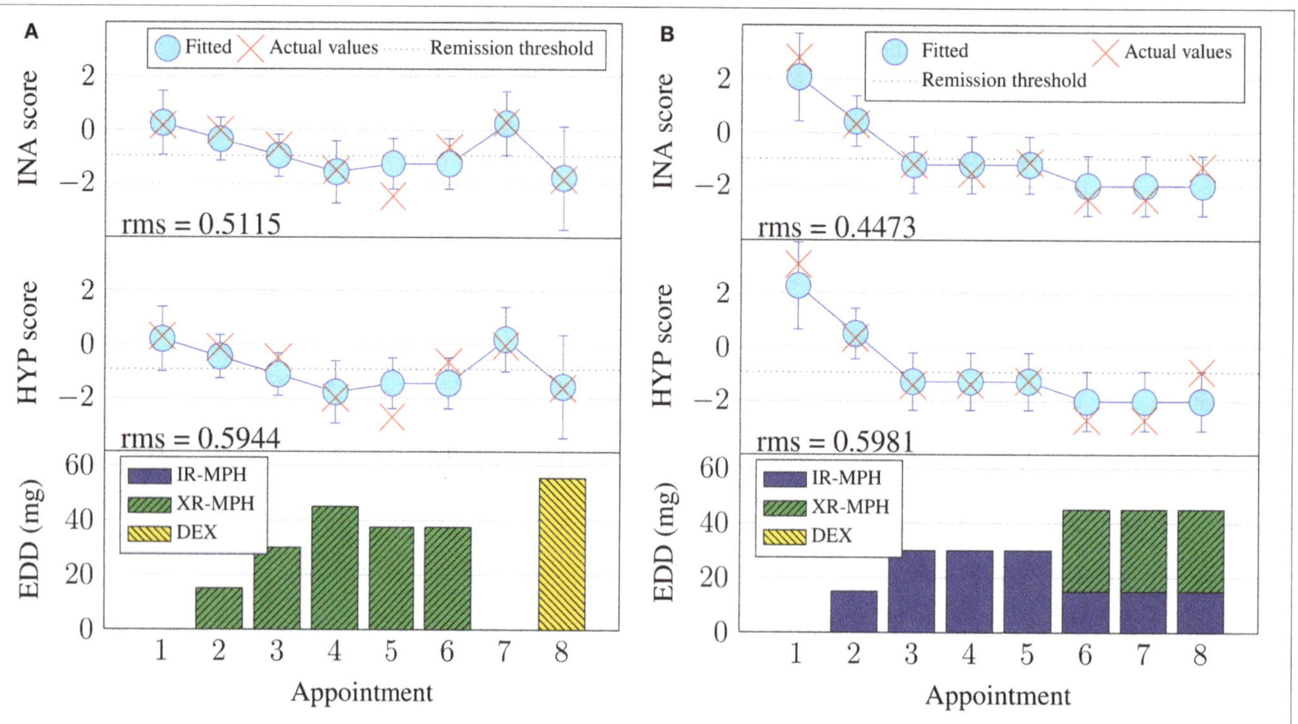

FIGURE 5 | Examples of Bayesian linear regression on continuous inattentiveness (INA) and hyperactivity (HYP) symptom scores with the Bayesian linear regression training dataset (BRR). (A) Subject #11, **(B)** Subject #74.

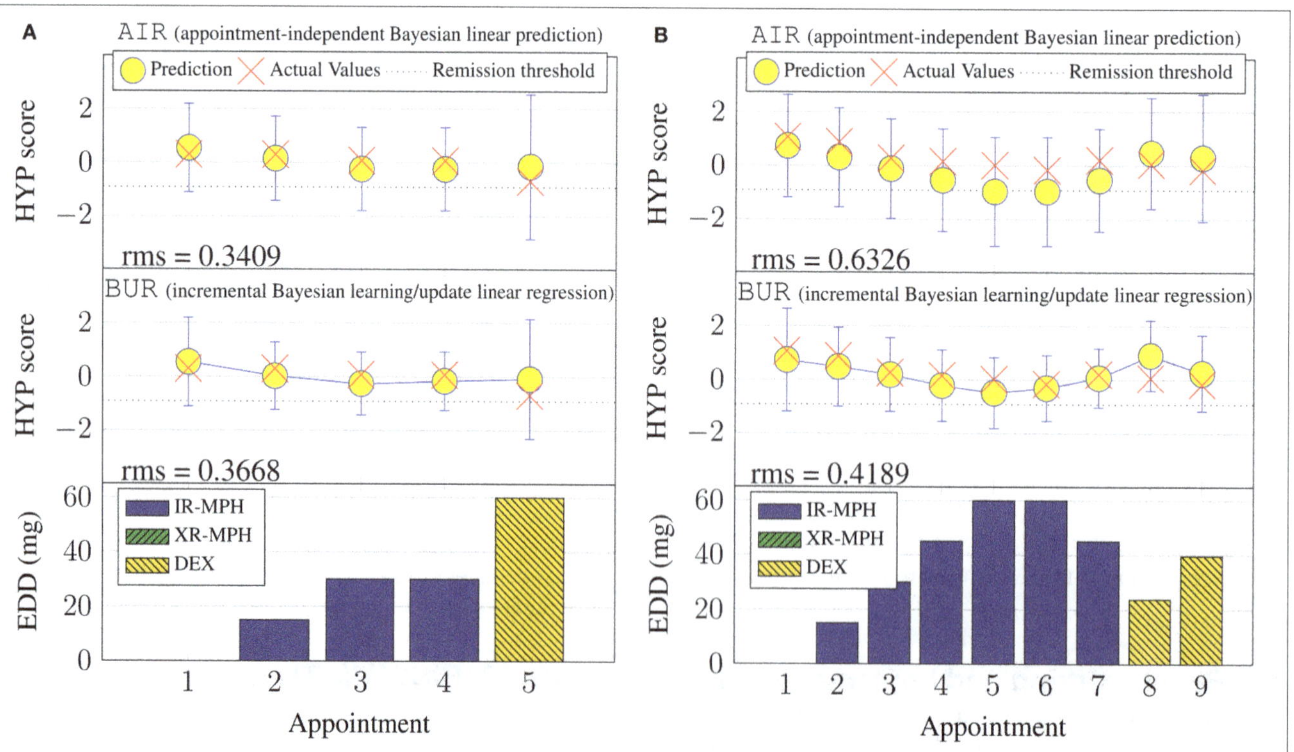

FIGURE 6 | Examples of continuous hyperactivity symptom score prediction with the validation set. (A) Subject #6. **(B)** Subject #148.

5.1. Mixed Effects Models

Linear mixed effects models (MEM) are widely used in many fields; for instance, biology (Rico et al., 2007), ecology (Stevens et al., 2007), linguistics (Nooteboom and Quené, 2008) and social sciences (Kliegl et al., 2009). They extend upon classical linear regression techniques to support data that have some form of grouping. For example, in this paper, each patient had one or more clinical appointments, and the data from each subject form a group. For each patient s with a number of appointments (from 1 to A_s), the severity symptom score vector $\mathbf{y}_s$ (as in Equation 2) is, for simplicity, assumed to have a linear relationship with the baseline and treatment effect via the following formulation:

$$\mathbf{y}_s = q_s + b_0 + \hat{\mathbf{X}}_s \boldsymbol{\omega}_s + \boldsymbol{\epsilon}, \tag{12}$$

where $\hat{\mathbf{X}}_s$ is similar in structure to $\mathbf{X}$ defined in Equation (4) but without the last column of ones; $\boldsymbol{\omega}_s \in \mathbb{R}_{P\times 1}$ is the subject-specific parameter vector for the fixed effects, $q_s \sim N(0, \sigma_q^2)$ is the random effect affecting only the intercept b_0, and $\boldsymbol{\epsilon} \sim N(0, \sigma_{\epsilon,s}^2) \in \mathbb{R}_{A_s\times 1}$ is an error term assumed to have a normal distribution. Observe that in this model, the scalar intercept b_0 is a fixed component for all of the patients in the population and the random effect a_s is a subject-specific scalar and is not grouped under any other parameters.

A linear mixed effects model was constructed within the R software (R Core Team, 2013), using the package "*lme4*" (Bates et al., 2015). Severity score predictions were produced by performing out-of-sample forecasts, i.e., on the validation data for each of the folds, using the "*predict*" function in the R software. To generate a prediction, the random effects are assumed to be zero and the population intercept was used. The continuous symptom score regression results for the MEM are prefixed `MER`.

For a dichotomized clinical remission classification, the symptom score thresholds 0.92 and 0.97 from Section 2.3 for hyperactivity and inattentiveness were used to generate the ground truths. Following the rationale in Section 4.3, the symptom scores predicted by the MEM are given thresholds at different levels to produce a set of classifiers trading off sensitivity against specificity. These threshold-adjusted classifiers are labeled `taMEC`. The best (in terms of Youden's statistic) threshold settings found using the training data were used for the validation data; the thresholds were 0.05 and -0.20, respectively, for the inattentiveness and hyperactivity symptom scores. In addition, the "*melogit*" function in the Stata software (StataCorp, 2015) was used to directly estimate a mixed effects logistic regression model—a MEM with a logistic link function that predicts the probability of the binary remission outcome. In this case, the threshold procedure was applied to the probabilities rather than the raw symptom scores. The resulting classifier is labeled `lrMEC`.

5.2. Support Vector Machines and Gaussian Processes

In addition to MEM, machine learning classification approaches using support vector machines (SVM) and Gaussian processes (GP) were benchmarked. Both the SVM and GP learning methods are kernel machines and were implemented using linear (dot product kernel: $k(\mathbf{x}_i, \mathbf{x}_j) = (\mathbf{x}_i.\mathbf{x}_j)$) and nonlinear kernels (the Gaussian kernel: $k(\mathbf{x}_i, \mathbf{x}_j) = \exp\left[-\gamma(\|\mathbf{x}_i - \mathbf{x}_j\|)^2\right]$). Readers are invited to refer to Burges (1998) for a detailed description of support vector machines and to Rasmussen and Nickisch (2010) for a detailed description of GP learning. Compared to MEM and the learning in the model space approach, the SVM and GP are non-parametric methods—there are no subject specific parameters to identify; the models map the subject-specific inputs, such as baseline characteristics and the medication dosage, to the output symptom scores.

For the SVM, nested cross-validation was employed to optimize the parameters in the model. A broad log range spanning $[10^{-3} : 10^{2}]$ was arbitrarily chosen as the search range for the regularization parameter C. Similarly, the gamma parameter of the Gaussian kernel was optimized in the log range spanning $[10^{-4} : 10^{1}]$. For the GP, the model parameters were optimized using conjugate gradient descent, avoiding the need for nested cross-validation. SVM and GP learning approaches were employed as regression models (support vector regression `SVR` and Gaussian process regression `GPR`) for the linear and nonlinear kernels to predict the clinical scores. Dichotomous remission predictions were obtained by thresholding the distance from the hyperplane for the SVM, and for thresholding the probabilistic predictions of class membership for GP. These binary classifiers are respectively labeled as `SVC` and `GPC`.

From Section 3.6, the number of remission cases outweighed non-remission cases by a ratio of roughly 1:7. For a classification task, this imbalance of data is problematic for many classification algorithms (He and Garcia, 2009). To help alleviate this, a downsampling approach was implemented for both linear and nonlinear kernels of the SVM and GP classifiers `dsSVC` and `dsGPC`. During the training phase, the non-responder class was downsampled randomly to match the number of training instances in the remission class. By repeating this downsampling procedure, an ensemble of 1,000 classifiers was trained. A classification prediction was generated by majority voting of the ensemble. Additionally, for the SVMs, an alternative is to learn the regularization parameters C on a per-class basis. The rationale is that a higher penalty for errors can be placed on the more abundant class (Osuna et al., 1997); this method is referred to as the weighted SVM (`rwSVC`). The per-class C parameters C^{+} and C^{-} were optimized using the ranges $\left[10^{-3} : 10^{1}\right]$ and $\left[10^{-2} : 10^{2}\right]$, respectively. Finally, for the Gaussian process classifier, it is possible to calibrate the probabilistic predictions in order to help account for imbalanced data (Bishop, 2006); this approach is referred to as a re-calibrated GP (`rcGPC`).

Using MATLAB software, the SVM was implemented using the `libsvm` toolbox (Chang and Lin, 2011), the Gaussian process learning was implemented using the `GPML` toolbox (Rasmussen and Nickisch, 2010).

6. RESULTS AND DISCUSSIONS

6.1. Continuous Symptom Score Prediction

The rms errors across all models are reported in **Table 1**.

TABLE 1 | Rms errors for predicting symptom scores for inattentiveness (INA) and hyperactivity (HYP) using the (A) learning in model space and (B) conventional approaches.

(1A) Learning in model space

	Inattentiveness		Hyperactivity	
	Method 1	**Method 2**	**Method 1**	**Method 2**
AIR*	0.98	0.84	0.97	0.85
BRR†	0.82	0.82	0.84	0.84
BUR‡	0.99	0.73	1.01	0.75

*AIR: *appointment-independent Bayesian linear prediction.*
†BRR: *retrospective Bayesian linear regression.*
‡BUR: *incremental Bayesian learning/update linear regression.*

(1B) Conventional approaches

Kernel	Inattentiveness		Hyperactivity	
	Linear	**Nonlinear**	**Linear**	**Non-linear**
SVR*	0.73	0.74	0.76	0.81
GPR†	0.72	0.77	0.76	0.84
MER‡	0.82		0.83	

*SVR: *support vector machine regression.*
†GPR: *Gaussian processes regression.*
‡MER: *mixed effects regression.*

6.1.1. Learning in the Model Space Approach

Looking at the learning in the model space approach, one can observe that the virtual patient profile construction method, labeled Method 2, resulted in lower errors overall compared to Method 1. In Method 1, the mappings were learnt using simple linear regression from baseline variables to the parameter space. In addition to this simple linear regression, low degree polynomial (quadratic to quartic) basis functions were tried; whilst degrees up to a cubic resulted in a slightly lower training error, there was worse generalizability (i.e., higher validation error). For Method 2, the incremental Bayesian learning (BUR) approach performed the best overall; its performance advantage over the appointment-independent prediction (AIR) approach is expected given that it allows the model to adapt to a new patient as the treatment continues. The performance advantage over the retrospective Bayesian linear regression (BRR) approach can be attributed to the fact that the virtual patient profile (Section 3.3) had utilized the prior whilst the Bayesian linear regression only uses the elicited prior (see Section 3.2.1). This supports the fact that the training population was able to add valuable information to the prediction task.

We recall that the BUR constructs virtual patient profiles while the BRR only uses the prior knowledge. It is interesting to note that the BRR outperforms the BUR using Method 1, suggesting that Method 1 was not an effective method for incorporating information from existing patient models.

Figure 7 shows the rms values averaged across all subjects during the validation phase and sorted by the clinical appointment (visit) number. Data above 15 visits are not shown as only a single patient had more than 15 visits. There is a slight downward trend visible with the BUR; suggesting that incremental Bayesian learning approach is able to reduce the prediction error as more data are known about a new patient through repeated appointments. The BRR also shows a downward trend, but the error is slightly higher than the BUR. This is because the BRR starts with only the elicited prior and performs learning (fitting) when more data are available, unlike the BUR which starts off with information from the training set in the form of a constructed/estimated virtual patient profile.

Figure 5 illustrates some examples of Bayesian linear regression performed during the training phase and their associated fitting rms errors. It can be seen that Bayesian linear regression fits similarly well for both the inattentiveness and hyperactivity symptom scores. Switching to a new type of medicine is usually associated with larger uncertainty (error bars). Looking at subject #74 (**Figure 5B**) in particular, it can be seen that, despite having the same input dosage from appointments 6–8, there were variations in the severity of the ADHD symptoms. It is not possible to know the exact reason for the variation for this subject during this particular period, without further information—perhaps this was due to adherence issues (the patient not taking their medication as prescribed), physiological factors, measurement "noise," or perhaps something else entirely. By design, Bayesian linear regression can only fit the same outcome given the same input. This does highlight the fact that the current model may not have enough information in the form of covariates to account for some of these factors.

A subset of results for the validation phase is plotted in **Figure 6**. For brevity, only prediction outcomes for hyperactivity

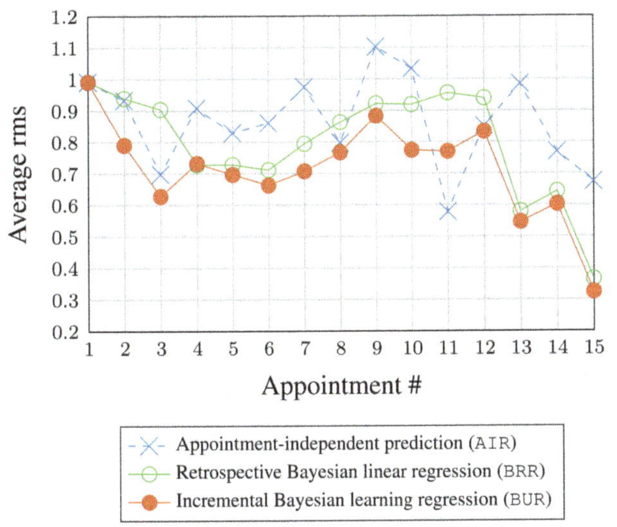

FIGURE 7 | Rms prediction (validation) error averaged across all subjects vs. appointment number.

symptom scores are shown and retrospective Bayesian linear regression results (BRR) are omitted. The lack of solid lines connecting the predictions in the topmost subplot serves as a reminder that the model does not incorporate temporal aspects for the case of appointment-independent (AIR) prediction, which treats each appointment as the first (new) appointment for a new patient. This is also why the 95% confidence intervals for AIR are larger (more uncertain) than those for the BUR. Also note that, by design, the prediction results for the first appointment are identical for both approaches.

The figures illustrate that, during prognosis, incremental learning does not always improve the prediction error compared to simply predicting at every appointment without updating the model using new information. However, based on the rms errors in **Table 1A**, one expects incremental learning to perform better overall across subjects, especially for subjects with a prediction offset, such as over- or under-estimates. This is illustrated by the results for subject #148 given in **Figure 6B**, where the virtual patient profile for this patient consistently underestimates the actual hyperactivity score. Here, the incremental Bayesian learning was able to adapt the parameter ω and shifted the prediction upwards, resulting in lower prediction errors over the subsequent appointments.

6.1.2. Conventional Machine Learning Approaches

Looking at the results in **Table 1B**, the conventional approaches yield similar performance, with linear SVR and GPR methods performing better than their nonlinear counterparts. The mixed effects model has slightly worse results. Errors of linear SCR and linear GPR are similar to each other, and to those for the learning in the model space approach BUR with Method 2. We conclude that for the task of predicting continuous symptom scores with the dataset investigated, the learning in the model space approach performs comparably with conventional approaches.

6.2. Dichotomous Remission Prediction

6.2.1. Learning in the Model Space Approach

For the learning in the model space approach, the ROC curves for the dichotomous predictor are plotted in **Figure 8** for both of the virtual patient profile (Section 3.3) construction methods. As the results for inattentiveness and hyperactivity scores were similar, only the ROC curves for inattentiveness are shown. The AUC values are given in the legend. The crosses on the lines mark the resulting classifier performance if one uses point estimates for the continuous symptom score from the model and simply applies the clinical remission thresholds. The squares mark the classifiers that have critical values based on maximizing the Youden's J-statistic (or the BAC, see Section 4.2.3) for the training set—this is equivalent to the sensitivity and specificity measures being maximized equally as a function of the critical values. Lastly, the circles mark the best classifier for the validation set in terms of the Youden's J-statistic. The closer the squares are to the circles, the better optimized the classifier is assuming no knowledge of the validation dataset. Those optimized classifiers marked by squares in the graph are used to generate various binary classifier performance metrics (see Section 4.2) in **Table 2**.

Looking at **Table 2**. The confusion matrices (CFM) show the number of true positives and false negatives in the first column, and false positives and true negatives in the second column. These may be used to calculate any classification performance metrics not included in this paper, such as the F-measure.

Similar to the continuous symptom prediction task, the BRC outperforms the AIC showing that posterior information is utilized effectively. As in the continuous task, the virtual patient profile construction method labeled Method 2 is better overall than Method 1, but the difference is much smaller in the classification task and the advantage is not universal across all metrics, especially for the AIC. Note that the virtual patient profile construction method has little effect on the BRC as it does not use it. The BRC achieves higher sensitivity values but a lower PPV compared to the BUC, meaning that the BRC is better at recalling remission cases, but the remission predictions by the BUC are more reliable. The SPC achieved by the BUC is notably higher, being better at ruling out false positives.

6.2.2. Conventional Machine Learning Approaches

Table 3 shows the binary classifier performance metrics for the conventional machine learning approaches. Apart from the AUC, all of the other metrics in the table were derived from classifier settings (set-points) that had optimized the balanced accuracy (BAC) during the training stage. Apart from rcGPC, the BAC values across the different approaches are similar. The MEC classifiers perform well compared with GPC and SVC, with consistently high AUC values for both inattentiveness and hyperactivity. However, the set-points of the MEC classifiers achieve lower sensitivity (but higher specificity) than the SVC. As mentioned in Section 4.2, a higher sensitivity is more important for this exercise. PPV is the other measure of interest; the lrMEC, in particular, achieved the highest PPV amongst all the conventional approaches—partially helped by its low sensitivity.

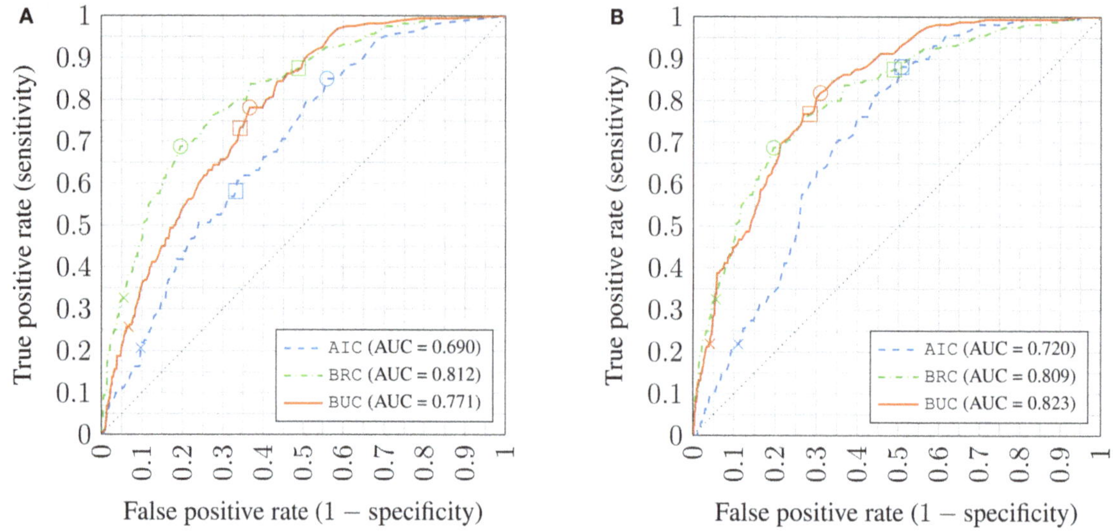

FIGURE 8 | Receiver operator characteristic (ROC) plots of inattentiveness prediction using virtual patient profile constructed by Methods 1 and 2 (A,B) in the learning in model space approach; crosses: no critical value adjustment (based on point estimates); squares: best performing critical values on training set; circles: best performing critical values on the validation set; AUC: Area under the ROC curve AIC: appointment-independent classifier; BRC: retrospective Bayesian linear regression classifier; BUC: incremental Bayesian learning/update classifier.

The ROC plot for the MEC is shown in **Figure 9**. The lrMEC variant fitted the training set better but both the lrMEC and the taMEC achieve similar validation performance. Tracing the ROC values, the lrMEC seems more suitable for high specificity settings while taMEC appears to be more suitable for high sensitivity classification.

For the machine learning approaches GPC and SVC, linear models work better. This was similarly observed in the continuous symptom score prediction task. Comparing methods in tackling data imbalance, the weighted SVM classifier rwSVC method performed better than the downsampled dsSVC method, while the downsampled Gaussian process classifier dsGPC method performed better than the re-calibrated rcGPC. Looking at both the BAC and AUC metrics, rwSVC and dsGPC perform similarly, with the former slightly better at classifying remission of hyperactivity, whereas the latter is slightly better for inattentiveness.

Overall for the conventional methods, the rwSVC achieves the best compromise between SEN and PPV, meaning that it can identify remission cases more readily and at the same time the remission predictions are more reliable. Comparing **Tables 2**, **3** it can be seen that the learning in the model space approach is superior overall. With respect to the BAC and AUC measures, the best performing BUC approach has an advantage of about 6–7%. This is interesting given the similar performance in the continuous symptom score prediction task amongst all approaches. The rms error measure in the symptom score prediction task was based on point estimate calculations, and thus used no information on the shape of the posterior distribution. The posterior predictive distribution (Section 3.4) for the BUC has a Student's t-distribution specific to each patient. The distributions were used to construct a probabilistic threshold in trading off specificity and specificity. This subject-specific nonlinear thresholding procedure may have contributed to its performance advantage over other approaches.

6.2.3. Comparison with Literature

As far as the authors are aware, Kim et al. (2015) is the only published literature on treatment response prediction of ADHD patients using machine learning techniques. Their best attempt achieved an AUC value of 0.84 and 86.4% classification accuracy (that is, the percentage of correct predictions, different from the BAC measure used in this paper) using a wide range of information types including demographical, clinical, genetic, environmental, neuropsychological and neuroimaging measures. In comparison, this paper includes only the more readily obtainable demographical and clinical information and is able to achieve best-case AUCs of 0.82–0.84. Restricting to demographical and clinical information, the highest performing method using SVMs in Kim et al. (2015) had an AUC of 0.69. Granted, the comparison is imprecise because the quality, quantity and sources of demographical and clinical information are different between this paper and Kim et al. (2015). Judging from the AUC values achieved by SVMs in this paper of about 0.71 (see **Table 3**), the results appears to be very close to those in Kim et al. (2015). Due to this similarity, the previous comparisons should be valid.

7. CLINICAL UTILITY AND FURTHER WORK

The proposed learning in the model space approach is capable of predicting, for an individual, the minimum dosage of a particular medication required to have a user-defined chance of achieving symptomatic remission. It is highly flexible and potentially can be extended to any disease or disorder where medication is used

TABLE 2 | Sensitivity, specificity, accuracy, and AUC of the remission classifier with critical values adjusted with respect to uncertainties in the predicted symptom scores.

		Inattentiveness				Hyperactivity			
		Method 1		Method 2		Method 1		Method 2	
CFM[a]	AIC*	93	327	141	501	82	314	121	448
		67	660	19	486	57	694	18	560
	BRC†	140	482	140	483	125	480	125	483
		20	505	20	504	14	528	14	525
	BUC‡	117	338	123	280	97	307	109	253
		43	649	37	707	42	701	30	755
SEN[b]	AIC	58.1%		88.1%		59.0%		87.1%	
	BRC	87.5%		87.5%		89.9%		89.9%	
	BUC	73.1%		76.9%		69.8%		78.4%	
SPC[c]	AIC	66.9%		49.2%		68.9%		55.6%	
	BRC	51.2%		51.1%		52.4%		52.1%	
	BUC	65.8%		71.6%		69.5%		74.9%	
BAC[d]	AIC	62.5%		68.7%		63.9%		71.3%	
	BRC	69.3%		69.3%		71.2%		71.0%	
	BUC	69.4%		74.3%		69.7%		76.7%	
PPV[e]	AIC	22.1%		22.0%		20.7%		21.3%	
	BRC	22.5%		22.5%		20.7%		20.6%	
	BUC	25.7%		30.5%		24.0%		30.1%	
NPV[f]	AIC	90.8%		96.2%		92.4%		96.9%	
	BRC	96.2%		96.2%		97.4%		97.4%	
	BUC	93.8%		95.0%		94.4%		96.2%	
AUC[g]	AIC	69.0%		72.0%		68.0%		73.8%	
	BRC	81.2%		80.9%		83.6%		83.3%	
	BUC	77.1%		82.3%		76.7%		84.4%	

[a]*CFM: confusion matrix.*
[b]*SEN: sensitivity.*
[c]*SPC: specificity.*
[d]*BAC: balanced accuracy.*
[e]*PPV: positive predictive value.*
[f]*NPV: negative predictive value.*
[g]*AUC: area under ROC curve.*
**AIC: appointment-independent classifier.*
†*BRC: retrospective Bayesian linear regression classifier.*
‡*BUC: incremental Bayesian learning/update classifier.*
Shaded values represent best performance amongst the compared methods.

in the course of treatment, speeding up and reducing the cost of the dose optimization/forced titration process, and potentially improving the quality of life for patients by ending the treatment sooner.

The current model, however, does not take into account adverse drug reactions (ADRs), minimization of which is another goal of a dose optimization titration process. To improve clinical utility, it is essential that ADRs are modeled. While data on this are available from the clinical notes accompaning the ADDUCE trial, a different modeling approach is required to incorporate the many different types of ADRs, with prevalence ranging from infrequent to very rare.

While the proposed approach achieves excellent performance in terms of treatment response classification, there is room for improvement. One obvious way to achieve this is to incorporate more data, especially covariates that are functions of time. In this exercise for example, the body mass index and age variables measured at baseline (first appointment) of the patients contribute to the latent factors, which in turn form the baseline variables. As such, they do not vary over time. It may be worth investigating whether the addition of temporal covariates, such as blood pressure, would improve the model.

Another venue for potential improvement is to extend the linear model to a nonlinear model—there is no guarantee that all

TABLE 3 | Sensitivity, specificity, accuracy, and AUC of the remission classifier with critical values adjusted with respect to uncertainties in the predicted symptom scores.

<table>
<tr><th></th><th></th><th colspan="2">Inattentiveness</th><th colspan="2">Hyperactivity</th></tr>
<tr><th></th><th></th><th>Linear</th><th>Non-linear</th><th>Linear</th><th>Non-linear</th></tr>
<tr><td rowspan="6">Sensitivity</td><td>dsSVC[a]</td><td>70.0%</td><td>70.6%</td><td>67.6%</td><td>66.9%</td></tr>
<tr><td>dsGPC[b]</td><td>68.1%</td><td>67.5%</td><td>67.6%</td><td>66.9%</td></tr>
<tr><td>rwSVC[c]</td><td>76.9%</td><td>33.8%</td><td>76.9%</td><td>69.1%</td></tr>
<tr><td>rcGPC[d]</td><td>43.1%</td><td>43.2%</td><td>71.3%</td><td>48.2%</td></tr>
<tr><td>taMEC[e]</td><td colspan="2">60.0%</td><td colspan="2">58.9%</td></tr>
<tr><td>lrMEC[f]</td><td colspan="2">53.1%</td><td colspan="2">56.0%</td></tr>
<tr><td rowspan="6">Specificity</td><td>dsSVC</td><td>61.9%</td><td>62.3%</td><td>67.8%</td><td>66.6%</td></tr>
<tr><td>dsGPC</td><td>67.9%</td><td>69.1%</td><td>67.8%</td><td>71.3%</td></tr>
<tr><td>rwSVC</td><td>55.9%</td><td>77.6%</td><td>62.1%</td><td>62.1%</td></tr>
<tr><td>rcGPC</td><td>59.0%</td><td>18.4%</td><td>50.7%</td><td>50.6%</td></tr>
<tr><td>taMEC</td><td colspan="2">73.3%</td><td colspan="2">73.8%</td></tr>
<tr><td>lrMEC</td><td colspan="2">81.4%</td><td colspan="2">81.4%</td></tr>
<tr><td rowspan="6">Balanced accuracy</td><td>dsSVC</td><td>66.0%</td><td>66.5%</td><td>67.7%</td><td>66.7%</td></tr>
<tr><td>dsGPC</td><td>68.0%</td><td>68.3%</td><td>67.7%</td><td>69.1%</td></tr>
<tr><td>rwSVC</td><td>66.4%</td><td>55.7%</td><td>69.5%</td><td>65.6%</td></tr>
<tr><td>rcGPC</td><td>51.1%</td><td>44.8%</td><td>46.9%</td><td>49.4%</td></tr>
<tr><td>taMEC</td><td colspan="2">66.6%</td><td colspan="2">66.4%</td></tr>
<tr><td>lrMEC</td><td colspan="2">67.2%</td><td colspan="2">70.6%</td></tr>
<tr><td rowspan="6">Positive predictive value</td><td>dsSVC</td><td>23.0%</td><td>23.3%</td><td>22.4%</td><td>21.6%</td></tr>
<tr><td>dsGPC</td><td>25.6%</td><td>26.2%</td><td>22.4%</td><td>24.3%</td></tr>
<tr><td>rwSVC</td><td>22.0%</td><td>19.6%</td><td>21.9%</td><td>20.1%</td></tr>
<tr><td>rcGPC</td><td>15.6%</td><td>12.4%</td><td>10.8%</td><td>11.9%</td></tr>
<tr><td>taMEC</td><td colspan="2">26.7%</td><td colspan="2">23.7%</td></tr>
<tr><td>lrMEC</td><td colspan="2">31.6%</td><td colspan="2">29.0%</td></tr>
<tr><td rowspan="6">Negative predictive value</td><td>dsSVC</td><td>92.7%</td><td>92.9%</td><td>93.8%</td><td>93.9%</td></tr>
<tr><td>dsGPC</td><td>92.9%</td><td>92.9%</td><td>93.8%</td><td>93.4%</td></tr>
<tr><td>rwSVC</td><td>93.7%</td><td>87.8%</td><td>95.1%</td><td>93.6%</td></tr>
<tr><td>rcGPC</td><td>86.5%</td><td>79.8%</td><td>86.6%</td><td>87.63%</td></tr>
<tr><td>taMEC</td><td colspan="2">91.8%</td><td colspan="2">92.9%</td></tr>
<tr><td>lrMEC</td><td colspan="2">91.5%</td><td colspan="2">94.1%</td></tr>
<tr><td rowspan="6">Area under ROC curve</td><td>dsSVC</td><td>71%</td><td>69%</td><td>73%</td><td>71%</td></tr>
<tr><td>dsGPC</td><td>75%</td><td>71%</td><td>73%</td><td>70%</td></tr>
<tr><td>rwSVC</td><td>71%</td><td>60%</td><td>76%</td><td>71%</td></tr>
<tr><td>rcGPC</td><td>49%</td><td>41%</td><td>46%</td><td>48%</td></tr>
<tr><td>taMEC</td><td colspan="2">74.8%</td><td colspan="2">77.5%</td></tr>
<tr><td>lrMEC</td><td colspan="2">75.8%</td><td colspan="2">77.2%</td></tr>
</table>

[a]dsSVC: down-sampled support vector machine classifier; [b]dsGPC: down-sampled Gaussian processes classifier; [c]rwSVC: regularization-weighted support vector machine classifier; [d]rwGPC: regularization-weighted support Gaussian processes classifier; [e]taMEC: threshold-adjusted mixed effects classifier; [f]lrMEC: logitic regression mixed effects classifier. Shaded values represent best performance amongst the compared methods.

the covariates have a linear relationship with treatment response. Identifying the level and nature of nonlinear relationships is the first challenge. In the current Bayesian framework, the introduction of nonlinearities increases computational complexity for Bayesian inference, requiring the use of techniques such as Gibbs sampling.

There are other areas of interest. For example, what is the optimum strategy, in terms of timing and requirements, for incorporating semi-new patient data to the model space to improve the generalizability of the model for other new patients? How can medical adherence/concordance be modeled? Does gender of the patient play a role in their treatment response?

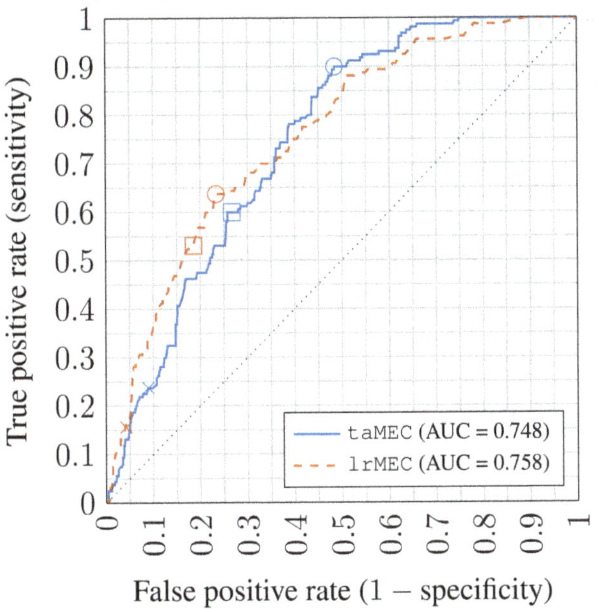

FIGURE 9 | Receiver operator characteristic (ROC) plot of inattentiveness prediction using mixed effects models; crosses: no threshold adjustment (based on point estimates); squares: best performing threshold setting on training set; circles: best performing threshold setting on the validation set; AUC: Area under the ROC curve; `taMEC`: threshold-adjusted mixed effects classifier; `lrMEC`: logitic regression mixed effects classifier.

8. CONCLUSION

A learning in the model space framework has been utilized to develop a personalized medicine approach to treatment response prediction. First of all, factor analysis was performed to extract latent factors from a large clinical dataset, collected from a UK sample of 157 patients suffering from attention-deficit hyperactivity disorder. The resulting reduced-order patient information was then encoded in a model parameter space resulting in a cloud of personalized models. Then, the patient-specific model space parameters were used to train a Bayesian linear regression model. New patients are then matched to existing patients most similar to themselves to obtain a virtual patient profile, which in turn forms a prior parameter set for the Bayesian linear regression model. Through a Bayesian update algorithm, new data are continuously integrated to improve the prediction performance for a given patient. In addition, the parameters of the "new" patients can be added to the model parameter space (once sufficient data are available) to improve the generalizability of the model for future patients.

Comparisons were made between the learning in the model space approach with conventional data-driven machine learning and regression approaches. In terms of the prediction of the continuous symptom factor scores, the performance of the learning in model space framework was on a par with conventional approaches. However, the new approach is shown to outperform support vector machines, Gaussian processes and linear mixed effects classifiers in the prediction of symptomatic remission. The effective gain in classification performance of the new model can potentially speed up and reduce the cost of a forced titration or dose optimization titration process, which is normally manually performed by the clinician to assess the effective dosage of medication. Further work includes incorporating the prediction of adverse drug reactions, which is also an important element in the dose optimization titration process.

ETHICS STATEMENT

This is secondary data analysis of the ADDUCE ADHD study. ADDUCE has received a favorable ethical opinion from the relevant UK National Health Service (NHS) Research Ethics Committee (REC reference 11/ES/0016). All subjects gave written informed consent in accordance with the Declaration of Helsinki. The protocol was approved by the REC. Consent to information access and secondary data analysis has been granted. All data were handled as per relevant guidelines and rules for such information, including the Data Protection Act 1998 and as per the initial ethical approval.

AUTHOR CONTRIBUTIONS

HW was responsible for performing feature extraction of the clinical data, literature search, causal factor model simplification, the software implementation of the learning in model space framework, and determined benchmarking protocol. PAT provided the clinical context of ADHD, constructed the prior knowledge used by the Bayesian framework, aided development on the causal factor model, and designed the literature search protocol. PW aided HW on the literature search and written the background on ADHD. OD implemented the SVM and GP machine learning algorithms. BL constructed the linear mixed effects models. YS helped with the technical implementation of the Bayesian linear regression algorithm. PT provided the mathematical formulation of the learning in model space approach and guidance on software implementation. MC and TN suggested candidate modeling approaches compatible with the framework, suggested a number of conventional approaches to be evaluated, appraised the output of the Bayesian approach, and provided recommendations for the improvement of the implementation and the methodology to ensure fair comparisons. SI coordinated access to the clinical data, facilitated digitization of paper-based clinical records, and maintained the database. DC provided the clinical context of ADHD, suggested causal mechanisms for treatment response, and defined the scope of clinical data collection.

FUNDING

We gratefully acknowledge the support from the UK Engineering and Physical Sciences Research Council (EPSRC), grant number EP/L000296/1. The ADDUCE project from which this piece of research borrowed clinical data is funded by the European Union's Seventh Framework Programme for research, technological development and demonstration under grant agreement number 260576.

ACKNOWLEDGMENTS

The authors would like to thank the staff of Ninewells Hospital and Medical School, the University of Dundee, and NHS Tayside; in particular trial database manager Emma McKenzie for coordinating the data exchange and clinical research nurse Jacqueline Paton for her meticulous data digitization efforts. The authors also would like to acknowledge the invaluable feedback by fellow POEMS (Predictive modeling for hEalthcare through MathS) network members and members in the ADDUCE consortium.

REFERENCES

Abdi, H. (2003). "Factor rotations," in *Encyclopedia for Research Methods for the Social Sciences*, eds M. Lewis-Beck, A. Bryman, and T. Futimg (Thousand Oaks, CA: Sage Publications), 978–982.

Alberg, A. J., Park, J. W., Hager, B. W., Brock, M. V., and Diener-West, M. (2004). The use of "overall accuracy" to evaluate the validity of screening or diagnostic tests. *J. Gen. Intern. Med.* 19, 460–465. doi: 10.1111/j.1525-1497.2004.30091.x

American Psychiatric Association (2000). *Diagnostic and Statistical Manual of Mental Disorders, 4th Edn.* Washington, DC: APA.

American Psychiatric Association (2013). *Diagnostic and Statistical Manual of Mental Disorders, 5th Edn.* Washington, DC: APA.

Angold, A., Costello, E. J., van Kämmen, W., and Stouthamer-Loeber, M. (1996). Development of a short questionnaire for use in epidemiological studies of depression in children and adolescents: factor composition and structure across development. *Int. J. Methods Psychiatr. Res.* 5, 251–262.

APS Group Scotland (2012). *Scottish Index of Multiple Deprivation.* Technical Report, The Scottish Government, Edinburgh.

Askland, K. D., Garnaat, S., Sibrava, N. J., Boisseau, C. L., Strong, D., Mancebo, M., et al. (2015). Prediction of remission in obsessive compulsive disorder using a novel machine learning strategy. *Int. J. Methods Psychiatr. Res.* 24, 156–169. doi: 10.1002/mpr.1463

Asparouhov, T., and Muthén, B. (2009). Exploratory structural equation modeling. *Struct. Equat. Model.* 16, 397–438. doi: 10.1080/10705510903008204

Atkins, M. S., Pelham, W. E., and Licht, M. H. (1985). A comparison of objective classroom measures and teacher ratings of attention deficit disorder. *J. Abnorm. Child Psychol.* 13, 155–167.

Banaschewski, T., Coghill, D., Santosh, P., Zuddas, A., Asherson, P., Buitelaar, J., et al. (2006). Long-acting medications for the hyperkinetic disorders. A systematic review and European treatment guideline. *Eur. Child Adolesc. Psychiatry* 15, 476–495. doi: 10.1007/s00787-006-0549-0

Barbaresi, W. J., Katusic, S. K., Colligan, R. C., Weaver, A. L., Leibson, C. L., and Jacobsen, S. J. (2006). Long-term stimulant medication treatment of attention-deficit/hyperactivity disorder: results from a population-based study. *J. Dev. Behav. Pediatr.* 27, 1–10. doi: 10.1097/00004703-200602000-00001

Bates, D., Mächler, M., Bolker, B., and Walker, S. (2015). Fitting linear mixed-effects models using lme4. *J. Stat. Softw.* 67, 1–48. doi: 10.18637/jss.v067.i01

Bishop, C. M. (2006). *Pattern Recognition and Machine Learning.* Berlin: Springer-Verlag.

Bradley, A. P. (1997). The use of the area under the roc curve in the evaluation of machine learning algorithms. *Pattern Recogn.* 30, 1145–1159.

Brodersen, K. H., Ong, C. S., Stephan, K. E., and Buhmann, J. M. (2010). "The balanced accuracy and its posterior distribution," in *Pattern Recognition (ICPR), 20th International Conference on* (Istanbul), 3121–3124.

Brodersen, K. H., Schofield, T. M., Leff, A. P., Ong, C. S., Lomakina, E. I., Buhmann, J. M., et al. (2011). Generative embedding for model-based classification of fMRI data. *PLoS Comput. Biol.* 7:e1002079. doi: 10.1371/journal.pcbi.1002079

Burges, C. J. (1998). A tutorial on support vector machines for pattern recognition. *Data Min. Knowl. Discov.* 2, 121–167. doi: 10.1023/A:1009715923555

Bussing, R., Fernandez, M., Harwood, M., Hou, W., Garvan, C. W., Swanson, J. M., et al. (2008). Parent and teacher SNAP-IV ratings of attention deficit/hyperactivity disorder symptoms: Psychometric properties and normative ratings from a school district sample. *Assessment* 15, 317–328. doi: 10.1177/1073191107313888

Chang, C.-C., and Lin, C.-J. (2011). LIBSVM: a library for support vector machines. *ACM Trans. Intell. Syst. Technol.* 2, 27. doi: 10.1145/1961189.1961199

Chen, H., Tiño, P., Rodan, A., and Yao, X. (2014). Learning in the model space for cognitive fault diagnosis. *IEEE Trans. Neural Netw. Learn. Syst.* 25, 124–136. doi: 10.1109/TNNLS.2013.2256797

Chou, W.-J., Chen, S.-J., Chen, Y.-S., Liang, H.-Y., Lin, C.-C., Tang, C.-S., et al. (2012). Remission in children and adolescents diagnosed with attention-deficit/hyperactivity disorder via an effective and tolerable titration scheme for osmotic release oral system methylphenidate. *J. Child Adolesc. Psychopharmacol.* 22, 215–225. doi: 10.1089/cap.2011.0006

Dopheide, J. A., and Pliszka, S. R. (2009). Attention-deficit-hyperactivity disorder: an update. *Pharmacotherapy* 29, 656–679. doi: 10.1592/phco.29.6.656

Doyle, O. M., Tsaneva-Atansaova, K., Harte, J., Tiffin, P. A., Tiño, P., and Diaz-Zuccarini, V. (2013). Bridging paradigms: hybrid mechanistic-discriminative predictive models. *IEEE Trans. Biomed. Eng.* 60, 735–742. doi: 10.1109/TBME.2013.2244598

Fawcett, T. (2006). An introduction to roc analysis. *Pattern Recogn. Lett.* 27, 861–874. doi: 10.1016/j.patrec.2005.10.010

Fletcher, R., and Fletcher, S. (2005). *Clinical Epidemiology: The Essentials.* Epidemiology/Biostatistics. Baltimore, MD: Lippincott Williams & Wilkins.

Goodman, R. (1997-07). The strengths and difficulties questionnaire: a research note. *J. Child Psychol. Psychiatry* 38, 581–586. doi: 10.1111/j.1469-7610.1997.tb01545.x

Goodman, R., Ford, T., Richards, H., Gatward, R., and Meltzer, H. (2000). The development and well-being assessment: description and initial validation of an integrated assessment of child and adolescent psychopathology. *J. Child Psychol. Psychiatry* 41, 645–655. doi: 10.1111/j.1469-7610.2000.tb02345.x

Goodfellow, I., Bengio, Y., and Courville, A. (2016). *Deep Learning.* Cambridge, MA: MIT Press.

Greenhill, L., Kollins, S., Abikoff, H., McCracken, J., Riddle, M., and Swanson, J. M. (2006). Efficacy and safety of immediate-release methylphenidate treatment for preschoolers with ADHD. *J. Am. Acad. Child Adolesc. Psychiatry* 45, 1284–1293. doi: 10.1097/01.chi.0000235077.32661.61

Guy, W. (1974). *ECDEU Assessment Manual for Psychopharmacology, Revised Edn.* Rockville, MD: US Department of Health, Education and Welfare, Public Health Service, Alcohol, Drug Abuse and Mental Health Administration, NIMH Psychopharmacology Research Branch, Division of Extramural Research Programs. DHEW No. ADM 76-338.

He, H., and Garcia, E. (2009). Learning from imbalanced data. *IEEE Trans. Knowledge Data Eng.* 21, 1263–1284. doi: 10.1109/TKDE.2008.239

Hechtman, L. (2005). Effects of treatment on the overall functioning of children with ADHD. *Can. Child Adolesc. Psychiatr. Rev.* 14, 10–15.

Holden, S. E., Jenkins-Jones, S., Poole, C. D., Morgan, C. L., Coghill, D., and Currie, C. J. (2013). The prevalence and incidence, resource use and financial costs of treating people with attention deficit/hyperactivity disorder (ADHD) in the United Kingdom (1998 to 2010). *Child Adolesc. Psychiatry Ment. Health* 7:34. doi: 10.1186/1753-2000-7-34

Horn, J. L. (1965). A rationale and test for the number of factors in factor analysis. *Psychometrika* 30, 179–185. doi: 10.1007/BF02289447

Khangura, S., Konnyu, K., Cushman, R., Grimshaw, J., and Moher, D. (2012). Evidence summaries: the evolution of a rapid review approach. *Syst. Rev.* 1:10. doi: 10.1186/2046-4053-1-10

Kim, J., Sharma, V., and Ryan, N. D. (2015). Predicting methylphenidate response in ADHD using machine learning approaches. *Int. J. Neuropsychopharmacol.* 18, 1–7. doi: 10.1093/ijnp/pyv052

Kliegl, R., Rolfs, M., Laubrock, J., and Engbert, R. (2009). Microsaccadic modulation of response times in spatial attention tasks. *Psychol. Res.* 73, 136–146. doi: 10.1007/s00426-008-0202-2

Leckman, J. F., Riddle, M. A., Hardin, M. T., Ort, S. I., Swartz, K. L., Stevenson, J., et al. (1989). The Yale Global Tic Severity Scale: initial testing of a clinician-rated scale of tic severity. *J. Am. Acad. Child Adolesc. Psychiatry* 28, 566–573. doi: 10.1097/00004583-198907000-00015

Lorenzo-Seva, U., and Ferrando, P. J. (2006). FACTOR: a computer program to fit the exploratory factor analysis model. *Behav. Res. Methods* 38, 88–91. doi: 10.3758/BF03192753

McCarthy, S., Asherson, P., Coghill, D., Hollis, C., Murray, M., and Potts, L. (2009). Attention-deficit hyperactivity disorder: treatment discontinuation in adolescents and young adults. *Br. J. Psychiatry* 273–277. doi: 10.1192/bjp.bp.107.045245

Muthén, B. O., du Toit, S. H. C., and Spisic, D. (1997). *Robust Inference using Weighted Least Squares and Quadratic Estimating Equations in Latent Variable Modeling with Categorical and Continuous Outcomes.* Available online at: https://www.statmodel.com/download/Article_075.pdf

Nooteboom, S. G., and Quené, H. (2008). Self-monitoring and feedback: a new attempt to find the main cause of lexical bias in phonological speech errors. *J. Mem. Lang.* 58, 837–861. doi: 10.1016/j.jml.2007.05.003

O'Hagan, A., and Forester, J. J. (eds.). (2004). "The linear model," in *Bayesian Inference, Vol. 2b, Kendall's Advanced Theory of Statistics*, 2nd Edn. (New York, NY: John Wiley & Sons), 305–339.

Osuna, E., Freund, R., and Girosi, F. (1997). *Support Vector Machines: Training and Applications.* Technical Report AIM-1602, Massachusetts Institute of Technology, Cambridge, MA.

Powers, D. M. W. (2011). Evaluation: from precision, recall and *F*-measure to ROC, informedness, markedness & correlation. *J. Mach. Learn. Technol.* 2, 37–63. doi: 10.9735/2229-3981

R Core Team (2013). *R: A Language and Environment for Statistical Computing.* Vienna: R Foundation for Statistical Computing.

Rasmussen, C. E., and Nickisch, H. (2010). Gaussian processes for machine learning (GPML) toolbox. *J. Mach. Learn. Res.* 11, 3011–3015.

Reeves, G., and Schweitzer, J. (2004). Pharmacological management of attention-deficit hyperactivity disorder. *Expert Opin. Pharmacother.* 5, 1313–1320. doi: 10.1517/14656566.5.6.1313

Rico, D., Martin-Diana, A. B., Bary-Ryan, C., Henehan, G. T. M., and Frias, J. M. (2007). Simultaneous modelling of the thermal degradation kinetics of pectin methylesterase in lettuce (*Lactuca sativa* L.) and carrot (*Daucus carota* L.) extracts: analysis of seasonal variation and tissue type. *Biosci. Biotechnol. Biochem.* 71, 2383–2392. doi: 10.1271/bbb.60484

Rutter, M., Bailey, A., and Lord, C. (2003). *SCQ: The Social Communication Questionnaire.* Torrance, CA: Western Psychological Services.

Safer, D. J., Zito, J. M., and Fine, E. M. (1996). Increased methylphenidate usage for attention deficit disorder in the 1990s. *Pediatrics* 98, 1084–1088.

Schachter, H. M., Pham, B., King, J., Langford, S., and Moher, D. (2001). How efficacious and safe is short-acting methylphenidate for the treatment of attention-deficit disorder in children and adolescents? A meta-analysis. *Can. Med. Assoc. J.* 165, 1475–1488.

Schulz, K. F., Altman, D. G., and Moher, D. (2010). CONSORT 2010 statement: updated guidelines for reporting parallel group randomised trials. *BMJ* 340:c332. doi: 10.1136/bmj.c332

Shaffer, D., Gould, M. S., Brasic, J., Ambrosini, P., Fisher, P., Bird, H., et al. (1983). A children's global assessment scale (CGAS). *Arch. Gen. Psychiatry* 40, 1228. doi: 10.1001/archpsyc.1983.01790100074010

Shen, Y., Tiño, P., and Tsaneva-Atanasova, K. (2016). A classification framework for partially observed dynamical systems. *ArXiv e-prints.*

Spencer, T., Biederman, J., Wilens, T., Harding, M., O'Donnell, D., and Griffin, S. (1996). Pharmacotherapy of attention-deficit hyperactivity disorder across the life cycle. *J. Am. Acad. Child Adolesc. Psychiatr.* 35, 409–432. doi: 10.1097/00004583-199604000-00008

StataCorp (2015). *Stata Statistical Software: Release 14.* College Station, TX: StataCorp LP.

Stevens, M. H. H., Sanchez, M., Lee, J., and Finkel, S. E. (2007). Diversification rates increase with population size and resource concentration in an unstructured habitat. *Genetics* 177, 2243–2250. doi: 10.1534/genetics.107.076869

Storebø, O. J., Ramstad, E., Krogh, H. B., Nilausen, T. D., Skoog, M., Holmskov, M., et al. (2015). Methylphenidate for children and adolescents with attention deficit hyperactivity disorder (ADHD). *Cochrane Database Syst. Rev.* 11:CD009885. doi: 10.1002/14651858.CD009885.pub2

Swanson, J. M. (1992). *School-Based Assessments and Interventions for ADD Students.* Irvine, CA: KC Publishing.

Swanson, J. M., Kraemer, H. C., Hinshaw, S. P., Arnold, L. E., Conners, C. K., and Abikoff, H. B. (2001). Clinical relevance of the primary findings of the MTA: success rates based on severity of ADHD and ODD symptoms at the end of treatment. *J. Am. Acad. Child Adolesc.* 40, 168–179. doi: 10.1097/00004583-200102000-00011

Swanson, J. M., Sandman, C. A., Deutsch, C., and Baren, M. (1983). Methylphenidate hydrochloride given with or before breakfast: I. Behavioral, cognitive, and electrophysiologic effects. *Pediatrics* 72, 49–55. doi: 10.1097/00004583-198311000-00019

The ADDUCE Consortium (2016). *The ADDUCE (Attention Deficit/Hyperactivity Disorder Drugs Use Chronic Effects) Project.* Availbale online at: http://www.adhd-adduce.org/

Thurstone, L. L. (1931). Multiple factor analysis. *Psychol. Rev.* 38, 406–427. doi: 10.1037/h0069792

van der Oord, S., Prins, P. J. M., Oosterlaan, J., and Emmelkamp, P. M. G. (2008). Efficacy of methylphenidate, psychosocial treatments and their combination in school-aged children with ADHD: a meta-analysis. *Clin. Psychol. Rev.* 28, 783–800. doi: 10.1016/j.cpr.2007.10.007

von Elm, E., Altman, D. G., Egger, M., Pocock, S. J., Gøtzsche, P. C., and Vandenbroucke, J. P. (2007). Strengthening the reporting of observational studies in epidemiology (STROBE) statement: guidelines for reporting observational studies. *BMJ* 335, 806–808. doi: 10.1136/bmj.39335.541782.AD

World Health Organization (2010). *ICD-10, 10 Edn.* Geneva: World Health Organization.

Youden, W. J. (1950). Index for rating diagnostic tests. *Cancer* 3, 32–35. doi: 10.1002/1097-0142(1950)3:1<32::AID-CNCR2820030106>3.0.CO;2-3

13

Electrode Position and Current Amplitude Modulate Impulsivity after Subthalamic Stimulation in Parkinsons Disease—A Computational Study

Alekhya Mandali[1], *V. Srinivasa Chakravarthy*[1]*, *Roopa Rajan*[2], *Sankara Sarma*[3] *and Asha Kishore*[2]

[1] *Department of Biotechnology, Bhupat and Jyoti Mehta School of Biosciences, Indian Institute of Technology Madras, Chennai, India,* [2] *Department of Neurology, Comprehensive Care Centre for Movement Disorders, Trivandrum, India,* [3] *Achutha Menon Centre for Health Science Studies, Sree Chitra Tirunal Institute for Medical Sciences and Technology, Trivandrum, India*

Edited by:
Krasimira Tsaneva-Atanasova, University of Exeter, UK

Reviewed by:
Paolo Melillo, Second University of Naples, Italy
Alok Ranjan Nayak, International Institute of Information Technology, India

****Correspondence:***
V. Srinivasa Chakravarthy
schakra@iitm.ac.in

Background: Subthalamic Nucleus Deep Brain Stimulation (STN-DBS) is highly effective in alleviating motor symptoms of Parkinson's disease (PD) which are not optimally controlled by dopamine replacement therapy. Clinical studies and reports suggest that STN-DBS may result in increased impulsivity and *de novo* impulse control disorders (ICD).

Objective/Hypothesis: We aimed to compare performance on a decision making task, the Iowa Gambling Task (IGT), in healthy conditions (HC), untreated and medically-treated PD conditions with and without STN stimulation. We hypothesized that the position of electrode and stimulation current modulate impulsivity after STN-DBS.

Methods: We built a computational spiking network model of basal ganglia (BG) and compared the model's STN output with STN activity in PD. Reinforcement learning methodology was applied to simulate IGT performance under various conditions of dopaminergic and STN stimulation where IGT total and bin scores were compared among various conditions.

Results: The computational model reproduced neural activity observed in normal and PD conditions. Untreated and medically-treated PD conditions had lower total IGT scores (higher impulsivity) compared to HC ($P < 0.0001$). The electrode position that happens to selectively stimulate the part of the STN corresponding to an advantageous panel on IGT resulted in de-selection of that panel and worsening of performance ($P < 0.0001$). Supratherapeutic stimulation amplitudes also worsened IGT performance ($P < 0.001$).

Conclusion(s): In our computational model, STN stimulation led to impulsive decision making in IGT in PD condition. Electrode position and stimulation current influenced impulsivity which may explain the variable effects of STN-DBS reported in patients.

Keywords: impulsivity, sub thalamic stimulation, Parkinson's disease, Iowa gambling task, reinforcement learning

INTRODUCTION

Deep brain stimulation (DBS) of the subthalamic nucleus (STN), is a surgical technique now widely applied for the treatment of Parkinson's disease (PD) when dopamine replacement therapy fails to provide sustained relief of motor symptoms or induces drug-induced dyskinesias (Benabid, 2003). Though the exact mechanism of action of DBS is not well-established, it is known that stimulation disrupts (Rosa et al., 2012) the increased synchrony and bursting activity in the β band (8–30 Hz; Kühn et al., 2006) of the STN neurons.

Several reports have highlighted the development of new onset, often transient, impulse control disorders (ICDs) following STN stimulation (Hershey et al., 2004; Smeding et al., 2007; Combs et al., 2015). This was thought to be, due to stimulation of the cognitive sub territory of STN or the spread of stimulation to adjacent parts of the cortico-limbic circuits. In support of this theory, stimulation parameters such as current spread and electrode position were shown to affect the outcome in cognitive tasks in PD patients (Sudhyadhom et al., 2007; York et al., 2009; Witt et al., 2013). Stimulation of the ventral STN decreased the number of correct hits and increased the number of errors on commission in Go-No Go task, when compared to stimulation to dorsal STN (Hershey et al., 2010). STN stimulation can also increase risk taking behavior in Iowa Gambling Task (IGT; Evens et al., 2015). Patients with STN-DBS tended to overestimate their performance with a preference toward competitive environments (Florin et al., 2013). On the other hand, pre-existing ICDs were reported to resolve following STN-DBS, as a result of reduction in dopaminergic medication (Castrioto et al., 2015). Thus STN-DBS may lead to varying net effects on impulsivity in PD through different mechanisms. Clinical studies aimed at dissecting out the effect of STN stimulation on impulsivity have been limited by the confounding effects of therapeutic reduction of dopaminergic medication following STN-DBS.

Computational models provide an opportunity to overcome this limitation by simulating the effect of variations in stimulation and medication protocols individually which cannot be easily applied in human subjects. We hypothesized that electrode position and stimulation parameters affect the decision making ability of PD patients who received STN-DBS. We used a spiking network model of basal ganglia (BG; Mandali et al., 2015) to test the performance under various conditions on the standard gambling task, IGT (Evens et al., 2015). It is known that IGT captures certain impulsive features such as the risk taking ability (Fukui et al., 2005) and lack of premeditation (Zermatten et al., 2005) during decision making. PD patients are known to have poor IGT performance, especially during dopaminergic medication (Poletti et al., 2011; Gescheidt et al., 2012).

Simulating IGT requires learning, which was incorporated in the proposed model using Reinforcement learning (RL; Chakravarthy et al., 2010). RL describes the manner in which an agent learns stimulus-response (S-R) relations based on action outcomes: S-R pairs associated with rewarding outcomes are reinforced while those associated with punishment are attenuated (Dayan and Abbott, 2001). Experimental evidence shows that dopamine (DA) codes for reward prediction error or the temporal difference error term ("δ") in RL (Niv, 2009).

Using the spiking network model of BG (described in Section Materials and Methods), we studied the performance of the model in IGT in normal, PD with and without medication [L-DOPA and Dopamine Agonists (DAA)] and STN-DBS conditions. Our results show that model in normal condition was able to learn from bad choices during the initial trials and improved its performance as the trials progressed. This was observed to be absent in PD with medication (both L-DOPA and DAA) condition.

We then studied the effect of STN stimulation alone on learning and performance by comparing it PD with and without medication. The simulation results show that during the initial trials, the stimulation current interferes with the learning which is reflected as poor performance. We then proceeded to study the factors of stimulation such as electrode position, amplitude of the current and spread that are specific to a patient. The simulation results indicate that electrode position played a significant role in altering performance in the model. We also observed that the model's performance improved for a narrow band of current amplitude.

MATERIALS AND METHODS

Spiking Neuron Model of the Basal Ganglia

The network model of BG (Mandali et al., 2015; **Figure 1A**) was built using 2-variable Izhikevich spiking neurons (Izhikevich, 2003) where each nucleus was modeled as a 2D array (=50 × 50). Parameters for each of the nuclei [STN, Globus Pallidus internus (GPi) and externus (GPe) were chosen such that the model neurons display firing patterns (in terms of firing rate and firing patterns such as rebound firing) of their biological counterparts (Mandali et al., 2015). STN and GPe neurons were bi-directionally connected (Plenz and Kital, 1999) where projections from GPe (STN) are inhibitory (excitatory). Each GPi neuron received both glutamatergic projections from STN and GABAergic projections from D1R-expressing medium spiny neurons (MSNs) of the striatum (Gerfen and Surmeier, 2011). The final action selection was done at thalamus which was simulated as a race model (Vickers, 1970). The activities of both D1- and D2-receptor expressing, striatal MSNs that receive input from cortex (Tritsch and Sabatini, 2012) were modeled as Poisson spike trains. The full set of equations and module sizes related to the model are described in Appendix A and Table A.I (Datasheet in Supplementary Material). The input from cortex to STN, also known as hyper-direct pathway and the GABAergic projection from GPe to GPi, were not included in the model as their functional significance has not been fully explored. The list of acronyms and parameters used in the model are listed in **Table 1** and Tables A.II, IV (Supplementary).

$$\frac{dv_{ij}^{x}}{dt} = 0.04(v_{ij}^{x})^2 + 5v_{ij}^{x} - u_{ij}^{x} + 140 + I_{ij}^{x} + I_{ij}^{syn} \tag{1}$$

$$\frac{du_{ij}^{x}}{dt} = a(bv_{ij}^{x} - u_{ij}^{x}) \tag{2}$$

$$if\ v_{ij}^{x} \geq v_{peak} \left\{ \begin{array}{l} v_{ij}^{x} \leftarrow c \\ u_{ij}^{x} \leftarrow u_{ij}^{x} + d \end{array} \right\} \tag{3}$$

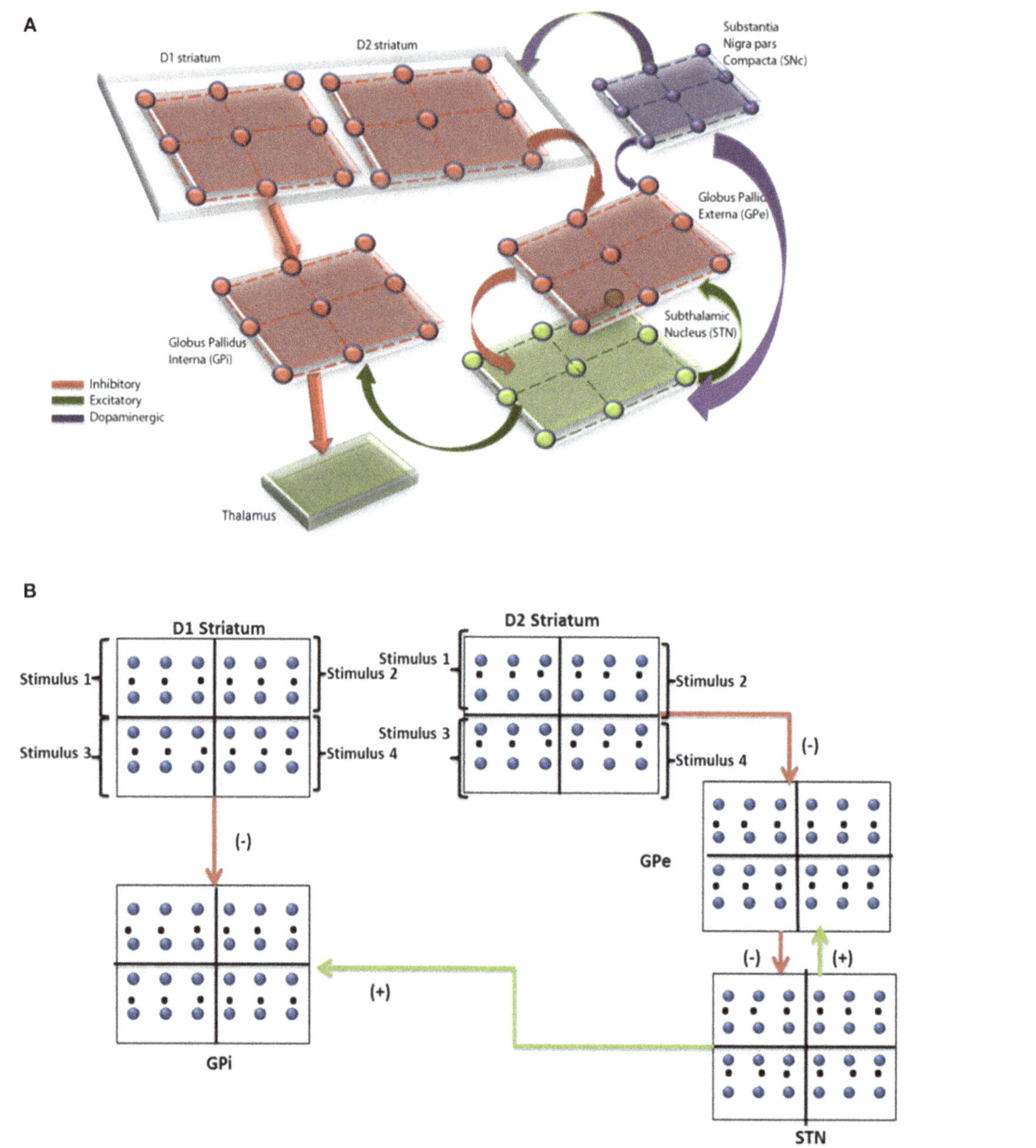

FIGURE 1 | (A) Shows the computational spiking basal ganglia model with key nuclei such as striatum (D1, D2), STN, GPe, GPi, and thalamus. Excitatory/inhibitory/modulatory glutamatergic/GABAergic/dopaminergic projections are shown by green/red/violet arrows. **(B)** Shows the BG model and the regions within each nuclei corresponding to the 4 decks are indicated.

where, v_{ij}^{x} = membrane potential, u_{ij}^{x} = membrane recovery variable, I_{ij}^{Syn} = total synaptic current received, I_{ij}^{x} = external current applied to neuron "x" at location (i, j), v_{peak} = maximum voltage (+30 mv) with x being STN/GPe/GPi neuron.

Behavioral Task-IGT

The task involved presentation of four decks of cards wherein each of the decks A/B/C/D is associated with a combination of reward and penalty. IGT was conducted for a total of 100 trials (5 bins of 20 trials each).The net outcome of a certain card selection (reward + penalty) in each trial was calculated. The probability and amount of penalty varied from deck to deck as explained in Table A.III (Datasheet in Supplementary Material). Over a few trials, one can observe that cards from the decks A and B (C and D) were disadvantageous (advantageous) as the corresponding expected value is negative (positive) The performance was measured as IGT total score (number of selections from "C," "D"—number of selections from "A," "B")

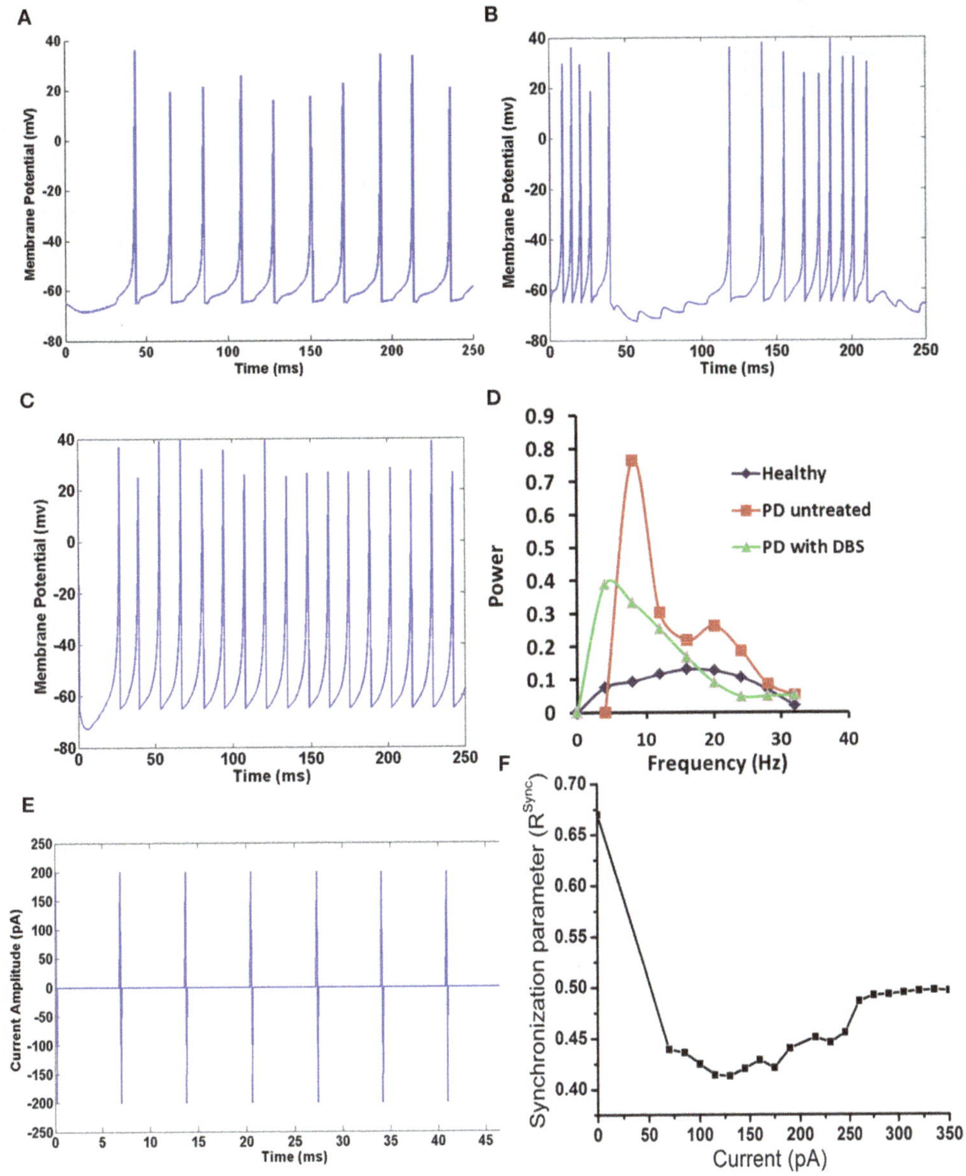

FIGURE 2 | The activity of STN neuron healthy, with and without DBS in PD. (A) Shows the activity of STN neurons in healthy condition, **(B)** shows the bursting activity of STN neurons in PD condition. **(C)** STN neurons resume to tonic firing after DBS, **(D)** the reduction in the frequency content at tremor frequency (4 Hz) in STN neurons in mentioned conditions, **(E)** shows the DBS current in biphasic mode (frequency = 130 Hz with amplitude of 200 pA), **(F)** shows the synchronization levels in STN neurons with increase in DBS current.

Simulating IGT Using Spiking Neuron Network Model

Since IGT consists of 4 decks, each nucleus [STN/GPe/GPi/Striatum (both D1 and D2)] in the network was divided equally into 4 quadrants, where each quadrant received input from one of the decks (**Figure 1B**).The expected value of each card was represented by the cortico-striatal weight which was modulated by DA term "δ." The input to GPe and GPi (i.e., the output of D2 and D1 striatum) was modeled as Poisson spike train (Reti, 2015), whose frequency was proportional to the cortico-striatal weight ($w_{i,k}^{D1}$, $w_{i,k}^{D2}$) of the corresponding card (i) and trial (k). The striatal neuronal firing rate was restricted to 2–40 Hz as per experimental literature (Kravitz et al., 2010).Since DA is known to modulate plasticity in cortico-striatal conditions, the error term "δ" (in the model) was used to update the cortico-striatal synapses (Surmeier et al., 2007). DA also modulated the synaptic strengths within various BG nuclei such as STN (Cragg et al., 2004), GPe (Smith and Kieval, 2000).

Cortico Striatal Weight Update and Temporal Difference Error

Each deck was associated with 2 cortico-striatal weights ($w_{i,0}^{D1}$, $w_{i,0}^{D2}$) which were initialized with random values from a uniform distribution over (0, 1). The two cortico-striatal weights were

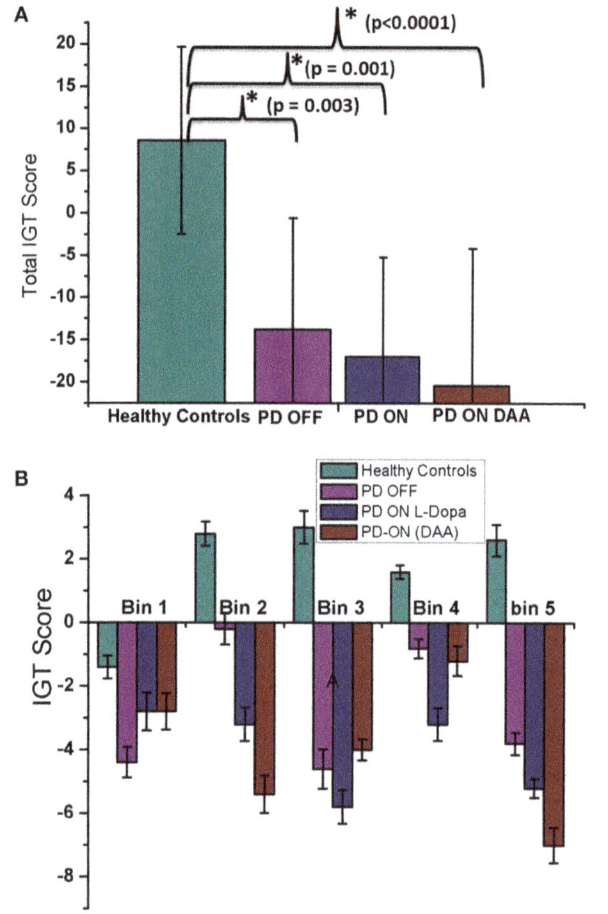

FIGURE 3 | Mean IGT score obtained from spiking model with standard error (SE) for four conditions, HC, PD OFF, PD ON (L-DOPA), and PD-ON (DAA) (A) Total IGT score **(B)** bin score.

trained as,

$$\Delta w_{i,k+1}^{D1} = \eta \delta_k x_{i,k}^{inp} \tag{4}$$

$$\Delta w_{i,k+1}^{D2} = -\eta \delta_k x_{i,k}^{inp} \tag{5}$$

The expected value (V_k) for kth trial was calculated as,

$$V_k = \sum_{i=1}^{4} w_{i,k}^{D1} * x_{i,k}^{sel} \tag{6}$$

The reward (Re_k) for kth trial was calculated as,

$$\text{Re}_k = \sum_{i=1}^{4} r_{i,k} * x_{i,k}^{sel} \tag{7}$$

The loss (L_k) for the kth trial was calculated as,

$$L_k = \sum_{i=1}^{4} l_{i,k} * x_{i,k}^{sel} \tag{8}$$

The error (δ) for kth trial was defined as,

$$\delta_k = \text{Re}_k + L_k - V_k \tag{9}$$

where, $w_{i,k+1}^{D1}(w_{i,k+1}^{D2})$, $w_{i,k}^{D1}(w_{i,k}^{D2})$ were the cortico-striatal weights of *D1* (*D2*) striatum for *i*th card in k+1th and kth trial, $r_{i,k}$ and $l_{i,k}$ were the reward and loss obtained for the selected ith card in kth trial, x^{inp} was the input binary vector representing the 4 decks, x^{sel} was the binary vector representing the selected card e.g., if the card "A" is selected $x^{sel} = [1\ 0\ 0\ 0]$.

Simulating Untreated PD and Medically Treated PD Conditions

Bearing in mind that "δ" is similar to DA activity (Schultz, 1998; Niv, 2009) and there is loss of DA neurons in PD, we simulated PD condition by clamping the "δ" value (Equation 9) to a low limit (δ_{lim}) which resembles the untreated PD condition (Equation 10).

$$\delta_{\lim} = \min(\delta, DA_{ceil}) \tag{10}$$

Where min (y,a) is defined as $\begin{matrix} z = y & if & y < a \\ a & if & y > a \end{matrix}$ and DA_{ceil} is the upper limit of "δ." Medically treated PD condition clinically involves external intake of dopamine precursors such as L-DOPA which was simulated by adding a positive "δ_{med}" term to the δ_{lim} (Mandali and Chakravarthy, 2016) (Equation 10).

$$\delta_{new} = \delta_{\lim} + \delta_{med} \tag{11}$$

Another class of medication prescribed to PD patients is DAA, which has differential affinity for dopamine receptors. We simulated DAA with preferential affinity for *D2* receptors, also known to be linked to impulsivity (MacMahon and Macphee, 2008). The δ_{new} in the Equation (11) was used to update only *D2* cortico-striatal weight (w^{D2}) unlike for L-DOPA where both w^{D1} and w^{D2} were updated.

DBS Current

An external current which mimics the clinically delivered DBS current was applied to the STN neurons in the model. The parameters (frequency, pulse duration, and amplitude) of the stimulation current were chosen to be similar to the typical values used in a clinical setting (Garcia et al., 2005) [Appendix A (Datasheet in Supplementary Material)]. The spread of current over network of neurons spatially is known to follow a Gaussian distribution (Lukasiewicz and Werblin, 1990). Apart from that, Foutz and McIntyre (2010) have simulated various stimulation waveforms and observed that non-rectangular waveforms are more efficient (Foutz and McIntyre, 2010). The stimulation current was applied to the entire/part of STN module in the form of Gaussian distribution (Foutz and McIntyre, 2010). The mean of the Gaussian coincides with the lattice position (i_c, j_c) which was assumed to be the center of the electrode and extent of current spread was controlled by the variance parameter (σ).

$$I_{ij}^{DBS} = A_{DBS} * e^{\frac{-((i-i_c)^2 + (j-j_c)^2)}{\sigma^2}} \tag{12}$$

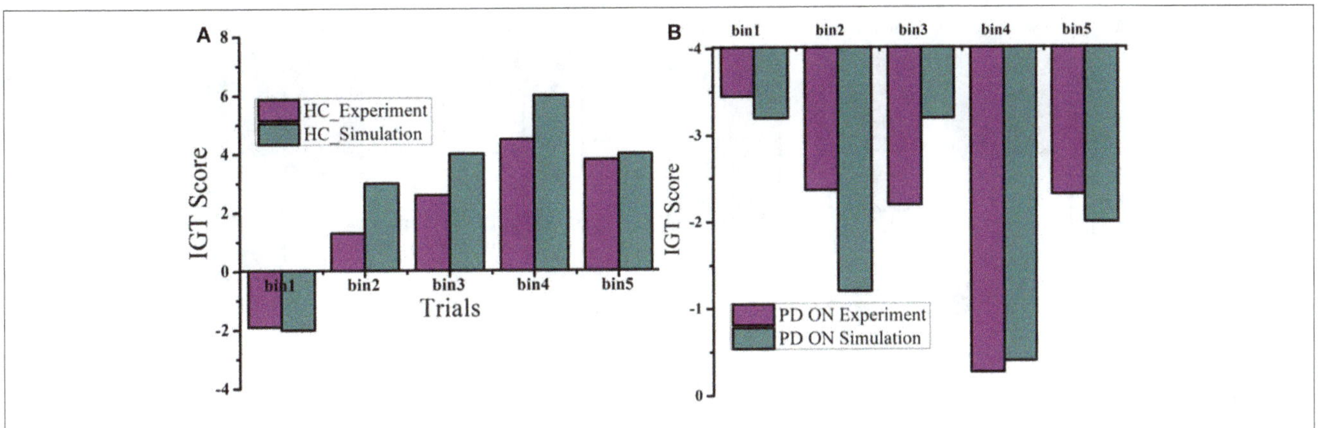

FIGURE 4 | IGT performance results were redrawn from Gescheidt et al. (2012). (A) HC from experiment and simulation. **(B)** Medically treated PD patients from experiment and simulation (L-DOPA).

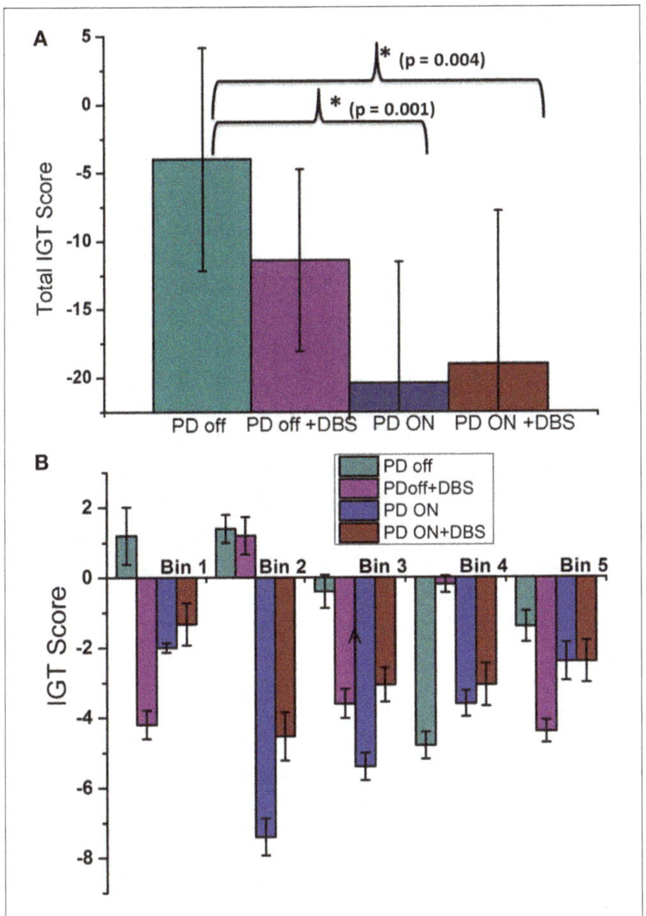

FIGURE 5 | The mean IGT score with SE obtained from spiking model calculated for four conditions; PD OFF, PD OFF+DBS, PD ON, PD ON+DBS (A) Total IGT score and **(B)** bin score.

where I_{ij}^{DBS} is the current received by the neuron at position (i, j) added to equation 1 of STN neurons, A_{DBS} is the amplitude of the current (pA), σ controls the current spread and (i_c, jc) is the mean/center point of the electrode. The effect of electrode position (i_c, j_c) and stimulation parameters A_{DBS} and σ on STN activity and on behavior was explored.

Values of the model parameters used for simulating various conditions (PD, medication and stimulation) are given in Appendix A (Datasheet in Supplementary Material).

Statistical Analysis

The IGT score obtained from each of the conditions [Healthy Controls (HC), untreated PD, medically treated PD, with and without STN stimulation] were compared using repeated measures of ANOVA. ANOVA (which stands for analysis of variance) is a popular statistical technique used to check if the means of two or more populations are equal and come from same distribution. It calculates the variance of means between the groups rather than intra group variance to determine whether the groups come from the same distribution.

Once the statistical test such as ANOVA has been performed, it is important to determine the underlying patterns in the data i.e., groups which could be representatives of different populations and the detects used are generally called as the a-posteriori tests. One such simple yet powerful test is the Bonferroni test which uses the "*p*-value" to determine the significance of the result. We have used the *post-hoc* Bonferroni test to study the effect of stimulation, medication on performance. All statistical tests were performed using IBM SPSS Statistics for Windows, Version 21.0, and Armonk, NY: IBM Corp., USA.

RESULTS

De-Synchronization by DBS Current

The membrane potential of STN neurons PD untreated condition (**Figures 2B,D**) showed bursting activity and frequency content showed a peak at around 4 Hz with high synchrony level (=0.67; Mandali et al., 2015) which was absent in healthy condition (**Figures 2A,D**). On stimulating the STN neurons in PD condition, the peak around tremor frequency (=4 Hz) was significantly reduced (**Figure 2D**; $P < 0.00001$). Similarly the bursting activity in **Figure 2B** was overridden and suppressed by the stimulating current (**Figure 2C**). The synchrony level, R^{sync} (Equation A.13) in the presence of DBS current (**Figure 2E**) decreased from 0.67 (in PD condition and stimulation-OFF) to

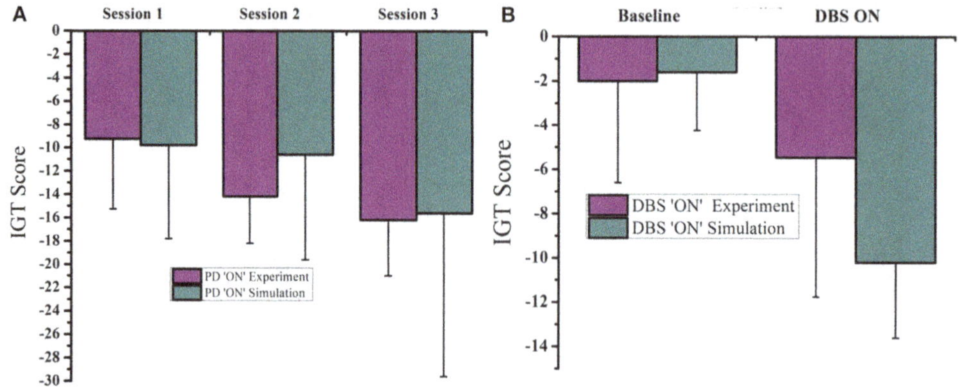

FIGURE 6 | Plots of IGT scores calculated for PD "ON" and PD ON+DBS condition. The experimental results are redrawn from Oyama et al. (2011). **(A)** PD "ON" controls from experiment and simulation for 3 sessions **(B)** DBS subjects for baseline and DBS ON.

0.42 (stimulation-ON; **Figure 2F**) but increased at higher current amplitudes.

IGT-Healthy vs. PD Condition

Total IGT scores for HC, untreated PD, medically treated PD (L-DOPA and DAA) were significantly different [$F_{(3, 36)}$ =9.813, $P < 0.0001$] [**Figure 3A**, Table A.V (Datasheet in Supplementary Material)]. *Post-hoc* analysis indicated a significant difference between HC and the other three conditions: untreated PD ($P = 0.003$), medically treated PD L-DOPA ($P = 0.001$) and DAA ($P < 0.0001$). No statistically significant difference was observed between the two medically treated PD (L-DOPA and DAA) conditions.

In HC condition, the score was negative in the 1st bin, but changed to positive at the 2nd bin and continued to increase, a trend that was absent in both untreated and medically treated PD (both L-DOPA and DAA) conditions (**Figure 3B**). The mean scores [Table A.V Datasheet in Supplementary Material)] for 1st bin of HC, untreated PD, medically treated PD L-DOPA and DAA were similar ($F = 0.684$, $P = 0.568$). For both 2nd and 3rd bins medically treated PD condition performed worse than healthy controls [bin 2 showed significant difference between HC, medically treated PD L-DOPA and medically treated PD DAA ($P < 0.005$); bin 3: The scores were significantly different between HC, medically treated PD L-DOPA ($P = 0.01$) and medically treated PD DAA, $P = 0.025$)]. Though untreated PD had lower scores compared to HC and higher scores compared to treated PD in both the bins, these differences did not reach statistical significance. The mean scores for 4th bin among the four conditions were not significantly different. For the 5th bin, except HC, all other conditions showed poor performance ($F = 8.744$, $P < 0.0001$). The individual bin scores obtained in each of the above described condition are given in **Table 2**.

The mean IGT score values (**Figure 4A**) for HC, obtained from experiment (Gescheidt et al., 2012) and simulation (**Figure 4B**) were not significantly different ($P = 0.19$). Similarly the mean IGT score values obtained from medically treated PD subjects (Gescheidt et al., 2012) and simulation were not significantly different ($P = 0.74$).

IGT-PD Condition with and without Stimulation

Total IGT score was negative for untreated PD, untreated PD with stimulation, medically treated PD and medically treated PD with stimulation (**Figure 5A**) with a significant difference among them [$F_{(3, 36)} = 7.24$, $P = 0.001$]. *Post-hoc* analysis revealed that medically treated PD ($P = 0.001$) and medically treated PD with stimulation ($P = 0.004$) had worse performance compared to untreated PD. No significant difference was observed between medically treated PD with and without stimulation.

A significant difference was observed among the 4 conditions (**Figure 5B**) for bin 1[$F_{(3, 36)} = 3.24$, $P = 0.033$] [Tables A.VI, A.VII (Datasheet in Supplementary Material)]. *Post-hoc* analysis indicated a significant difference only between untreated PD and untreated PD with stimulation ($P = 0.039$). For the 2nd bin [$F_{(3, 36)} = 5.58$, $P = 0.003$], medically treated PD performed worse compared to untreated PD ($P < 0.008$), no significant effect of stimulation was noted. No significant difference was observed for IGT score for the last three bins. The individual bin scores obtained in each of the above described condition are given in **Table 3**.

The mean IGT score values obtained from medically treated PD (PD "ON") subjects from experiment (Oyama et al., 2011) and simulation (**Figures 6A,B**) were statistically similar ($P = 0.42$). Similarly the mean IGT score values for PD with STN-DBS (DBS "ON") experiment and simulation were similar ($P = 0.55$).

Effect of DBS Electrode Parameters on IGT Score

When the electrode position (positions explained in **Figure 7** legend) was changed such that stimulation (for untreated PD) is given selectively to a part of the STN module corresponding to each deck in IGT, we observed a significant variation in the IGT score (**Figure 7B**) ($P < 0.0001$).On changing the spread of DBS current (**Figure 7C**), there was a trend toward better performance with lower radius of spread ($\sigma = 10$), which, however, did not reach statistical significance ($P = 0.67$). We also observed a lower IGT score at higher (=300 pA) and lower (=70 pA)

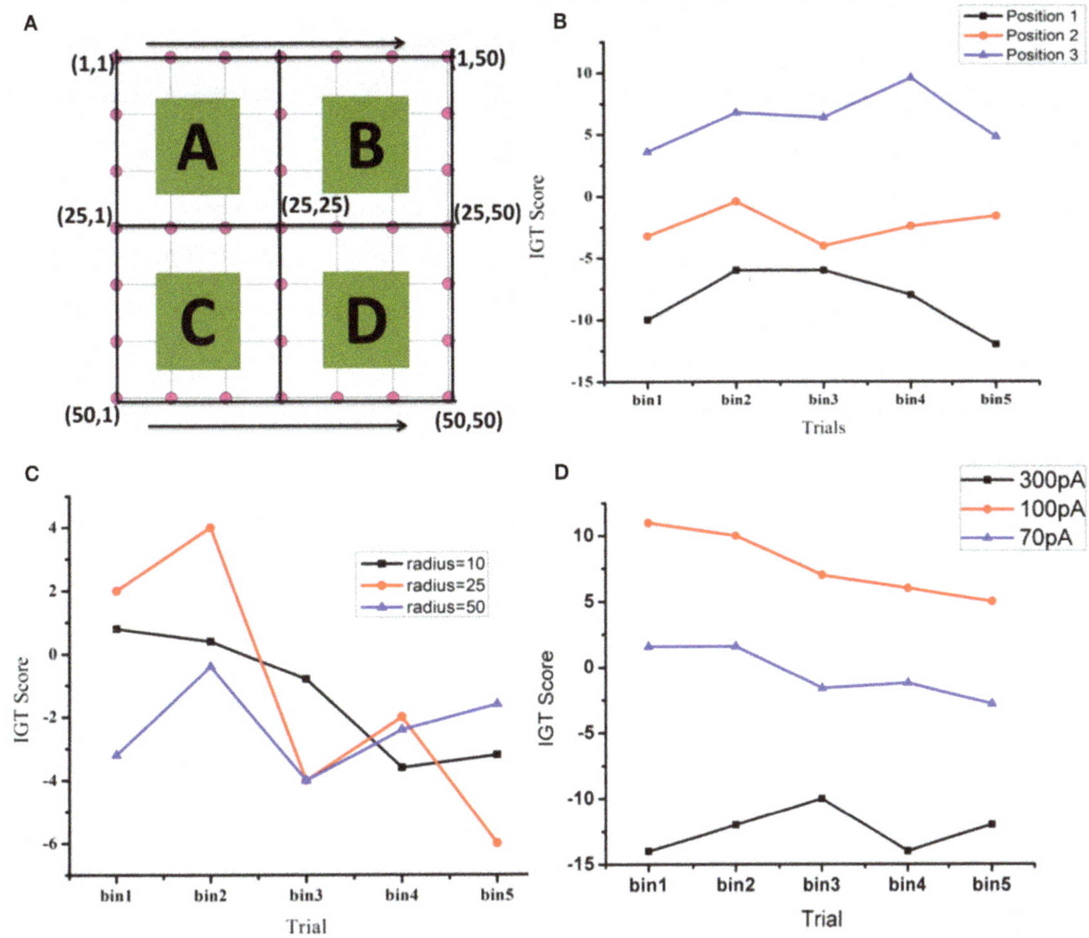

FIGURE 7 | The IGT score. (A) Shows the STN network (=50 × 50) with quadrants that receive input from each of the corresponding decks **(A–D) (B)** The IGT score for three electrode positions [Position 1—in first quadrant with electrode center at lattice point (13,13), Position 2—center of the electrode at the lattice point (25,25), and Position 3- center of the electrode at the lattice point (38,38) in the fourth quadrant]. **(C)** For the electrode at position-2, the spread of the current (σ) was varied **(D)** the effect of DBS current amplitude (70, 100, and 300 pA) on IGT scores when the electrode is placed in position 3.

currents compared to that obtained from optimal current (=100 pA) ($P < 0.001$) (**Figure 8A**). Interestingly, the current range where the highest IGT score was obtained was nearly the same range where the synchrony in STN neurons was observed to be lowest (**Figure 8B**). The underlying cause for such an effect (**Figures 7D, 8A,B**), was investigated by observing the spiking activity of STN for both optimal (=100 pA) and high (=300 pA) current scenarios. At optimal current levels, DBS desynchronized the activity of the STN neurons (**Figure 2F**) that received the stimulation, leading to the selection of the stimulated panel (**Figure 2D**; Figure A1; purple line) and a positive IGT score. On the contrary, at non-optimal (=300 pA) current amplitudes, the corresponding STN neuron activity is increased leading to de-selection of the panel "D" (Figure A1).

DISCUSSION

We built a computational spiking basal ganglia network model to understand the effects of STN stimulation on impulsivity. We first tested the spiking network model, by comparing its simulated activity with the known pathophysiological alterations in PD. It has been observed that desynchronized, irregular STN activity observed in HC changes into synchronized bursting behavior in PD (Wilson and Bevan, 2011), which is also observed in STN neurons of MPTP treated primates (Bergman et al., 1994). This bursting oscillatory activity (Plenz and Kital, 1999) was also observed in our spiking model which correlated with tremor frequency, suggesting that the spiking model has the ability to reflect pathological STN activity.

Effect of PD and Dopaminergic Therapy on IGT Performance

Overall IGT performance was poor in untreated and medically treated PD conditions compared to HC. Medically treated PD condition did worse than HC and untreated PD, regardless of the type of medication used. Analysis of bin scores revealed that learning of the task was poor in all PD conditions compared to HC. Medically treated PD condition resulted in lower bin scores

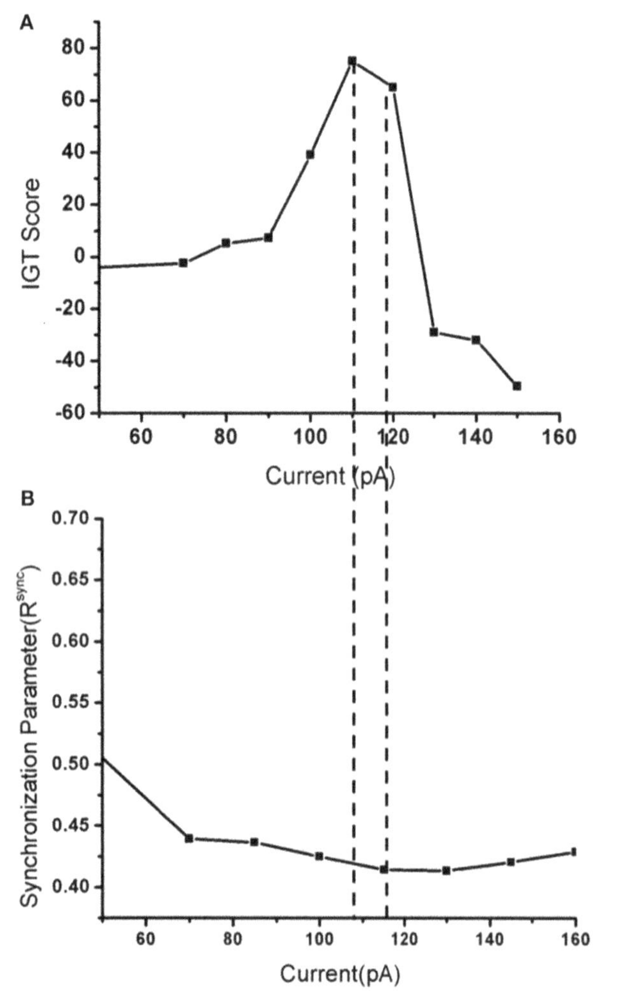

FIGURE 8 | Shows (A) The IGT score obtained when the DBS current was increased from 0 (only medication) to up to 150 pA with frequency 130 Hz, biphasic mode. **(B)** The corresponding synchrony levels observed in STN neurons for that stimulation current.

TABLE 1 | Lists the Acronyms and parameters used in this article.

Variable/Acronym	Full form
STN	Sub Thalamic Nucleus
PD	Parkinson's Disease
DBS	Deep Brain Stimulation
IGT	Iowa Gambling Task
BG	Basal Ganglia
RL	Reinforcement Learning
DA	Dopamine
L-DOPA	Leva Dopa
DAA	Dopamine Agonist
GPe	Globus Pallidus externus
GPi	Globus Pallidus internus
Untreated PD/PD OFF	PD condition without medication
Medically treated PD/PD ON	PD condition with medication
HC	Healthy controls
MPTP	1-methyl-4-phenyl-1,2,3,6-tetrahydropyridine
ICD	Impulse Control disorder
GABA	Γ-Aminobutyric acid
δ	Error term similar to dopamine
σ	Parameter that controls the spread of the current
$w_{i,k}^{D1}$	Cortico-striatal weight to D1 striatum
$w_{i,k}^{D2}$	Cortico-striatal weight to D2 striatum
A_{DBS}	Parameter that controls the amplitude of the DBS current
δ_{lim}	Clamped DA-value resembling PD condition
δ_{med}	DA medication
η	(=0.1) Learning rate of the model
R^{sync}	Synchronization measure

from the 2nd bin onwards, suggesting significantly impaired learning of the task, while in untreated PD; a significantly low bin score compared to HC was seen only in the last bin. The model in treated PD condition does not learn from its action outcomes (rewards/punishments) and wanders among the decks, which is reflected in the negative IGT score (**Figure 3**). This behavior is similar to that previously reported in a probabilistic action selection paradigm where PD "ON" subjects fail to avoid punitive choices (Frank et al., 2007). Physiologically this behavior is attributed to excess DA levels in striatum (Frank et al., 2007). In the model, striatal weights were positively updated even in punishment situation due to dopaminergic medication (δ_{med}), leading to the selection of wrong choice. Clinical studies have identified dopamine agonists to be associated with higher risk of impulsivity, mediated through their D3 receptor affinity. We did not specifically model the D3 receptor activity and did not observe a higher risk with DAA (which selectively increased the D2 weight) compared to L-DOPA in our model.

Effect of STN Stimulation

Stimulation applied to the entire STN module did not result in a significant deterioration of overall IGT performance (**Figure 5A**). However, stimulation significantly impaired performance in the 1st bin when applied to untreated PD condition (**Figure 5B**). It is during these early trials that most of the learning regarding deck's reward pattern happens and this learnt information is used for future card selection. Stimulation seems to affect this learning ability, making the model performance worse as trials progress. Similar behavior is also reported in clinical experiments, where it was observed that PD patients with stimulation tend to overestimate their choices (Florin et al., 2013). No significant change was observed between medically treated PD with and without stimulation. This could be due to an overriding effect of dopaminergic medication over stimulation.

A few aspects of DBS that are specific to a PD patient who receives stimulation are the active contact point in the electrode, the amplitude of the current and the spread due to the current. Keeping this mind, we first changed the position of electrode within the STN nucleus (**Figures 7A,B**) and observed

TABLE 2 | The mean IGT scores with standard deviation for HC, untreated PD, medically treated PD (L-DOPA and DAA) at total and individual bin levels and total IGT score.

Condition	Score bin 1	Score bin 2	Score bin 3	Score bin 4	Score bin 5	Total score
HC	−1.40 ± 3.65	2.8 ± 3.8	3.0 ± 5.2	1.6 ± 2.1	2.6 ± 4.99	8.60 ± 11.07
PD OFF	−4.4 ± 4.78	−2 ± 4.85	−4.6 ± 6.2	−0.8 ± 3.01	−3.8 ± 3.45	−13.8 ± 13.21
PD ON-LDOPA	−1.8 ± 5.96	−2.8 ± 5.26	−4.8 ± 3.3	−3.2 ± 5.1	−4.4 ± 2.95	−17 ± 11.78
PD ON-DAA	−2.8 ± 5.67	−5.4 ± 5.9	−2.6 ± 5.9	−1.2 ± 4.733	−7 ± 5.5	−20.4 ± 16.27

TABLE 3 | The mean IGT scores with standard deviation for untreated PD, medically treated PD (L-DOPA) with and without stimulation at total and individual bin levels.

Condition	Score bin 1	Score bin 2	Score bin 3	Score bin 4	Score bin 5	Total score
PD OFF	1.2 ± 4.02	1.4 ± 4.7	−0.4 ± 3.86	−4.8 ± 4.4	−1.4 ± 5.73	−4 ± 8.17
PD OFF + stim	−4.2 ± 4.05	1.2 ± 5.3	−3.6 ± 4.19	−0.2 ± 2.3	−4.4 ± 3.23	−11.4 ± 6.67
PD ON	−2 ± 1.33	−7.4 ± 5.33	−5.4 ± 3.89	−3.6 ± 3.7	−2.4 ± 5.4	−20.4 ± 8.98
PD ON + stim	−3.4 ± 5.96	−3.4 ± 2.1	−4.2 ± 4.8	−4.2 ± 6.07	−2.9 ± 6	−19 ± 11.20

a significant change in IGT score. Physiologically, it might be possible that stimulation of different active points (which are 0.5–1 mm apart in the electrode) could lead to differential activation of neurons. This activation not only leads to overall difference in the current spread (McIntyre et al., 2004) but also in the activity of neurons that receive input from a different sources. This controlled activation might eventually changes the behavior (Witt et al., 2013).

Apart from position, another DBS parameter thought to significantly influence cognition is current amplitude. Various computational and experimental studies showed that the volume of tissue activated is dependent on the stimulating current amplitude (McIntyre et al., 2004; Arle et al., 2007; Yousif et al., 2010). We observed that a high stimulation current can increase the firing rate of STN neurons that received the stimulation whereas sufficiently low amplitude current would just desynchronize the activity. This change in STN, in terms of spiking activity during optimal and non-optimal (Figure A1; for a fixed position) stimulation currents might be the reason behind conflicting results observed at behavioral level (**Figure 7D**). For instance, de-selection of panel "D" increases probability of selection of other panels. In Figure A1, the selected panel was "A" (blue line) is a highly rewarding panel when viewed at a shorter time scale but punitive in long term. The model selects this panel due to its inability to learn the punishments associated with it. This selection of other panels (A and B are disadvantageous, and only C is rewarding) gave rise to a negative IGT score with high current.

With the above results one can consider the possibility that stimulation current when applied to the corresponding topographical areas of the panels within STN might lead to inhibition/facilitation of the corresponding panel selection depending on the current amplitude. To relate the above results physiologically, we suggest a role for the parallel functional loops within BG nuclei (Alexander and Crutcher, 1990) (motor, cognitive, and the limbic loops) and topographical mapping in impulsive behavior. A coarse functional organization is also observed within STN nucleus with motor and cognitive areas being adjacent to each other (Temel et al., 2005). We suspect that the highly variable cognitive outcomes in experimental studies could be correlated to electrode position and current spread (**Figures 7B–D, 8A**). These results are similar to those observed in the clinical study where a decrease in performance (hit rate) was observed when the position of the electrode was changed (Hershey et al., 2010).

Limitations of our model include the connectivity pattern within BG nuclei (GPe-GPi connection and hyper direct pathway is not included) and neuronal number. More elaborate modeling studies are required to further explore the effect of electrode position and stimulus waveform on motor and cognitive aspects of the PD patients. Although the human STN is known to be organized into motor and cognitive sub territories, a further topographical division into separate areas for various choices is yet unconfirmed. Nevertheless, our model reflects the pathophysiology of STN in PD and predicts behavioral changes similar to clinical data. Our results yield valuable information on the effect of electrode position and current amplitude on behavioral and cognitive outcomes of STN stimulation in PD that may help in the development of optimal stimulation protocols in a clinical setting.

AUTHOR CONTRIBUTIONS

AM: Designed the study, built the computational model and performed the simulations, Manuscript Preparation and Analysis. VSC: Designed the study, built the computational model and performed the simulations, Manuscript Preparation and Analysis. RR: Designed the study, Manuscript Preparation and Analysis. AK: Designed the study, Manuscript Preparation and Analysis. SS: Analysis.

REFERENCES

Alexander, G. E., and Crutcher, M. D. (1990). Functional architecture of basal ganglia circuits: neural substrates of parallel processing. *Trends Neurosci.* 13, 266–271. doi: 10.1016/0166-2236(90)90107-L

Arle, J., Mei, L., and Shils, J. (2007). Modeling Parkinsonian circuitry and the DBS electrode. *Stereotact. Funct. Neurosurg.* 86, 1–15. doi: 10.1159/000108584

Benabid, A. L. (2003). Deep brain stimulation for Parkinson's disease. *Curr. Opin. Neurobiol.* 13, 696–706. doi: 10.1016/j.conb.2003.11.001

Bergman, H., Wichmann, T., Karmon, B., and Delong, M. (1994). The primate subthalamic nucleus. II. Neuronal activity in the MPTP model of parkinsonism. *J. Neurophysiol.* 72, 507–520.

Castrioto, A., Funkiewiez, A., Debû, B., Cools, R., Lhommée, E., Ardouin, C., et al. (2015). Iowa gambling task impairment in Parkinson's disease can be normalised by reduction of dopaminergic medication after subthalamic stimulation. *J. Neurol. Neurosurg. Psychiatry* 86, 186–190. doi: 10.1136/jnnp-2013-307146

Chakravarthy, V., Joseph, D., and Bapi, R. S. (2010). What do the basal ganglia do? A modeling perspective. *Biol. Cybern.* 103, 237–253. doi: 10.1007/s00422-010-0401-y

Combs, H. L., Folley, B. S., Berry, D. T., Segerstrom, S. C., Han, D. Y., Anderson-Mooney, A. J., et al. (2015). Cognition and depression following deep brain stimulation of the subthalamic nucleus and globus pallidus pars internus in Parkinson's disease: a meta-analysis. *Neuropsychol. Rev.* 25, 439–454. doi: 10.1007/s11065-015-9302-0

Cragg, S. J., Baufreton, J., Xue, Y., Bolam, J. P., and Bevan, M. D. (2004). Synaptic release of dopamine in the subthalamic nucleus. *Eur. J. Neurosci.* 20, 1788–1802. doi: 10.1111/j.1460-9568.2004.03629.x

Dayan, P., and Abbott, L. F. (2001). *Theoretical Neuroscience.* Cambridge, MA: MIT Press.

Evens, R., Stankevich, Y., Dshemuchadse, M., Storch, A., Wolz, M., Reichmann, H., et al. (2015). The impact of Parkinson's disease and subthalamic deep brain stimulation on reward processing. *Neuropsychologia.* 75, 11–19. doi: 10.1016/j.neuropsychologia.2015.05.005

Florin, E., Müller, D., Pfeifer, J., Barbe, M. T., Fink, G. R., and Timmermann, L. (2013). Subthalamic stimulation modulates self-estimation of patients with Parkinson's disease and induces risk-seeking behaviour. *Brain.* 136, 3271–3281. doi: 10.1093/brain/awt241

Foutz, T. J., and McIntyre, C. C. (2010). Evaluation of novel stimulus waveforms for deep brain stimulation. *J. Neural Eng.* 7:066008. doi: 10.1088/1741-2560/7/6/066008

Frank, M. J., Samanta, J., Moustafa, A. A., and Sherman, S. J. (2007). Hold your horses: impulsivity, deep brain stimulation, and medication in parkinsonism. *Science* 318, 1309–1312. doi: 10.1126/science.1146157

Fukui, H., Murai, T., Fukuyama, H., Hayashi, T., and Hanakawa, T. (2005). Functional activity related to risk anticipation during performance of the Iowa Gambling Task. *Neuroimage* 24, 253–259. doi: 10.1016/j.neuroimage.2004.08.028

Garcia, L., D'alessandro, G., Bioulac, B., and Hammond, C. (2005). High-frequency stimulation in Parkinson's disease: more or less? *Trends Neurosci.* 28, 209–216. doi: 10.1016/j.tins.2005.02.005

Gerfen, C. R., and Surmeier, D. J. (2011). Modulation of striatal projection systems by dopamine. *Annu. Rev. Neurosci.* 34, 441. doi: 10.1146/annurev-neuro-061010-113641

Gescheidt, T., Czekóová, K., Urbánek, T., Mareček, R., Mikl, M., Kubíková, R., et al. (2012). Iowa Gambling Task in patients with early-onset Parkinson's disease: strategy analysis. *Neurol. Sci.* 33, 1329–1335. doi: 10.1007/s10072-012-1086-x

Hershey, T., Campbell, M. C., Videen, T. O., Lugar, H. M., Weaver, P. M., Hartlein, J., et al. (2010). Mapping Go-No-Go performance within the subthalamic nucleus region. *Brain* 133, 3625–3634. doi: 10.1093/brain/awq256

Hershey, T., Revilla, F. J., Wernle, A., Gibson, P. S., Dowling, J. L., and Perlmutter, J. S. (2004). Stimulation of STN impairs aspects of cognitive control in PD. *Neurology* 62, 1110–1114. doi: 10.1212/01.WNL.0000118202.19098.10

Izhikevich, E. M. (2003). Simple model of spiking neurons. *IEEE Trans. Neural Netw.* 14, 1569–1572. doi: 10.1109/TNN.2003.820440

Kravitz, A. V., Freeze, B. S., Parker, P. R., Kay, K., Thwin, M. T., Deisseroth, K., et al. (2010). Regulation of parkinsonian motor behaviours by optogenetic control of basal ganglia circuitry. *Nature* 466, 622–626. doi: 10.1038/nature09159

Kühn, A. A., Kupsch, A., Schneider, G. H., and Brown, P. (2006). Reduction in subthalamic 8–35 Hz oscillatory activity correlates with clinical improvement in Parkinson's disease. *Eur. J. Neurosci.* 23, 1956–1960. doi: 10.1111/j.1460-9568.2006.04717.x

Lukasiewicz, P. D., and Werblin, F. S. (1990). The spatial distribution of excitatory and inhibitory inputs to ganglion cell dendrites in the tiger salamander retina. *J. Neurosci.* 10, 210–221.

MacMahon, D. G., and Macphee, G. J. (2008). Dopamine agonists and impulse control disorders in Parkinson's disease. *Prog. Neurol. Psychiatry* 12, 5–9. doi: 10.1002/pnp.100

Mandali, A., and Chakravarthy, V. (2016). Probing the role of medication, DBS electrode position and antidromic activation on impulsivity using a computational model of Basal Ganglia. *Front. Hum. Neurosci.* 10:450. doi: 10.3389/fnhum.2016.00450

Mandali, A., Rengaswamy, M., Chakravarthy, V. S., and Moustafa, A. A. (2015). A spiking Basal Ganglia model of synchrony, exploration and decision making. *Front. Neurosci.* 9:191. doi: 10.3389/fnins.2015.00191

McIntyre, C. C., Mori, S., Sherman, D. L., Thakor, N. V., and Vitek, J. L. (2004). Electric field and stimulating influence generated by deep brain stimulation of the subthalamic nucleus. *Clin. Neurophysiol.* 115, 589–595. doi: 10.1016/j.clinph.2003.10.033

Niv, Y. (2009). Reinforcement learning in the brain. *J. Math. Psychol.* 53, 139–154. doi: 10.1016/j.jmp.2008.12.005

Oyama, G., Shimo, Y., Natori, S., Nakajima, M., Ishii, H., Arai, H., et al. (2011). Acute effects of bilateral subthalamic stimulation on decision-making in Parkinson's disease. *Parkinsonism Relat. Disord.* 17, 189–193.

Plenz, D., and Kital, S. T. (1999). A basal ganglia pacemaker formed by the subthalamic nucleus and external globus pallidus. *Nature* 400, 677–682. doi: 10.1038/23281

Poletti, M., Cavedini, P., and Bonuccelli, U. (2011). Iowa gambling task in Parkinson's disease. *J. Clin. Exp. Neuropsychol.* 33, 395–409. doi: 10.1080/13803395.2010.524150

Reti, I. (2015). *Brain Stimulation: Methodologies and Interventions.* Hoboken, NJ: John Wiley & Sons.

Rosa, M., Giannicola, G., Marceglia, S., Fumagalli, M., Barbieri, S., and Priori, A. (2012). "Emerging horizons in neuromodulation new Frontiers in brain and spine stimulation," in *International Review of Neurobiology*, eds C. Hamani and E. Moro (San Diego, CA: Elsevier), 2–380.

Schultz, W. (1998). Predictive reward signal of dopamine neurons. *J. Neurophysiol.* 80, 1–27.

Smeding, H. M. M., Goudriaan, A. E., Foncke, E. M. J., Schuurman, P. R., Speelman, J. D., and Schmand, B. (2007). Pathological gambling after bilateral subthalamic nucleus stimulation in Parkinson disease. *J. Neurol. Neurosurg. Psychiatry* 78, 517–519. doi: 10.1136/jnnp.2006.102061

Smith, Y., and Kieval, J. Z. (2000). Anatomy of the dopamine system in the basal ganglia. *Trends Neurosci.* 23, S28–S33. doi: 10.1016/S1471-1931(00)00023-9

Sudhyadhom, A., Bova, F. J., Foote, K. D., Rosado, C. A., Kirsch-Darrow, L., and Okun, M. S. (2007). Limbic, associative, and motor territories within the targets for deep brain stimulation: potential clinical implications. *Curr. Neurol. Neurosci. Rep.* 7, 278–289. doi: 10.1007/s11910-007-0043-1

Surmeier, D. J., Ding, J., Day, M., Wang, Z., and Shen, W. (2007). D1 and D2 dopamine-receptor modulation of striatal glutamatergic signaling in striatal medium spiny neurons. *Trends Neurosci.* 30, 228–235. doi: 10.1016/j.tins.2007.03.008

Temel, Y., Blokland, A., Steinbusch, H. W., and Visser-Vandewalle, V. (2005). The functional role of the subthalamic nucleus in cognitive and limbic circuits. *Prog. Neurobiol.* 76, 393–413. doi: 10.1016/j.pneurobio.2005.09.005

Tritsch, N. X., and Sabatini, B. L. (2012). Dopaminergic modulation of synaptic transmission in cortex and striatum. *Neuron* 76, 33–50. doi: 10.1016/j.neuron.2012.09.023

Vickers, D. (1970). Evidence for an accumulator model of psychophysical discrimination. *Ergonomics* 13, 37–58. doi: 10.1080/00140137008931117

Wilson, C. J., and Bevan, M. D. (2011). Intrinsic dynamics and synaptic inputs control the activity patterns of subthalamic nucleus neurons in health and in Parkinson's disease. *Neuroscience* 198, 54–68. doi: 10.1016/j.neuroscience.2011.06.049

Witt, K., Granert, O., Daniels, C., Volkmann, J., Falk, D., van Eimeren, T., et al. (2013). Relation of lead trajectory and electrode position to neuropsychological outcomes of subthalamic neurostimulation in Parkinson's disease: results from a randomized trial. *Brain* 136, 2109–2119. doi: 10.1093/brain/awt151

York, M. K., Wilde, E. A., Simpson, R., and Jankovic, J. (2009). Relationship between neuropsychological outcome and DBS surgical trajectory and electrode location. *J. Neurol. Sci.* 287, 159–171. doi: 10.1016/j.jns.2009.08.003

Yousif, N., Purswani, N., Bayford, R., Nandi, D., Bain, P., and Liu, X. (2010). Evaluating the impact of the deep brain stimulation induced electric field on subthalamic neurons: a computational modelling study. *J. Neurosci. Methods* 188, 105–112. doi: 10.1016/j.jneumeth.2010.01.026

Zermatten, A., Van Der Linden, M., d'acremont, M., Jermann, F., and Bechara, A. (2005). Impulsivity and decision making. *J. Nerv. Ment. Dis.* 193, 647–650. doi: 10.1097/01.nmd.0000180777.41295.65

14

A Pointwise Method for Identifying Biomechanical Heterogeneity of the Human Gallbladder

Wenguang Li[1], Nigel C. Bird[2] and Xiaoyu Luo[3]*

[1] School of Engineering, University of Glasgow, Glasgow, UK, [2] Academic Surgical Unit, Royal Hallamshire Hospital, Sheffield, UK, [3] School of Mathematics and Statistics, University of Glasgow, Glasgow, UK

Edited by:
Vanessa Diaz,
University College London, UK

Reviewed by:
Haoxiang Luo,
Vanderbilt University, USA
Andrew James Narracott,
University of Sheffield, UK

***Correspondence:**
Wenguang Li
wenguang.li@glasgow.ac.uk

Identifying the heterogeneous biomechanical property of human gallbladder (GB) walls from non-invasive measurements can have clinical significance in patient-specific modeling and acalculous biliary pain diagnosis. In this article, a pointwise method was proposed to measure the heterogeneity of ten samples of human GB during refilling. Three different points, two on the equator of GB body 90° apart and one on the apex of GB fundus, were chosen to represent the typical regions of interest. The stretches at these points were estimated from ultrasound images of the GB during the bile emptying phase based on an analytical model. The model was validated against the experimental data of a lamb GB. The material parameters at the different points were determined inversely by making use of a structure-based anisotropic constitutive model. This anisotropic model yielded much better accuracy when compared to a number of phenomenologically-based constitutive laws, as demonstrated by its significantly reduced least-square errors in stress curve fitting. The results confirmed that the human GB wall material was heterogeneous, particularly toward the apex region. Our study also suggested that non-uniform wall thickness of the GB was important in determining the material parameters, in particular, on the parameters associated with the properties of the matrix and the longitudinal fibers—the difference could be as large as 20–30% compared to that of the uniform thickness model.

Keywords: gallbladder, strain energy function, heterogeneity, anisotropic property, constitutive law, optimization, inverse problem

INTRODUCTION

Human gallbladder (GB) is a small pear-shaped organ that is attached to the underside of the right lobe of the liver. Its function is to store and concentrate bile produced continuously by the liver. Induced by cholecystokinin (CCK), bile can be expelled from the GB to the gut to aid the digestion of fat. Cholecystitis, often due to blockage of the cystic duct by gallstones, and acalculous biliary pain are common GB diseases that affect both women and men (Cozzolino et al., 1963; Williamson, 1988). The symptoms in acalculous biliary pain disease vary widely from discomfort to severe pain, which usually follows food intake. However, the painful symptoms remain in nearly 50% patients following gallbladder removal (Cholecystectomy) (Smythe et al., 1998, 2004). This is in part due to the lack of understanding of the underlying mechanism for acalculous biliary pain.

Interestingly, many human tissues, such as artery, breast, liver, and pancreas, can develop local disease, examples include vulnerable plaque (Baldewsing et al., 2004a; Trivedi et al., 2007), atheroma in coronary and femoral arteries (Chandran et al., 2003; Baldewsing et al., 2004b; Hamilton et al., 2005), arterial stenosis (Franquet et al., 2011), and cerebral aneurysms (Zhao et al., 2011a,b). Biomechanical properties of the diseased soft tissue are different to those of the healthy ones and are often heterogeneous. Inverse methods have been developed to identify isotropic biomechanical properties (Chandran et al., 2003; Baldewsing et al., 2004a,b; Hamilton et al., 2005; Trivedi et al., 2007; Franquet et al., 2011) in terms of the Young's modulus. In studies by Zhao, Raghavan, and Lu, pointwise inverse approaches were used to reveal the anisotropic heterogeneous biomechanical properties of cerebral aneurysms (Zhao et al., 2011a,b), ascending thoracic aneurysms (Davis et al., 2015) and murine aortas (Bersi et al., 2016) on a membrane mechanic model.

Healthy human GB wall is commonly regarded as a homogenous anisotropic non-linear material in passive state, i.e., bile refilling phase (Li et al., 2012, 2013; Xiong et al., 2013). However, recent work based on *in vitro* test of a healthy lamb GB suggested that this might not be true (Genovese et al., 2014). In addition, human acalculous biliary disease can lead to increased material heterogeneity in the GB wall. In this paper, we addressed this issue by extending the homogenous anisotropic non-linear biomechanical model for human GB wall proposed in Li et al. (2013) to a heterogeneous anisotropic case. We used an inverse pointwise method to identify the heterogeneous anisotropic property at three different points on the GB wall. The method was based on an ellipsoid membrane model and an in-house developed program using MATLAB.

COMPUTATIONAL MODELS

Geometrical Model and Stresses under Internal Pressure

A series of ultrasonic images of acalculous human GB had been scanned in 10 min interval for 60 min during the emptying phase at the Sheffield Hallamshire Teaching Hospital. A typical example is illustrated in **Figure 1**, marked by the three axes D_1, D_2, and D_3 ($D_1 \leq D_2 \leq D_3$). From these images we generated the corresponding ellipsoid models, as shown in **Figure 2A**, which are used to estimate the GB volume.

The passive biomechanical property of GB wall exhibits in the refilling phase only, hence we will focus this process. The refilling phase is the reverse process of the emptying phase. This means that for the same volume, the refilling and the emptying phases share the same ellipsoid (Li et al., 2013). The heterogeneous anisotropic biomechanical property of human GB wall in the refilling phase will be determined inversely at points 1, 2, and 3 in a GB wall. Point 1 is an intersected point of two ellipses, one is along the equator and the other is in the longitudinal direction in a meridian plane, point 2 is also on the equator but 90° apart from point 1, and point 3 is at the apex as shown in **Figure 2A**. In the spherical coordinate system (r, ϕ, θ), the coordinates of points 1, 2, and 3 are $(D_1/2, 0, \pi/2)$, $(D_2/2, \pi/2, \pi/2)$, and $(D_3/2, 0, \pi)$, respectively.

Observing that the volume of the GB model was reduced by 50% by the end of emptying (Li et al., 2013), we chose the end of the emptying configuration of the GB as the reference configuration.

We interpolated the GB model with fifteen time moments throughout the refilling phase based on GB images using the geometrical similarity (Li et al., 2011). The GB wall circumferential and longitudinal in-plane stretches at point 1 at a time instant t_j is calculated with (Ragab and Bayoumi, 1998).

$$\begin{cases} \lambda_{1j}^{\phi} = 1 + \frac{u_r}{D_{11}/2} + \frac{D_{11}}{2}\frac{\partial u_\phi}{\partial \phi} \\ \lambda_{1j}^{\theta} = 1 + \frac{u_r}{D_{11}/2} + \frac{D_{11}}{2}\frac{\partial u_\theta}{\partial \theta} \end{cases} \tag{1}$$

where the radial displacement $u_r = 0.5\left(D_{1j} - D_{11}\right)$, and $j = 1,2,3...,N$, here $N = 15$, D_{11} is the length of the principal axis D_1 at time t_1, D_{1j} is the length of the principal axis D_1 at time t_j. Since point 1 is on the axis of the ellipsoid, symmetry requires that $\partial u_\phi/\partial \phi = \partial u_\theta/\partial \theta = 0$. Hence, Equation (1) can be simplified to:

$$\begin{cases} \lambda_{1j}^{\phi} = 1 + \frac{0.5(D_{1j} - D_{11})}{0.5D_{11}} = D_{1j}/D_{11} \\ \lambda_{1j}^{\theta} = 1 + \frac{0.5(D_{1j} - D_{11})}{0.5D_{11}} = D_{1j}/D_{11} \end{cases} \tag{2}$$

The incompressibility of the GB wall means that the stretch component of GB thickness must satisfy:

$$\lambda_{1j}^{h} = 1/\left(\lambda_{1j}^{\phi}\lambda_{1j}^{\theta}\right). \tag{3}$$

Similarly, the stretch components are for point 2:

$$\begin{cases} \lambda_{2j}^{\phi} = \lambda_{2j}^{\theta} = D_{2j}/D_{21} \\ \lambda_{2j}^{h} = 1/\left(\lambda_{2j}^{\phi}\lambda_{2j}^{\theta}\right), \end{cases} \tag{4}$$

and for point 3:

$$\begin{cases} \lambda_{3j}^{\phi} = \lambda_{3j}^{\theta} = D_{3j}/D_{31} \\ \lambda_{3j}^{h} = 1/\left(\lambda_{3j}^{\phi}\lambda_{3j}^{\theta}\right). \end{cases} \tag{5}$$

where D_{21} and D_{2j} are the length of principal axis D_2 at time t_1 and t_j, while D_{31} and D_{3j} are the length of principal axis D_3 at time t_1 and t_j.

These stretch components at t_j ($j = 1,...,N$) and point i ($i = 1,2,3$) can be presented simply:

$$\begin{cases} \lambda_{ij}^{\phi} = \lambda_{ij}^{\theta} = D_{ij}/D_{i1} \\ \lambda_{ij}^{h} = 1/\left(\lambda_{ij}^{\phi}\lambda_{ij}^{\theta}\right). \end{cases} \tag{6}$$

The stretches during the refilling phase are plotted against the GB volume in **Figure 2B** at points 1, 2, and 3 for a typical GB sample. The GB volume changed with time exponentially based on an earlier model in Li et al. (2011): $V = Ge^{Ht} + M$, where

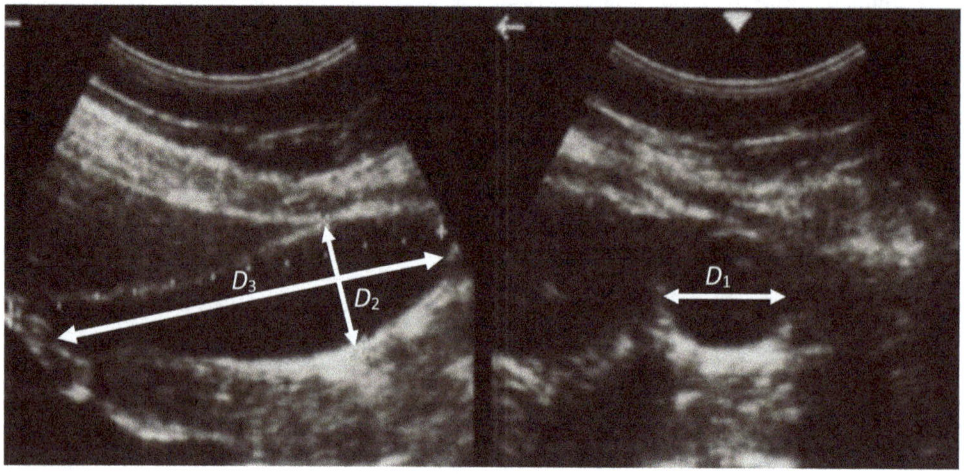

FIGURE 1 | A typical ultrasonic image of human gallbladder during the emptying phase.

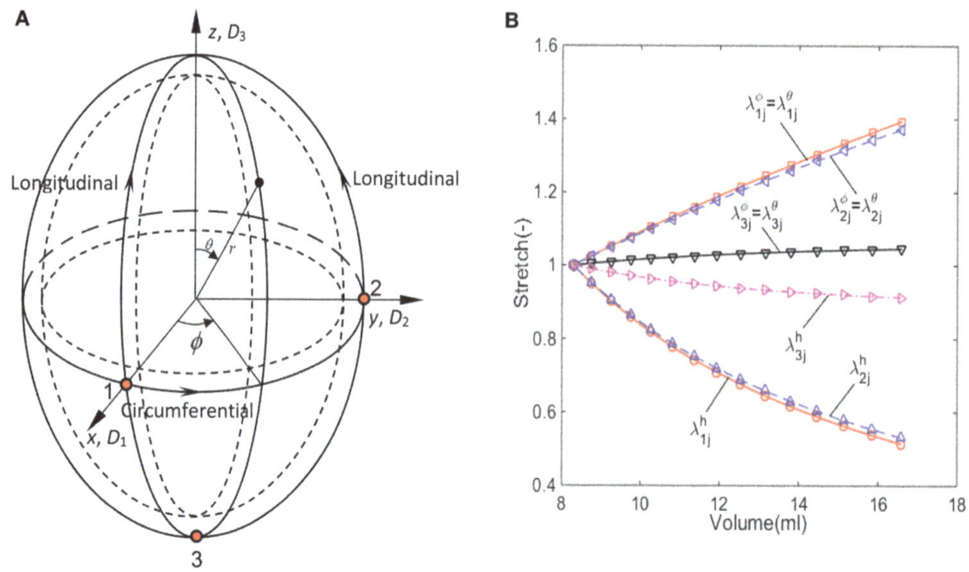

FIGURE 2 | The imaged-based ellipsoid model for GB during the refilling phase, (A) ellipsoid model with three control points on the surface, **(B)** the stretch components estimated from the ellipsoid model for GB sample No.1 listed in **Table 1**.

G, H, and M are parameters determined analytically using the measured GB volume and pressure at the moments t_1 and t_N.

The expressions of in-plane stress components in the GB wall during the refilling phase were the same as these in the emptying phase (Li et al., 2011) since we assumed the GB material is an elastic membrane:

$$\begin{cases} \sigma_{ij}^{\theta} = p_j F_\theta F_n \\ \sigma_{ij}^{\phi} = p_j \frac{F_\phi}{F_n}, \end{cases} \tag{7}$$

where p_j is the refilling pressure at time t_j, and F_θ, F_ϕ, and F_n are the functions describing the geometry of the GB:

$$\begin{cases} F_\theta = \frac{D_{3j}K_{1j}K_{2j}}{4h_{ij}} \left(1 - \frac{K_{1j}^2 - K_{2j}^2}{K_{1j}^2 K_{2j}^2} \cos 2\phi_i\right) \\ F_\phi = \frac{D_{3j}}{4K_{1j}K_{2j}h_{ij}} \begin{bmatrix} K_{1j}^2 K_{2j}^2 + \left(K_{1j}^2 + K_{2j}^2 - 2K_{1j}^2 K_{2j}^2\right) \sin^2\theta_i \\ + \left(K_{1j}^2 - K_{2j}^2\right) \cos^2\theta_i \cos 2\phi_i \end{bmatrix} \\ F_n = \frac{\sqrt{K_{1j}^2 \cos^2\theta_i \cos^2\phi_i + K_{2j}^2 \cos^2\theta_i \sin^2\phi_i + \sin^2\theta_i}}{\sqrt{K_{1j}^2 \sin^2\phi_i + K_{2j}^2 \cos^2\phi_i}} \end{cases} \tag{8}$$

where $K_{1j} = D_{1j}/D_{3j}$, $K_{2j} = D_{2j}/D_{3j}$, h_{ij} is the GB wall thickness at point i, and time t_j, and $D_{ij} = \lambda_{ij}^{\theta} D_{i1}$, $h_{ij} = \lambda_{ij}^{h} h_{i1}$. The internal pressure p_j is given by Li et al. (2013):

TABLE 1 | Parameters of ten human GB samples, these parameters are for one dataset rather than an average of the whole set.

Moment	Parameter	GB sample No.									
		1	3	4	17	19	21	29	37	39	43
At end of refilling	p_N (Pa)	1,466.5									
	D_{1N} (mm)	23.4	26.8	32.9	27.2	34.7	28.2	28.1	30.2	33.2	37.6
	D_{2N} (mm)	25.0	27.9	35.2	27.2	35.7	30.1	28.9	30.8	33.5	38.0
	D_{3N} (mm)	54.1	70.7	57.5	55.9	92.3	74.5	56.1	53.8	53.9	82.1
At start of refilling	p_1 (Pa)	466.6									
	D_{11} (mm)	16.8	21.0	24.8	21.4	26.8	20.8	20.1	24.7	24.2	28.1
	D_{21} (mm)	18.2	21.2	25.8	20.7	29.2	24.2	22.7	24.7	26.1	29.7
	D_{31} (mm)	51.7	59.3	54.9	46.7	72.9	62.9	49.9	41.2	47.5	70.3
Ejection Fraction (EF) in 30 min (%)		4.5	11.4	13.3	32.4	49.4	66.3	37.8	77.0	60.1	2.7

A uniform thickness, i.e., $h_{11} = h_{21} = h_{31} = 2.5$ mm, is assumed.

$$p_j = p_1 \left(\frac{p_N}{p_1} \right)^{t_j/t_N}, t_j \in [0, t_N] \tag{9}$$

where p_N is the mean final bile pressure in a GB after the refilling phase chosen to be 1,466.5 Pa (11 mmHg), p_1 is the bile pressure when the refilling starts, $p_1 = 466.6$ Pa (3 mmHg), and t_N is the total time of the refilling phase. These values are estimated from *in vivo* measurements (Li et al., 2013).

The Constitutive Model

To determine the heterogeneous material parameters of human GB wall, the structure-based anisotropic constitutive model used in Li et al. (2013) was extended so that the material parameters are location dependent. At each point, the GB wall is assumed to be composed of homogeneous matrix and two families of fibers along the circumferential and longitudinal directions, respectively, as shown in **Figure 3**. The strain energy functions are:

$$\psi_i = c^i (I_1 - 3) + \frac{\kappa_1^i}{2\kappa_2^i} \left[e^{\kappa_2^i \left((\lambda_i^{\phi})^2 - 1 \right)^2} - 1 \right] + \frac{\kappa_3^i}{2\kappa_4^i} \left[e^{\kappa_4^i \left((\lambda_i^{\theta})^2 - 1 \right)^2} - 1 \right]. \tag{10}$$

where the parameters c^i, κ_m^i ($i = 1,2,3$, $m = 1$–4) are location dependent. The total number of material property constants in Equation (10) for points 1–3 should be 15 in general. However, at point 3, there are no circumferential fibers, so the second term in Equation (10) disappears, i.e., κ_1^3 and κ_2^3 vanish. Hence, there are a total of 13 parameters to be determined.

The in-plane Cauchy stress components at t_j are:

$$\begin{cases} \sigma_{ij}^{\prime\phi} = 2c^i \left(\lambda_{ij}^{\phi 2} - \lambda_{ij}^{h2} \right) + 2\lambda_{ij}^{\phi 2} \kappa_1^i \left(\lambda_{ij}^{\phi 2} - 1 \right) \exp\left(\kappa_2^i \left(\lambda_{ij}^{\phi 2} - 1 \right)^2 \right) \\ + \sigma_{i1}^{\phi} \\ \sigma_{ij}^{\prime\theta} = 2c^i \left(\lambda_{ij}^{\theta 2} - \lambda_{ij}^{h2} \right) + 2\lambda_{ij}^{\theta 2} \kappa_3^i \left(\lambda_{ij}^{\theta 2} - 1 \right) \exp\left(\kappa_4^i \left(\lambda_{ij}^{\theta 2} - 1 \right)^2 \right) \\ + \sigma_{i1}^{\theta} \end{cases} \tag{11}$$

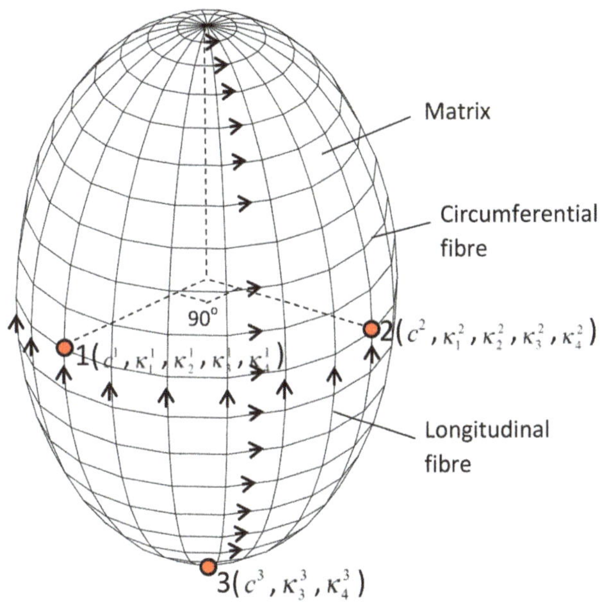

FIGURE 3 | The GB wall is composed of matrix and two families of fibers, the thirteen material parameters are location dependent, changing from points 1–3.

where σ_{i1}^{ϕ} and σ_{i1}^{θ} ($i = 1,2,3$) are interpreted as the initial stresses imbedded in the GB wall, which are estimated using Equations (7–9) with the internal pressure p_1.

Comparison with Other Constitutive Models

Several phenomenological anisotropic strain energy functions have been proposed for soft tissues. Here we did not intend to be exhaustive but will choose three commonly used strain energy functions for comparisons. These include the Fung strain energy function (Ferruzzi et al., 2011):

$$\psi = \frac{c}{2}\left\{e^{\frac{1}{4}\left[a_1(\lambda^{\phi 2}-1)^2 + a_2(\lambda^{\theta 2}\ 1)^2 + 2a_3(\lambda^{\phi 2}-1)(\lambda^{\theta 2}-1)\right]} - 1\right\}, \tag{12}$$

The Choi-Vito strain energy function (Ferruzzi et al., 2011):

$$\psi = c\left\{e^{\frac{1}{4}a_1(\lambda^{\phi 2}-1)^2} + e^{\frac{1}{4}a_2(\lambda^{\theta 2}-1)^2} + e^{\frac{1}{4}a_3(\lambda^{\phi 2}-1)(\lambda^{\theta 2}-1)} - 3\right\} \tag{13}$$

and the Zhou-Fung strain energy function (Zhou and Fung, 1997):

$$\psi = \frac{c}{2}\left\{\begin{array}{l} e^{\frac{1}{4}\left[a_1(\lambda^{\phi 2}-1)^2 + a_2(\lambda^{\theta 2}-1)^2 + 2a_3(\lambda^{\phi 2}-1)(\lambda^{\theta 2}-1)\right]} \\ -\frac{1}{4}\left[a_1(\lambda^{\phi 2}-1)^2 + a_2(\lambda^{\theta 2}-1)^2\right. \\ \left. +\,2a_3(\lambda^{\phi 2}-1)(\lambda^{\theta 2}-1)\right] - 1 \end{array}\right\} + \frac{1}{8}\left[b_1(\lambda^{\phi 2}-1)^2 + b_2(\lambda^{\theta 2}-1)^2 + 2b_3(\lambda^{\phi 2}-1)(\lambda^{\theta 2}-1)\right] \tag{14}$$

Inverse Estimate of the Material Parameters

The material parameters in Equations (10) or (12) or (13) or (14) are selected to minimize the objective function:

$$f = \sum_{i=1}^{3}\sum_{j=1}^{N}\left[\left(\sigma_{ij}^{\phi} - \sigma_{ij}^{\prime\phi}\right)^2 + \left(\sigma_{ij}^{\theta} - \sigma_{ij}^{\prime\theta}\right)^2\right]. \tag{15}$$

The minimization was performed using the Trust-Region-Reflective algorithm in MATLAB (More and Sorensen, 1983) which terminated when the objective function value is less than 10^{-5}. In addition, the following RMS error (RMSE) is calculated to assess the curve-fitting quality:

$$\varepsilon = \frac{\sqrt{\frac{1}{6N}\sum_{i=1}^{3}\sum_{j=1}^{N}\left[\left(\sigma_{ij}^{\phi} - \sigma_{ij}^{\prime\phi}\right)^2 + \left(\sigma_{ij}^{\theta} - \sigma_{ij}^{\prime\theta}\right)^2\right]}}{\frac{1}{6N}\sum_{i=1}^{3}\sum_{j=1}^{N}\left[\sigma_{ij}^{\phi} + \sigma_{ij}^{\theta}\right]} \times 100\%. \tag{16}$$

It should be pointed out that the optimization process was conducted at points 1, 2, and 3 simultaneously rather than separately at each point. To secure a global minimum, the initial guesses of the parameters were chosen randomly within a suitable range, such as [0.01, 10] for c^1, κ_1^1, κ_2^1, κ_4^1, c^2, κ_1^2, κ_2^2, κ_4^2, c^3, and κ_4^3, but [0.01, 3] for κ_3^1, κ_3^2, and κ_3^3. In those ranges, the optimized material parameters did not occur at the boundaries, and the curve fitting error was in the minimum. 80 initial guesses were generated randomly, followed by 80 optimization processes. The mean property constants and curve fitting errors were chosen to be the results. The detail of initial guesses on property constants optimization is given in Section Effects of Material Heterogeneity.

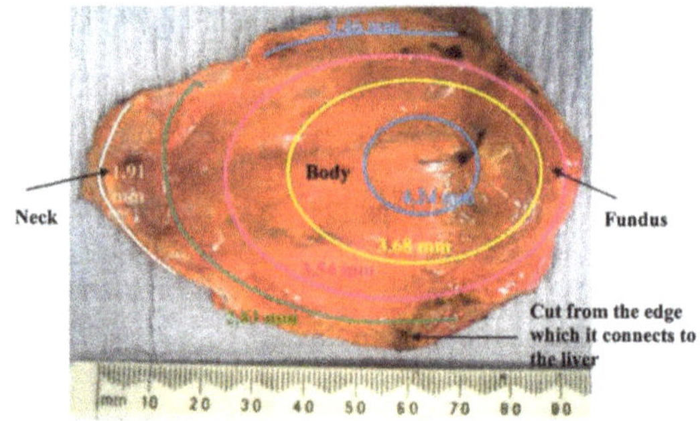

FIGURE 4 | A GB wall thickness profile, showing the thickest wall at the GB apex and thinnest wall near the neck, adopted from Su (2005) with permission.

The Variable Wall Thickness

The GB wall thickness is related to the stress magnitude determined from the experimental images, as shown in Equations (7) and (8). This means that even when the pressure is the same, stresses can be different due to a varied thickness. This leads to different material parameters in the strain energy function in Equation (10). We now address the issue of the variable wall thickness of GB. A three-dimensional *in vivo* measurement of wall thickness of the GB was not yet available (Engel et al., 1980; Sanders, 1980; Prasad et al., 2008; Mohammed et al., 2010; Ugwu and Agwu, 2010); however, varying thickness of GB wall was measured *in vitro* with a digital slide caliper (Su, 2005; Khan et al., 2012). A contour of GB wall thickness is illustrated in **Figure 4** (Su, 2005). It is observed that the thickness of the GB apex in the fundus increases to around 5 mm maximum and the wall of the neck is as thin as 2 mm. The ratio of the maximum thickness over the thickness of the body is 1.2.

In Khan et al. (2012), 62 GB samples were divided into three age groups; (10–20) years, (21–40) years, and (41–70) years. The thicknesses of these GBs were measured manually at the fundus, body and neck. It was identified that the maximum thickness was found on the neck, and the thinnest wall is located at the fundus. For the (41–70) years group, which coincides with the patient's age group for GB surgery in the paper, the ratio of the thickness at the fundus over the thickness of the body is 0.9. This is contrary to the finding in Su (2005). These ratios are used to examine the effect of a varying thickness.

GB Samples

The input data for our model were the geometrical parameters based on ultrasound images, and internal pressures at the start and end of emptying/refilling phase of ten GB samples from a previous study (Li et al., 2013). These geometrical parameters and internal pressures are shown in **Table 1**. Additionally, the geometrical parameters and pressure profile at 15 or more moments between the start and the end of refilling phase were interpolated according to the method in Li et al. (2011) and Equation (9). A uniform wall thickness was assumed at the start of refilling phase, i.e., $h_{11} = h_{21} = h_{31} = 2.5\,\text{mm}$ (Li et al., 2011,

2012, 2013). The stretch components of GB No. 1 over time are shown in **Figure 2B**. Note that the stretch-volume profiles are patient-specific.

In **Table 1**, the ejection fraction (EF) of a GB is defined as the ratio of the difference between the initial and emptied GB volumes at 30 min after venous injection of stimulator-CCK. Ethical approval for the use of data in **Figure 4** and (Li et al., 2013) were approved by the ethical committees in the hospital where the studies were conducted, and the subjects gave informed consent to these studies.

The Solution of Inverse Problem

From **Table 1**, we had the ellipsoid model geometrical parameters at the beginning and the end of the refilling phase, which were the same as the end (30 min after CCK) and beginning of the emptying phase from the routine ultrasound images taken in hospital, as shown in **Figure 1**. As the least squares method required more scattered points than the number of parameters to be estimated, the ellipsoid model (Li et al., 2011) was interpolated over 15 or more time points for the emptying phase. The internal bile pressure is then given by Equation (9). These data were used to obtain the initial guess for the optimization process. The stretch and stress components were then computed, and the objective function and the RMSE were evaluated and compared to a given criterion of 10^{-6}. If the criterion was not satisfied, a new guess based on the Trust-Region-Reflective algorithm would be generated, and the procedure repeated until the convergent result is reached.

RESULTS

Effects of the Initial Guesses on Material Property Constants

GB No.1 shown in **Table 1** was randomly chosen to identify effects of initial guesses on the repeatability of inversely determined GB wall biomechanical property constants at points 1, 2, and 3. The initial guesses of the constants were generated randomly in the ranges for search of property constants mentioned in Section Inverse Estimate of the Material Parameters by normal distribution function in MATLAB and the numbers of initial guesses were specified 10, 20, 30, ..., 130, respectively. The means of the determined property constants and RMSE as well as their standard error at 95% confidence level are illustrated in **Table 2**. The true value of these property constants should be equal to the mean ± its standard error at 95% confidence level.

Note that the parameters at points 1–3 determined by the least squares method based on the Trust-Region-Reflective algorithm could not be repeated from one initial guess to another due to the complexity of the inverse problem. Considering the property constants determined from statistics point of view, however, the material parameters and their standard error at 95% confidence level inversely determined remained unchanged basically, especially when the number of initial guesses was 80 or more. This suggests that the biomechanical property constant values are repeatable in a statistical sense and are globally optimum. In the following sections, the property constants are extracted with 80 initial guesses at a computational cost of around

TABLE 2 | Mean material property constants and standard error at 95% confidence level of GB No.1 with 2.5 mm uniform initial thickness at various initial guests.

No of initial guesses	c^1 (kPa)	κ_1^1 (kPa)	κ_2^1 (–)	κ_3^1 (kPa)	κ_4^1 (–)	c^2 (kPa)	κ_1^2 (kPa)	κ_2^2 (–)	κ_3^2 (kPa)	κ_4^2 (–)	c^3 (kPa)	κ_3^3 (kPa)	κ_4^3 (–)	ε (%)	Time costed (min)
10	0.2509 ± 0.0340	2.1467 ± 0.0531	0.3263 ± 0.0170	0.5875 ± 0.0518	0.6255 ± 0.0782	0.0155 ± 0.0036	2.4547 ± 0.0058	0.2858 ± 0.0017	0.1048 ± 0.0075	1.5144 ± 0.0937	1.4855 ± 0.0088	2.4611 ± 0.0476	4.4470 ± 1.4226	7.2492 ± 0.0487	4.2
20	0.2169 ± 0.0222	2.1999 ± 0.0352	0.3105 ± 0.0109	0.6411 ± 0.0340	0.5532 ± 0.0419	0.0156 ± 0.0022	2.4547 ± 0.0036	0.2858 ± 0.0010	0.1042 ± 0.0047	1.5207 ± 0.0571	1.4808 ± 0.0080	2.4712 ± 0.0336	5.2825 ± 0.9006	7.1797 ± 0.0601	8.5
30	0.2136 ± 0.0239	2.2064 ± 0.0372	0.3089 ± 0.0112	0.6467 ± 0.0376	0.5531 ± 0.0478	0.0172 ± 0.0039	2.4519 ± 0.0064	0.2867 ± 0.0019	0.1030 ± 0.0066	1.5481 ± 0.0933	1.4798 ± 0.0058	2.4768 ± 0.0213	5.2257 ± 0.6993	7.1762 ± 0.0440	13.0
40	0.2234 ± 0.0219	2.1906 ± 0.0344	0.3136 ± 0.0105	0.6313 ± 0.0340	0.5734 ± 0.0429	0.0172 ± 0.0034	2.4520 ± 0.0055	0.2866 ± 0.0016	0.1031 ± 0.0055	1.5465 ± 0.0798	1.4802 ± 0.0067	2.4812 ± 0.0294	4.7908 ± 0.6233	7.1914 ± 0.0513	17.2
50	0.2268 ± 0.0185	2.1847 ± 0.0289	0.3154 ± 0.0090	0.6260 ± 0.0284	0.5794 ± 0.0395	0.0176 ± 0.0028	2.4512 ± 0.0045	0.2868 ± 0.0013	0.1018 ± 0.0048	1.5598 ± 0.0658	1.4831 ± 0.0046	2.4671 ± 0.0194	4.8035 ± 0.5842	7.2110 ± 0.0376	21.1
60	0.2269 ± 0.0146	2.1856 ± 0.0227	0.3147 ± 0.0069	0.6253 ± 0.0227	0.5765 ± 0.0299	0.0162 ± 0.0022	2.4537 ± 0.0036	0.2861 ± 0.0011	0.1042 ± 0.0038	1.5275 ± 0.0525	1.4826 ± 0.0050	2.4617 ± 0.0223	5.3255 ± 0.5797	7.2025 ± 0.0368	25.2
70	0.2231 ± 0.0117	2.1906 ± 0.0184	0.3131 ± 0.0057	0.6310 ± 0.0181	0.5667 ± 0.0242	0.0161 ± 0.0018	2.4536 ± 0.0030	0.2861 ± 0.0009	0.1043 ± 0.0034	1.5248 ± 0.0453	1.4842 ± 0.0042	2.4603 ± 0.0190	4.9036 ± 0.4887	7.2082 ± 0.0310	28.8
80	0.2180 ± 0.0131	2.1992 ± 0.0206	0.3105 ± 0.0064	0.6391 ± 0.0203	0.5603 ± 0.0267	0.0160 ± 0.0016	2.4538 ± 0.0025	0.2861 ± 0.0008	0.1041 ± 0.0030	1.5253 ± 0.0385	1.4801 ± 0.0040	2.4807 ± 0.0176	4.8695 ± 0.4743	7.1827 ± 0.0310	32.7
90	0.2182 ± 0.0129	2.1981 ± 0.0203	0.3113 ± 0.0062	0.6397 ± 0.0201	0.5592 ± 0.0249	0.0153 ± 0.0013	2.4550 ± 0.0022	0.2857 ± 0.0007	0.1056 ± 0.0025	1.5064 ± 0.0338	1.4815 ± 0.0044	2.4764 ± 0.0200	4.6928 ± 0.4266	7.1907 ± 0.0318	37.1
100	0.2114 ± 0.0124	2.2094 ± 0.0194	0.3075 ± 0.0060	0.6493 ± 0.0192	0.5494 ± 0.0248	0.0159 ± 0.0018	2.4541 ± 0.0030	0.2860 ± 0.0009	0.1049 ± 0.0030	1.5217 ± 0.0440	1.4815 ± 0.0041	2.4714 ± 0.0175	5.0405 ± 0.4098	7.1845 ± 0.0317	40.8
110	0.2182 ± 0.0134	2.1982 ± 0.0209	0.3115 ± 0.0064	0.6397 ± 0.0207	0.5646 ± 0.0283	0.0175 ± 0.0024	2.4515 ± 0.0039	0.2867 ± 0.0012	0.1029 ± 0.0035	1.5565 ± 0.0580	1.4799 ± 0.0033	2.4794 ± 0.0154	5.0577 ± 0.4053	7.1850 ± 0.0269	46.0
120	0.2184 ± 0.0101	2.1977 ± 0.0158	0.3112 ± 0.0048	0.6391 ± 0.0157	0.5578 ± 0.0194	0.0168 ± 0.0020	2.4526 ± 0.0033	0.2864 ± 0.0010	0.1033 ± 0.0029	1.5477 ± 0.0508	1.4797 ± 0.0032	2.4829 ± 0.0141	4.8604 ± 0.3984	7.1803 ± 0.0233	48.5
130	0.2153 ± 0.0105	2.2033 ± 0.0164	0.3094 ± 0.0051	0.6432 ± 0.0162	0.5564 ± 0.0228	0.0170 ± 0.0019	2.4523 ± 0.0030	0.2865 ± 0.0009	0.1033 ± 0.0029	1.5451 ± 0.0457	1.4815 ± 0.0039	2.4729 ± 0.0177	4.9616 ± 0.3444	7.1894 ± 0.0284	51.1

TABLE 3 | Heterogeneous material parameters of ten GB samples determined inversely with model Equation (10) and compared with homogeneous model (Li et al., 2013) in uniform thickness $h_{11} = h_{21} = h_{31} = 2.5$ mm.

GB No.	Model	Point i	c^i (kPa)	κ_1^i (kPa)	κ_2^i (–)	κ_3^i (kPa)	κ_4^i (–)	ε (%)
1	Heterogeneous	1	0.2180 ± 0.0131	2.1992 ± 0.0206	0.3105 ± 0.0064	0.6391 ± 0.0203	0.5603 ± 0.0267	7.1827 ± 0.0310
		2	0.0160 ± 0.0016	2.4538 ± 0.0025	0.2861 ± 0.0008	0.1041 ± 0.0030	1.5253 ± 0.0385	
		3	1.4801 ± 0.0040	–	–	2.4807 ± 0.0176	4.8695 ± 0.4743	
	Homogeneous	1, 2, 3	2.3349	0.5977	0.8512	1.3952	1.0430	2.2
3	Heterogeneous	1	0.1742 ± 0.0076	3.6549 ± 0.0142	0.7242 ± 0.0053	0.8486 ± 0.0141	1.4082 ± 0.0281	1.9123 ± 0.0152
		2	0.0885 ± 0.0084	3.3645 ± 0.0150	0.4526 ± 0.0047	0.1987 ± 0.0144	0.9714 ± 0.1228	
		3	0.3487 ± 0.0001	–	–	0.9275 ± 0.0012	4.2149 ± 0.0085	
	Homogeneous	1, 2, 3	1.8375	4.7385	1.3538	0.8694	0.9568	3.9
4	Heterogeneous	1	0.2229 ± 0.0086	4.0107 ± 0.0159	0.4920 ± 0.0045	1.1588 ± 0.0160	0.8603 ± 0.0185	3.2065 ± 0.0092
		2	0.1669 ± 0.0091	3.9481 ± 0.0168	0.4622 ± 0.0047	0.5402 ± 0.0162	1.2017 ± 0.0454	
		3	0.9313 ± 0.0004	–	–	0.6766 ± 0.0043	6.2768 ± 0.0462	
	Homogeneous	1, 2, 3	2.1817	2.9539	0.7230	0.6578	1.0458	3.0
17	Heterogeneous	1	0.1484 ± 0.0080	3.3757 ± 0.0151	0.8418 ± 0.0065	0.6389 ± 0.0144	1.8043 ± 0.0455	5.1699 ± 0.0131
		2	0.2152 ± 0.0116	2.9225 ± 0.0205	0.5591 ± 0.0074	0.5387 ± 0.0204	0.9941 ± 0.0516	
		3	0.8548 ± 0.0004	–	–	0.1702 ± 0.0017	5.1634 ± 0.1110	
	Homogeneous	1, 2, 3	1.6810	2.9213	0.1161	0.4784	1.5530	2.5
19	Heterogeneous	1	0.5583 ± 0.0195	4.4503 ± 0.0362	0.4628 ± 0.0094	0.6047 ± 0.0349	1.0724 ± 0.1010	3.6488 ± 0.0141
		2	0.0978 ± 0.0078	6.7567 ± 0.0169	0.6573 ± 0.0049	0.0512 ± 0.0025	3.2080 ± 0.0823	
		3	0.5965 ± 0.0003	–	–	0.6651 ± 0.0042	1.1060 ± 0.0216	
	Homogeneous	1, 2, 3	2.2772	6.2427	0.1106	0.2182	0.8042	3.0
21	Heterogeneous	1	0.2321 ± 0.0088	6.1282 ± 0.0227	3.2180 ± 0.0189	1.7915 ± 0.0157	7.2733 ± 0.0529	7.8681 ± 0.0058
		2	0.0576 ± 0.0002	2.7506 ± 0.0005	0.0628 ± 0.0002	0.0101 ± 0.0001	0.1280 ± 0.0228	
		3	0.0188 ± 0.0005	–	–	1.8847 ± 0.0046	5.3105 ± 0.0167	
	Homogeneous	1, 2, 3	2.2309	3.0375	0.0176	0.2213	0.7755	3.4
29	Heterogeneous	1	0.3411 ± 0.0158	2.3990 ± 0.0247	0.2708 ± 0.0066	0.5445 ± 0.0249	0.4547 ± 0.0363	5.1186 ± 0.0165
		2	0.1083 ± 0.0081	3.5589 ± 0.0156	0.8243 ± 0.0065	0.4838 ± 0.0135	2.6977 ± 0.0584	
		3	1.2416 ± 0.0013	–	–	0.5071 ± 0.0059	3.1976 ± 0.2470	
	Homogeneous	1, 2, 3	2.0624	1.6658	0.7148	0.8237	1.1547	2.6
37	Heterogeneous	1	0.2382 ± 0.0126	4.1246 ± 0.0250	1.2410 ± 0.0120	1.0673 ± 0.0257	1.7882 ± 0.0579	6.3186 ± 0.0054
		2	0.2411 ± 0.0117	4.1189 ± 0.0233	1.2435 ± 0.0113	1.0612 ± 0.0236	1.7999 ± 0.0555	
		3	0.9575 ± 0.0002	–	–	0.0149 ± 0.0005	6.7694 ± 0.0517	
	Homogeneous	1, 2, 3	1.9243	4.3563	0.4350	0.1451	1.3890	2.5
39	Heterogeneous	1	0.3117 ± 0.0121	2.9031 ± 0.0199	0.2749 ± 0.0051	0.8384 ± 0.0197	0.5063 ± 0.0218	5.5799 ± 0.0092
		2	0.2700 ± 0.0127	3.6645 ± 0.0238	0.6873 ± 0.0084	0.8535 ± 0.0221	1.8349 ± 0.0444	
		3	1.8285 ± 0.0019	–	–	0.1972 ± 0.0062	3.0987 ± 0.2607	
	Homogeneous	1, 2, 3	2.4066	1.7295	0.5803	0.8437	1.1167	2.5

30 min. This time consumption is much less than 3.5–7.0 h based on the approach of ABAQUS 3D FEA plus MATLAB optimization solver in Li et al. (2013).

Effects of Material Heterogeneity

The material parameters of heterogeneity inversely determined are listed in **Table 3** and compared with those from the corresponding homogenous model in Li et al. (2013). The error ε in the homogenous model reflects the error in GB volume between image observation and homogenous model prediction.

For all the GBs, the material parameter associated with the matrix in the heterogeneous model is around 10 times that of the homogenous model. For GB 3, 4, 17, 29, 37, 39, and 43, the mean values of fibers-related material parameters at points 1 and 2, κ_1^2 and κ_3^2, basically agree with κ_1 and κ_3 in the homogenous model, i.e., $\kappa_1^2 \approx (1\text{–}2)\,\kappa_1$ and $\kappa_3^2 \approx (1\text{–}2)\,\kappa_3$. For GB 1, 19, and 21, κ_1^2 and

TABLE 4 | The first principal stresses in 10 GB samples wall estimated by using the homogenous model in Li et al. (2013) and the heterogeneous model Equation (10) in the present paper.

Stress (kPa)	GB sample No.									
	1	3	4	17	19	21	29	37	39	43
Homogenous (Li et al., 2013)	9.75	12.38	13.66	11.89	17.11	14.41	12.17	12.78	13.26	17.09
Heterogeneous	10.86	12.40	14.49	11.40	16.25	14.09	14.41	11.41	14.96	17.85
Pain due to CCK	No	No	No	Yes	Yes	No	No	No	Yes	Yes

TABLE 5 | Material parameters of Fung strain energy function Equation (14) inversely determined with uniform thickness $h_{11} = h_{21} = h_{31} = 2.5$ mm.

GB No.	Point i	c^i (kPa)	a_1^i	a_2^i	a_3^i	ε (%)
1	1	5.9207 ± 0.0444	1.0182 ± 0.0060	0.0102 ± 0.0002	0.6373 ± 0.0006	7.4833 ± 0.0119
	2	6.2282 ± 0.0398	1.4220 ± 0.0106	0.0755 ± 0.0064	0.0600 ± 0.0062	
	3	5.6380 ± 0.0916	–	5.6863 ± 0.0900	–	
3	1	4.4149 ± 0.0299	2.4513 ± 0.0135	0.0225 ± 0.0020	0.4580 ± 0.0013	6.8817 ± 0.0102
	2	6.6937 ± 0.0337	1.9242 ± 0.0123	0.1144 ± 0.0087	0.0725 ± 0.0086	
	3	2.5090 ± 0.0910	–	3.2397 ± 0.1120	–	
4	1	5.4869 ± 0.0489	1.9630 ± 0.0145	0.0102 ± 0.00010	0.5777 ± 0.0003	13.6283 ± 0.0140
	2	6.8674 ± 0.0568	2.0333 ± 0.0252	0.2729 ± 0.0233	0.2018 ± 0.0237	
	3	3.1520 ± 0.0934	–	3.5454 ± 0.0952	–	
17	1	3.6600 ± 0.0395	2.865 ± 0.0244	0.0419 ± 0.0034	0.3753 ± 0.0020	17.5786 ± 0.0106
	2	6.0800 ± 0.0578	1.8191 ± 0.0262	0.3547 ± 0.0227	0.1892 ± 0.0227	
	3	2.4674 ± 0.1469	–	2.7210 ± 0.1183	–	
19	1	7.6011 ± 0.0673	1.9387 ± 0.0144	0.0103 ± 0.0004	0.3693 ± 0.0002	12.3314 ± 0.0231
	2	6.6126 ± 0.0695	3.5333 ± 0.0413	0.2148 ± 0.0265	0.3362 ± 0.0264	
	3	3.2850 ± 0.1789	–	1.9863 ± 0.1098	–	
21	1	3.3914 ± 0.0674	4.8971 ± 0.0613	0.8129 ± 0.0153	0.5357 ± 0.0100	11.7094 ± 0.0132
	2	9.3404 ± 0.0119	0.9968 ± 0.0010	0.0100 ± 0.0001	0.0934 ± 0.0001	
	3	1.5754 ± 0.0672	–	7.3935 ± 0.1600	–	
29	1	6.324 ± 0.0650	1.1187 ± 0.0095	0.0100 ± 0.0001	0.5099 ± 0.0004	16.0779 ± 0.0202
	2	4.2754 ± 0.0757	2.7660 ± 0.0531	0.3330 ± 0.0414	0.5028 ± 0.0414	
	3	3.4407 ± 0.0716	–	3.5408 ± 0.0713	–	
37	1	3.5278 ± 0.0919	3.5628 ± 0.0604	0.1031 ± 0.0094	0.4602 ± 0.0055	24.8315 ± 0.0129
	2	3.4863 ± 0.0700	3.7051 ± 0.0844	0.7644 ± 0.0690	0.7056 ± 0.0684	
	3	2.6134 ± 0.2148	–	1.9942 ± 0.1109	–	
39	1	5.9501 ± 0.0708	1.2697 ± 0.0124	0.0100 ± 0.0001	0.6480 ± 0.0005	20.5731 ± 0.0203
	2	5.5450 ± 0.0627	2.1687 ± 0.0566	0.5208 ± 0.0539	0.4967 ± 0.0539	
	3	3.7173 ± 0.0586	–	3.9577 ± 0.0636	–	
43	1	7.7685 ± 0.0518	1.6277 ± 0.0086	0.0100 ± 0.0001	0.3810 ± 0.0002	17.0028 ± 0.0199
	2	7.6550 ± 0.0566	2.2885 ± 0.0319	0.3060 ± 0.0297	0.3350 ± 0.0300	
	3	3.1778 ± 0.0862	–	3.1777 ± 0.0862	–	

κ_3^2 are different from κ_1 and κ_3 in the homogenous model. On one hand, the material parameters in the heterogeneous model at point 1 and point 2 are similar, implying the heterogeneity of the GB wall along the circumference is small. This is especially true for GB 37 which has $D_1 \approx D_2$; the points 1 and 2 share the same parameters. On the other hand, the parameters at point 3 differ substantially from the other two, suggesting a strong heterogeneity from GB body to fundus.

The first principal stresses of all the GB samples are compared in **Table 4** with those predicted by the homogenous model in

TABLE 6 | Material parameters of Choi-Vito strain energy function Equation (13) inversely determined with uniform thickness $h_{11} = h_{21} = h_{31} = 2.5$ mm.

GB No.	Point i	c^i (kPa)	a_1^i	a_2^i	a_3^i	ε (%)
1	1	6.5709 ± 0.0467	1.0196 ± 0.0103	0.1209 ± 0.0070	0.9865 ± 0.0068	7.2885 ± 0.0086
	2	6.2533 ± 0.0550	1.3934 ± 0.0102	0.0112 ± 0.0003	0.3090 ± 0.0030	
	3	5.6432 ± 0.1665	–	5.7457 ± 0.1631	–	
3	1	6.7592 ± 0.0606	2.8159 ± 0.0111	0.6914 ± 0.0309	1.0173 ± 0.0838	6.8731 ± 0.0170
	2	6.1800 ± 0.0362	2.0092 ± 0.0094	0.0245 ± 0.0023	0.4406 ± 0.0054	
	3	2.4858 ± 0.1920	–	3.4989 ± 0.2282	–	
4	1	6.3626 ± 0.0556	2.0862 ± 0.0127	0.4669 ± 0.0211	1.2218 ± 0.04994	13.2913 ± 0.0049
	2	6.0338 ± 0.0369	2.1638 ± 0.0116	0.0814 ± 0.0056	1.1084 ± 0.0120	
	3	3.2084 ± 0.2091	–	3.6669 ± 0.19996	–	
17	1	3.5904 ± 0.0691	3.5174 ± 0.0342	0.8399 ± 0.0354	0.9713 ± 0.1038	17.4151 ± 0.0056
	2	4.4094 ± 0.0362	2.2944 ± 0.0147	0.2778 ± 0.0170	1.2093 ± 0.0349	
	3	2.7237 ± 0.2346	–	2.7146 ± 0.2294	–	
19	1	8.3497 ± 0.0547	1.9024 ± 0.0123	0.1462 ± 0.0066	1.1344 ± 0.0123	11.8413 ± 0.0103
	2	6.3147 ± 0.0600	3.6558 ± 0.0276	0.0840 ± 0.0110	1.1832 ± 0.0276	
	3	4.0103 ± 0.1681	–	1.6285 ± 0.0616	–	
21	1	3.0469 ± 0.0164	8.9908 ± 0.0453	4.3924 ± 0.0317	0.6240 ± 0.0482	12.1795 ± 0.0184
	2	9.2992 ± 0.0231	1.0055 ± 0.0020	0.0026 ± 0.0001	0.0277 ± 0.0002	
	3	1.4123 ± 0.0302	–	8.0059 ± 0.1113	–	
29	1	6.8779 ± 0.0807	1.1716 ± 0.0138	0.1738 ± 0.0076	0.8076 ± 0.0109	15.8654 ± 0.0173
	2	3.1266 ± 0.0826	3.7813 ± 0.0497	0.4507 ± 0.0607	1.8447 ± 0.1480	
	3	3.5246 ± 0.1540	–	3.5659 ± 0.1611	–	
37	1	3.4254 ± 0.0991	4.8591 ± 0.1016	1.5987 ± 0.0496	0.9439 ± 0.1021	24.6237 ± 0.0101
	2	2.7583 ± 0.1002	5.3099 ± 0.1160	1.2692 ± 0.0877	2.5030 ± 0.2218	
	3	2.5060 ± 0.2000	–	2.0923 ± 0.1252	–	
39	1	6.9593 ± 0.0467	1.3088 ± 0.0127	0.2592 ± 0.0121	1.0019 ± 0.0184	20.2444 ± 0.0071
	2	3.9188 ± 0.0587	2.9796 ± 0.0358	0.6190 ± 0.0436	2.1451 ± 0.0932	
	3	3.6297 ± 0.1896	–	4.2443 ± 0.2282	–	
43	1	8.5402 ± 0.0387	1.5659 ± 0.0074	0.0782 ± 0.0040	1.0716 ± 0.0060	16.2154 ± 0.0084
	2	6.9974 ± 0.0440	2.3804 ± 0.0143	0.0865 ± 0.0070	1.4685 ± 0.0132	
	3	3.5295 ± 0.2111	–	3.2640 ± 0.1878	–	

Li et al. (2013). These stresses are extracted at point 1 since the length of an ellipsoid major axis is the shortest through that point, resulting in the highest stress level there based on Equations (7) and (8) in the ϕ direction. It is shown that the homogenous model underestimate the stresses in the wall of GB1, 3, 4, 29, 39, and 43, and overestimates them for the remaining GBs. As a result, the relative error in the first principal stresses varies in a range of $-11.4\%\sim +10.8\%$ in comparison with the stresses in the homogenous model.

Comparison with Other Constitutive Models

The inversely estimated parameters are shown in **Tables 5–7** for the Fung, Choi-Vito, and Zhou-Fung strain energy functions, respectively. The parameters in the Zhou-Fung model were as many as 17 in total at the three points; thus the number of time instants was increased to 30 in the optimization procedure.

Our results show that even though the model parameters using these strain energy functions can also be inversely determined, the errors in stress are quite large. For instance, the mean errors are 14.8, 14.6, and 14.0% for the Fung, Choi-Vito and Zhou-Fung strain energy functions, respectively, while the structure-based model Equation (10) yields a mean error of 5.0% only (**Figure 5**).

Variation of the GB Wall Thickness

We notice from **Table 3** that there are some large errors ranged between 5.2 and 7.8% for the parameters estimated for five GB samples: 1, 17, 21, 37, and 39. To identify the cause of the errors, the stress-volume curves of GB 3 and 39 at points 1, 2, and 3

TABLE 7 | Material parameters of Zhou-Fung strain energy function Equation (14) inversely determined with uniform thickness $h_{11} = h_{21} = h_{31} = 2.5$ mm.

GB No.	Point i	c^i (kPa)	a_1^i	a_2^i	a_3^i	b_1^i	b_2^i	b_3^i	ε (%)
1	1	0.8073 ± 0.1082	2.0643 ± 0.1619	0.6341 ± 0.0806	0.5654 ± 0.0920	8.0082 ± 0.1667	1.6571 ± 0.1635	2.3635 ± 0.1627	13.8872 ± 0.0363
	2	0.7743 ± 0.0974	2.5808 ± 0.1633	0.7208 ± 0.0566	0.5077 ± 0.0877	9.8112 ± 0.0320	0.2040 ± 0.0241	0.2899 ± 0.0234	
	3	8.7721 ± 0.2652	–	9.2796 ± 0.2362	–	–	9.9707 ± 0.0095	–	
3	1	1.6195 ± 0.1581	3.1320 ± 0.2319	0.9319 ± 0.1349	0.9728 ± 0.1115	12.8219 ± 0.1550	1.4955 ± 0.1487	3.3015 ± 0.1511	6.4490 ± 0.0178
	2	1.2704 ± 0.1266	3.6118 ± 0.1468	0.1869 ± 0.0356	0.2755 ± 0.0562	13.4187 ± 0.0686	0.4733 ± 0.0459	0.8521 ± 0.0497	
	3	4.4095 ± 0.1615	–	4.1456 ± 0.0669	–	–	6.3141 ± 0.0108	–	
4	1	1.1620 ± 0.1257	3.2066 ± 0.3284	0.9681 ± 0.0987	0.7732 ± 0.0997	13.5437 ± 0.2035	2.0458 ± 0.2009	4.3685 ± 0.2090	13.0342 ± 0.0101
	2	1.0549 ± 0.1577	3.7930 ± 0.3425	0.8367 ± 0.1157	0.7100 ± 0.0833	14.7885 ± 0.1161	0.9880 ± 0.1008	2.4757 ± 0.1257	
	3	5.3204 ± 0.2107	–	4.8034 ± 0.0878	–	–	8.4977 ± 0.0886	–	
17	1	1.2488 ± 0.1413	3.7198 ± 0.2387	1.0768 ± 0.1045	1.1118 ± 0.1301	12.3554 ± 0.1548	1.2096 ± 0.1320	2.5024 ± 0.1405	17.0760 ± 0.0062
	2	0.7638 ± 0.1137	3.8420 ± 0.2851	0.9075 ± 0.1263	0.8016 ± 0.1500	11.2526 ± 0.1456	1.2561 ± 0.1235	2.3117 ± 0.1159	
	3	4.3245 ± 0.2086	–	4.0995 ± 0.1127	–	–	4.5930 ± 0.0455	–	
19	1	0.7828 ± 0.1307	4.6629 ± 0.3631	0.4719 ± 0.1150	0.4482 ± 0.1186	17.2832 ± 0.1329	1.2589 ± 0.1232	4.9771 ± 0.1244	11.0323 ± 0.0067
	2	1.4217 ± 0.1348	4.2640 ± 0.3540	1.5276 ± 0.1483	1.3113 ± 0.1283	25.9704 ± 0.1174	0.9578 ± 0.1076	2.0420 ± 0.1098	
	3	2.2564 ± 0.2448	–	1.3357 ± 0.1134	–	–	6.8376 ± 0.0216	–	
21	1	0.6073 ± 0.1682	5.5783 ± 0.7122	0.3904 ± 0.1108	0.3309 ± 0.0958	11.6934 ± 0.2064	2.6685 ± 0.2035	3.1046 ± 0.2028	5.5210 ± 0.0255
	2	1.1479 ± 0.1651	5.7567 ± 0.4519	1.3359 ± 0.1117	1.0352 ± 0.1403	18.4767 ± 0.0373	0.0749 ± 0.0118	0.1278 ± 0.0175	
	3	4.1435 ± 0.2240	–	3.3335 ± 0.0929	–	–	9.4489 ± 0.0439	–	
29	1	0.5637 ± 0.1207	2.9603 ± 0.3071	0.5336 ± 0.1119	0.4604 ± 0.0878	9.7022 ± 0.1465	1.8628 ± 0.1405	2.2909 ± 0.1386	15.0643 ± 0.0104
	2	1.0935 ± 0.1195	3.6787 ± 0.2249	1.6821 ± 0.1013	1.2663 ± 0.1381	13.5427 ± 0.1392	1.0954 ± 0.1270	1.7467 ± 0.1368	
	3	6.5934 ± 0.1729	–	6.5834 ± 0.1186	–	–	8.8734 ± 0.1268	–	
37	1	0.9666 ± 0.1439	3.1314 ± 0.1468	0.3768 ± 0.1183	0.4541 ± 0.1581	10.8332 ± 0.1713	2.1080 ± 0.1282	3.4023 ± 0.1355	22.8530 ± 0.0189
	2	4.3316 ± 0.1941	6.3448 ± 0.1610	1.5052 ± 0.1046	2.8789 ± 0.1302	18.7453 ± 0.1423	0.8814 ± 0.1296	5.9936 ± 0.1298	
	3	1.8696 ± 0.2859	–	1.0830 ± 0.1926	–	–	6.0539 ± 0.0198	–	
39	1	0.8969 ± 0.1399	1.8939 ± 0.1194	0.6789 ± 0.1009	0.6465 ± 0.1432	10.6963 ± 0.1726	2.1998 ± 0.1675	3.1981 ± 0.1706	19.7368 ± 0.0085
	2	1.3454 ± 0.1326	2.4035 ± 0.1253	1.5162 ± 0.0992	1.2377 ± 0.1315	13.2262 ± 0.2109	1.8452 ± 0.1971	3.6513 ± 0.1950	
	3	7.3471 ± 0.1801	–	7.3861 ± 0.0921	–	–	9.1861 ± 0.1025	–	
43	1	1.0615 ± 0.1756	2.5564 ± 0.1709	0.3917 ± 0.0843	0.4254 ± 0.1054	15.9433 ± 0.1504	1.5190 ± 0.1458	4.0383 ± 0.1455	15.3788 ± 0.0088
	2	1.3838 ± 0.1592	2.8815 ± 0.1915	1.2693 ± 0.0918	0.8113 ± 0.1182	18.9654 ± 0.0929	0.7135 ± 0.0829	4.3854 ± 0.0848	
	3	4.8098 ± 0.2783	–	4.4178 ± 0.1503	–	–	9.2673 ± 0.0868	–	

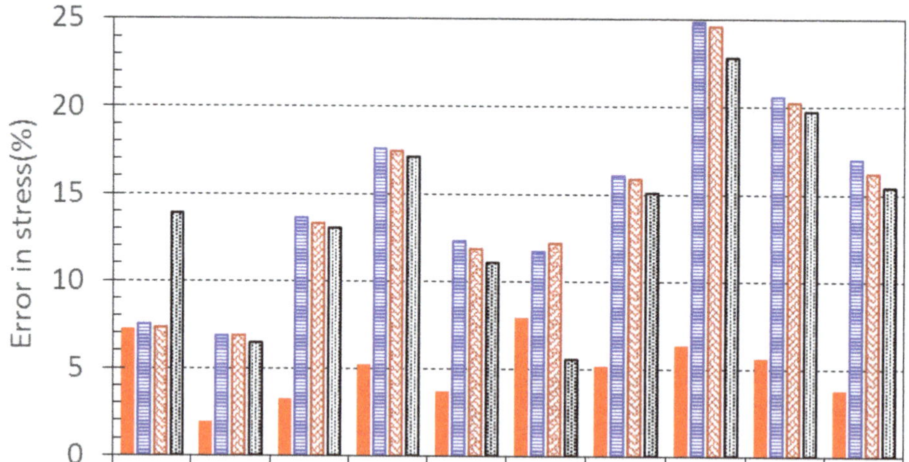

	1	3	4	17	19	21	29	37	39	43
Present	7.1827	1.9123	3.2065	5.1699	3.6488	7.8681	5.1186	6.3186	5.5799	3.7013
Fung	7.4833	6.8817	13.628	17.579	12.331	11.709	16.078	24.832	20.573	17.003
Choi-Vito	7.2885	6.8731	13.291	17.415	11.841	12.18	15.865	24.624	20.244	16.215
Zhou-Fung	13.887	6.449	13.034	17.076	11.032	5.521	15.064	22.853	19.737	15.379

GB No.

FIGURE 5 | A comparison of errors in the least-squares stress curve fitting between the present constitutive law and existing laws proposed by Fung, Choi-Vito, and Zhou-Fung, respectively.

are shown in **Figure 6**. The predicted stress agrees well with the observations at points 1 and 2, but not so well at point 3. In the following, we show that this is due to the uniform wall thickness assumption used in the model.

In Section Comparison with Other Constitutive Models, the GB wall heterogeneity mainly occurs in the apex region, resulting in poor agreement in the stress, as shown in **Figure 6**. Therefore, we altered the GB wall thickness at the apex to examine the effect of varying thickness. First, the apex thickness was changed to 3.0 mm, based on the ratio of 1.2 found in Su (2005), and kept at 2.5 mm at points 1 and 2. The extracted pointwise mechanical properties in Equation (10) are shown in **Table 8**.

The relative changes in these 13 parameters are tabulated in **Table 9**. The increased thickness at the apex by 20% has a considerable effect on the material parameters, with changes up to 30%, in particular, on κ_4^1, c^2, κ_4^2, c^3, κ_3^3, and κ_4^3, which are associated with the properties of the matrix and the longitudinal fibers. This is very different to the membrane theory, in which the Young's modulus is independent of the membrane thickness (Timoshenko and Woinowsky-Krieger, 1959).

We also found that an increased h_{31} could lower the error in the stress between the model production and the observation. If we increase h_{31} to 5.0 mm, the error reduces by 2.5%. Further increase in thickness does not decrease the error much, see **Figure 7**. Interestingly, 5.0 mm apex thickness seems to agree with measurement in Su (2005). Finally, if h_{31} is reduced to be 10% thinner than h_{11} and h_{21}, 2.25 m, according to Khan et al. (2012), then the errors in the stresses are greater, as shown in **Figure 7**. Thus, the observation that apex thickness was thinner than the GB body in Khan et al. (2012) did not agree with the results from the cohort of GB samples used here.

Note that in Khan et al. (2012) post-mortem samples from "unclaimed bodies" were used and so would not have been fresh. When left *in situ* the bile will start to break down the gallbladder wall–a process known as autolysis. Therefore, the results for wall thickness might not be reliable. Samples in Su (2005), on the other hand were obtained fresh from the operating theater and washed immediately. So we would have more faith in the results in Su (2005).

Comparison with Animal Test

Our ellipsoid model is different from the patient specific GB geometries. One may ask if such a simple model is of any practical use. To answer this question, we compared our model prediction with the *in vitro* measurements of a lamb GB (Genovese et al., 2014). In Genovese et al. (2014), a lamb GB was harvested and inflated at a pressure up to 50 mmHg, then a series points on the GB outside surface were tracked optically, and the strain fields were estimated from fitting curves of these points. The tension/stress fields were then calculated by using the elastic membrane model and solved numerically. In our model, we only used the diameters from Genovese et al. (2014) as the two minor axis lengths D_1 and D_2, respectively. The stress field, hence tension, are obtained analytically from Equation (7). The results are shown in **Figure 8**, where the comparisons of the second Piola-Kirchhoff surface tensions are plotted for pressure $p =$ 20 and 50 mmHg (we couldn't compare the results at $p = 3.5$ mmHg, as the tension profile in Genovese et al. (2014) seems to

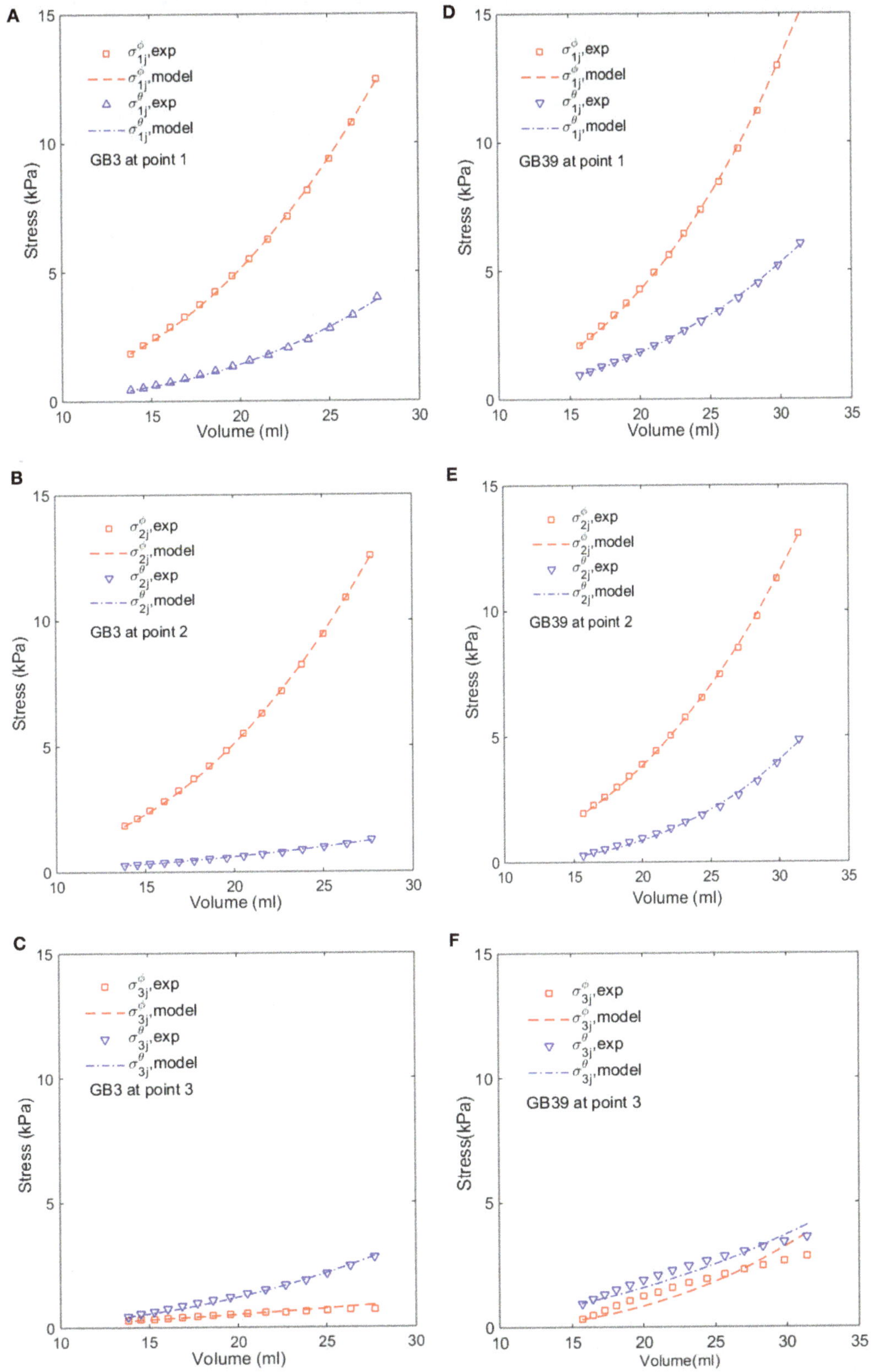

FIGURE 6 | Comparison of the modeled (lines) and estimated (symbols) circumferential and longitudinal stresses with the image-based ellipsoid membrane mechanic model at points 1, 2, 3, for GB 3 (A–C), and GB 39 **(D–F)**.

TABLE 8 | **Material parameters inversely determined with model Equation (10) and variable thicknesses: $h_{11} = h_{21} = 2.5$ mm, and $h_{31} = 3.0$ mm.**

GB No.	Point i	c^i (kPa)	κ_1^i (kPa)	κ_2^i	κ_3^i (kPa)	κ_4^i	ε (%)
1	1	0.2146 ± 0.0118	2.2040 ± 0.0186	0.3090 ± 0.0056	0.6441 ± 0.0185	0.5514 ± 0.0218	5.9256 ± 0.0297
	2	0.0106 ± 0.0029	2.4626 ± 0.0047	0.2836 ± 0.0014	0.1089 ± 0.1089	1.5081 ± 0.0662	
	3	1.1918 ± 0.0046	–	–	2.2841 ± 0.0212	4.8171 ± 0.4371	
3	1	0.1739 ± 0.0077	3.6555 ± 0.0144	0.7239 ± 0.0053	0.8493 ± 0.0142	1.4068 ± 0.0291	1.8261 ± 0.0173
	2	0.0926 ± 0.0072	3.3569 ± 0.0130	0.4549 ± 0.0040	0.1907 ± 0.0123	0.9927 ± 0.1021	
	3	0.2897 ± 0.0002	–	–	0.7816 ± 0.0026	4.1440 ± 0.0214	
4	1	0.2299 ± 0.0087	3.9979 ± 0.0161	0.4956 ± 0.0046	1.1460 ± 0.0161	0.8751 ± 0.0196	2.8756 ± 0.0119
	2	0.1633 ± 0.0095	3.9544 ± 0.0174	0.4604 ± 0.0049	0.5469 ± 0.0169	1.1846 ± 0.0468	
	3	0.7727 ± 0.0003	–	–	0.6013 ± 0.0036	5.8020 ± 0.0544	
17	1	0.1572 ± 0.0086	3.3594 ± 0.0160	0.8488 ± 0.0070	0.6231 ± 0.0152	1.8528 ± 0.0515	4.6031 ± 0.0195
	2	0.2326 ± 0.0138	2.8920 ± 0.0241	0.5705 ± 0.0089	0.5087 ± 0.0238	1.0820 ± 0.0671	
	3	0.7162 ± 0.0007	–	–	0.1522 ± 0.0020	4.2194 ± 0.1741	
19	1	0.5454 ± 0.0152	4.4740 ± 0.0284	0.4564 ± 0.0073	0.6271 ± 0.0276	0.9859 ± 0.0699	3.2060 ± 0.0129
	2	0.0918 ± 0.0076	6.7692 ± 0.0165	0.6541 ± 0.0049	0.0493 ± 0.0025	3.1356 ± 0.0818	
	3	0.4974 ± 0.0003	–	–	0.5500 ± 0.0042	1.1336 ± 0.0261	
21	1	0.2266 ± 0.0094	6.1367 ± 0.0238	3.2163 ± 0.0194	1.8135 ± 0.0201	7.1830 ± 0.0671	7.5871 ± 0.0058
	2	0.0577 ± 0.0002	2.7500 ± 0.0006	0.0631 ± 0.0002	0.0101 ± 0.0002	0.1368 ± 0.0252	
	3	0.0157 ± 0.0005	–	–	1.5921 ± 0.0060	5.2078 ± 0.0272	
29	1	0.3272 ± 0.0156	2.4211 ± 0.0244	0.2648 ± 0.0065	0.5663 ± 0.0245	0.4261 ± 0.0358	4.4857 ± 0.0208
	2	0.1032 ± 0.0063	3.5691 ± 0.0120	0.8197 ± 0.0049	0.4906 ± 0.0105	2.6659 ± 0.0461	
	3	1.0348 ± 0.0013	–	–	0.4320 ± 0.0059	2.6961 ± 0.2625	
37	1	0.2469 ± 0.0126	4.1076 ± 0.0252	1.2490 ± 0.0122	1.0492 ± 0.0253	1.8288 ± 0.0601	5.5412 ± 0.0075
	2	0.2414 ± 0.0140	4.1183 ± 0.0280	1.2440 ± 0.0136	1.0606 ± 0.0283	1.8085 ± 0.0664	
	3	0.7965 ± 0.0002	–	–	0.0153 ± 0.0005	6.3465 ± 0.0628	
39	1	0.2999 ± 0.0131	2.9229 ± 0.0217	0.2699 ± 0.0054	0.8575 ± 0.0216	0.4888 ± 0.0222	4.8833 ± 0.0117
	2	0.2650 ± 0.0123	3.6739 ± 0.0230	0.6839 ± 0.0081	0.8623 ± 0.0215	1.8161 ± 0.0420	
	3	1.5221 ± 0.0023	–	–	0.1718 ± 0.0083	2.8898 ± 0.2615	
43	1	0.3508 ± 0.0156	4.3353 ± 0.0275	0.2502 ± 0.0054	0.7968 ± 0.0275	0.5590 ± 0.0371	3.2537 ± 0.0129
	2	0.2305 ± 0.0099	5.3712 ± 0.0192	0.4119 ± 0.0044	0.7883 ± 0.0181	1.5736 ± 0.0399	
	3	1.1491 ± 0.0005	–	–	0.2174 ± 0.0031	3.0536 ± 0.2349	

be unrealistically large, which is possibly due to a typo in the color map scale).

The overall agreement is encouraging; in both our model and experiments, the highest tension is found near the GB equator, and the minimum tension occurs at the apex. The values of the predicted surface tension are also in good agreement with the experimental data except some isolated tension spots due to the rapid change in the wall curvature of the lamb GB. The predicted surface tension magnitude near the GB body/equator is in a range of 0.023–0.027 N/mm, compared with 0.03–0.04 N/mm in the experiments at 20 mmHg pressure. Likewise, the predicted tension is in a range of 0.063–0.073 N/mm, compared to the range of 0.06–0.08 N/mm in the experiments at 50 mmHg.

DISCUSSION

Ultrasonography is a common method for monitoring GB volume variations in daily diagnosis (Dodds et al., 1985; Portincasa et al., 2003; Ugwu and Agwu, 2010). Although a detailed 3D model is more accurate, simplified geometry models are fast and therefore frequently used in clinical assessment. When an ellipsoidal model is used to estimate GB volume based on the images scanned during emptying phase, the error of the model in GB volume is about 0.8 ± 0.1 ml. This compares better to the error of 2.1 ± 0.2 ml if using sum-of-cylinder method (Dodds et al., 1985). To assess if the simplified model could predict the stress distribution of a realistic GB sample, we also

TABLE 9 | Relative changes in the parameters due to varied wall thickness.

GB No.	Point i	$\Delta c^i/c^i$ (%)	$\Delta\kappa_1^i/\kappa_1^i$ (%)	$\Delta\kappa_2^i/\kappa_2^i$ (%)	$\Delta\kappa_3^i/\kappa_3^i$ (%)	$\Delta\kappa_4^i/\kappa_4^i$ (%)	$\Delta\varepsilon$ (%)
1	1	−1.5229	0.2183	−0.4831	0.7824	−1.5884	−1.2571
	2	−33.7500	0.3586	−0.8738	4.61095	−1.12765	
	3	−19.4784	–	–	−7.9252	−1.0761	
3	1	−0.1722	0.01642	−0.0414	0.0825	−0.0994	−0.0862
	2	4.6328	−0.2259	0.5082	−4.0262	2.1927	
	3	−16.9200	–	–	−15.7305	−1.6821	
4	1	3.14042	−0.3192	0.7317	−1.1046	1.7203	−0.3309
	2	−2.1570	0.1596	−0.38944	1.2403	−1.4230	
	3	−17.0300	–	–	−11.1292	−7.5644	
17	1	5.9299	−0.4829	0.8316	−2.4730	2.6880	−0.5668
	2	8.0855	−1.0436	2.0390	−5.5690	8.8422	
	3	−16.2143	–	–	−10.5758	−18.2825	
19	1	−2.3106	0.5325	−1.3829	3.7043	−8.0660	−0.4428
	2	−6.1350	0.1850	−0.4868	−3.7109	−2.2569	
	3	−16.6136	–	–	−17.3057	2.4955	
21	1	−2.3697	0.1387	−0.0528	1.2280	−1.2415	−0.2810
	2	0.1736	−0.0218	0.47771	0	6.8750	
	3	−16.4894	–	–	−15.5250	−1.9339	
29	1	−4.0751	0.9212	−2.2157	4.0037	−6.2898	−0.6329
	2	−4.7091	0.2866	−0.5581	1.4055	−1.1788	
	3	−16.6559	–	–	−14.8097	−15.6836	
37	1	3.6524	−0.4122	0.6446	−1.6959	2.2704	−0.7774
	2	0.1244	−0.0146	0.0402	−0.0565	0.4778	
	3	−16.8146	–	–	2.6846	−6.2472	
39	1	−3.7857	0.6820	−1.8188	2.2781	−3.4565	−0.6966
	2	−1.8519	0.2565	−0.4947	1.0310	−1.0246	
	3	−16.7569	–	–	−12.8803	−6.7415	
43	1	−3.7322	0.56367	−1.8439	3.0656	−5.6381	−0.4476
	2	−3.2326	0.28192	−0.8426	1.8476	−1.9197	
	3	−16.5808	–	–	−13.4554	−17.0601	

Δc^i, $\Delta\kappa_1^i$, $\Delta\kappa_2^i$, $\Delta\kappa_3^i$, $\Delta\kappa_4^i$ and $\Delta\varepsilon$ are the differences of these parameters and error between the case of $h_{31} = 3.0$ mm thick apex wall and the case of 2.5 mm uniform GB wall, c^i, κ_1^i, κ_2^i, κ_3^i κ_4^i are the parameters for the 2.5 mm uniform GB wall.

compared our model prediction with the surface tension data for a lamb GB, and the overall agreement was surprisingly good.

We comment that the segmentation error of estimating the GB diameter from a GB image is usually around 4.31–7.21% (Bocchi and Rogai, 2011). To address the effect of this error on the GB wall material parameters, we introduce a random error (or noise) of 4.31–7.21% in the major axis lengths, i.e.:

$$\varepsilon_{seg} = 0.0431 + rand \times (0.0721 - 0.0431) \quad (17)$$

where *rand* is the inner random function in MATLAB to generate a random number in value 0–1. Then we run the inverse heterogeneous problem code with these noisy data for a number of GB sample, say No. 1, 3, 17, and 21. The parameters estimated are compared in **Table 10** against those without the noise.

When noise is considered, the error in the curve fitting increased mostly by 3.1–6.7%, some can go as high as 12.3%, in comparison with the case without the noise. The influence of segmentation error on the parameters varies from one GB to another, however, the parameters at points 1, 2 are mostly likely affected by the segmentation error, particularly, changes in c^1, κ_2^1, κ_3^1, κ_4^1, c^2, κ_3^2, are κ_4^2 can be large. Hence, the segmentation error should be reduced as much as possible to

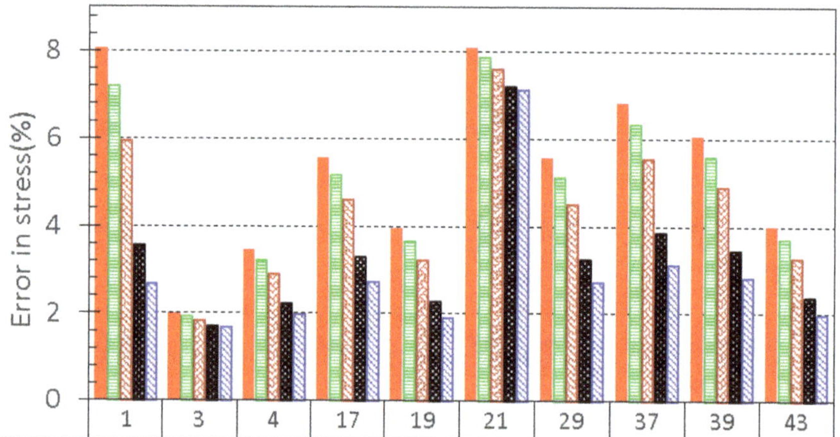

GB No.	1	3	4	17	19	21	29	37	39	43
2.25mm	8.0445	1.9781	3.4335	5.5471	3.9388	8.0780	5.5589	6.8100	6.0426	3.9921
2.5mm	7.1827	1.9123	3.2065	5.1699	3.6488	7.8681	5.1186	6.3186	5.5799	3.7013
3.0mm	5.9388	1.8291	2.8760	4.6031	3.2060	7.5871	4.4857	5.5412	4.8833	3.2537
5.0mm	3.5485	1.6973	2.2237	3.2913	2.2785	7.1909	3.2229	3.8510	3.4330	2.3598
7.0mm	2.6627	1.6842	1.9679	2.7225	1.9055	7.1110	2.7059	3.1107	2.8236	1.9813

GB No.

FIGURE 7 | Effect of GB wall thickness at the apex on the error in stress, the thickness at the apex is varied to be 2.25, 2.5, 3.0, 5.0, and 7.0 mm, respectively, while the thickness at the other two points 1 and 2 is kept to be 2.5 mm.

improve the accuracy of the inverse estimation. In future, using an automatic segmentation method with small segmentation error as introduced in Bocchi and Rogai (2011) may be the way forward.

In our previous work in Li et al. (2013), the human GB wall was considered to be a non-linear composite material of matrix and two orthogonal families of fibers in the circumferential and longitudinal directions, respectively, and the material parameters were assumed to be constant. These parameters were determined inversely in Li et al. (2013) by using the FEM software-ABAQUS with a user subroutine and a MATLAB code. However, such an inverse approach is extremely time-consuming (~7 h) and unsuitable for clinical applications. In this work, we have developed a simpler approach using analytical or simpler forward solvers, which makes it possible for clinical assessment of in GB human wall disease in real time.

In addition, we extended the previous model from homogenous membrane model in Li et al. (2013) to heterogeneous model, which has significantly improved the fitting accuracy. The heterogeneity of the GB has been confirmed in the experimental work on lamb GB (Genovese et al., 2014). The inverse estimation of the heterogeneous property constants had an error less than 7% for the ten human GB samples, and the computational time was reduced by 20 times (~30 min). Further, allowing the wall thickness variation following experiments (Su, 2005), reduced the error to be less than 4%.

One potential clinical use of the model is to assessing the GB pain. In **Table 4**, we compare the first principal stress with the pain score associated with the CCK venous injection. It is clear that there is a strong correlation between the magnitude of the stress and the pain score. Although given the limited sample size, the homogeneous and heterogeneous model seem to do equally well in terms of pain prediction.

The limitations of our study should also be mentioned. In the study, the stretches at the points 1–3 were determined analytically during GB emptying phase. The analytical method was based on the GB volume change from images. To our best knowledge, no speckle tracking echocardiography on GB has been reported to validate our model, unlike extensive measured on human left ventricle (Edvardsen et al., 2002; Marwick, 2006; Crosby et al., 2009; Maffessanti et al., 2009; Marwick et al., 2009; Tanaka et al., 2010; Hoit, 2011; Kleijn et al., 2011). In addition, there were also no *in vitro* passive tensile tests on the specimens harvested from the body and fundus of human GB. In future, we may be able to utilize the measured strain/stretch to validate our analytical method for stretch extraction, this will make our regional GB biomechanical property identification more accurate.

Further, we only used one *in vitro* observation to determine the reference configuration of human GBs. In reality, the size of a reference configuration may not be exactly 50% of the size of totally refilled GB. It is possible that the GB reference configuration can be estimated using GB ejection fraction (EF) in cholecystokinin-cholescintigraphy (CCK-CS) (Ozden and DiBaise, 2003) or fatty meal Cholescintigraphy (FM-CS) (Al-Muqbel et al., 2010) examination for GB patients' *in vivo* clinical diagnosis.

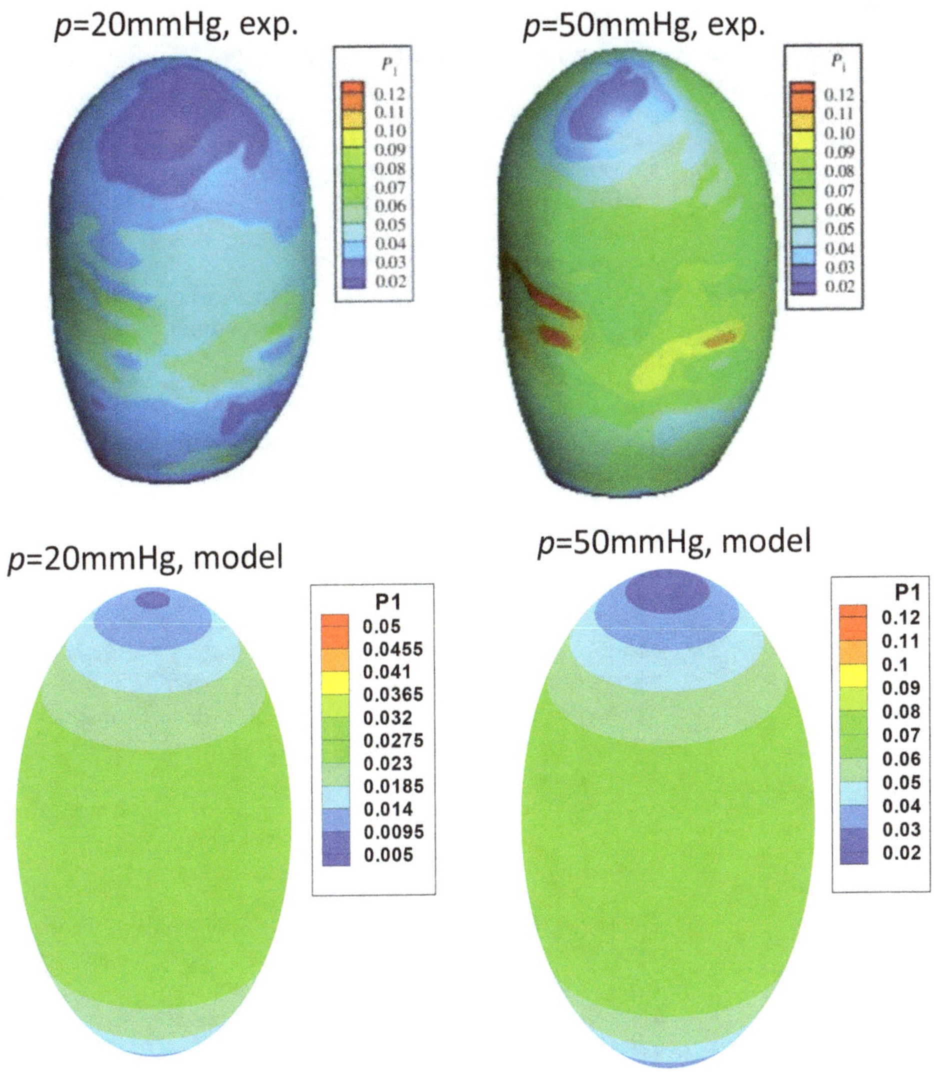

FIGURE 8 | Comparison of the peak tension from the ellipsoid model with the *in vitro* experimental tension of a lamb GB (Genovese et al., 2014) at the internal pressure of 20 and 50 mmHg, respectively, two plots in the top row are from Genovese et al. (2014), with permission.

Although we have investigated the thickness variation in our model, the values we used were applicable only for a healthy GB. When human GBs suffer from diseases such as acute cholecystitis, acalculous cholecystitis and ascites (Sanders, 1980; Runner et al., 2014), the GB thickness can increase significantly. Indeed, diseased GB body wall thickness was ranged in 3–5 mm (Sanders, 1980; Mohammed et al., 2010; Runner et al., 2014). How to estimate the GB wall thickness change in disease will be an important challenge for future studies.

CONCLUSIONS

The heterogeneity of ten samples of human GB is investigated theoretically in refilling phase using a structure-based constitutive model, ellipsoidal GB and membrane in-plane mechanic model. Three different points, two on the equator of GB body with 90° apart and one on the apex of GB fundus, are chosen to evaluate the variation of the material properties. The stretches at these points are tracked over time from the routine ultrasonic images scanned at the Sheffield Hallamshire Hospital during the emptying phase. The material parameters at the three different points are determined inversely using a MATLAB code. The human GBs are found to exhibit heterogeneity, especially from GB body to its apex region. It is found that using a homogeneous model underestimate the peak stresses in the GB wall, and that a strong heterogeneity occurs from GB body to fundus. Finally, our model results indicate that the GB wall is much thicker at the apex, which clarify the contrary findings reported in the literature.

TABLE 10 | Heterogeneous material parameters of four GB samples inversely determined with Equation (10) and uniform thickness $h_{11} = h_{21} = h_{31} = 2.5$ mm when the error in segmentation is considered.

GB No.	Point i	With segmentation error	c^i (kPa)	κ_1^i (kPa)	κ_2^i (–)	κ_3^i (kPa)	κ_4^i (–)
1	1	No	0.2180 ± 0.0131	2.1992 ± 0.0206	0.3105 ± 0.0064	0.6391 ± 0.0203	0.5603 ± 0.0267
		Yes	0.2179 ± 0.0069	2.2538 ± 0.0108	0.4583 ± 0.0037	0.5776 ± 0.0097	1.0080 ± 0.0154
	2	No	0.0160 ± 0.0016	2.4538 ± 0.0025	0.2861 ± 0.0008	0.1041 ± 0.0030	1.5253 ± 0.0385
		Yes	0.0681 ± 0.0063	2.7813 ± 0.0109	0.1910 ± 0.0028	0.1074 ± 0.0100	0.7072 ± 0.1303
	3	No	1.4801 ± 0.0040	–	–	2.4807 ± 0.0176	4.8695 ± 0.4743
		Yes	0.7986 ± 0.0033	–	–	2.6040 ± 0.0094	4.9530 ± 0.2609
	ε (%)	No			7.1827 ± 0.0310		
		Yes			10.2758 ± 0.0171		
3	1	No	0.1742 ± 0.0076	3.6549 ± 0.0142	0.7242 ± 0.0053	0.8486 ± 0.0141	1.4082 ± 0.0281
		Yes	0.2905 ± 0.0183	4.4777 ± 0.0349	0.4755 ± 0.0099	1.1143 ± 0.0357	0.7544 ± 0.0501
	2	No	0.0885 ± 0.0084	3.3645 ± 0.0150	0.4526 ± 0.0047	0.1987 ± 0.0144	0.9714 ± 0.1228
		Yes	0.0546 ± 0.0041	3.6458 ± 0.0074	0.5561 ± 0.0022	0.1301 ± 0.0063	2.4910 ± 0.0864
	3	No	0.3487 ± 0.0001	–	–	0.9275 ± 0.0012	4.2149 ± 0.0085
		Yes	0.3173 ± 0.0004	–	–	1.4950 ± 0.0064	1.0760 ± 0.0253
	ε (%)	No			1.9123 ± 0.0152		
		Yes			8.6650 ± 0.0106		
17	1	No	0.1484 ± 0.0080	3.3757 ± 0.0151	0.8418 ± 0.0065	0.6389 ± 0.0144	1.8043 ± 0.0455
		Yes	0.0207 ± 0.0021	4.3780 ± 0.0042	0.3579 ± 0.0013	1.1763 ± 0.0069	0.4943 ± 0.0133
	2	No	0.2152 ± 0.0116	2.9225 ± 0.0205	0.5591 ± 0.0074	0.5387 ± 0.0204	0.9941 ± 0.0516
		Yes	0.4705 ± 0.0048	2.8344 ± 0.0085	0.7250 ± 0.0035	0.1470 ± 0.0060	3.4810 ± 0.0704
	3	No	0.8548 ± 0.0004	–	–	0.1702 ± 0.0017	5.1634 ± 0.1110
		Yes	0.8166 ± 0.0001	–	–	0.1426 ± 0.0018	6.6020 ± 0.0852
	ε (%)	No			5.1699 ± 0.0131		
		Yes			8.9515 ± 0.0048		
21	1	No	0.2321 ± 0.0088	6.1282 ± 0.0227	3.2180 ± 0.0189	1.7915 ± 0.0157	7.2733 ± 0.0529
		Yes	0.1136 ± 0.0068	5.5528 ± 0.0200	4.1576 ± 0.0214	1.6702 ± 0.0104	7.9210 ± 0.0451
	2	No	0.0576 ± 0.0002	2.7506 ± 0.0005	0.0628 ± 0.0002	0.0101 ± 0.0001	0.1280 ± 0.0228
		Yes	0.0402 ± 0.0002	2.9211 ± 0.0005	0.0536 ± 0.0002	0.0101 ± 0.0001	0.0990 ± 0.0208
	3	No	0.0188 ± 0.0005	–	–	1.8847 ± 0.0046	5.3105 ± 0.0167
		Yes	0.0161 ± 0.0006	–	–	1.7578 ± 0.0036	4.4671 ± 0.010
	ε (%)	No			7.8681 ± 0.0058		
		Yes			12.3111 ± 0.0049		

AUTHOR CONTRIBUTIONS

WL developed math model, designed numerical method, composed and validated code, carried out case studies, analyzed the results and drafted the manuscript. NB checked medical issues in human gallbladder, revised the draft. XL checked the math model, method, result explanation and conducted English text editing.

ACKNOWLEDGMENTS

The project was supported by EPSRC (Grant no. EP/G015651 EP/G028257, and EP/N014642/1).

REFERENCES

Al-Muqbel, K. M., Bani Hani, M, N., Elheis, M. A., and Al-Omari, M. H. (2010). Reproducibility of gallbladder ejection fraction measured by fatty meal cholescintigraphy. *Nucl. Med. Mol. Imaging* 44, 246–251. doi: 10.1007/s13139-010-0046-8

Baldewsing, R. A., de Korte, C. L., Schaar, J. A., Mastik, F., and van der Steen, A. F. (2004a). Finite element modelling and intravascular ultrasound elastrography of vulnerable plaques: parameters variation. *Ultrasonics* 42, 723–729. doi: 10.1016/j.ultras.2003.11.017

Baldewsing, R. A., de Korte, C. L., Schaar, J. A., Mastik, F., and van der Steen, A. F. (2004b). Finite element model for performing intravascular ultrasound elastography of human atherosclerotic coronary arteries. *Ultrasound Med. Biol.* 30, 803–813. doi: 10.1016/j.ultrasmedbio.2004.04.005

Bersi, M. R., Bellini, C., Di Achille, P., Gumphrey, J. D., Genovese, K., and Avril, S. (2016). Novel methodology for characterizing regional variations in the material properties of murine aortas. *J. Biomech. Eng.* 138, 071005. doi: 10.1115/1.4033674

Bocchi, L., and Rogai (2011). *Segmentation of Ultrasound Breast Images: Optimization of Algorithm Parameters*. Torino: Applications of Evolutionary Computation-LNCS 6624.

Chandran, K. B., Mun, J. H., Choi, K. K., Chen, J. S., Hamilton, A., Nagaraj, A., et al. (2003). A method for *in vivo* analysis for regional arterial wall property alternations with atherosclerosis: preliminary results. *Med. Eng. Phys.* 25, 289–298. doi: 10.1016/S1350-4533(02)00224-2

Cozzolino, H. J., Goldstein, F., and Green, R. R. (1963). The cystic duct syndrome. *JAMA* 185, 920–104. doi: 10.1001/jama.1963.03060120030017

Crosby, J., Amundsen, B. H., Hergum, T., Remme, E. W., Langeland, S., and Torp, H. (2009). 3-D speckle tracking for assessment of regional left ventricular function. *Ultrasound Med. Biol.* 36, 458–471. doi: 10.1016/j.ultrasmedbio.2008.09.011

Davis, F. M., Luo, Y., Avril, A., Duprey, A., and Lu, J. (2015). Pointwise characterization of the elastic properties of planar soft tissues; application to ascending thoracic aneurysms. *Biomech. Model. Mechanobiol.* 14, 967–978. doi: 10.1007/s10237-014-0646-9

Dodds, W. J., Groh, W. J., Darweesh, R. M. A., Lawson, T. L., Kishk, S. M. A., and Kern M. K. (1985). Sonographic measurement of gallbladder volume. *Am. J. Roentgenol.* 145, 1009–1011. doi: 10.2214/ajr.145.5.1009

Edvardsen, T., Gerber, B. L., Garot, J., Bluemke, D. A., Lima, J. A. C., and Smiseth, O. A. (2002). Quantitative assessment of intrinsic regional myocardial deformation by Doppler strain rate echocardiography in humans. *Circulation* 106, 50–56. doi: 10.1161/01.CIR.0000019907.77526.75

Engel, J. M., Deitch, E. A., and Sikkema, W. (1980). Gallbladder wall thickness: sonographic accuracy and relation to disease. *Am. J. Roentgenol.* 134, 907–909. doi: 10.2214/ajr.134.5.907

Ferruzzi, J., Vorp, D. A., and Humphrey, J. D. (2011). On constitutive descriptors of the biaxial mechanical behaviour of human abdominal aorta and aneurysms. *J. R. Soc. Interface* 8, 435–450. doi: 10.1098/rsif.2010.0299

Franquet, A., Avril, S., Le Riche, R., and Badel, P. (2011). Identification of heterogeneous elastic properties in stenosed arteries; a numerical plane strain study. *Comput. Methods Biomech. Biomed. Engin.* 15, 49–58. doi: 10.1080/10255842.2010.547192

Genovese, K., Casaletto, L., Humphrey, J. D., and Lu, J. (2014). Digital image correlation-based point-wise inverse characterization of heterogeneous material properties of gallbladder *in vitro*. *Pro. R. Soc. A.* 470:20140152. doi: 10.1098/rspa.2014.0152

Hamilton, A. J., Kim, H., Nagaraj, A., Mun, J. H., Yan, L. L., Roth, S. I., et al. (2005). Regional material property alterations in porcine femoral with atheroma development. *J. Biomech.* 38, 2354–2364. doi: 10.1016/j.jbiomech.2004.10.018

Hoit, B. D. (2011). Strain and strain rate echocardiography and coronary artery disease. *Circulation* 4, 179–190. doi: 10.1161/circimaging.110.959817

Khan, L. F., Naushaba, H., Paul, U. K., Banik, S., Ahmed, M., and Al-Safri, M. A. (2012). Gross and histomorphological study of thickness of the gallbladder wall. *J. Dhaka Natl. Med. Coll. Hosp.* 18, 34–38. doi: 10.3329/jdnmch.v18i1.12238

Kleijn, S. A., Aly, M. F. A., Terwee, C. B., van Rossum, A. C., and Kamp, O. (2011). Three-dimensional speckle tracking echocardiography for automatic assessment of global and regional left ventricular function based on area strain. *J. Am. Soc. Echocardiogr.* 24, 314–321. doi: 10.1016/j.echo.2011.01.014

Li, W. G., Hill, N. A., Ogden, R. W., Smythe, A., Majeed, A. W., Bird, N., et al. (2013). Anisotropic behaviour of human gallbladder walls. *J. Mech. Behav. Biomed. Mater.* 20, 363–375. doi: 10.1016/j.jmbbm.2013.02.015

Li, W. G., Luo, X. Y., Hill, N. A., Ogden, R. W., Smythe, A., Majeed, A. W., et al. (2011). A mechanical model for CCK-induced acalculous gallbladder pain. *Ann. Biomed. Eng.* 39, 786–800. doi: 10.1007/s10439-010-0205-1

Li, W. G., Luo, X. Y., Hill, N. A., Ogden, R. W., Smythe, A., Majeed, A. W., et al. (2012). A quasi-nonlinear analysis on anisotropic behaviour of human gallbladder wall. *J. Biomech. Eng.* 134:0101009. doi: 10.1115/1.4007633

Maffessanti, F., Nesser, H. J., Weinert, L., Steringer-Mascherbauer, R., Niel, J., Gorissen, W., et al. (2009). Quantitative evaluation of regional left ventricular function using three-dimensional speckle tracking echocardiography in patients with and without heart disease. *Am. J. Cardiol.* 104, 1755–1762. doi: 10.1016/j.amjcard.2009.07.060

Marwick, T. H. (2006). Measurement of strain and strain rate by echocardiography. *J. Am. Coll. Cardiol.* 47, 1313–1327. doi: 10.1016/j.jacc.2005.11.063

Marwick, T. H., Leano, R. L., Brown, J., Sun, J. P., Hoffmann, R., Lysyansky, P., et al. (2009). Myocardial strain measurement with 2-dimensional speckle-tracking echocardiography. *J. Am. Coll. Cardiol.* 2, 80–84. doi: 10.1016/j.jcmg.2007.12.007

Mohammed, S., Tahir, A., Ahidjo, A., Mustapha, Z., Franza, O., Okoye, I., et al. (2010). Sonographic gallbladder wall thickness in normal adult population in Nigeria. *South Afr. J. Radiol.* 14, 84–87. doi: 10.4102/sajr.v14i4.450

More, J. J., and Sorensen, D. C. (1983). Computing a trust region step. *SIAM J. Sci. Stat. Comput.* 4, 553–572. doi: 10.1137/0904038

Ozden, N., and DiBaise, J. K. (2003). Gallbladder ejection fraction and symptom outcome in patients with acalculous biliary-like pain. *Dig. Dis. Sci.* 48, 890–897. doi: 10.1023/A:1023039310574

Portincasa, P., Moschetta, A., Colecchia, A., Festi, D., and Palasciano, G. (2003). Measurements of gallbladder motor function by ultrasonography: towards standardization. *Digest. Liver Dis.* 35, S56–S61. doi: 10.1016/S1590-8658(03)00096-3

Prasad, M. N., Brown, M. S., Ni, C., Douek, M., Raman, S., Lu, D., et al. (2008). Three-dimensional mapping of gallbladder wall thickness on computed tomography using Laplace's equation. *Acad. Radiol.* 15, 1075–1081. doi: 10.1016/j.acra.2008.02.006

Ragab, A. R., and Bayoumi, S. E. (1998). *Engineering Solid Mechanics: Fundamentals and Applications*. Boca Raton, FL: CRC Press.

Runner, G. J., Corwin, M. T., Siewert, B., and Eisenberg, R. L. (2014). Gallbladder wall thickening. *Am. J. Roentgenol.* 202, W1–W12. doi: 10.2214/AJR.12.10386

Sanders, R. (1980). The significance of sonographic gallbladder wall thickening. *J. Clin. Ultrasound* 8, 143–146. doi: 10.1002/jcu.1870080210

Smythe, A., Ahmed, R., Fitzhenry, M., Johnson, A. G., and Majeed, A. W. (2004). Bethanechol provocation testing does not predict symptom relief after cholecystectomy for acalculous biliary pain. *Digest. Liver Dis.* 36, 682–686. doi: 10.1016/j.dld.2004.05.009

Smythe, A., Majeed, A., Fizhenry, M., and Johnson, A. G. (1998). A requiem for the cholecystokinin provocation test? *Gut* 43, 571–574. doi: 10.1136/gut.43.4.571

Su, Y. (2005). *The Mechanical Properties of Human Gallbladder*. Thesis of BEng in Mechanical Engineering, University of Sheffield, Sheffield, UK.

Tanaka, H., Hara, H., Saba, S., and Gorcsan, J. III. (2010). Usefulness of three-dimensional speckle tracking strain to quantify dyssynchrony and the site of latest mechanical activation. *Am. J. Cardiol.* 105, 235–242. doi: 10.1016/j.amjcard.2009.09.010

Timoshenko, S., and Woinowsky-Krieger, S. (1959). *Theory of Plates and Shells, 2nd Edn*. New York, NY: McGraw-Hill Book Company.

Trivedi, R. A., Li, Z. Y., U-King-Im, J., Graves, M. J., Kirkpatrick, P. J., and Gillard, J. H. (2007). Identifying vulnerable carotid plaques *in vivo* using high resolution magnetic resonance imaging-based finite element analysis. *J. Neurosurg.* 107, 536–542. doi: 10.3171/JNS-07/09/0536

Ugwu, A. C., and Agwu, K. K. (2010). Ultrasound quantification of gallbladder volume to establish baseline contraction indices in healthy adults: a pilot study. *South Afr. Radiogr.* 48, 9–12.

Williamson, R. (1988). Acalculous disease of the gall bladder. *Gut* 29, 860–872. doi: 10.1136/gut.29.6.860

Xiong, L., Chui, C. K., and Teo, C. L. (2013). Reality based modelling and simulation of gallbladder shape deformation using variational methods. *Int. J. Comput. Assist. Radiol. Surg.* 8, 857–865. doi: 10.1007/s11548-013-0821-y

Zhao, X., Raghavan, M. L., and Lu, J. (2011a). Identifying heterogeneous anisotropic properties in cerebral aneurysms; a pointwise approach. *Biomech. Model. Mechanobiol.* 10, 177–189. doi: 10.1007/s10237-010-0225-7

Zhao, X., Raghavan, M. L., and Lu, J. (2011b). Characterizing heterogeneous anisotropic properties of cerebral aneurysms with unknown stress-free geometry: a precursor to *in vivo* identification. *J. Biomech. Eng.* 133, 051008. doi: 10.1115/1.4003872

Zhou, J., and Fung, Y. C. (1997). The degree of nonlinearity and anisotropic of blood vessel elasticity. *Proc. Natl. Acad. Sci. U.S.A.* 94, 14255–14260. doi: 10.1073/pnas.94.26.14255

Permissions

All chapters in this book were first published in MHPCM, by Frontiers Media SA; hereby published with permission under the Creative Commons Attribution License or equivalent. Every chapter published in this book has been scrutinized by our experts. Their significance has been extensively debated. The topics covered herein carry significant findings which will fuel the growth of the discipline. They may even be implemented as practical applications or may be referred to as a beginning point for another development.

The contributors of this book come from diverse backgrounds, making this book a truly international effort. This book will bring forth new frontiers with its revolutionizing research information and detailed analysis of the nascent developments around the world.

We would like to thank all the contributing authors for lending their expertise to make the book truly unique. They have played a crucial role in the development of this book. Without their invaluable contributions this book wouldn't have been possible. They have made vital efforts to compile up to date information on the varied aspects of this subject to make this book a valuable addition to the collection of many professionals and students.

This book was conceptualized with the vision of imparting up-to-date information and advanced data in this field. To ensure the same, a matchless editorial board was set up. Every individual on the board went through rigorous rounds of assessment to prove their worth. After which they invested a large part of their time researching and compiling the most relevant data for our readers.

The editorial board has been involved in producing this book since its inception. They have spent rigorous hours researching and exploring the diverse topics which have resulted in the successful publishing of this book. They have passed on their knowledge of decades through this book. To expedite this challenging task, the publisher supported the team at every step. A small team of assistant editors was also appointed to further simplify the editing procedure and attain best results for the readers.

Apart from the editorial board, the designing team has also invested a significant amount of their time in understanding the subject and creating the most relevant covers. They scrutinized every image to scout for the most suitable representation of the subject and create an appropriate cover for the book.

The publishing team has been an ardent support to the editorial, designing and production team. Their endless efforts to recruit the best for this project, has resulted in the accomplishment of this book. They are a veteran in the field of academics and their pool of knowledge is as vast as their experience in printing. Their expertise and guidance has proved useful at every step. Their uncompromising quality standards have made this book an exceptional effort. Their encouragement from time to time has been an inspiration for everyone.

The publisher and the editorial board hope that this book will prove to be a valuable piece of knowledge for researchers, students, practitioners and scholars across the globe.

List of Contributors

Antonella Iuliano, Claudia Angelini and Italia De Feis
Istituto per le Applicazioni del Calcolo "Mauro Picone," Consiglio Nazionale delle Ricerche, Naples, Italy

Annalisa Occhipinti and Pietro Lió
Computer Laboratory, University of Cambridge, Cambridge, UK

Mona Alimohammadi, Cesar Pichardo-Almarza and Vanessa Díaz-Zuccarini
Mechanical Engineering, University College London, London, UK

Obiekezie Agu
Vascular Unit, University College Hospital, London, UK

Kyle C.A. Wedgwood
Center for Biomedical Modelling and Analysis, University of Exeter, Exeter, UK

Sarah J.Richardson and Noel G. Morgan
University of Exeter Medical School, University of Exeter, Exeter, UK

Krasimira Tsaneva- Atanasova
College for Engineering, Mathematics and Physical Sciences, University of Exeter, Exeter, UK
Engineering and Physical Sciences Research Council Centre for Predictive Modelling in Healthcare, University of Exeter, Exeter, UK

Felicity Kendrick, Neil D. Evans and Michael J. Chappell
School of Engineering, University of Warwick, Coventry, UK

Bertrand Arnulf and Stéphanie Harel
Hôpital Saint-Louis, Paris, France

Hervé Avet-Loiseau
Unité de Génomique du Myélome, Institut Universitaire du Cancer de Toulouse Oncopole, Toulouse, France

Olivier Decaux
Centre Hospitalier Universitaire de Rennes, Rennes, France

Thomas Dejoie , Cyrille Touzeau and Philippe Moreau
Centre Hospitalier Universitaire de Nantes, Nantes, France

Guillemette Fouquet and Susanna Schraen
Centre Hospitalier Régional Universitaire de Lille, Lille, France

Stéphanie Guide , Vincent Javaugue and Xavier Leleu
Centre Hospitalier Universitaire de Poitiers, Poitiers, France

Benjamin Hebraud
Centre Hospitalier Universitaire de Toulouse, Toulouse, France

Valentine Richez
Centre Hospitalier Universitaire de Nice, Nice, France

Stephen Harding
Department of Research and Development, The Binding Site Group Limited, Birmingham, UK

Kavinda Jayawardana and Nicholas C. Ovenden
Department of Mathematics, University College London (UCL), London, UK

Alan Cottenden
Department of Bioengineering and Medical Physics, University College London (UCL), London, UK

Dalila Hamami
Department of Computing Science, University of Oran1 Ahmed Benbella,Oran, Algeria

Ross Cameron and Kevin G. Pollock
Health Protection Scotland, Glasgow, UK

Carron Shankland
Department of Computing Science and Mathematics, University of Stirling, Stirling, UK

Laurent Mackay and Anmar Khadra
Department of Physiology, McGill University, Montreal, QC, Canada

Nicholas Mikolajewicz and Svetlana V. Komarova
Faculty of Dentistry, McGill University, Montreal, QC, Canada
Shriners Hospital for Children-Canada, Montreal, QC, Canada

Neville J. Boon and Reiko J. Tanaka
Department of Bioengineering, Imperial College London, London, UK

Elisa Domínguez-Hüttinger
Department of Bioengineering, Imperial College London, London, UK
Instituto de Ecología, Universidad Nacional Autónoma de México, Mexico City, Mexico

Thomas B. Clarke
Department of Medicine, Imperial College London, London, UK

Caroline Colijn, Nick Jones and Mauricio Barahona
Department of Mathematics, Imperial College London, London, UK
EPSRC Centre for Mathematics of Precision Healthcare, Imperial College London, London, UK

Iain G. Johnston
EPSRC Centre for Mathematics of Precision Healthcare, Imperial College London, London, UK
School of Biosciences, University of Birmingham, Birmingham, UK

Sophia Yaliraki
EPSRC Centre for Mathematics of Precision Healthcare, Imperial College London, London, UK
Department of Chemistry, Imperial College London, London, UK

Sherry-Ann Brown
Department of Cardiovascular Diseases, Mayo Clinic, Rochester, MN, USA

Neil D. Evans and Michael J. Chappell
Biomedical and Biological Systems Laboratory, School of Engineering, University of Warwick, Coventry, UK

Linnéa Bergenholm
Biomedical and Biological Systems Laboratory, School of Engineering, University of Warwick, Coventry, UK
DrugMetabolism and Pharmacokinetics, Cardiovascular and Metabolic Diseases, iMED, AstraZeneca, Gothenburg, Sweden

Mats Jirstrand
Fraunhofer-Chalmers Centre, Chalmers Science Park, Gothenburg, Sweden

David L. I. Janzén
Biomedical and Biological Systems Laboratory, School of Engineering, University of Warwick, Coventry, UK
DrugMetabolism and Pharmacokinetics, Cardiovascular and Metabolic Diseases, iMED, AstraZeneca, Gothenburg, Sweden
Fraunhofer-Chalmers Centre, Chalmers Science Park, Gothenburg, Sweden

Joanna Parkinson
Early Clinical Development, Quantitative Clinical Pharmacology, iMED, AstraZeneca, Gothenburg, Sweden

James Yates
Oncology, iMED, AstraZeneca, Cambridge, UK

Hin K. Wong, Thomas E. Nichols and Boryana C. Lopez-Kolkovska
Warwick Manufacturing Group, Institute of Digital Healthcare, Unversity of Wanwick, Coventry, UK

Paul A. Tiffin
Mental Health and Addiction Research Group, Department of Health Sciences, Unversity of York, York, UK

Michael J. Chappell
School of Engineering, Unversity of Wanwick, Coventry, UK

Patrick R. Welsh
School of Psychology, Newcastle University, Newcastle upon Tyne, UK

Orla M. Doyle
Center for Neuroimaging Sciences, King's College London, London, UK

Sarah K. Inglis
Division of Maternal and Child Health Sciences, Ninewells Hospital and Medical School, University of Dundee, Dundee, UK

David Coghill
Departments of Paediatrics and Psychiatry, University of Melbourne, Melbourne, VIC, Australia

Yuan Shen and Peter Tiño
School of Computer Science, University of Birmingham, Birmingham, UK

Alekhya Mandali and V. Srinivasa Chakravarthy
Department of Biotechnology, Bhupat and Jyoti Mehta School of Biosciences, Indian Institute of Technology Madras, Chennai, India

Roopa Rajan and Asha Kishore
Department of Neurology, Comprehensive Care Centre for Movement Disorders, Trivandrum, India

Sankara Sarma
Achutha Menon Centre for Health Science Studies, Sree Chitra Tirunal Institute for Medical Sciences and Technology, Trivandrum, India

Wenguang Li
School of Engineering, University of Glasgow, Glasgow, UK

Nigel C. Bird
Academic Surgical Unit, Royal Hallamshire Hospital, Sheffield, UK

Xiaoyu Luo
School of Mathematics and Statistics, University of Glasgow, Glasgow, UK

Index

R

S

T

U

V

W

www.ingramcontent.com/pod-product-compliance
Lightning Source LLC
La Vergne TN
LVHW060559210926
848485LV00005B/1343